Rapid Thermal Processing of Semiconductors

MICRODEVICES
Physics and Fabrication Technologies

Series Editors: Ivor Brodie and Arden Sher
SRI International
Menlo Park, California

COMPOUND AND JOSEPHSON HIGH-SPEED DEVICES
Edited by Takahiko Misugi and Akihiro Shibatomi

ELECTRON AND ION OPTICS
Miklos Szilagyi

ELECTRON BEAM TESTING TECHNOLOGY
Edited by John T. L. Thong

GaAs DEVICES AND CIRCUITS
Michael Shur

ORIENTED CRYSTALLIZATION ON AMORPHOUS SUBSTRATES
E. I. Givargizov

PHYSICS OF HIGH-SPEED TRANSISTORS
Juras Požela

THE PHYSICS OF MICRO/NANO-FABRICATION
Ivor Brodie and Julius J. Muray

PHYSICS OF SUBMICRON DEVICES
David K. Ferry and Robert O. Grondin

THE PHYSICS OF SUBMICRON LITHOGRAPHY
Kamil A. Valiev

RAPID THERMAL PROCESSING OF SEMICONDUCTORS
Victor E. Borisenko and Peter J. Hesketh

SEMICONDUCTOR ALLOYS
Physics and Materials Engineering
An-Ban Chen and Arden Sher

SEMICONDUCTOR LITHOGRAPHY
Principles, Practices, and Materials
Wayne M. Moreau

SEMICONDUCTOR PHYSICAL ELECTRONICS
Sheng S. Li

A Continuation Order Plan is available for this series. A continuation order will bring delivery of each new volume immediately upon publication. Volumes are billed only upon actual shipment. For further information please contact the publisher.

Rapid Thermal Processing of Semiconductors

Victor E. Borisenko
Belarussian State University of Informatics and Radioelectronics
Minsk, Republic of Belarus

Peter J. Hesketh
The University of Illinois at Chicago
Chicago, Illinois

Plenum Press • New York and London

Library of Congress Cataloging-in-Publication Data

Borisenko, V. E. (Viktor Evgen´evich)
Rapid thermal processing of semiconductors / Victor E. Borisenko, Peter J. Hesketh.
p. cm. -- (Microdevices)
Includes bibliographical references and index.
ISBN 0-306-45054-2
1. Semiconductors--Design and construction. 2. Rapid thermal processing. I. Hesketh, P. J. (Peter J.) II. Title. III. Series.
TK7871.85.B64 1997
621.3815'2--dc21 97-3783
CIP

ISBN 0-306-45054-2

A Division of Plenum Publishing Corporation
233 Spring Street, New York, N. Y. 10013

http://www.plenum.com

10 9 8 7 6 5 4 3 2 1

Printed in the United States of America

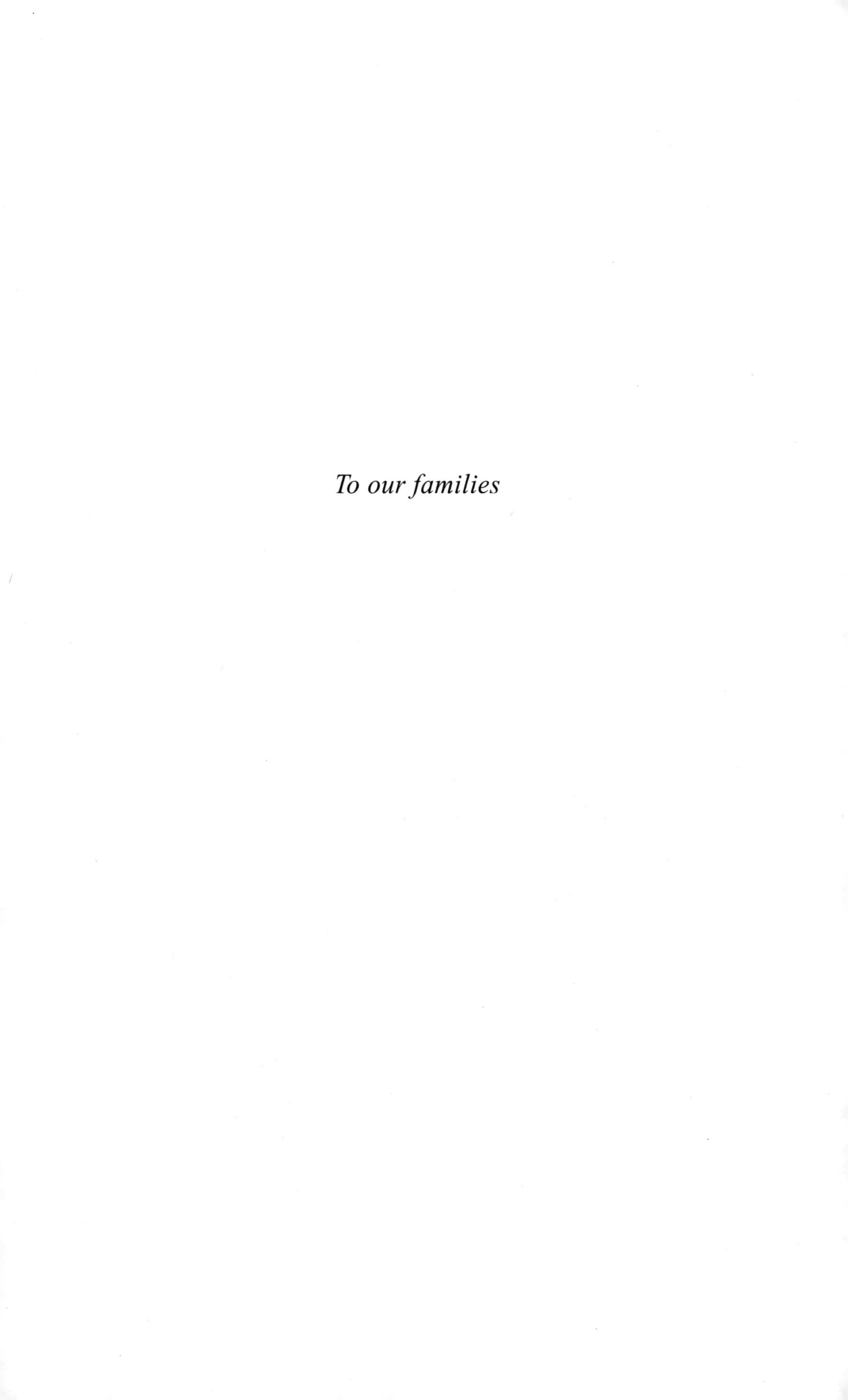

To our families

Preface

Thermal processing is one of the most effective ways to control the phase-structure, properties, and electrophysical parameters of materials for semiconductor devices and integrated circuits. The duration of conventional thermal processing ranges from tens of minutes to several hours. However, rapid thermal processing has shorter time scales, usually measured in seconds. Transient heating for a few seconds appears to have promising advantages: (1) the opportunity to modify material properties of the semiconductor in the solid phase combined with minimal negative effects at high temperature, (2) reduced temperature gradients causing a reduction in the elastic stresses and consequent defect generation in semiconductor wafers, and (3) simplicity and efficiency of technical implementation based on the use of incoherent light from halogen and arc lamps.

The most important current trends in the development of microelectronics are the decrease in the size of integrated circuit elements and fabrication trends to submicron structures. A key requirement in achieving successful fabrication processes is the control of the thermal budget. This has stimulated a considerable interest in transient methods of thermal processing, which minimize the negative influence of high-temperature processes on semiconductor crystals. Such negative effects include diffusion redistribution of dopants, generation of defects, and the generation of carrier traps both in the oxide and at the semiconductor–oxide interface. Over the past 5 to 10 years there has been rapid growth of research activity in this area for all types of processes, ranging from annealing of ion implant damage, dopant diffusion, gettering, silicide formation, metal contact alloying, glass reflow, oxidation, nitridation, and chemical vapor deposition of a range of materials. Integrated circuit manufacturing environments are constantly evolving to take advantage of new processing technologies. With an increasing number of processes that can be carried out by rapid thermal processing, the development of single wafer cluster processing tools will become increasingly more important in the future.

The aim of this book is to review the field of rapid thermal processing for both scientists and students interested in the field. We have attempted to cover as wide a range of materials, processes, and conditions as possible, allowing the reader to look up details for a particular material of interest. Chapter 1 analyzes the transient heating of semiconductors by radiation, related elastic stresses, and thermoplastic effects. Recrystallization phenomena, defect annealing in implanted layers, and impurity behavior in monocrystalline silicon are discussed in Chapter 2. Impurity segregation and diffusion in polycrystalline silicon are presented in Chapter 3. Impurity behavior in III–V semiconductor compound crystals is discussed in Chapter 4. Chapter 5 presents the main features of silicide formation in thin-film metal–silicon structures subjected to rapid thermal processing. In Chapter 6 the rapid thermal oxidation and nitridation of silicon, which has an important technological impact for ULSI and VLSI integrated circuits, particularly DRAMs and EPROMs, is reviewed. Chapter 7 looks forward to exciting developments in rapid thermal chemical vapor deposition, which is emerging as an important processing technique for a range of materials.

The background of the book has been formed by theoretical and experimental investigations carried out by Professor Borisenko at the Microelectronics Department at the Belarussian State University of Informatics and Radioelectronics and at the Physics Institute of Arhus University. Results of other researchers are also demonstrated and summarized.

ACKNOWLEDGMENTS

I express my deepest gratitude to my principal advisor Professor V. A. Labunov for brilliant ideas and current support. I am grateful to my colleagues from the former Soviet Union and foreign colleagues for the fruitful discussions we shared which contributed to the viewpoints presented in this book. I thank personally Professors I. A. Abroyan, F. F. Komarov, F. P. Korshunov, R. S. Malkovich, V. N. Mordkovich, L. S. Smirnov, I. B. Khaibullin, G. Carter, and A. Nylandsted Larsen. I am also indebted to my co-workers, especially Drs. V. P. Bondarenko, A. M. Dorofeev, V. V. Gribkovskii, N. T. Kvasov, and S. G. Yudin for their contribution. Charming D. E. P. Nesvetailova is acknowledged for her help in the preparation of the English version of the first five chapters of this book.

Victor E. Borisenko

I express my thanks to Drs. I. Brodie and A. Sher for the opportunity to contribute to the Microdevices series. I am indebted to my colleagues in the Electrical Engineering and Computer Science Department at the University of Illinois in Chicago, Professors G. J. Maclay, M. J. McNallan, D. Arnold, and D. Babic for fruitful discussions and their assistance in reviewing drafts of the manuscript. Thanks to S. Steward, J. Butler, E. Conwell, M. Yaffee, L. St. Clair, J. Cunneen, S. Hardeman, D. Beirne, and Jaihui Ji for assistance with typing, proofreading, graphics preparation, and literature searches. Finally, special thanks to my wife, Ann Marie, for her constant support throughout the preparation of this book.

Peter J. Hesketh

Contents

CHAPTER 5. Diffusion Synthesis of Silicides in Thin-Film Metal–Silicon Structures

CHAPTER 6. Rapid Thermal Oxidation and Nitridation

CHAPTER 7. Rapid Thermal Chemical Vapor Deposition

Nomenclature

ENGLISH SYMBOLS

a	lattice constant
$a_e(x, x_i)$	thermal generation parameter
a_v	evaporation coefficient
A	oxidation constant
$A(E, r, T, t)$	absorbed power density
A^*	dimensionless factor
Abs_{max}, Abs_{min}	maximum and minimum infrared absorbance
$A_i(E, x, T, t)$	absorbed energy in the ith film fraction
A_{Me}	atomic weight of metal atom
b	diffusivity ratio
b'	ion implantation range parameter
B	oxidation parabolic rate constant
B_e	ion implantation range parameter
B_{eff}	effective parabolic rate constant
c	velocity of light
c^*	material-dependent constant
C^*	oxygen solubility
$C(T)$	specific heat
$C_i(T)$	specific heat of ith layer
C_{LF}	low-frequency capacitance
C_o	oxidant concentration
C_{ox}	oxide capacitance
d	wafer thickness
$\dot{d}$	strain rate in s^{-1}
D	impurity diffusivity, diffusion constant
D^0	natural impurity diffusivity
D^*	constant parameter
D^+	+1 impurity diffusivity
D^-	−1 impurity diffusivity
$D^=$	−2 impurity diffusivity

D_1	Si diffusivity in $MeSi_2$
D_2	Si diffusivity in MeSi
D_a	impurity diffusivity in amorphous silicon
D_{eff}	effective diffusivity
D_i	intrinsic impurity diffusion constant
D_{it}	interface state density
$D_{it\text{-}m}$	midgap interface state density
D_o	diffusion constant preexponential factor
D_{ox}	diffusivity of oxidant in the growing film
D_{t0}	diffusivity defined at the initial moment of processing
D_v	vacancy diffusion coefficient
e	electron charge
E	energy
$\overline{E}$	average energy of photons
$E(r)$	field of lattice deformation
E^*	Young's modulus
E_1, E_2	oxidation parameters
E_a	activation energy
E_{BD}	breakdown field
E_c	conduction band energy
ED1, ED2, ED3	traps in epitaxial layer
E_g	band gap energy
EG6, EG7, EG8	traps
EN1, EL2, EL3, EL5, EL6	traps
E_o	free exciton threshold energy
E_r	specific heat of new phase formed
E_s	sublimation energy
E_{th}	threshold energy
E_v	valence band energy
$f(N)$	stepwise function represented by $f(N) = 0$ at $N < N_s$ and $f(N) = 1$ at $N > N_S$
f_T	unity gain maximum frequency
$f(t)$	dimensionless radiation intensity–time profile
F_B	barrier height
F^I	force driving all impurity atoms characterized by U^I
F_{ox}	parameter in model for oxide trapped charge
F^V	force driving all impurity atoms characterized by U^V
$g\ (T, d)$	factor related to the radiation transmitted through the wafer
g_{am}	Gibbs energy per atom in the amorphous phase
g_{cr}	Gibbs energy per atom in the crystalline phase
g_m	transconductance
G	Gibbs energy
G^*	oxidation parameter
h	Planck's constant

H	enthalpy
$\hbar$	$h/2\pi$
h_c	convection coefficient
h_e	internal electric field enhancement factor
i	integer
$I(\hbar\omega)$	photoluminescence intensity
I_0	modified Bessel function of the first kind of order zero
I_1	modified Bessel function of the first kind of order one
I_g	gate current
$I_{g,0}$	gate current at time zero
j	ion current density
k	Boltzmann constant
$k'(t)$	segregation coefficient
k^*	modulus of volume compression
k_{eq}	equilibrium clustering coefficient
$K(T)$	thermal conductivity
K^*	oxidation parameter
K_1, K_2, K_3, K_4, K_5	reaction rate constants
K_H	constant in two-factor model for oxide charge
$K_i(T)$	thermal conductivity of ith layer
K_N	constant in two-factor model for oxide charge
K_s	oxidation reaction rate
l	recrystallized layer thickness
ℓ	layer thickness
ℓ_d	length of atomic diffusion
ℓ_g	average grain size
L, L_1, L_2	oxidation parameters
m	empirical factor
m^*	material-dependent constant
m_1, m_2	fit parameters
M	mass of evaporated atoms or particles
M1, M2, M3, M4	electron traps
n	electron concentration
n_A	Avogadro's number (6.02×10^{23} atoms/mol)
n_i	intrinsic carrier concentration
n_0	initial impurity concentration
N	impurity concentration
N_1	number of oxidant molecules incorporated into 1 cm^3
N_1^*	volume of oxide produced per reaction between oxidant and vacancy
N'_a	number of atoms
N'_c	number of atoms in crystallite of critical radius
N'_g	number of impurity atoms inside grains
$N_{gb}(t)$	concentration of electrically inactive impurity segregated at grain boundaries

$N_i(t)$	impurity concentration in the nondissociated solid solution
N_{Me}	concentration profile of metal atoms in the silicide layer
N_0	initial impurity concentration in the nondissociated solid solution
N_0^I	equilibrium concentration of impurity interstitials
N_p	density of the precipitate nucleation centers
N_r	concentration of reacting components
N_S	steady-state impurity solubility
N_{Si}	atom density of silicon
N_{Si}^*	silicon host atom concentration
N_t	total impurity concentration
N'_{tg}	total number of impurity atoms inside grains and at grain boundaries
N1, N2, N3	traps
p	hole concentration
P	ambient pressure
$P(A)$	partial pressure of species A
P^*	vapor pressure
P_{eq}	equilibrium pressure
P_m	ratio of partial pressure of CVD process gases
P_0	preexponential factor in vapor pressure
q	electron charge
Q	integral absorbed power density
Q_{BD}	charge to breakdown
Q_f	fixed oxide charge at the interface
Q_i	interaction parameter
Q_{it}	oxide interface trap charge
Q_m	mobile ionic charge in the bulk oxide
Q_{ot}	bulk oxide trap charge
Q_S	the surface density of vacant sites for impurity atoms at the grain boundary
r	radial coordinate
(r,θ)	polar coordinate representation
$\overline{r}$	vector coordinate
r_g	radius of grain
r_i	impurity atom radius
r_o	ion implantation radius
r_s	host atom radius
r_w	wafer radius
R	reflectivity
R^*	universal gas constant
R_i	reflectivity of ith layer
R_p	projected range of electrons or ions
S	entropy
s	sputtering yield
$\lvert S_i \rvert$	stresses resolved in independent slip direction

S_{ox} sticking coefficient
t time
t_{bd} time to breakdown
t_i mean time of impurity atom diffusion inside a grain
t_0 initial time displacement to represent the initial oxide thickness
t_p pulse duration
t_{pr} processing time
T absolute temperature (in Kelvin)
$T(r)$ stress components
$T(x,t)$ temperature time profile
T_a ambient temperature
T_c temperature of black body
T_e steady-state temperature induced in wafer center
T_m temperature for melting
T_{max} steady-state maximum temperature
T_r temperature of recrystallization
T_s temperature of surrounding surfaces involved in a radiation exchange
u equation variable
U parabolic rate constant
U_1 the volume of one atom or a molecule in the crystal
U^I dilation volume of an atom in the interstitial position
U_o preexponential factor of Arrhenius expression
U^V dilation volume of an atom in the substitutional position
v drift velocity
$v_c(t)$ regrowth rate, the rate at which the amorphous–crystalline interface moves
v_o preexponential factor
V linear rate of Arrhenius expression
V_D drain voltage
V_{DS} drain source voltage
V_{fb} flat band voltage
V_G gate voltage
V^I drift velocity of interstitial impurity atoms
V_0 preexponential factor
V^V drift velocity of substitutional impurity atoms
w_a power consumption for evaporation or sputtering
w_c convection power
w_e power carried away by thermally evaporated or sputtered atoms
w_i possible heat losses
w_r consumption and release of power in chemical reactions and phase transitions
w_t power losses by thermal radiation
W^* impurity dose
$W_A(T,t)$ absorbed power density

W_c power loss through convection of heat
W_I incident power density
W_i possible power losses
W_R reflected radiation power density
W_T transmitted power
x rectangular coordinate
x_a thickness of absorbing layer
x_{Ge} germanium stoichiometry
$x_{initial}$ initial oxide thickness in oxidation model
x_{j1}, x_{j2}, x_{j3} phase boundaries
x_{ox} oxide thickness
X_a initial amorphous layer thickness
X solid-state mole fraction of the impurity
z impurity charge state
Z atomic number

SUBSCRIPT AND SUPERSCRIPTS

i subscript describing ith layer of film
I superscript representing fast-diffusing nonequilibrium impurity interstitials
V superscript representing equilibrium impurity atoms in substitutional positions

GREEK SYMBOLS

$\alpha(x)$ piecewise linear absorption coefficient
α' oxidation parameter
$\alpha_i(E, T)$ components of the temperature-dependent absorption coefficient
α_v thermal expansion coefficient
β injection rate of impurity atoms from grain boundaries into electrically active positions inside grains
β^* oxidation parameter
β_i impurity evaporation rate
χ_{min} minimum backscattering yield of channeled ions (measured just after the surface peak)
δ jump distance of an atom at the interface
"δ" x coordinate for the right side of the interface
"$-\delta$" x coordinate on left side of interface
Δ change in
Δa_i change in lattice constant due to interstitial impurity
Δa_s change in lattice constant due to substitutional impurity
Δa_v change in lattice constant due to vacancy
Δg Gibbs free energy for crystallization

Δg_{m}	Gibbs energy for migration
ΔG	change in Gibbs energy
ΔG_{c}	Gibbs energy for the formation of a crystallite having critical size
ΔH	variation of enthalpy
ΔH_{m}	enthalpy for migration
ΔR_{p}	standard deviation of ion or electron projected range
Δs	variation of entropy
ΔS	difference between oscillation entropy of impurity atoms at the grain boundary and inside the grain
ΔS_{m}	entropy for migration
ΔV_{t}	threshold voltage shift
Δx	heat redistribution depth
ε	dielectric constant
ε_{e}	effective emissivity
$\Phi(E,x,T)$	radiation absorption function
γ	rate of impurity atom capture in electrically inactive positions at grain boundaries
η	geometry factor for a sphere
ψ	normalizing factor for wafer radius
Ψ	surface free energy
Ψ_{s}	surface potential
κ	thermal diffusivity
λ	wavelength
$\lambda_{i}(E, T)$	absorption coefficient
$\lambda_{i}(X^{1})$	piecewise continuous function made up of partial $\lambda_{i}(E, T)$
λ_{v}	determines relationship between the generation of vacancies at the Si/SiO_2 surface
υ	Poisson's ratio
υ_{o}^{*}	atomic vibrational frequency
π	3.142
θ	polar angle
ρ	material density
ρ_{Me}	density of metal
σ	Stefan–Boltzmann constant
$\sigma_{xx}\ \sigma_{yy}\ \sigma_{zz}$	normal stress components
$\sigma_{xy}\ \sigma_{xz}\ \sigma_{zy}$	shear stress components
$\sigma_{\theta\theta}\ \sigma_{rr}\ \sigma_{r\theta}$	stress components in polar coordinates
τ	relaxation time
τ^{*}	transmittance
τ_{i}	lifetime of nonequilibrium interstitials
τ_{0}	relaxation time preexponential factor
τ_{s}	supersaturation relaxation time
ω	angular frequency
ω_{e}	evaporation rate
ω_{r}	nucleation rate

$\omega_r(T)$ reaction rate
Ω Ohm
ξ effective planar charge density

1

Transient Heating of Semiconductors by Radiation

Photon, electron, and ion beam interactions with solids provide a wide range of physical and chemical effects which depend on the nature of the irradiated materials as well as the characteristics of the radiation. Energy, mass, charge state of the particles, and intensity of the beam appear to be the characteristics that have the greatest impact. While photons, electrons, and ions differ considerably in their characteristics, there are some common features in their interaction with solids. At low and moderate energy the main effects are produced in a surface layer extending several microns into the wafer. As the beam intensity is increased heating becomes the dominant effect. For ions, the heaviest particles, in the energy range between tens of electron volts to hundreds of mega-electron-volts, surface energy absorption occurs. A higher penetrability is observed with the lighter particles, which includes electrons in the energy range from tens to thousands of electron volts, and photons in the ultraviolet to infrared spectral range.

This chapter describes heating of semiconductor wafers with intensive beams that produce noticeable heating of a sample. The temperature profiles, induced thermal stresses, and thermoplastic effects are analyzed. Particular attention is paid to photon heating by incoherent light because of its practical applications.

1.1. ENERGY BALANCE DURING IRRADIATION AND HEATING

The interaction of intensive beams with semiconductors is a complicated process. We will consider only the effects responsible for heating. Consider radiation incident on a sample of finite thickness which is partly reflected by the surface, absorbed, and transmitted through the sample (Fig. 1.1). The balance of incident radiation (power density) W_I comprises reflected W_R, absorbed W_A, and transmitted W_T components

$$W_I = W_R + W_A + W_T \tag{1.1}$$

With the reflectivity, R, and transmittance, τ^*, the absorbed fraction amounts to

$$W_A = W_I(1 - R - \tau^*) \tag{1.2}$$

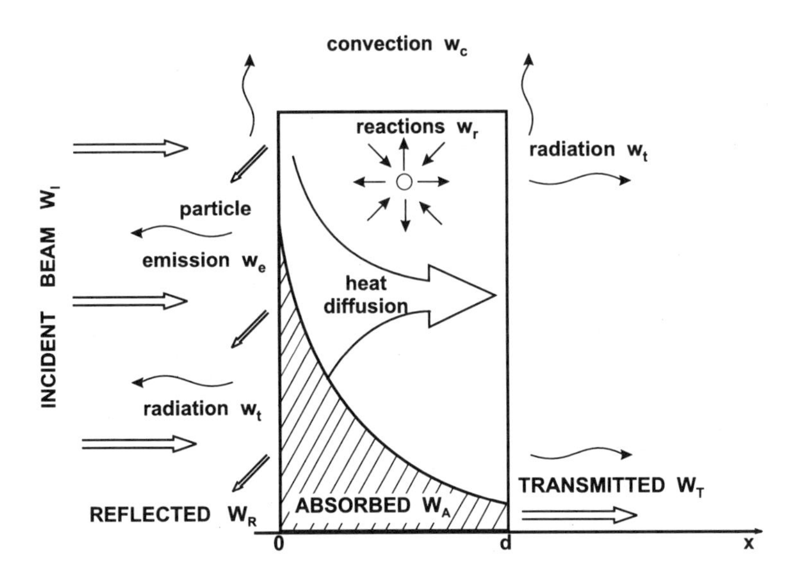

FIGURE 1.1. Incident energy distribution in a semiconductor wafer subjected to transient heating with radiation.

In the case of intensive irradiation the correlation between the components varies in time because of heating-induced changes of the semiconductor material parameters. Therefore, Eq. (1.1) has to be considered as an instant energy balance. It is written for the incident, reflected, absorbed, and transmitted radiation power density.

Incident electron and photon energy are converted to thermal energy by excitation and relaxation of electrons within the solids [1, 2]. In the case of ion bombardment, heating appears as a result of direct excitation of atom oscillations and ionization [3]. The excited electron and atom subsystems relax within 10^{-10}–10^{-14} s [1, 4], resulting in an increase of lattice temperature. It is fast enough to consider this process instantaneous in the majority of practically important cases.

The absorbing layer acts as an intrinsic heat source, whose location is related to the characteristics of the radiation and the irradiated material properties. Accounting for variation with energy, E, coordinate, $\overline{r}$, temperature, T, and time, t, the absorbed power density in the intrinsic heat source is

$$A(E,\overline{r},T,t) = \text{grad } W_A \tag{1.3}$$

Further evolution of thermal processes is a function of heat diffusion through the sample and energy losses. The absorbed energy balance includes the energy carried away by thermally evaporated or sputtered atoms, w_e, losses by thermal radiation, w_t, convection, w_c, consumption and release of energy in chemical reactions and phase transitions, w_r (see Fig. 1.1).

The energy carried by atoms being evaporated or sputtered has two components: (1) the potential energy of interatomic bonds and (2) the kinetic energy of the atoms leaving the sample's surface. The potential energy component may be estimated with the use of the known specific sublimation energies, E_s, for evaporation and by the threshold energies, E_{th}, for sputtering [5]. Energy consumption for evaporation according to the Hertz–Knudsen equation is

$$w_a = \frac{E_s\, a_v\, P^*}{(2\pi M kT)^{1/2}} \tag{1.4}$$

where a_v is the evaporation factor, P^* the vapor pressure, M the mass of evaporated atoms, k the Boltzmann constant, and T the absolute temperature. For sputtering

$$w_e = \frac{S\, j\, E_{th}}{q} \tag{1.5}$$

where S is the sputtering yield, j the ion current density, and q the electron charge. The kinetic energy of atoms evaporated or sputtered by low-energy ions (50 eV–5 keV) lies in the range of 0–0.3 eV, while the sublimation and threshold energies for sputtering are 10–20 eV [5–8]. In comparison the energy contribution from secondary electrons is negligibly small.

Radiative heat losses take place on the sample surface and are described by the Stefan–Boltzmann law:

$$w_t = \sigma \varepsilon_e (T^4 - T_s^4) \tag{1.6}$$

where σ is the Stefan–Boltzmann constant, ε_e the effective emissivity, and T_s the temperature of the surrounding surfaces involved in the radiation exchange.

Convective heat losses are proportional to the difference between the sample temperature, T, and the temperature of the surrounding ambient, T_a [9]:

$$w_c = h_c (T - T_a) \tag{1.7}$$

where h_c is the convection coefficient.

Intrinsic chemical reactions and phase transitions can both consume thermal energy and release it [10]. The nature of the reactions or phase transitions, their specific heat, temperature, and the quantity of matter involved in these transformations are important. This energy is given by

$$w_r = \omega_r(T)\, N_r\, E_r \tag{1.8}$$

where $\omega_r(T)$ is the reaction rate, N_r the concentration of the reacting component, and E_r the specific heat of the new phase formed.

The efficiency of different heat loss mechanisms depends to a great extent on the time duration of the irradiation through its influence on the absorbed energy redistribution. The heat redistribution depth, Δx, achieved within the radiation pulse duration, t_p, is a good criterion to classify transient heating regimes [11]:

$$\Delta x = (\kappa t_p)^{1/2} \tag{1.9}$$

where κ is the thermal diffusivity. According to this criterion there are three important cases:

1. The heat redistribution depth is much less than the thickness of the absorbing layer x_a ($\Delta x << x_a$).

2. It is somewhere in between the absorbing layer depth and wafer thickness, d ($x_a < \Delta x < d$).
3. It is comparable to or greater than the wafer thickness ($\Delta x > d$).

These three thermal processing regimes [11–13] corresponding to (1) adiabatic, (2) thermal flux, and (3) heat balance are shown schematically in Fig. 1.2. The thermal diffusivities for Si, Ge, GaAs, InSb and wafer thicknesses of 200 to 800 μm have been considered [10, 11].

The adiabatic regime is produced at very short pulses (10^{-11} s $< t_p < 10^{-6}$ s). The absorbed energy is conserved in the absorbing layer and an extremely rapid temperature rise occurs which results in melting the surface layer. The melted depth depends only on the energy density while both radiative and convective losses are negligible.

In the thermal flux regime (10^{-6} s $< t_p < 10^{-2}$ s) temperature profiles are determined by heat diffusion from the radiation-absorbing region. There is a depth–temperature gradient in the wafer and lateral gradients appear when a point or line source is used. Heat loss via radiation and heat diffusion into the wafer holder begin to play a role as the pulse duration increases, but convection still remains negligible. Solid-phase and surface melting modes may occur in this regime.

The heat balance regime ($t_p > 10^{-2}$ s) occurs when the pulse duration is long enough to have a uniform temperature distribution in the wafer. For scanned continuous wave (cw) sources a laterally uniform region can locally exist inside the irradiated area. All of the above-mentioned heat loss mechanisms influence the wafer temperature and its temperature evolution. Two modes in the heat balance regime are defined, a transient heat balance and a steady-state one. In the starting transient mode the power density absorbed is higher than that consumed by heat loss mechanisms although an instantaneous heat balance exists. In general, the wafer temperature increases with irradiation time and this enhances radiative heat loss that is proportional to T^4. An equilibrium between deposited and lost power can be achieved resulting in a steady-state heat balance. This is known as the isothermal mode and is characterized by the constant temperature of the irradiated wafer.

The time scale depicted in Fig. 1.2 for the regimes discussed above are quite general, reflecting those that are most frequently encountered in practical situations. For a specific material, wafer diameter, beam spot geometry, heat sink configuration, thermal conductivity of the wafer holder, and pulse duration, the boundaries between regimes can change.

Transient thermal processing in the heat balance regime has been widely applied because of its simple technical implementation with halogen or arc lamps [14] and the minimal temperature gradients in the semiconductor wafer [15].

Summarizing, the temperature profile in a semiconductor wafer being transiently heated by radiation is defined by

1. The radiation adsorption and instant heat source formation
2. The kinetics of heat diffusion
3. The energy losses related to heat exchange with the surroundings and intrinsic consumption or generation of heat in the wafer

These factors have to be included in procedures for simulating the temperature profiles in the wafer.

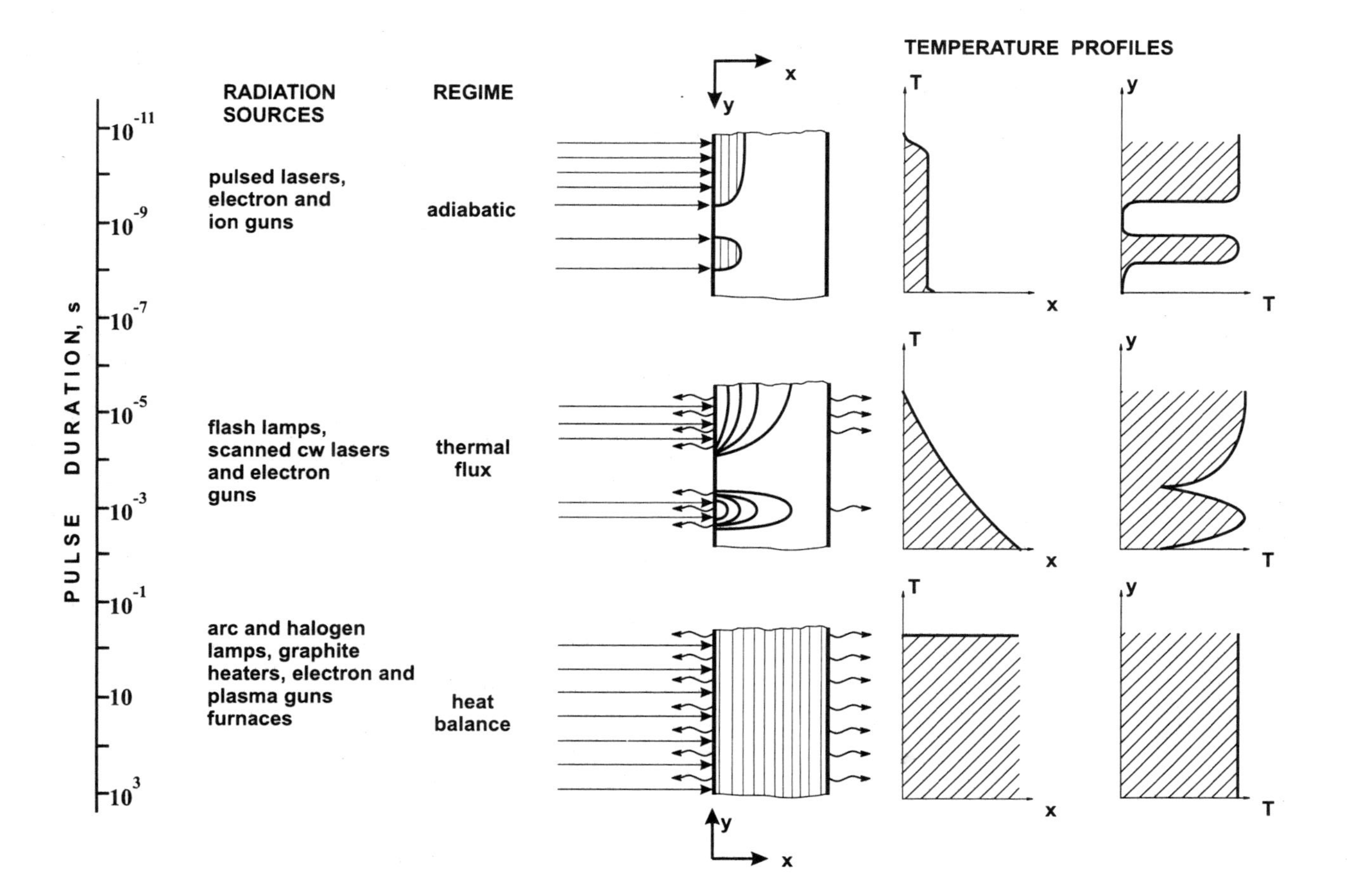

FIGURE 1.2. Transient heating regimes.

1.2. INSTANT HEAT SOURCE

In the thermal flux and heat balance regimes the duration of the processes involved in the radiation energy absorption is much less than the shortest incident radiation pulse. The deposited energy may be considered converted into heat immediately. The absorbed power depth distribution in the instant heat source, $A(E,x,T,t)$, assumed energy spectrum of radiation, $\partial W_I/\partial E$, temperature and spectral dependence of the semiconductor material parameters in a one-dimensional case is described by

$$A(E,x,T,t) = g(T,d)\,f(t)\int_0^\infty \frac{\partial W_I}{\partial E}(1-R)\,\Phi(E,x,T)\,dE \tag{1.10}$$

where $f(t)$ is the dimensionless radiation intensity–time profile and $\Phi(E,x,T)$ the absorption function representing radiation absorption. The factor $g(T,d)$ is related to the radiation transmitted through the wafer at d_w thickness and is defined by the normalization of the power absorbed in the wafer

$$\int_0^d A(E,x,T,t)\,dx = W_A = W_I(1-R-\tau^*) \tag{1.11}$$

Then,

$$A(E,x,T,t) = \frac{W_I(1-R-\tau^*)\,f(t)\int_0^\infty \frac{\partial W_I}{\partial E}(1-R)\,\Phi(E,x,T)\,dE}{\int_0^d dx\int_0^\infty \frac{\partial W_I}{\partial E}(1-R)\Phi(E,x,T)\,dE} \tag{1.12}$$

The energy spectrum of incident radiation is closely related to the radiation source. Lasers, electron guns, and ion guns produce beams of practically monoenergetic particles, and are therefore characterized by $\partial W_I/\partial E$ equal to a delta function.

For incoherent light generated by halogen and arc lamps the radiation energy spectrum must be taken into account. The radiation energy spectra of a ruby laser, arc xenon lamp [16], and halogen lamp [17] are sketched in Fig. 1.3. In spite of the differences that exist, Planck's formula for radiation from a blackbody heated to temperature T_c approximates the incoherent light spectral distributions well:

$$\frac{\partial W_I}{\partial E} = \frac{2\pi E^3}{c^2 h^3\,[\exp(E/kT_c)-1]} \tag{1.13}$$

where c is the velocity of light, h is Planck's constant, $E = hc/\lambda$, and λ is the wavelength.

The absorption function describes the depth profile of the absorbed energy. In general, it depends both on the nature of the radiation and on the properties of the irradiated material. Photon absorption obeys Bouguer's law:

$$\Phi(E,x,T) = \Phi_o(E,T)\exp[-\alpha(E,T)\,x] \tag{1.14}$$

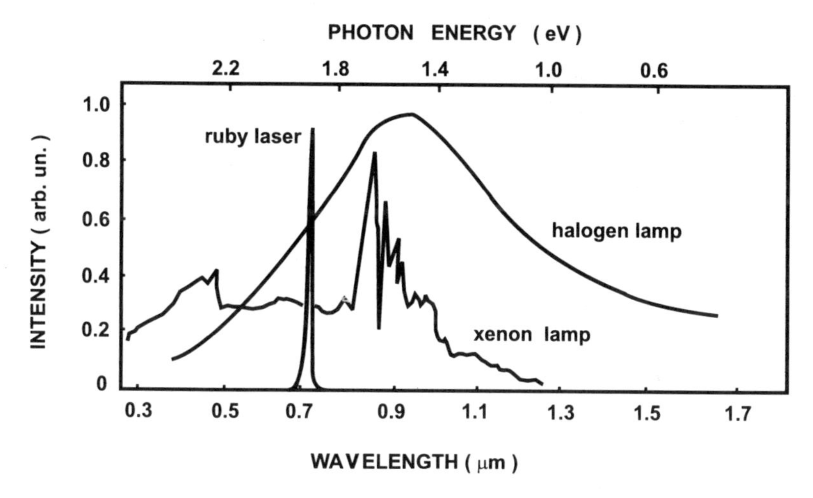

FIGURE 1.3. Emission spectra of different light sources.

where $\alpha(E,T)$ is the temperature-dependent spectral absorption coefficient. In the case of electrons and ions the absorption function may be approximated by the Gaussian distribution

$$\Phi(E,x,T) = \frac{1}{\sqrt{2\pi}\,\Delta R_{\mathrm{p}}} \exp\left[-\frac{1}{2}\left(\frac{x-R_{\mathrm{p}}}{\Delta R_{\mathrm{p}}}\right)^2\right] \tag{1.15}$$

or more precisely by the high-moment Pearson IV functions [18]. The parameters R_{p} and ΔR_{p} represent the range and standard deviation, respectively, of the absorbed energy profile following the generated defect distribution in the sample. They are functions of particle energy and mass, and their temperature dependence can be neglected. These parameters are tabulated for ion implantation by Burenkov *et al.* [18]. Electron beam energy absorption in materials of atomic number, Z, from 6 to 50 (carbon to tin) is described by [19]

$$R_{\mathrm{p}} = (0.143\,Z + 0.622)\,r_{\mathrm{o}} \tag{1.16}$$

$$\Delta R_{\mathrm{p}} = (-0.0538\,\ln Z + 0.374)\,r_{\mathrm{o}} \tag{1.17}$$

where $r_{\mathrm{o}} = B_{\mathrm{e}}E^{b'}$ is in g/cm^2, $B_{\mathrm{e}} = 3.92 \times 10^{-6} + Z\,1.562 \times 10^{-7}$; $b' = 1.777 - Z\,2.165 \times 10^{-3}$, and the energy of electrons is in keV.

Temperature and wavelength dependencies of thermophysical and optical parameters of semiconductors for irradiation-induced temperature calculations are collected in Table 1.1. The empirical formulas shown were obtained by fitting to experimental data with the precision required for numerical simulation of real processes.

As a first approximation, for irradiation with light of higher energy than the semiconductor band gap, electrons, or ions, the transmittance is considered to be equal to zero. However, for incoherent light with a considerably longer wavelength tail in the spectral range where the semiconductor is transparent, transmission must be included.

TABLE 1.1. Thermophysical and Optical Properties of Semiconductors

Material	Density, ρ (g/cm^3)	Melting point, T_m (K)	Melting enthalpy, H_m (J/g)	Thermal conductivity, $K(T)$ ($W\ cm^{-1}\ K^{-1}$)	Specific heat, $C(T)$ ($J\ g^{-1}\ K^{-1}$)	Band gap, E_g (eV)	Reflectivity, $R(T)$	Absorption coefficient, $\alpha(E,T)$ (cm^{-1})
Crystalline silicon	2.33 [20]	1685 [23,25]		$802.99\ T^{-1.12}$, 300–1683 K [23] $\frac{299}{(T-99)}$ [27] $2.722 \exp(-2.338 \times 10^{-3}\ T)$, 300–900 K [28] $0.648 \exp(-7.275 \times 10^{-4}\ T)$, > 900 K [28]	$0.204 + 2.2 \times 10^{-5}\ T - 3.6 \times 10^{-4}\ T^2$ [20] $0.641 + 2.473 \times 10^{-4}\ T$, > 300 K [23] $0.863 + 8.345 \times 10^{-5}\ T + 1.624 \times 10^{4}\ T^{-2}$ [21]	$1.21 - 3.6 \times 10^{-4}\ T$ [20] $1.167 - 8.6 \times 10^{-5}\ T$ $2.2 \times 10^{-7}\ T^2$ [23] $1.16 - \frac{7.02 \times 10^{-4}}{(T + 1108)}$ [28]	0.33 [23] $0.267 + 1.68 \times 10^{-4}\ T - 1.12 \times 10^{-7}\ T^2 - 5.25 \times 10^{-11}\ T^3 + 5.65 \times 10^{-14}\ T^4$, 1373–1673 K [30]	$2500 \exp[2.48(E' - 1.79)]$, $1.6 < E' < 3.3$[a] [23] $6000[\frac{E - E_g(T) - 0.0575)^2}{1 - \exp(-670/T)} + \frac{(E - E_g(T) + 0.0575)^2}{\exp(670/T) - 1}$ [29]
Amorphoussilicon	—	1400 [25] 1420 [26]	1250 [25]	—	—	—	0.44 [23,29]	—
Liquidsilicon	2.5 [21]	—	—	0.075 [21]	0.914 [21]	—	0.7 – 0.73 [24,31,32]	—
Crystalline germanium	5.32 [20]	1214 [23]	467 [20]	$\frac{180}{T}$, 200 – 1214 K [23]	$0.303 + 6.133 \times 10^{-5}\ T$, >300 K [23]	$0.785 - 3.5 \times 10^{-4}\ T$ [20] $0.918 - 3.9 \times 10^{-4}\ T$ [23]	0.44 [23] 0.60 [31]	$\frac{1.4 \times 10^4}{\exp[2.81(E' - 1.17)]}$, $0.95 < E' < 2.0$ [23]
Liquid germanium	—	—	—	—	—	—	0.65 [31] 0.76 [33]	—

Indium antimonide	5.77 [20]	798 [22]	—	$20.48\ T^{-0.83}$, > 300 K [23]	$0.18 + 1.0 \times 10^{-4}\ T$, > 300 K [23]	$0.235 - 1.363 \times 10^{-4}\ T - 1.622 \times 10^{-7}\ T^2$ [23] $0.235 - \frac{2.7 \times 10^{-4}\ T^2}{T + 106}$, 300 – 700 K [24]	0.45 [23]	22.7 exp[164(E' – 0.16)], $E' < 0.185$ 2050 exp[6.52(E' – 0.238)], $0.185 < E' < 0.364$ 4.34×10^4 exp[2.77 (E' – 1.17)], $0.364 < E' < 1.55$ 1.51×10^5 exp[0.824(E' – 1.79)], $1.55 < E' < 2.30$ [23]
Liquid indium antimonide	6.48 [22]	—	—	—	—	—	—	—
Crystalline gallium arsenide	5.31 [22]	1511 [22]	729 [20]	$225.54\ T^{-1.1}$, > 300 K [23] $\frac{91}{T - 90.9}$ [28]	0.303 + $5.0 \times 10^{-5}\ T$, > 300 K [23]	$1.575 - 5.0 \times 10^{-4}\ T$, > 300 K [23]	0.33 [23]	10 exp[149(E' – 1.38)], $E' < 1.43$ 2.91×10^4 exp[3.22(E' – 1.79)], $1.43 < E' < 1.83$ 3.48×10^4 exp[1.71 (E' – 1.83)] $1.83 < E' < 2.50$ [23]
Amorphous gallium arsenide	—	—	—	—	—	—	0.40 [32]	—
Liquid gallium arsenide	—	—	—	—	—	—	0.70 [32]	—

[a] $E' = E + E_g(300\ \text{K}) - E_g(T)$, where E is the photon energy in eV and T is the absolute temperature.

Transmittance τ^* is a complex function of thickness, temperature of the sample, and energy of the incident photons. Reliable experimental data as well as simple analytic expressions for the temperature range of practical interest (from room temperature to 600–800 °C) are not available. Analytical integration of (1.12) with normalization (1.11) for a slight wavelength dependence of the absorption coefficient gives an approximate formula for the transmittance [11]:

$$\tau^* = \exp\left[-\frac{d}{\bar{E}}\int_0^{\bar{E}} \Phi(E,T)\, dE\right] \tag{1.18}$$

where

$$\bar{E} = \frac{\int_0^{\infty} \frac{\partial W_{\mathrm{I}}}{\partial E} E\, dE}{\int_0^{\infty} \frac{\partial W_{\mathrm{I}}}{\partial E}\, dE}$$

$\bar{E}$ being the average energy.

Absorbed power depth profiles in Si, Ge, GaAs, InSb were computed in Ref. 11 with temperature- and wavelength-dependent material parameters as shown in Fig. 1.4. It is evident that monochromatic laser radiation is absorbed in a surface layer about 3 μm thick, but the absorption of incoherent light occurs over a region 20 times greater so that the surface absorption is reduced. The absorption layer becomes thinner as the temperature of the semiconductor is raised. This is related to an increase in the absorption coefficient produced by narrowing of the band gap and free carrier generation. In the temperature range of 400–700 °C the silicon absorption coefficient for visible and infrared light grows by a factor of 3–5, which leads to practically complete photon energy absorption in the irradiated wafers [34–36].

It is worth noting that the integral absorbed power calculated for the halogen lamp radiation spectrum, as tabulated, deviates by only 3–5% from that calculated with Planck's radiation formula. However, for a xenon arc lamp the calculated absorbed power appears to be 7–9% higher than with Planck's formula.

The model presented in this section for instantaneous heat generation appears to be the most general because it accounts for temperature and spectral variations of the semiconductor material parameters.

1.3. TEMPERATURE PROFILES IN SEMICONDUCTOR WAFERS

Diffusion of the absorbed radiant energy, which is converted into heat, occurs when the pulse duration becomes longer than 10^{-6} s. In addition, radiative and convective heat losses start to play a role in defining the temperature distribution. For a semiconductor wafer that is uniformly irradiated, the one-dimensional heat conduction equation takes the form [9, 37]

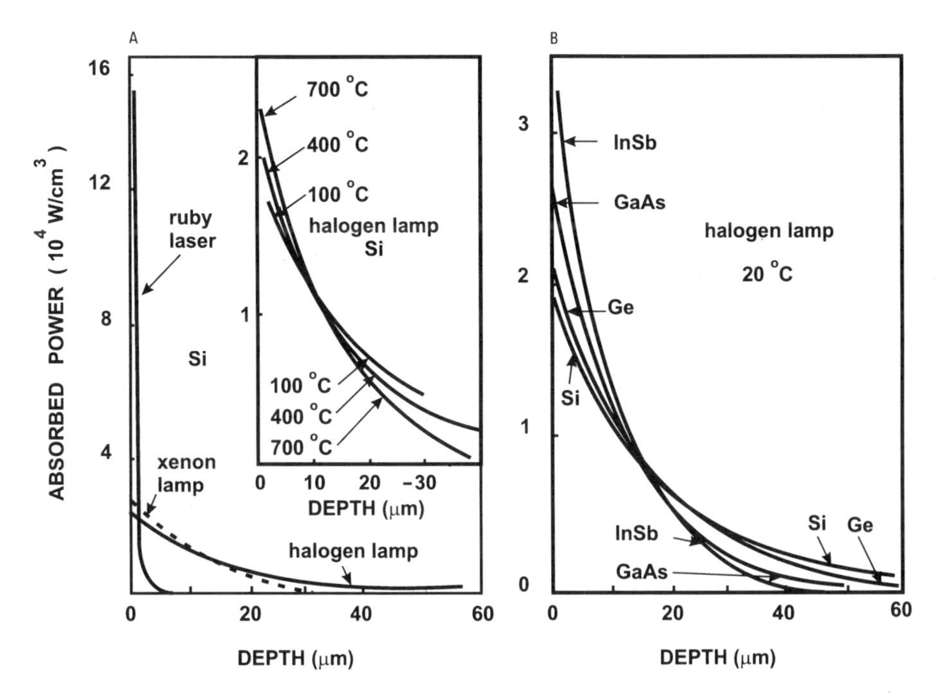

FIGURE 1.4. Absorbed power depth profiles in (a) Si with temperature as a parameter and (b) InSb, GaAs, and Ge at 20 °C.

$$C(T)\,\rho\frac{\partial T(x,t)}{\partial t}=\frac{\partial}{\partial x}\left[K(T)\frac{\partial T(x,t)}{\partial x}\right]+A(E,x,T,t) \tag{1.19}$$

where $C(T)$ is the specific heat, ρ the material density, and $K(T)$ the thermal conductivity. Possible heat losses, w_i, are introduced through the boundary conditions as follows:

$$-K(T)\frac{\partial T}{\partial x}\bigg|_{x=0}+\sum_{i=e,t,c,r} w_i=0 \tag{1.20}$$

$$K(T)\frac{\partial T}{\partial x}\bigg|_{x=d}+\sum_{i=e,t,c,r} w_i=0 \tag{1.21}$$

where the subscript i represents the heat losses related to evaporation ($i = e$), thermal radiation ($i = t$), convection ($i = c$), reactions ($i = r$).

Solution of (1.19) with the boundary conditions (1.20), (1.21) is available only by numerical methods. The Kirchhoff transform followed by numerical integration is generally used. Finite-difference schemes [38, 39], in particular the Crank–Nicolson method, are also widely employed. Herein, the initial equation nonlinearity is eliminated by iteration [29, 40]. Moreover, the Laplace–Carson method of numerical transform [41] or integral transforms [42] can also be applied.

An analytical solution of the heat conduction equation (1.19) becomes possible with the simplifications implying adiabatic boundary conditions, temperature- and spectral-independent semiconductor material parameters [29, 40–44]. One of them for the pulse duration $t_p = 15$ s is [44]

$$T(x,t) = \frac{W_I(1-R)}{K}\Big\{(\kappa/\omega)^{1/2}\exp[-x\,(\omega/2\kappa)^{1/2}]\sin\left[\omega t - \pi/4 - x\,(\omega/2\kappa)^{1/2}\right] + 2\,\kappa/\pi\int_0^{\infty}\frac{\omega\cos(ux)}{\omega^2 + 2\,\kappa^2 u^4}\exp(+\kappa\, u^2\, t)\,du\Big\} \tag{1.22}$$

where the thermal diffusivity $\kappa = K/\rho C$, $\omega = \pi/t_p$, and $0 < \omega\, t < \pi$.

Analytical solutions provide a good approximation for calculation of the temperature distribution only in the near-surface region of a semiconductor wafer, for the pulses less than 10^{-4} s [37]. Therefore, numerical integration of the heat conduction equation is most often applied.

Computations for different radiation sources [11, 21, 29, 45, 46] demonstrate that the semiconductor wafer's thickness and initial temperature have a considerable influence on the heating process. Thinner wafers and higher preheating temperatures result in a higher overall processing temperature. Pulse duration and incident power density contribute more directly to the temperature profile formation. Figure 1.5 shows calculated temperature depth profiles induced in a silicon wafer by intense coherent and incoherent light [11]. Here the volume temperature is normalized to the surface temperature and plotted versus wafer depth. The two light sources that produce pulse durations in the range of 10^{-6}–10^{-2} s, i.e., ruby laser and xenon flash lamp, are presented for comparison. The temperature profiles shown demonstrate the existence of considerable temperature gradients, especially at short pulses 10^{-6}–10^{-4} s. An increase of the pulse duration to 10^{-3}–10^{-2} s leads to an almost uniform temperature distribution with depth through the wafer.

Semiconductor layers used for device fabrication are normally very thin, less than 1 μm, and therefore the surface temperature is important in defining the transient heating. Figure 1.6 presents the computed surface temperature for Si, Ge, GaAs, InSb wafers exposed to xenon flash lamp irradiation at pulse duration, energy density, and a wafer thickness range of practical interest [11]. Within the experimental uncertainty there is good agreement between the calculated (Fig. 1.6) and experimental values.

Processing times of 10^{-2} s and longer produce a uniform temperature distribution through an irradiated semiconductor wafer. Radiative heat loss becomes dominant at high temperatures. The heat balance equation should be solved to calculate the wafer temperature. An account of the other heat losses supposes this equation in the form

$$dC(T)\,\rho\,\frac{dT}{dt} = W_A(T,t) - 2\,\sigma\,\varepsilon_e\,(T^4 - T_s^4) - \sum_{i=e,c,r} w_i \tag{1.23}$$

where the absorbed power density is

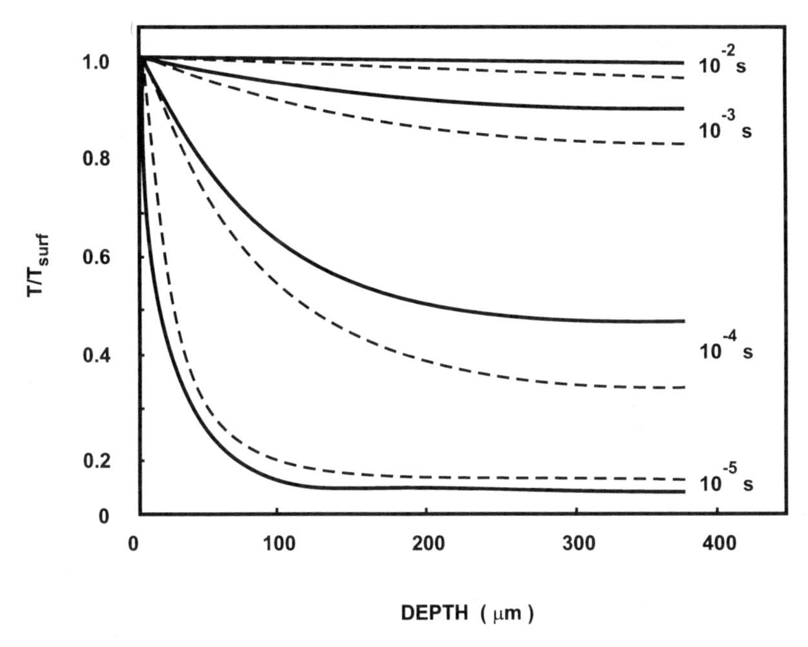

FIGURE 1.5. Normalized temperature as a function of depth in a 380-μm-thick silicon wafer heated with ruby laser (solid lines) and xenon flash lamp (dashed lines) at an energy density of 5 J/cm^2 and pulse duration of 10^{-5} to 10^{-2} s.

$$W_A(T,t) = \int_0^d A(E, x, T, t)\, dx \tag{1.24}$$

The parameters in (1.23) and (1.24) are traditionally assumed to be temperature independent [47–49] and the accuracy of the calculations is limited. Since the temperature dependence of the absorbed power density has no adequate analytical expression, an exact solution of (1.23) can be defined only by numerical integration. However, neglecting the temperature dependence, useful approximate formulas have been obtained [49] to estimate a steady-state maximum temperature T_{max} and time interval, t, at which the wafer temperature, T, is attained, as follows:

$$T_{max} = \left[\frac{W_I(1-R)}{2\sigma\varepsilon_e}\right]^{1/4} \tag{1.25}$$

$$t = \frac{C(T)\rho d}{8T_{max}^3\sigma\varepsilon_e}\left\{\left[\ln\left(\frac{T_{max}+T}{T_{max}-T}\right) - \ln\left(\frac{T_{max}+Ts}{T_{max}-Ts}\right)\right] + 2\left[\text{arctg}\left(\frac{T}{T_{max}}\right) - \text{arctg}\left(\frac{Ts}{T_{max}}\right)\right]\right\} \tag{1.26}$$

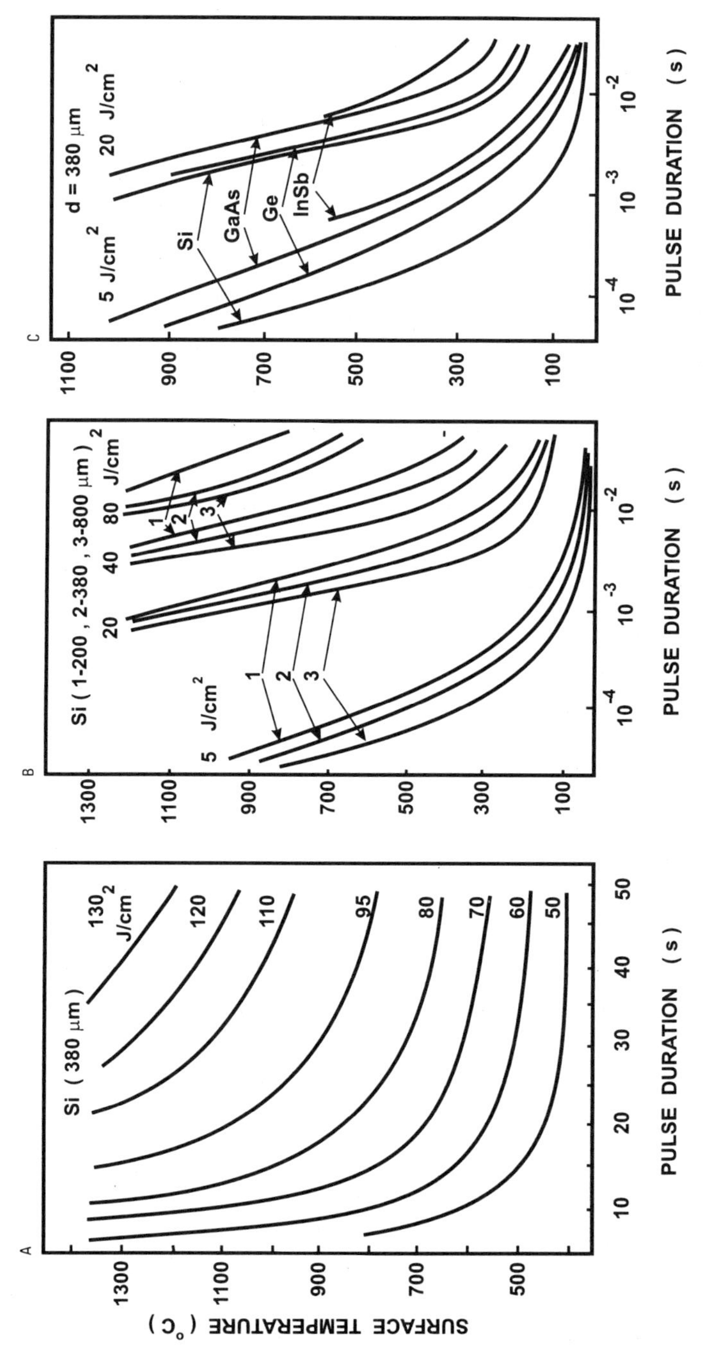

FIGURE 1.6. Surface temperature of semiconductor wafers as a function of xenon flash lamp pulse duration and energy density ($T_0 = 20$ °C).

A rectangular-shaped radiation pulse with time is assumed.

For the temperature–time calculations in a real transient heating apparatus at an incident power density, W_I, the effective emissivity, ε_e, has to be determined experimentally. There is no problem measuring accurately the incident power density with either standard or specially designed calorimeters, which are widely used for this purpose. However, the effective emissivity is determined from the experimental wafer temperature measurements because theoretical calculations are difficult.

A diagram connecting incident power density, effective emissivity, and induced silicon wafer temperature in a 15-s heating cycle (Fig. 1.7) was computed by numerical integration of the heat balance equation (1.23) with only the radiative heat loss term [11]. Thus, the effective emissivity can be easily derived if the two other parameters are known.

In the heat balance regime there are insignificant depth temperature gradients. They can be numerically estimated by [46]

$$\frac{\Delta T}{d} = \frac{\sigma \varepsilon_e T^4}{K} \tag{1.27}$$

Here the value of thermal conductivity, K, corresponds to the processing temperature. Calculations made for silicon wafers 300–500 μm thick show temperature differences

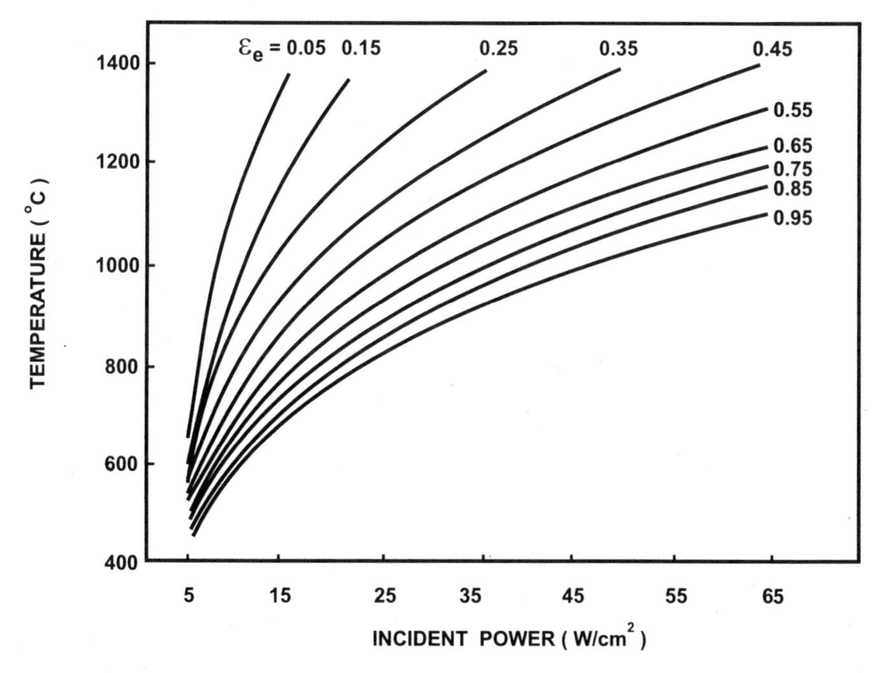

FIGURE 1.7. Induced temperature/incident radiation power/effective emissivity diagram for a 380-μm-thick silicon wafer heated for 15 s with incoherent light from a halogen lamp.

between the irradiated and back surfaces of 7–10 °C, even at the melting point. This is much less than the radial temperature drop across the wafer.

Radial gradients can be calculated with the thin wafer approximation ($d << r_w$, where r_w is the wafer radius) and with the steady-state heat conduction equation in the form [50, 51]

$$K \frac{d}{r} \frac{d}{dr}\left(r \frac{dT}{dr} \right) = -Q + 2\,\sigma \varepsilon_e\, T^4(r) \tag{1.28}$$

where $Q = 2\,\sigma \varepsilon_e\, T_e^4$ is the integral absorbed power density in the wafer center ($r = 0$), T_e the steady-state temperature induced in the wafer center, and r the radial coordinate. Uniform irradiation of the wafer surface and only radiative heat losses are assumed. The factor of 2 in the radiation term of (1.28) represents the fact that both surfaces of the wafer radiate heat.

The boundary condition for the radiative heat loss at the wafer edge is

$$K \left. \frac{dT}{dr} \right|_{r = r_w} = -\sigma \varepsilon_e\, T^4(r_w) \tag{1.29}$$

The solution of (1.28) satisfying (1.29) has been expressed in the form [51]

$$T(r) = T_e - \frac{\sigma \varepsilon_e T_e}{K} \psi \frac{I_0(r/\psi)}{I_1(r_w/\psi)} \tag{1.30}$$

where $\psi = [Kd/(8\,\sigma \varepsilon_e\, T_e^3)]^{1/2}$, I_0 and I_1 are the modified Bessel functions of the first kind of order zero and one, respectively. For the argument $u >> 1$ these functions may by asymptotically approximated by

$$I_0(u) = I_1(u) = \exp(u)/(2\,\pi\, u)^{1/2} \tag{1.31}$$

It should be noted that the term with a minus sign on the right-hand side of (1.30) corresponds to additional radiative heat loss from the wafer edge.

Radial temperature gradients reach the maximum at the wafer edge. Their location is mainly determined by the wafer diameter spreading over a distance of $(0.2–0.3)r_w$ from the wafer edge [51–53].

The above simulation of depth and radial temperature profiles in a transiently heated semiconductor wafer provides a basis for calculating thermally induced elastic stresses and related defect generation in the wafers.

1.4. HEATING OF THIN-FILM STRUCTURES

Transient thermal processing of thin-film microelectronic structures is promising for silicide formation and metal contact alloying on semiconductor substrates, as well as for oxide or nitride film growth and the annealing of interfaces. Induced temperature profiles produced by heating with coherent and incoherent radiation in thin-film structures are defined by both the properties of the materials and radiation parameters. Because experimental recording of the temperature profiles is extremely difficult, theoretical simulations are more frequently applied.

Simulations involve taking into account differences in temperature and spectral dependencies of optical and thermophysical parameters of the films and substrate materials [54–61]. For thin films additional terms need to be taken into consideration in the incident energy balance and heat diffusion equations. These are:

1. The radiative energy exchange at the inner interfaces especially when emissivities of the contacting films (or film and substrate) differ
2. The transmittance increase produced by interference effects, particularly in dielectric film structures
3. The absorption or release of heat through chemical reactions or phase transitions

The latter processes are supposed to be instant in the range of thermal processing time, i.e., seconds. However, in general, the transient time interval for these processes to reach steady state in the film depends on thickness and the induced temperature.

For a structure with linear dimensions greater than its thickness, and with the assumption of uniform irradiation, a one-dimensional temperature distribution may be assumed [61]. Let us analyze the structure composed of $n - 1$ films shown in Fig. 1.8. Each ith film ($i = 1, 2, \ldots n - 1$) and the substrate ($i = n$) is characterized by thickness $x_{i+1} - x_i$, density ρ_i, specific heat $C_i(T)$, reflectivity R_i, absorption coefficient $\alpha_i(E,T)$, and thermal conductivity $K_i(T)$. Temperature profiles in the structure can be calculated by numerical integration of equations in the form of (1.19) with the boundary conditions (1.20), (1.21) written for each film, and for the substrate.

Accounting for the incident light energy redistribution at the inner interfaces, the absorbed energy in the ith film is

$$A_i(E,x,T,t) = W_{i,i+1}\, a_e(x_i,x) + W_{i+1,i}\, a_e(x,x_i) \tag{1.32}$$

Here the thermal generation parameter is determined by

$$a_e(x,x_i) = \exp\left[-\int_x^{x_i} \alpha(x')\, dx'\right] \tag{1.33}$$

where $\alpha(x')$ is a piecewise continuous function made up of partial $\alpha_i(E,T)$ for the whole structure. $W_{i,k}$ is transmitted ($i < k$) and reflected ($i > k$) radiation power at the ith interface, expressed explicitly:

$$W_{0,1} = W_\mathrm{I} \tag{1.34}$$

$$W_{i,i+1} = W_{i-1,i}(1 - R_i) + W_{i+1,i}\, a_e(x_i,x_i+1)\, R_i \tag{1.35}$$

$$W_{i+1,i} = W_{i,i+1}\, a_e(x_i,x_{i+1})\, R_{i+1} + W_{i+2,i+1}\, a_e(x_{i+1},x_{i+2})\,(1 - R_{i+1}) \tag{1.36}$$

$$W_{i+1,i+2} = W_{i,i+1}\, a_e(x_i,x_{i+1})(1 - R_{i+1}) + W_{i+2,i+1}\, a_e(x_{i+1},x_{i+2})\, R_{i+1} \tag{1.37}$$

$$W_{i+2,i+1} = W_{i+1,i+2}\, a_e(x_{i+1},x_{i+2})\, R_{i+2} \tag{1.38}$$

The spectral composition of the radiation, thermal spectral dependencies of optical and thermophysical parameters of the semiconductor substrate and film are included in

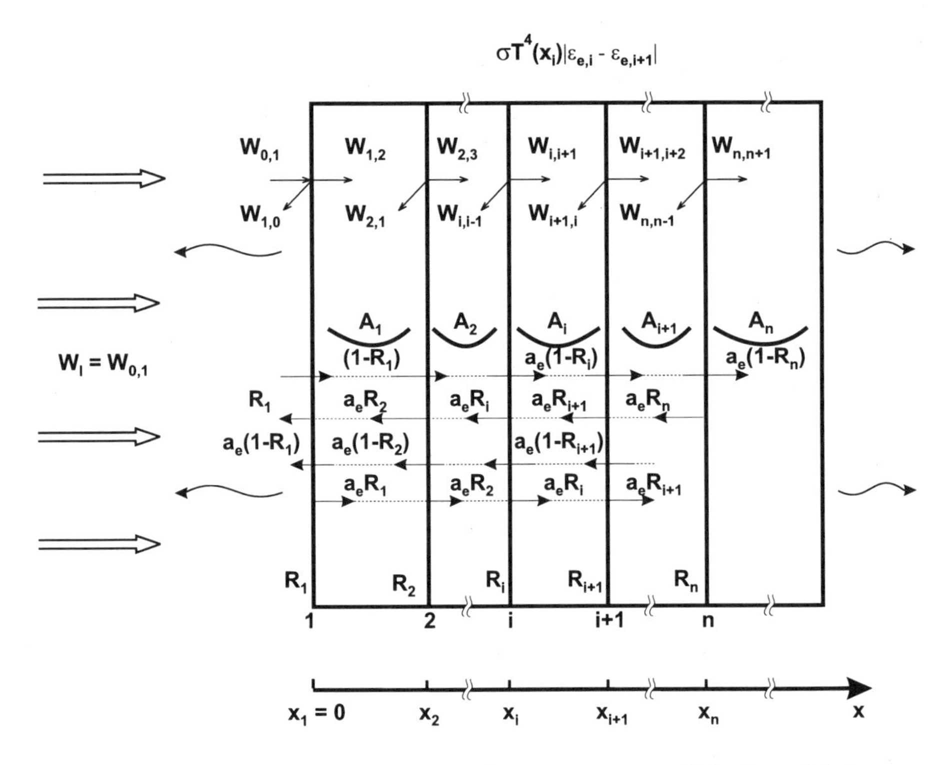

FIGURE 1.8. Energy balance diagram in a thin-film multilayer structure heated with incoherent light from one side of the film.

TABLE 1.2. Thermophysical and Optical Properties of Silicon Dioxide

Property	Value[a]		Reference
Density, ρ (g/cm^3)	2.2		62
Thermal conductivity, $K(T)$ (W cm^{-1} K^{-1})	$2.0 \times 10^{-3} T^{0.336}$	300–696 K	60
	$2.3 \times 10^{-6} T^{1.37}$	696–1149 K	60
	$7.3 \times 10^{-11} T^{2.84}$	> 1149 K	60
	$0.0115 + 1.343 \times 10^{-5}(T - 300)$	300–1000 K	63
	0.0209	> 1000 K	63
Specific heat, $C(T)$ (J g^{-1} K^{-1})	$931.3 + 0.256T - 24 \times 10^{6} T^{-2}$	300–2000 K	64
Reflectivity, R	0.05 – 0.10	300 K	17
Absorption coefficient	~0.01 α_{Si}		62

[a] T is absolute temperature in K and α_{Si} is the silicon absorption coefficient.

TABLE 1.3. Integral Reflectivity of Metal and Silicide Thin Films at Room Temperature

Metal	Metal reflectivity	Silicide	Silicide reflectivity	Wavelength (μm)	Refs.
Al	0.92 – 0.98			0.2–10	1
Ti	0.55			0.555	65
V	0.49	VSi_2	0.30	0.644	66
Cr	0.66			0.560	67
Co	0.68	CoSi	0.47	0.644	66
Ni	—	$NiSi_2$	0.50	0.4–0.9	68
Mo	0.60			0.560	65, 67
Pd	0.81	Pd_2Si	0.28	0.644	66
Ta	0.56			0.560	65, 67
	0.59	$TaSi_2$	0.33	0.644	66
	0.40	—	0.50	0.3–0.6	69
Re	—	$ReSi_2$	0.15	1.0–2.0	70
Pt	0.83			0.644	66
		Pt_2Si	0.44–0.56	0.3–1.3	71
		PtSi	0.44	0.3–1.3	66, 71
Au	0.97			0.7–10	1

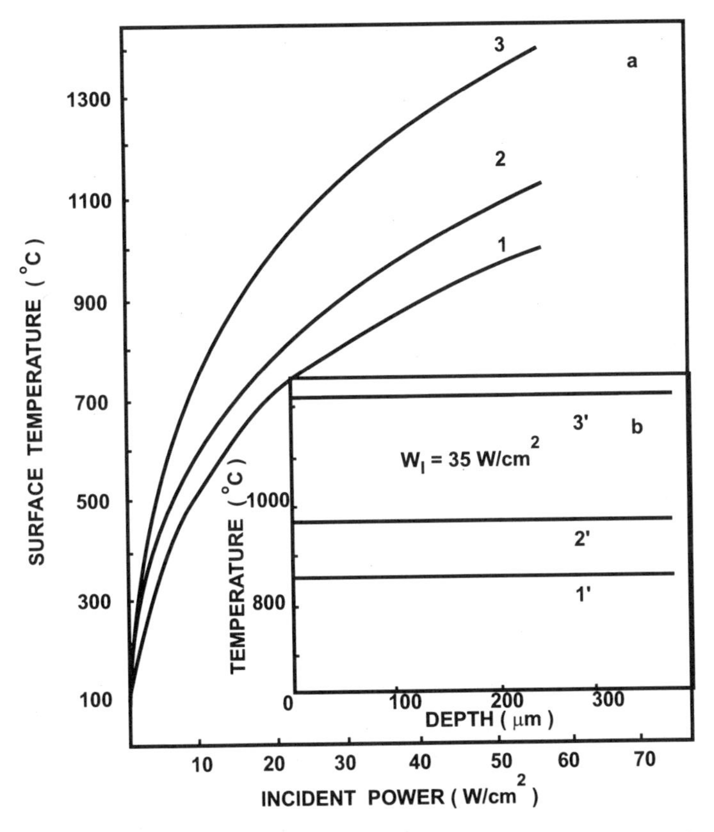

FIGURE 1.9. Temperature profiles for a 380-μm-thick silicon substrate coated with layers of polycrystalline silicon (0.5 μm)/silicon dioxide (0.3 μm) and heated with incoherent light from (1, 1′) the film side, (2, 2′) the substrate side, and (3, 3′) the bare silicon substrate: (a) steady-state surface temperature and (b) depth–temperature profiles.

the above description. The possible energy losses by evaporation, thermal radiation, convection, chemical reactions, and phase transitions are introduced in the boundary conditions. The energy losses from chemical reactions enable one to investigate the effect of oxidation, silicide formation, and nitridation on the temperature profiles. The calculations can be performed only by numerical methods. The appropriate parameters for thin-film materials are listed in Tables 1.2 and 1.3.

Numerical simulation performed for conventional microelectronic structures composed of a light-transparent dielectric or absorbing metallic film on a silicon substrate showed that the efficiency of energy deposition depends essentially on the optical properties of the outer film on the surface being irradiated [56, 57, 60, 61, 72].

Figure 1.9 presents the results of temperature calculations for a polycrystalline silicon/silicon dioxide thin-film sandwich on a single-crystal silicon substrate [61]. In this case 71% of the incident power is reflected, while the absorbed portion amounts to 20% in the polycrystalline silicon film and 9% in the silicon substrate.

Temperature gradients produced for silicon wafers of thickness 380 μm at a fixed incident power density of 35 W/cm^2 were characterized by $\Delta T = 7$ °C for irradiation of

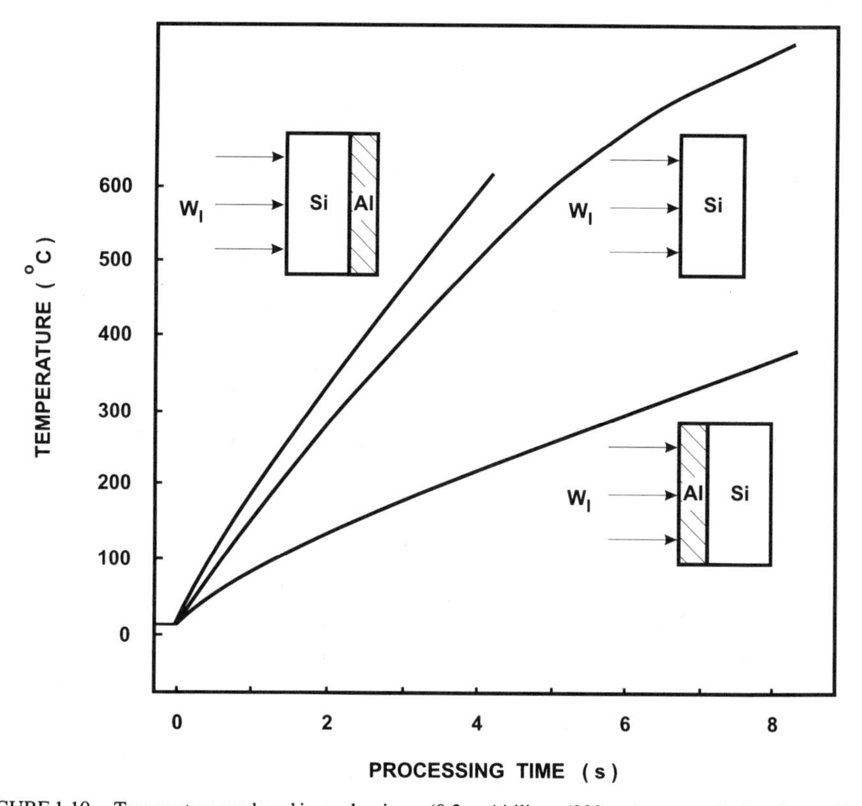

FIGURE 1.10. Temperature produced in an aluminum (0.3 μm)/silicon (380 μm) structure by incoherent light at an incident power density of $W_I = 14$ W/cm^2.

the film surface and $\Delta T = 15$ °C for heating from the other side of the wafer. However, absolute temperature values in the first case were almost 100 °C lower than in the second case. Moreover, the silicon wafer with a film had a much lower temperature than that for a bare silicon wafer. Evidently the presence of transparent or semitransparent films on the silicon wafer changes the absorbed power dramatically.

In the case of metal thin-film structures, light reflection becomes important too. The temperature induced can diminish up to four times in comparison to the one reached for the bare silicon substrate or if the irradiation was incident from the uncoated side of the silicon wafer [61]. The heating depends on the material reflectivity and it may be derived from the numerical estimations for a silicon wafer covered by gold, aluminum, platinum, or vanadium films. Figure 1.10 shows the results for an aluminum film. Here the light reflection by the film decreases the induced temperature by a factor of 2–2.5. However, the efficiency of heating by irradiation from the silicon substrate side is increased 18–20% over that for the bare silicon wafer. The results are identical for thin-film structures on gallium arsenide transiently heated by incoherent light [72].

The temperature calculations performed allow one to draw the conclusion that heating of the thin-film structures by irradiation from the substrate side improves the efficiency of energy deposition and minimizes temperature gradients.

1.5. THERMAL STRESSES AND THERMOPLASTIC EFFECTS

Thermoelastic stresses in a semiconductor wafer are mainly induced by the depth–temperature gradients produced when one side is heated by radiation in the thermal flux regime. The components of deformation in a wafer, located in the z–y plane, can be determined assuming the stressed state to be flat, and the silicon crystal planes perpendicular before and during the heating [28]. In this case the stress components

$$\sigma_{zy} = \sigma_{xz} = \sigma_{xy} = \sigma_{xx} = 0$$

and

$$\sigma_{zz} = \sigma_{yy} = f(x)$$

The equation of compatibility for the latter is

$$\frac{d^2}{dx^2}\left[f(x) + \frac{\alpha_V E^*}{1-\nu} T(x,t) \right] = 0 \tag{1.39}$$

where α_V is the thermal expansion coefficient, E^* Young's modulus, ν Poisson's ratio, and $T(x,t)$ the temperature–time profile. Assuming the force and its moment are equal to zero at the wafer surface, the expression for the stress components are

$$\sigma_{zz} = \sigma_{yy} = \frac{\alpha_V E^*}{1-\nu}$$

$$\left[-T(x,t) + \frac{1}{d}\int_0^d T(x,t)\,dx + \frac{12}{d^3}\left(x - \frac{d}{2}\right)\int_0^d T(x,t)\left(x - \frac{d}{2}\right)dx \right] \tag{1.40}$$

Once the temperature profile as a function of depth is known, (1.40) permits the numerical estimation of the related stresses. The calculated depth distributions of thermoelastic stresses induced in the silicon wafer subjected to one-side transient heating by incoherent light for 50–500 μs [28] are presented in Fig. 1.11. The incident power densities used ensure heating up to 1000–1300 °C. The calculations indicate that there is greater asymmetry in the thermal stress profile with an increase of incident power, and a decrease of pulse duration.

A comparative analysis of the heating regime and induced thermoelastic stresses permit the optimal choice of heating parameters for minimizing stress and related defect generation. Figure 1.12 depicts the relationship between pulse duration, radiation energy

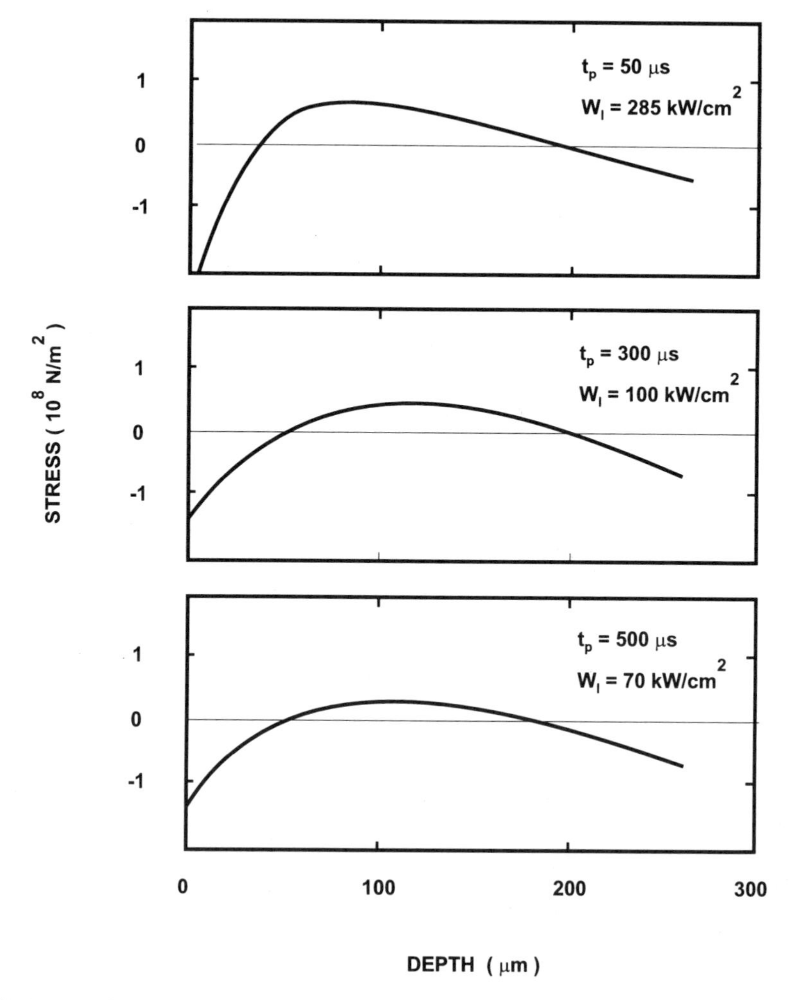

FIGURE 1.11. Calculated stress–depth profiles in a silicon wafer irradiated with incoherent light from one side.

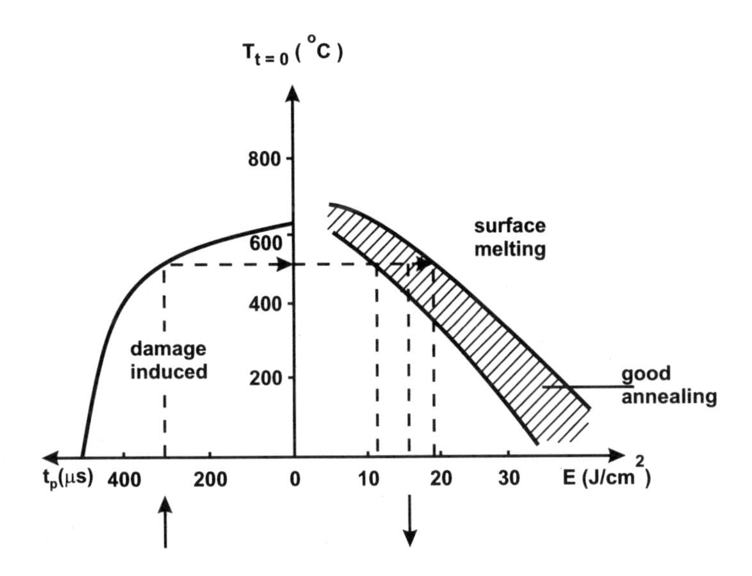

FIGURE 1.12. Relationship between pulse duration, silicon wafer preheating temperature, and pulse energy density for good annealing conditions.

density, and preheating temperature as computed for silicon in Ref. 28. For example, by applying the diagram to a solid-phase regrowth of a 100-nm implanted layer, stress is minimized with a 300-μs pulse, incident energy density in the range 13–19 J/cm^2, and preheating temperature of about 500 °C.

Temperature gradients through the thickness of the wafer appear to be small in the heat balance regime. Radiative heat loss is dominant and the edges of the wafers become cooler than the center as a result of the extra heat loss from the perimeter area of the wafer. Therefore, radial temperature gradients and related thermoelastic stresses are produced. In the steady-state for induced temperature profiles, $T(r)$, stress components in polar coordinates (r,θ) are given by [51, 73]

$$\sigma_{rr} = \frac{\alpha_v E^*}{1-\nu}\left[\frac{1}{r_w^2}\int_0^{r_w} T(r)\, r\, dr - \frac{1}{r^2}\int_0^{r} T(r)\, r\, dr\right] \tag{1.41}$$

$$\sigma_{\theta\theta} = \frac{\alpha_v E^*}{1-\nu}\left[\frac{1}{r_w^2}\int_0^{r_w} T(r)\, r\, dr + \frac{1}{r^2}\int_0^{r} T(r)\, r\, dr - T(r)\right] \tag{1.42}$$

$$\sigma_{r\theta} = 0 \tag{1.43}$$

where $2r_w$ is the wafer diameter. Substitution of the temperature profile in the form of (1.30) leads to

$$\sigma_{rr} = -\frac{\alpha_v E^* \sigma \varepsilon_e T_e^4 \psi^2}{K(1-\nu)} \left[\frac{1}{r_w} - \frac{1}{r} \frac{I_1(r/\psi)}{I_1(r_w/\psi)} \right] \tag{1.44}$$

$$\sigma_{\theta\theta} = -\frac{\alpha_v E^* \sigma \varepsilon_e T_e^4 \psi^2}{K(1-\nu)} \left[\frac{1}{r_w} + \frac{1}{r} \frac{I_1(r/\psi)}{I_1(r_w/\psi)} - \frac{1}{\psi} \frac{I_0(r/\psi)}{I_1(r_w/\psi)} \right] \tag{1.45}$$

$$\sigma_{r\theta} = 0 \tag{1.46}$$

The appropriate data for numerical estimation of stresses in single-crystal silicon wafers are presented in Table 1.4.

Figure 1.13 shows calculated steady-state σ_{rr} and $\sigma_{\theta\theta}$ stress components as a function of the distance from the center of the silicon wafer. The component $\sigma_{\theta\theta}$ appears to be the most valuable. It reaches a maximum value at the wafer edge ($r = r_w$). For the temperature profile described by (1.30) σ_{max} is given by [51]

$$\sigma_{\theta\theta}^{max} = \frac{\alpha_v E^* \sigma \varepsilon_e T_e^4 \psi^2}{K(1-\nu)} \left(\frac{1}{\psi} - \frac{2}{r_w} \right) \tag{1.47}$$

There are compressive stresses in the center of a wafer while tensile ones are dominant at the wafer edge.

Stresses in a crystal can relax by shear which occurs only along certain well-defined crystallographic directions [77]. For silicon, with a diamond cubic lattice, the slip planes are (111), and slip directions are ⟨110⟩. There are 12 possible slip systems relevant to different orientations of the (111) planes. However, because of the crystallographic symmetry, only 5 of them are independent. The stress components in the rectangular coordinates are [51]

$$\sigma_{zz} = 0.5\,[\sigma_{rr} + \sigma_{\theta\theta} + (\sigma_{rr} - \sigma_{\theta\theta})\cos(2\theta)] \tag{1.48}$$

$$\sigma_{yy} = 0.5\,[\sigma_{rr} + \sigma_{\theta\theta} - (\sigma_{rr} - \sigma_{\theta\theta})\cos(2\theta)] \tag{1.49}$$

TABLE 1.4. Silicon Parameters Used for Thermoelastic Stress Calculations

Parameter	Value	Crystal orientation	Refs.
Young's modulus ($\times 10^{11}$ N/m^2)	1.6	—	51
	1.30	(100)	73, 74
	1.70	(100)	75
	1.9	(111)	73, 74
Poisson's ratio	0.18	—	51
	0.2	—	73
	0.25	—	74, 76
Thermal expansion coefficient ($\times 10^{-6}$ deg^{-1})	2.33	—	51
	4	—	73
	4.68	—	76
	15	—	74

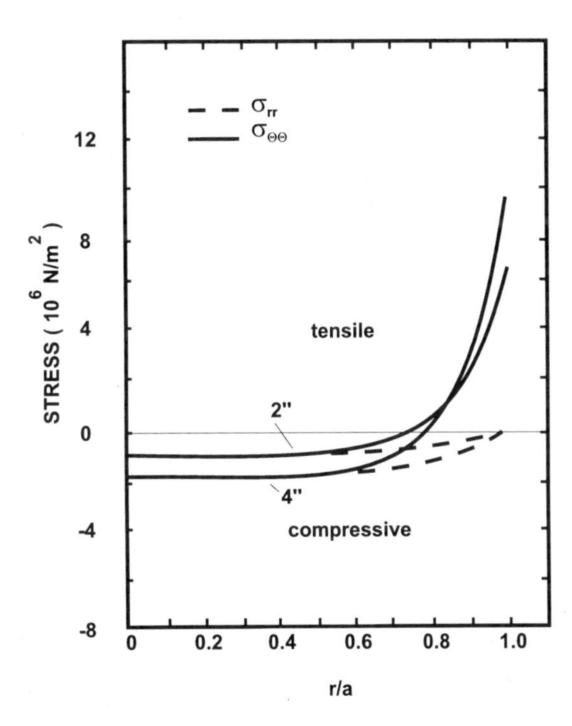

FIGURE 1.13. Computed radial profiles of the σ_{rr} stress (dashed line) and $\sigma_{\theta\theta}$ stress (solid line) in silicon wafers of 50-mm (2 inch) and 100-mm (4 inch) diameter ($d/a = 0.01$, where d is the wafer thickness and a is the wafer diameter) heated to 1300 °C with uniform irradiation.

$$\sigma_{zy} = 0.5\,(\sigma_{rr} - \sigma_{\theta\theta})\,\sin(2\theta) \tag{1.50}$$

where θ is measured counterclockwise starting from the z-axis, as illustrated in Fig. 1.14. For a (100) silicon wafer the stresses resolved in the independent slip directions are [51]

$$\begin{aligned} |S_1| &= (2/3)^{1/2}\left|\sigma_{zy}\right|, \qquad & |S_2| &= (1/6)^{1/2}\left|\sigma_{zz} + \sigma_{zy}\right| \\ |S_3| &= (1/6)^{1/2}\left|\sigma_{zz} - \sigma_{zy}\right|, \qquad & |S_4| &= (1/6)^{1/2}\left|\sigma_{yy} + \sigma_{zy}\right| \\ |S_5| &= (1/6)^{1/2}\left|\sigma_{yy} - \sigma_{zy}\right| \end{aligned} \tag{1.51}$$

The overall stress angle distribution has $\pi/2$ (i.e., fourfold) periodicity and is symmetrical about $\theta = \pi/4$. The maximum stresses occur for $\theta = (2i + 1)\pi/8$, $i = 0, 1, 2, \ldots$, and at the wafer edge

$$|S_{\max}| = (1/6)^{1/2}\,[1/2 + (1/2)^{1/2}]\,\sigma_{\theta\theta}^{\max} \approx 0.5\,\sigma_{\theta\theta}^{\max} \tag{1.52}$$

Therefore, there are eight equally spaced topographic positions in the wafer where stresses are concentrated and the elastic limit will be exceeded first. Thermoplastic effects appear where the induced thermal stresses are greater than the yield stress for the material, given by [51, 66]

$$\sigma_c(T,d) = c^*\,(\dot{d})^{1/m^*}\exp\left(\frac{E_a}{m^* kT}\right) \tag{1.53}$$

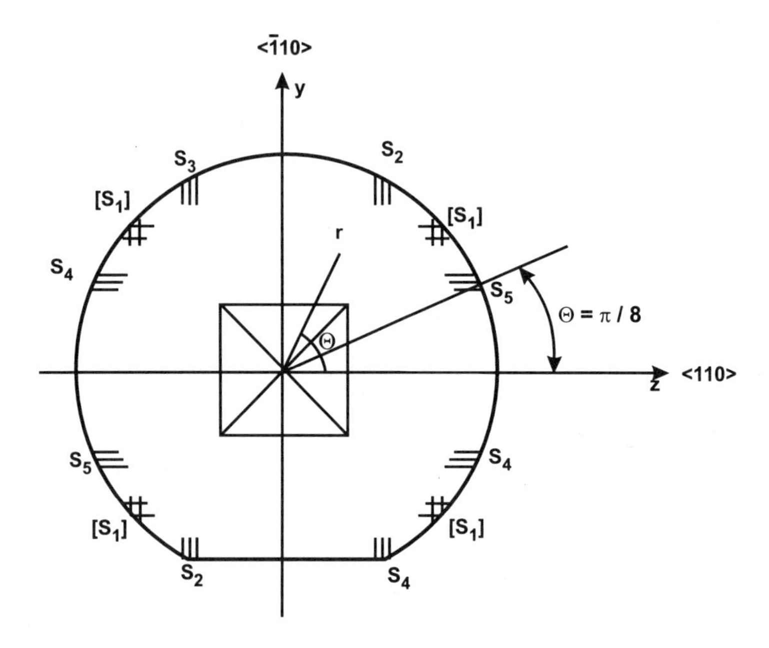

FIGURE 1.14. Azimuthal positions of the maxima in the resolved stresses and associated slip directions for a (100) silicon wafer.

where E_a is the activation energy of glide movement, $\dot{d}$ the strain rate in s^{-1}, and c^* and m^* the material-dependent constants. A wide spectrum of parameter values for silicon are summarized in Table 1.5. Among them, $c^* = 4.7 \times 10^4$ N/m^2, $m^* = 2.9$, $E_a = 2.32$ eV [83] are the best fit to follow the experimental results for undoped (100) silicon. Glide activation energy and constant c^* are changed essentially by impurity atoms in the crystal.

TABLE 1.5. Yield Stress Parameters for Silicon

Temperature range (° C)	Activation energy (eV)	Material constants		Strain rate, $\dot{d}$ (s^{-1})	Refs.
		m^*	c^* (N/m^2)		
800	2.14	—	—	10^{-3}	78
800	2.17	3.4	—	10^{-3}	79
800	2.30	—	—	10^{-3}	80
800	2.4	3.5	—	10^{-3}	81
700–900	2.1	3.0	—	$6 \times 10^{-5} - 6 \times 10^{-4}$	82
700–1050	2.19	2.3	2.3×10^4	$2 \times 10^{-5} - 2 \times 10^{-4}$	83
750–1025	2.4	3.0	4.8×10^4	$2 \times 10^{-6} - 1 \times 10^{-3}$	84
750–1100	2.3	2.9	4.7×10^4	$1 \times 10^{-4} - 1 \times 10^{-2}$	85
800–1000	2.25	—	—	$7 \times 10^{-5} - 7 \times 10^{-3}$	86
800–1000	2.64	3.3	2.4×10^4	$2 \times 10^{-5} - 6 \times 10^{-4}$	87
850–1030	2.35	2.9	3.5×10^4	$4 \times 10^{-5} - 2 \times 10^{-4}$	88
900–1300	2.32	2.9	4.7×10^4	$5 \times 10^{-4} - 5 \times 10^{-3}$	83

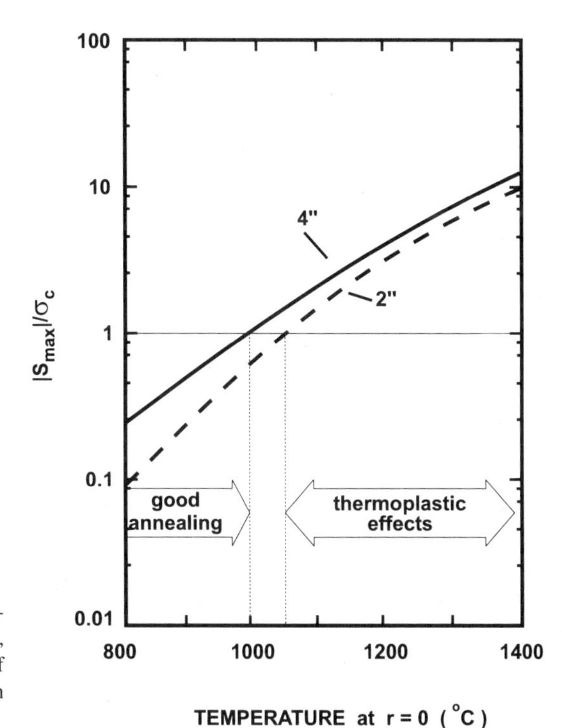

FIGURE 1.15. Ratio between the maximum resolved stress and the yield stress, computed at the wafer edge as a function of the temperature for silicon wafers of 50-mm (2 inch) and 100-mm (4 inch) diameter.

Thus, for the silicon doped by phosphorus up to the concentration of 10^{20} atoms/cm^3 the parameters have the values $E_a = 1.74$ eV and $c^* = (4.3 \pm 0.3) \times 10^5$ N/m^2 [83].

Thermoelastic stresses that exceed the yield strength will relax producing dislocations in the wafer. Figure 1.15 shows the ratio between the maximum resolved stress induced in silicon wafers of 50-mm (2 inch) and 100-mm (4 inch) diameter and corresponding yield stress as a function of the temperature in the center of the wafer [51]. The threshold temperature for defect formation is in the range of 1000–1050 °C depending on the silicon wafer diameter. Large-diameter wafers undergo plastic deformation at a lower temperature. This conclusion is supported by numerous experimental studies [51, 53, 74, 89].

Calculations combined with the experimental results [73] allow a determination of the range of the radial temperature gradients where the thermal stresses induce thermoplastic effects. The results are presented in Fig. 1.16 for a silicon wafer of 150-mm (6 inch) diameter. On the basis of Fig. 1.16 the maximum radial temperature gradient should not exceed 6–11 °C/cm over the temperature range 950–1200 °C. At lower wafer temperatures the higher-temperature gradients do not produce dislocations.

Radial temperature gradients in semiconductor wafers heated by incoherent light can be reduced if the energy absorption at the wafer edge is increased. A ring of silicon dioxide formed at the wafer edge is useful for this purpose [90]. The reflectivity of the SiO_2/Si structure is a function of the incident light wavelength and the oxide film thickness. The

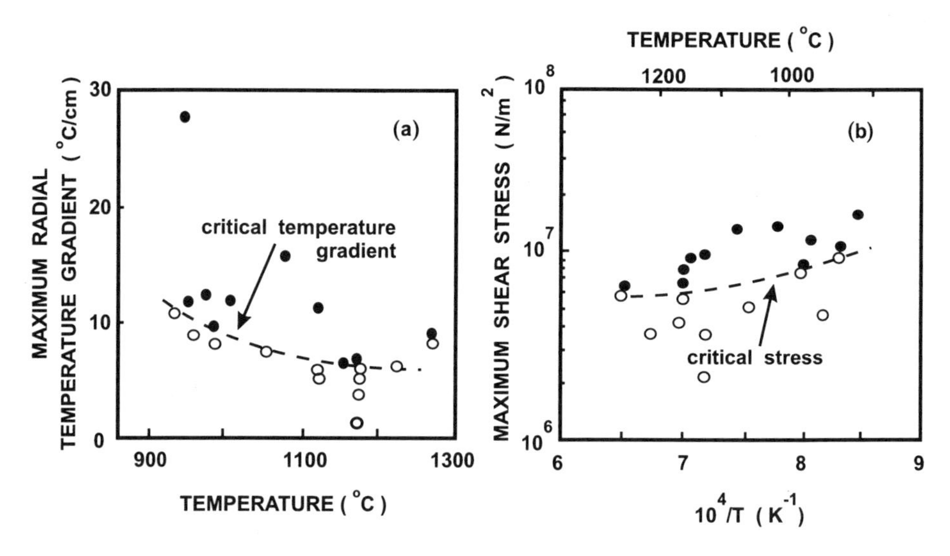

FIGURE 1.16. Temperature dependence of (a) the critical radial temperature gradient and (b) the maximum stress for inducing slip dislocations in 150-mm (6 inch) silicon wafers. ●, slip, ○, no slip.

film thickness and the ring width may be adjusted to reduce light reflection, thus compensating for the energy loss at the edge.

Another way to minimize the nonuniformity in the temperature distribution is to use a radiation shield [91]. Placing a cone-shaped shield at the edge of the wafer reduces heat loss from the wafer's edges by reflecting radiative energy back into the wafer. By varying the angle of the conic section, an optimum shield configuration can be found to minimize thermal stresses induced.

A better improvement of radial temperature uniformity might be provided by a supplementary heat source surrounding the wafer. Both a passive guard ring and independently heated ring [92, 93] or an annular lamp [92, 94] can be used. A passive guard ring around the wafer reflects and emits radiation laterally to the edge. The ring is heated simultaneously with the wafer and the heat radiation from the ring is absorbed by the wafer edge. In order to achieve the same temperature rate for the ring as for the wafer, the ring has to be designed in such a way that the ratio $\sigma\varepsilon_e/(C\rho d)$ for the ring matches the corresponding value for the wafer, where ε_e is the effective emissivity, C specific heat, ρ density, and d the thickness. Guard rings reduce the temperature gradient-induced wafer deformation compared to a freestanding wafer by a factor of ten [93].

In the case of a guard ring, the problem of radial temperature gradients is not solved but only transferred to the ring. Therefore, warping of the ring can occur. Application of an independently powered radiation source around a wafer [94] might be a better approach, although more complicated.

Design features of rapid thermal processing equipment also play a role determining the amount of plastic deformation of the semiconductor wafer. The most important factors are radiation source geometry and wafer holder configuration. For incoherent light tube-like sources (halogen and arc xenon lamps), which are widely used, the distance between the processed semiconductor wafer and the light sources is critical to provide a uniform irradiation, particularly for large-area wafers. It has been shown that for a block consisting of tube lamps of 10- to 15-mm diameter, the distance to the processed wafer must not be less than 20–25 mm [95, 96]. For conventional three-point wafer holders the distance between the point supports must also be optimized for compensation of mechanical tensile and compressive stresses in the wafer.

In practice, processing time and induced temperatures have the most influence on the thermal stress and related defect generation. Transient thermal processing in the heat balance regime is the most favorable, providing the lowest temperature gradients and induced stresses. It was found [93, 97] that wafers deform during the complete annealing cycle and moreover that their deformation is greatest during the heating and cooling transients. The optimal conditions for silicon include time duration over a range of seconds at induced temperatures up to 1000–1050 °C, provided that radial gradients are less than 8–9 °C/cm, and the rate of cooling in the high-temperature region does not exceed 100 °C/s. These limits may be exceeded if specially matched compensating thin-film coatings are utilized.

2

Recrystallization of Implanted Layers and Impurity Behavior in Silicon Crystals

Rapid thermal processing (RTP) is traditionally used for defect annealing, recrystallization, and activation of doped implanted layers. Most previous studies using RTP have operated in the thermal flux or heat balance regimes (as defined in Section 1.1). Moreover, the heat balance regime is often preferred with incoherent light and for short heating times (seconds). Solid-state processes initiated in semiconductors by RTP are mainly of thermal origin and hence there are many similarities with conventional furnace annealing. However, the much shorter processing times, rapid heating and cooling produce distinct features. The nonequilibrium fast transient processes occurring can introduce structural changes during the initial moment of the thermal processing. In this chapter the general features of point defect generation and annealing, implanted layer recrystallization, and impurity behavior in silicon are presented and discussed. In particular, the distinct features based on the nonequilibrium conditions created in RTP are emphasized. These phenomena have been investigated for the most part in the solid state for silicon.

2.1. POINT DEFECT GENERATION AND ANNEALING

Thermal processing is traditionally employed to control qualitative composition and quantitative balance of the defects in semiconductor crystals. A furnace process of long duration is frequently followed by point defect annealing. RTP can both anneal and generate point defects at the same time. For silicon these defects appear at processing times extending from nanoseconds, in the liquid phase [1–6], to seconds, in the solid state [7–22].

The electronic properties of point defects generated during solid-state RTP are strongly influenced by the crystal growing technique, type of silicon, and contamination of the sample with fast-diffusing impurities. The concentration of the electrically active point defects in Czochralski-grown single-crystal silicon is 10^{13}–10^{16} defects/cm^3. In float zone silicon crystals the concentration is usually lower than the sensitivity limit of deep-level transient spectroscopy (about 10^{11} defects/cm^3). This difference is related to unwanted impurities, such as oxygen, carbon, and fast-diffusing metals which can be at a relatively high concentration in Czochralski-grown silicon [23] or are deposited during chemical cleaning of the wafers [24, 25]. An extensive spectrum of the point defects associated with deep levels in *p*- and *n*-type silicon subject to RTP is summarized in Table 2.1.

2.1.1. *Traps*

In general, hole traps are observed in *p*-silicon and electron traps in *n*-silicon. They usually appear after rapid heating to 850–1200 °C over 3–10 s and rapid cooling at a rate of 5–50 °C/s. These conditions correspond to the transient mode of the heat balance regime (defined in Section 1.1). The origin of the traps and their nature are interpreted as follows.

TABLE 2.1. Deep Levels and Related Point Defects Generated by Rapid Thermal Processing in Czochralski-Grown Silicon

Processing conditions		Deep-level parameters				
Temperature, T (°C)	Time, t (s)	Activation energy (eV)	Capture cross section (cm^2)	Concentration (cm^{-3})	Defect[a]	Refs.
		p-Silicon (boron doped)				
900–1150	20	$E_v + 0.30$	2×10^{-15}	$10^{13} - 10^{15}$	v–B	7, 8
1010	0.5–14	$Ev + 0.23$	—	~10^{15}	v–v	9
		$Ev + 0.34$		~5×10^{14}	v–C–O	
900	10	$E_v + 0.32$	9×10^{-18}	$(2–5) \times 10^{13}$	Fe_i	10
1100		$E_v + 0.45$	5×10^{-19}	$(2–5) \times 10^{13}$	Fe_i	
850–1050	5	$E_v + 0.29$	1.4×10^{-15}	—	v–B, Cr_i–B	11
		$E_v + 0.30$	1.5×10^{-17}	—	V_1, Si_i–Si_i	
		$E_v + 0.45$	1.5×10^{-16}	$(1–10) \times 10^{13}$	Fe_i	
1050	40	$E_v + 0.28$	~10^{-15}	—	Cr_i–B	12
		n-Silicon (phosphorus doped)				
550	0.16	$E_c - 0.17$	>10^{-15}	$(2–5) \times 10^{14}$	v–O	15
1000	5, 10	$E_c - 0.18$	—	—	v–O	14, 18, 19, 22
300	0.07	$E_c - 0.39$	3.7×10^{-16}	$(5–10) \times 10^{14}$	v–P	15
550	0.16	$E_c - 0.40$	2.0×10^{-16}	$(5–50) \times 10^{14}$	v–P	15
1169–1172	9–12	$E_c - 0.21$	—	—	Co_i	16
		$E_c - 0.26$	—	—	Co–B	
		$E_c - 0.48$	—	—	Co–B	
1000	10	$E_c - 0.25$	—	—	—	22
		$E_c - 0.36$	—	2×10^{13}	—	18, 19, 22
		$E_c - 0.52$	—	4×10^{12}	—	22
800, 1000	10	$E_c - 0.53$	—	$(1–10) \times 10^{13}$	Au	17
1000	5	$E_c - 0.56$	2×10^{-15}	—	Au–O,Au–P	14
1000	10	$E_c - 0.58$	1.5×10^{-15}	~10^{15}	the surface	13

[a]v = vacancy; i = interstitial.

The level $E_v + 0.29$ (0.28, 0.30) eV with a capture cross section of about 10^{-15} cm^2 is normally ascribed to a hypothetical vacancy–boron complex which is stabilized by quenching [7, 8, 11]. However, this is not the only interpretation because there can be a pairing reaction between substitutional boron and interstitial chromium [26, 27] giving rise to the same energy level [12]. The maximum concentration of the defects producing the $E_v + 0.29$ level is observed after processing at about 1000 °C for 5 s, while an increase of the processing time to 20 s results in their complete annealing. This indicates the nonequilibrium nature of the defects and their direct connection with the rearrangement of intrinsic defects in the crystal. Conventional furnace annealing at temperatures higher than 400 °C provides an exponential decrease in the concentration of the traps with increasing temperature. It is characterized by an activation energy of 0.3 eV [7].

The levels $E_v + 0.23$ eV and $E_v + 0.34$ eV were found in *p*-silicon heated with an electron beam for a few seconds [9]. These levels correspond to a divacancy [28, 29] and vacancy–carbon–oxygen complexes [29], as identified by radiation experiments. Their origin is connected with thermally stimulated rearrangement of the system of vacancy point defects.

The level $E_v + 0.30$ eV with the capture cross section of about 10^{-17} cm^2 corresponds to two possible defects, namely, those of an interstitial vanadium, or silicon di-interstitial [11]. They both have close energy levels in the silicon band gap, a small capture cross section, and they are not stable at room temperature [30, 31]. This instability is typical for the defects discussed above induced by RTP [11].

The levels $E_v + 0.32$ eV and $E_v + 0.45$ eV are attributed to the presence of iron atoms in silicon [10, 11]. In particular, the appearance of level $E_v + 0.45$ eV is explained by dissociation of complexes formed by interstitial iron and substitutional boron. Interstitial iron, as shown by Chantre *et al.* [26], creates the hole trap level $E_v + 0.4$ eV, which is unstable at room temperature. The defects generated by RTP disappear after 60 days of storing the sample at room temperature.

The level $E_c - 0.17$ (0.18) eV corresponds to an extensively studied vacancy–oxygen complex (A-center) [32, 33]. It is formed by RTP at the relatively low temperature of about 550 °C, and at high temperatures reaching 1000–1200 °C [14, 18, 22]. Subsequent furnace annealing at 550 °C for 20 min eliminates this defect.

The levels $E_c - 0.39$ eV and $E_c - 0.40$ eV correspond to divacancy complexes. The researchers who observed these traps tried to refer them to vacancy–phosphorus complexes [15]. In fact, vacancy–phosphorus complex (E-center) forms the level $E_c - 0.44$ eV, while neutral and negatively charged divacancies produce levels $E_c - 0.39$ eV and $E_c - 0.40$ eV, respectively [33].

The family of levels $E_c - 0.21$ eV, $E_c - 0.26$ eV, $E_c - 0.48$ eV was found in *n*-silicon contaminated by cobalt [16]. The level $E_c - 0.21$ eV is assumed to belong to interstitial cobalt [27, 34]. The outer two traps appear only in the samples that are also doped with boron and they are consequently attributed to a negatively charged complex consisting of interstitial cobalt and boron. It forms two energy levels in the silicon band gap becoming neutral or twice negatively charged. Such an interpretation should be reexamined in light of the later investigation of cobalt-related traps in silicon [35], in which $E_c - 0.23$ eV and $E_c - 0.40$ eV levels were detected. Moreover, the level $E_c - 0.48$ eV (capture cross section 5×10^{-16} cm^2) has been attributed to iron [36].

The level E_c – 0.36 eV has been suggested [19] to be related to a residual impurity present in the starting material. It is probably a complex including metallic impurities activated during quenching of the RTP cycle.

Deep levels with energies from E_c – 0.52 eV to E_c – 0.58 eV are associated with heavy metal impurities, and in particular with gold [14].

2.1.2. *Effect of Defects*

Silicon processed by RTP is much more sensitive to contamination than a conventional heat treatment with a slow cooling rate [20, 21]. Some quenched-in defects are caused by the activation of isolated metallic impurities. Other contributing factors include structural defects generated during rapid heating and cooling [13, 14], damage in the surface produced by mechanical polishing [37, 38], dislocations [39, 40], as well as metal impurities absorbed at the surface [41]. Surface passivation with a 3-nm layer of silicon dioxide or additional chemical polishing removes these defects. The majority of the defects introduced by RTP are impurity complexes and interstitial impurities. They can be reduced by annealing the sample at a high temperature (1000 °C) for 20 s or longer [10], or at 550–650 °C for 60 min [13, 14]. Experiments show that these defects are primarily located in a surface layer 500–700 nm thick. Surface structures and the process of energy deposition through radiant heating are important in defining defect generation and migration. Work with the rapid annealing of oxygen thermal donors [42, 43], laser-induced defects [5, 6], defects in boron difluoride low-dose-implanted layers [44], and nuclear reaction-doped silicon [45–47] provide illustrative examples.

Other studies include RTP of thermal donors in Czochralski-grown silicon with a scanned electron beam at 650 °C in 10 s [42]. Conventional furnace annealing at the same temperature requires 1–3 h.

Similar behavior was observed in the RTP of defects generated by pulse laser melting of silicon surfaces [5, 6]. The annealing process was monitored via control of the behavior of the E_c – 0.32 eV trap over the temperature range of 500–650 °C. Defect annealing was about 30 times faster for RTP than annealing in a conventional furnace. Rapid thermal annealing (RTA) has an activation energy of 0.81 eV, while the classical furnace anneal (FA) requires 1.12 eV. This result lends strong support to the hypothesis that ionization-induced enhancement of point defect annealing takes place during rapid radiation heating.

An enhanced electrical activation of boron introduced in silicon by low-dose implantation (3×10^{13} ions/cm^2) of boron difluoride ions (BF_2^+) was observed. However, an amorphous layer was not formed [44]. Samples with different free charge carrier diffusion lengths heated with incoherent light, from the implanted and the back surfaces, demonstrate nonequilibrium charge carrier generation in the energy-absorbing layer. This is important for the electrical activation of boron and point defect annealing. There was also an improvement in electrophysical parameters for nuclear reaction-doped *p–n* silicon structures, formed within seconds after RTP by incoherent light [43, 45]. This confirms enhanced solid-state annealing of point defects.

The observed enhancement is explained consistently by the influence of ionization processes in the radiation-absorbing layer. Indeed, the stability of point defects in semiconductors essentially depends on their charge state [48]. Ionization of a semiconductor material with radiation changes the charge states of the atom through recombination of nonequilibrium free charge carriers at the defects. The relative position of the Fermi level and the energy of the traps created by point defects appears to be critical in this case. Generation of nonequilibrium charge carriers causes electron traps above the Fermi level to become empty and traps below the Fermi level to become full. For holes the situation is reversed.

Changing a defect charge state in the energy-absorbing layer results in a violation of thermodynamic equilibrium for the defects and causes rapid dissociation of the recharged defects. Hence, ionization-controlled effects may be as important in the radiation heating process. Moreover, photons have been observed [49] to play a key role in the atomic migration and rearrangement of the silicon lattice when surface chemical reactions are involved.

In summary, the concentration of point defects generated in single-crystal silicon by radiation heating within seconds is not higher than 10^{15}–10^{16} defects/cm^3. Point defects influence electrophysical parameters of the semiconductor where there is a comparable concentration of ionized substitutional impurities, such as in lightly doped layers and at the *p–n* junction.

2.2. RECRYSTALLIZATION OF IMPLANTED LAYERS

The main criterion for estimation of the effectiveness of the annealing is the degree of lattice perfection and residual defects in an implanted doped layer. Lattice imperfections and residual defects mainly determine electrical activation of implanted impurities. Annealing of ion implantation-amorphized silicon crystals by processing in the thermal flux [50] and heat balance [51–53] regimes has been shown to be a fundamentally heat-controlled process.

Recrystallization starts at an interface between the amorphous and the single-crystal phases, moving the interface to the surface. Ionization has practically no influence on the regrowth rate [54–56]. Therefore, we can apply the models developed for conventional furnace processing to the recrystallization process initiated by rapid heating radiation [57–66].

2.2.1. Epitaxial Regrowth

Recrystallization of an implanted layer can be considered as an increase in the crystal-phase volume through the addition of new epitaxially aligned layers. It occurs through nucleation followed by epitaxial growth. Nucleation of crystallites at the amorphous–crystalline interface is the initial rate-determining step for the regrowth process. The interface between the amorphous and crystalline phases is usually created at the surface at a specific energy. Nucleation of a crystallite consisting of N'_a atoms leads to a change in the thermodynamic potential, which is the sum of the bulk and the surface terms

$$\Delta G = N'_a \Delta g + s \Psi \tag{2.1}$$

where Ψ is the surface free energy, $\Delta g = g_{cr} - g_{am}$ (<0) is the Gibbs free energy for crystallization, i.e., the difference between the Gibbs energies per atom in crystalline and amorphous phases [51]. The crystallite surface area is [52]

$$s = \eta N_a'^{2/3} \tag{2.2}$$

where $\eta = (36\,\pi\,U_1^2)^{1/3}$ is the geometry factor for a sphere and U_1 the volume of one atom or a molecule in the crystal.

The crystalline phase is more stable, having a Gibbs energy lower than that for the amorphous state. The lower the Gibbs free energy, the more stable the corresponding state is. This is why the free energy of crystallization is negative providing an extremum of ΔG as a function of N_a' .The change of thermodynamic potential reaches a maximum when the number of atoms in the nucleus becomes critical. The function (2.1) accounting for (2.2) has the extremum at

$$N_a' = N_c' = -\left[\frac{2\eta\Psi}{3\,\Delta g}\right]^3 \tag{2.3}$$

A nucleus consisting of N_c' atoms is of critical size. Further growth of such a crystallite appears to be easy because the energy barrier is decreased. The Gibbs energy for the formation of a crystallite having critical size is

$$\Delta G_c = 4\,\frac{(\eta\Psi)^3}{27\,\Delta g^2} \tag{2.4}$$

Thus, the minimum number of atoms that create a stable crystallite and the Gibbs energy are correlated by the expression

$$N_c' = 2\,\frac{\Delta G_c}{\Delta g} \tag{2.5}$$

Crystallites consisting of 840 silicon atoms and of 53 germanium atoms are estimated to be stable, when parameters of the materials are taken from Germain *et al.* [58]. This confirms the experimental fact that crystallization in silicon requires more energy expense, and as a consequence is initiated at higher temperatures than with germanium.

The growth kinetics of the subsequent crystalline phase are controlled by the atomic rearrangement near the interface. The atoms have to overcome the potential barrier, corresponding to the Gibbs energy for migration

$$\Delta g_m = \Delta H_m - T\,\Delta S_m \tag{2.6}$$

where ΔH_m is the enthalpy and ΔS_m the entropy for migration. The nucleation rate is then [58]

$$\omega_r = N_a'\,\upsilon_o^*\exp\left(-\frac{\Delta g_m}{kT}\right)\exp\left(-\frac{\Delta G_c}{kT}\right)\left[1-\exp\left(\frac{\Delta g}{kT}\right)\right] \tag{2.7}$$

where N_a' is the number of atoms and υ_o^* the atomic vibrational frequency. Thus, the regrowth rate, i.e., the rate at which the amorphous–crystalline interface moves, is usually written as

$$v_c = \delta\, \upsilon_o^*\, \exp\left(\frac{\Delta S_m}{k}\right) \exp\left(-\frac{\Delta H_m}{kT}\right)\left[1 - \exp\left(\frac{\Delta g}{kT}\right)\right] \tag{2.8}$$

where δ is the jump distance of an atom at the interface.

The above approach allows us to simulate amorphized silicon recrystallization using estimates of the thermodynamic parameters: $\delta = 0.5$ nm, $\upsilon_o^* = 10^{13}\ \text{s}^{-1}$, $\Delta S_m = 37$ K, $\Delta H_m = 4.9$ eV, and $\Delta g = 0.12$ eV [58, 59, 62–66].

Numerical simulation based on these parameters may be considered as a zero-order approximation because of the limited accuracy of the parameters. Moreover, there is no account taken of the crystal orientation, impurities, precipitates, and other factors influencing the recrystallization process.

Amorphous layers are known to regrow epitaxially on (100) silicon and germanium crystals with a constant linear rate at a fixed temperature [58, 59]. Recrystallization is fastest for the (100) orientation. The experimentally observed temperature dependence of the recrystallization rate is approximated by the Arrhenius expression

$$v_c = v_o \exp\left(-\frac{E_a}{kT}\right) \tag{2.9}$$

The experimentally determined activation energy obtained for self-ion-amorphized (100) silicon, both at low temperatures (500–600 °C) [67–69] and at high temperatures (600–950 °C) [65, 70, 71], range from 2.24 to 2.85 eV. Within the experimental uncertainty, all of the experimental data belong to the region defined by the activation energy $E_a = 2.54 \pm 0.02$ eV and preexponential factor $v_o = (5.2 \pm 0.2) \times 10^{14}$ nm/s. Recrystallization rates can be estimated for the other principal crystal directions by considering the epitaxial regrowth for that direction relative to the ⟨100⟩ crystal direction. Epitaxial regrowth occurs 4 and 24 times slower for ⟨110⟩ and ⟨111⟩, respectively, than for ⟨100⟩ [60].

For a transient temperature–time profile $T(t)$ and a processing time t_{pr} the recrystallized layer has a thickness [51]

$$\ell_r = v_o \int_0^{t_{pr}} \exp\left[-\frac{E_a}{kT(t)}\right] dt \tag{2.10}$$

Sample heating for 10^{-2} s and longer may be described by the heat balance equation in the form

$$dt = \frac{C\rho d}{W_I(1-R)f(t) - 2\sigma\varepsilon_e(T^4 - T_s^4)}\, dT \tag{2.11}$$

Substitution of (2.11) into (2.10) leads to the integral expression for the recrystallized layer thickness

$$\ell_r = C\rho d\, v_o \int_{T_0}^{T_{max}} \frac{\exp(-E_a/kT)}{W_I(1-R)f(t) - 2\sigma\varepsilon_e(T^4 - T_s^4)}\, dT \tag{2.12}$$

Calculations made for (100) silicon for the slowest regrowth of self-ion-amorphized layers ($E_a = 2.54$ eV, $v_0 = 5.2 \times 10^{14}$ nm/s) are presented in Fig. 2.1. These calculations assume instant heating and cooling, and the absence of substitutional impurities, which would accelerate the process. For complete recrystallization of a typical layer thickness of 100 nm within milliseconds, a temperature of 1200–1300 °C is required. To obtain a sufficiently high temperature, a high radiation energy density will be needed.

For processing times in the range of seconds, such layers are recrystallized within the first second when the temperature reaches 750–800 °C. However, because of secondary extended defect formation, higher temperatures are needed for complete lattice restoration in an implanted layer. The degree of perfection attained depends on the as-implanted layer characteristics and annealing regimes. A dramatic difference is observed for implantation-disordered and -amorphized layers.

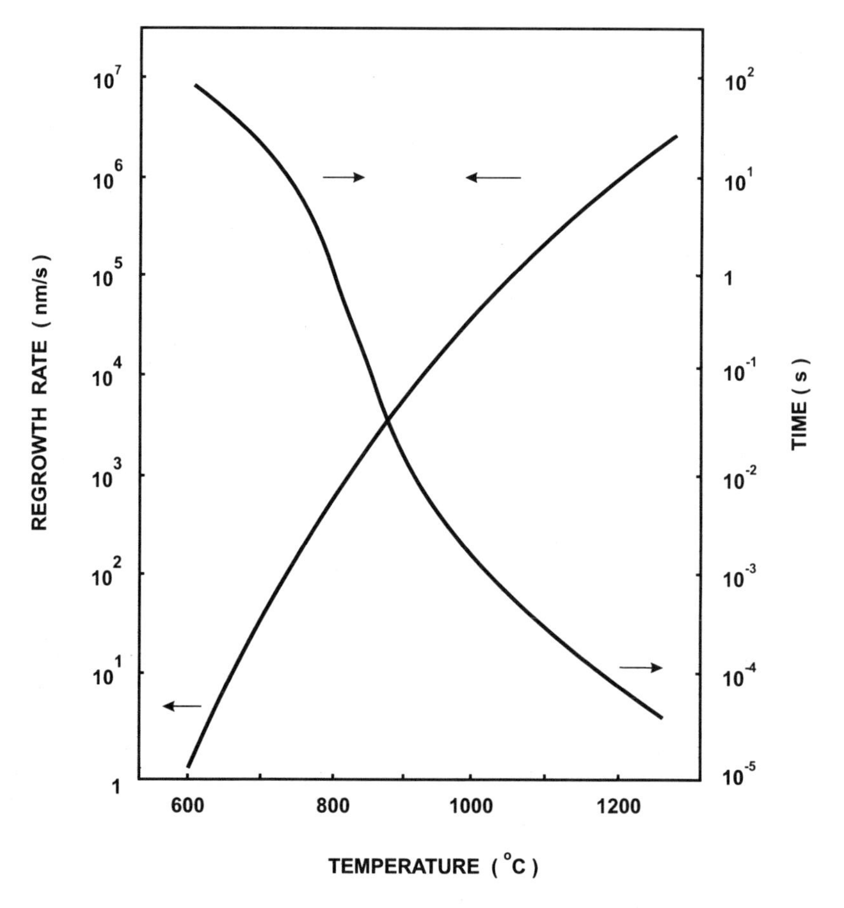

FIGURE 2.1. Solid-phase regrowth rate of self-ion-amorphized (100) silicon and the time required to regrow the amorphous layer to a thickness of 100 nm.

2.2.2. *Effect of Impurities*

Recrystallization is also controlled by implanted impurities. At concentrations above 0.1 at. %, impurities have a major influence on the regrowth kinetics. Substitutional impurities enhance recrystallization [69–77] and the regrowth rate increases linearly with substitutional impurity concentration for both conventional furnace annealing [78, 79] and RTA [73]. This effect is most pronounced for the concentration range close to the steady-state solubility of the impurity. However, oxygen, carbon, or inert gas atoms in the implanted layers retard the recrystallization process [77, 80].

Experimental data characterizing the recrystallization of (100) epitaxial silicon with different implanted impurities are collected in Table 2.2. There is considerable variation of activation energies and preexponential factors. These variations appear to be related to different experimental conditions, particularly the difficulties of temperature control in rapid heating and cooling experiments. Nevertheless, recrystallization rate estimates based on the parameters from Table 2.2 allow more or less accurate calculation of the implanted layer thickness recrystallized during RTP in the most practical temperature range of 700–1100 °C.

2.2.3. *Perfection of Annealed Layers*

Implantation of substitutional impurities is widely used for the doping of silicon and traditionally performed with ion energies of 10–100 keV and doses in the range 10^{14}–10^{16} ions/cm^2 at a temperature not higher than 100 °C. This normally amorphizes the implanted layer [81, 82]. However, light boron ions are the exception, even with doses above 10^{16} ions/cm^2. Although disorder is produced, no amorphous layer forms [82, 83].

Figure 2.2 shows 2-MeV He$^+$ ion channeling spectra for silicon crystals implanted with boron and boron difluoride molecular ions with the same equivalent energy as the boron atoms. The sample implanted by boron has disorder in the depth corresponding to

TABLE 2.2. Solid-Phase Recrystallization of Implanted (100) Silicon

Temperature range (°C)	Implanted impurity, N_{max} (atoms/cm^3)	Preexponential factor, v_0 (nm/s)	Activation energy, E_a (eV)	Refs.
450–500	B (2.0×10^{20})	8.93×10^{11} [a]	1.9	72
500–575	B (3.0×10^{20})	8.99×10^{16} [a]	2.6	73
550–850	B	3.18×10^{15}	2.52	70, 74
475–575	P (1.7×10^{20})	1.64×10^{15}	2.5	69
475–575	P + B (compensated)	1.74×10^{16}	2.8	69
500–550	P(2.5×10^{20})	1.0×10^{17}	2.7	75
550–850	P	6.92×10^{15}	2.68	70
550–850	P + B (compensated)	2.89×10^{15}	2.68	70
475–525	P or As [$(1.0 - 2.5) \times 10^{20}$]	2.37×10^{14}	2.35	71, 72, 74
450–900	As (4.3×10^{19})	2.21×10^{14}	2.52	76
1000–1150	As (1.9×10^{20})	5.07×10^{11} [a]	1.56	54
525–600	0 (2.5×10^{20})	4.65×10^{15}	2.8	77

[a] Calculated from the experimental data presented in the reference.

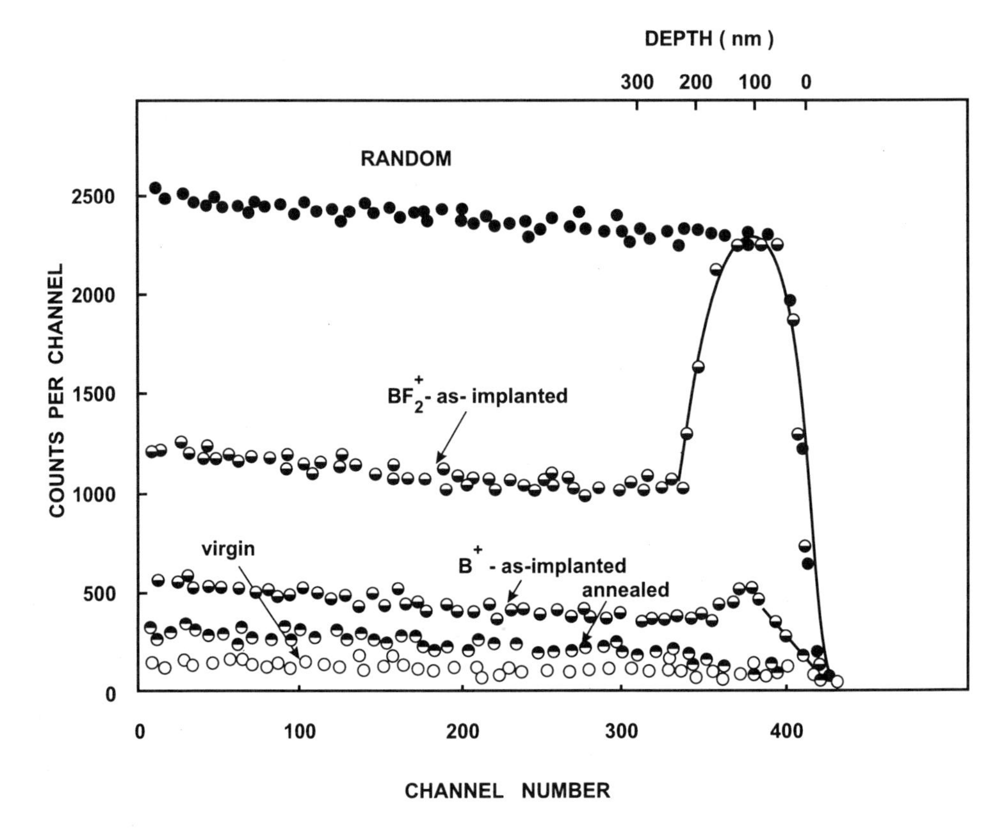

FIGURE 2.2. RBS spectra of 2-MeV He^+ channeled ions for (100) silicon implanted by boron difluoride (100 keV/1 × 10^{16} ions/cm^2) or boron (23 keV/1 × 10^{16} ions/cm^2) ions and transiently annealed at 1100 °C for 10 s.

the ions' projected range. Implantation of the heavier molecular ions leads to an amorphous layer formation of approximately the same thickness. Channeling spectra for both samples after RTP by incoherent light at 1100 °C for 10 s demonstrate evidence of lattice restoration in the implanted layers. Their perfection approaches that of the virgin crystal.

Crystal structure perfection may be quantitatively characterized by the minimum backscattering yield of channeled ions χ_{min} measured just after the surface peak. That is 3.1 and 3.4%, respectively, for (100) and (111) perfect silicon crystals analyzed with 2-MeV He^+ ions [84]. The upper limit for χ_{min} evaluated according to the criteria of the doped layer application in semiconductor devices is 5–6%. RTP provides χ_{min} = 3.5% (Fig. 2.2), while only χ_{min} = 4.5% was achieved by conventional furnace annealing at 900 °C for 30 min.

The presented spectra are typical for RTA of implanted silicon. Experimental χ_{min} varies as a function of annealing parameters for boron, phosphorus, arsenic, and antimony implants [85–99] and is summarized in Fig. 2.3. A temperature rise in the range of 950–1100 °C is accompanied by monotonic lowering of χ_{min} down to the level of a perfect silicon crystal. RTA is most effective in the epitaxial layers implanted with phosphorus and arsenic compared to antimony and boron.

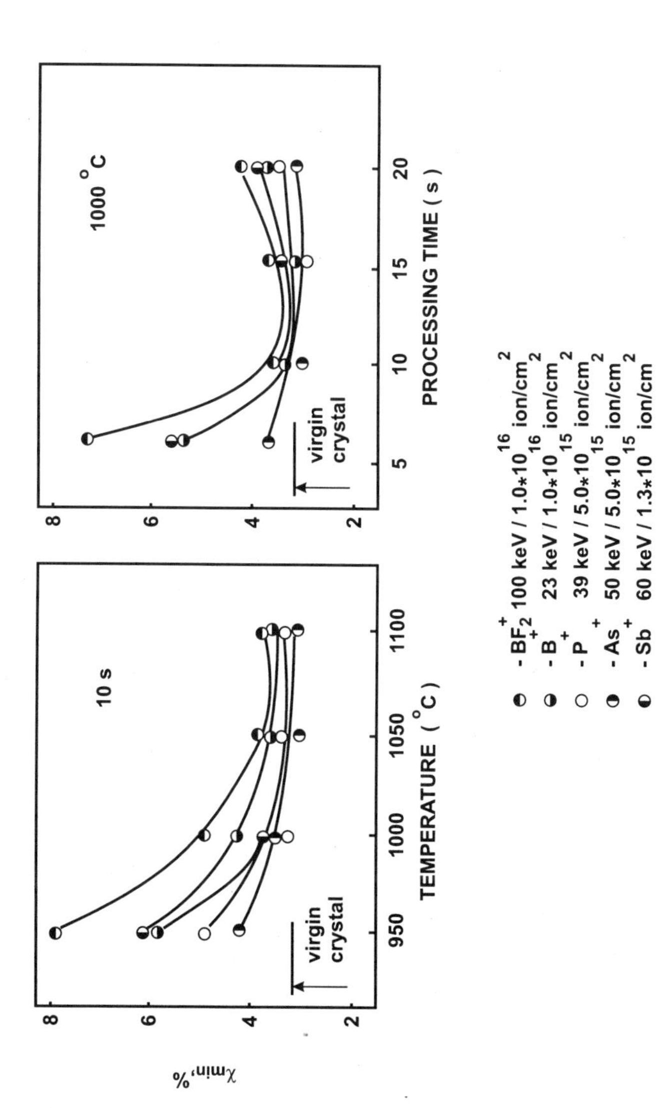

FIGURE 2.3. Minimum backscattering yield of 2-MeV He^+ channeled ions for substitutional impurity-implanted (100) silicon after rapid thermal annealing.

The results for antimony are explained by its known behavior in silicon [60, 90–93]. It has much greater tendency to form thermally stable defect complexes and precipitates, much more than other impurities. By comparing boron and boron difluoride implants, the activating role of fluorine in the residual thermally stable lattice defect formation can be seen. This phenomenon was observed earlier for solid-state recrystallization of such layers in a furnace [94, 95] and during laser processing [96, 97]. The same effects were noted in rapid thermal recrystallization of silicon implanted with phosphorus fluoride molecular ions [98].

The crystalline perfection of implanted layers is a monotonic function of the processing. There is an optimal processing duration, 8–15 s, which provides a minimum value of χ_{min}. Longer annealing produces a slow increase in χ_{min}, especially at temperatures above 1050 °C. This is the result of thermally stimulated generation of residual extended defects in the crystal. Although silicon crystals of different orientations behave identically, χ_{min} values are somewhat higher for the closer-packed (111) plane than for the (100) plane.

In practice, there is a limit to the degree of crystalline perfection that can be reached in implanted and RTP or furnace annealed layers. The degree of perfection is controlled by the residual extended defects [99, 101]. The type and location are determined by the as-implanted layer characteristics, which are very different for amorphized and disordered layers.

2.2.4. *Residual Defects*

There is a correlation between damage density in implanted layers and the type and location of annealing-produced extended defects. This is schematically illustrated in Fig. 2.4, which summarizes a wide range of experimental data [100, 101] within the classification proposed by Jones *et al.* [101].

2.2.4.1. Category I Defects. These are extrinsic dislocation loops. They appear for room-temperature implantation in disordered layers when amorphization is not achieved. The defects are typically associated with light ions such as boron. Their concentration increases in proportion with the dose. The defects are located around the projected range of the implanted species and a threshold boron dose of 1×10^{14} ions/cm^2 has been observed.

Annealing of light ion (boron, phosphorus)-implanted silicon is also followed by formation of rodlike or {311} defects [102]. They consist of interstitials precipitating in five- and seven-membered rings on {311} planes as a single monolayer at hexagonal silicon. During processing above 600 °C these defects emit interstitials providing an enhanced impurity diffusion. Defect-free boron-implanted layers can be distinctively obtained for doses not exceeding 1×10^{13} ions/cm^2. RTP for seconds was found experimentally to be more effective in disordered layer annealing than furnace processing [103–105]. The RTP is characterized by a heating rate of 200–300 °C/s, while 3–5 °C/s is usually realized during conventional furnace annealing. The high heating and cooling rates suppress nucleation of secondary extended defects.

2.2.4.2. Category II Defects. These defects, also known as "end of range" defects, arise whenever an amorphous layer is formed. They are located beneath the amorphous–

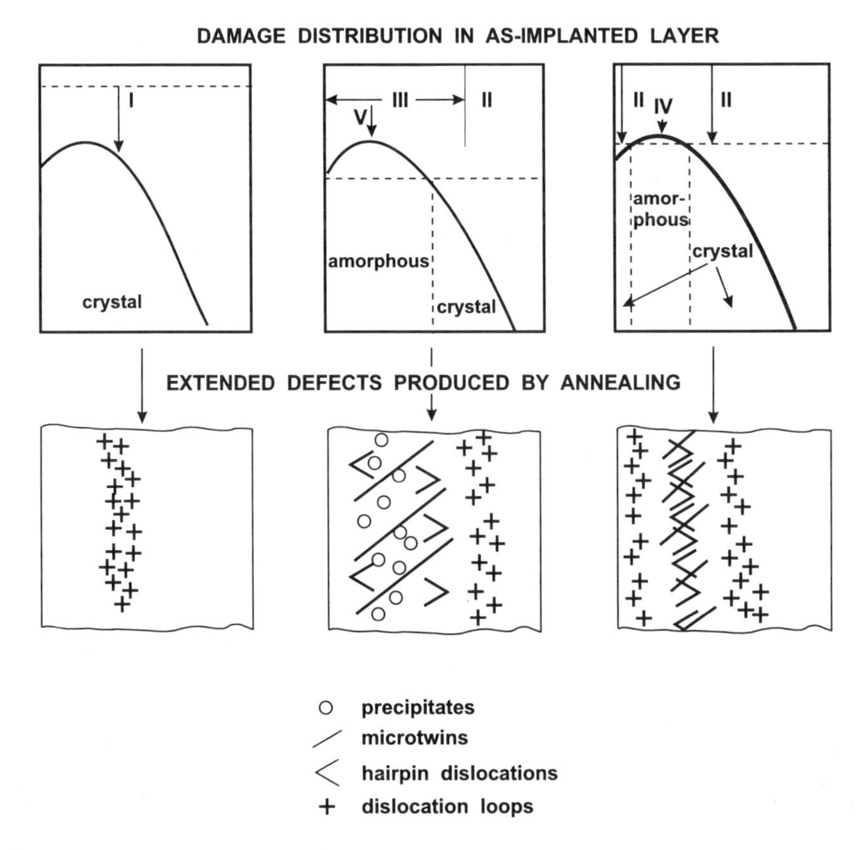

FIGURE 2.4. The relationship between the damage density distribution (solid lines) and amorphization threshold (dashed line) leading to different category defects.

crystalline interface in the heavily damaged but still crystalline material. They are interstitial dislocation loops with a concentration that is correlated to the number of implanted atoms located in the crystal under the interface (typically up to 5% of the implanted dose). It is worth noting that heavy ions such as As^+, Sb^+, Ge^+, Sn^+, and others produce a sharp amorphous–crystalline interface. A small amount of impurity atoms located in the crystalline region causes a reduction in the defect concentration.

2.2.4.3. Category III Defects. Category III defects arise from solid-state regrowth of amorphized layers. They are located in the region previously occupied by the amorphous layer. There are several forms of these defects, including hairpin dislocations, microtwins, and segregation defects. Hairpin dislocations nucleate when the regrowing amorphous–crystalline interface encounters pockets of misoriented crystalline material within the amorphous layer. Twinning is most frequently observed with (111) crystals. There have also been reports of defect formation just beneath the surface associated with impurities such as fluorine segregated at the advancing amorphous–crystalline interface [103, 104].

2.2.4.4. Category IV Defects. These defects are generated during recrystallization of buried amorphous layers formed by heavy ion implantation in the dose range of 5×10^{13}

to 5×10^{14} ions/cm^2. Regrowth of a buried amorphous layer occurs by the motion of amorphous–crystalline interfaces and leads to defect creation at a depth where these two interfaces meet. There can also be microtwin and hairpin dislocation networks. Moreover, category II interstitial dislocation loops are formed in such annealed structures at both interfaces. Category IV defects may be avoided by decreasing the energy of the implant, increasing the dose, or decreasing the implantation temperature. All of these alter the relationship between the damage density distribution and the amorphization threshold in favor of producing a surface amorphous layer.

2.2.4.5. Category V Defects. Category V defects are created in heavily doped layers when the solid solubility level of the impurity is exceeded. These defects comprise precipitates and related dislocation loops. The defects associated with the precipitation phenomena are formed at depths corresponding to the projected range of the implanted impurity. Growth and annealing of the secondary defects are diffusion controlled. An appropriate choice of RTP regimes allows one to get practically defect-free ($< 10^7$ defects/cm^2) doped layers. Figure 2.5 shows the temperature dependence of the time necessary to dissolve clusters and category II dislocations. It summarizes experimental data for both RTP [106–113] and conventional furnace [106, 111, 114, 115] annealing.

For all impurities investigated, namely, boron, aluminum, gallium, phosphorus, arsenic, and antimony, activation energies for secondary defect annealing are 4.5–5.0 eV. This gives support to the decisive role of self-diffusion in reconstruction and annihilation

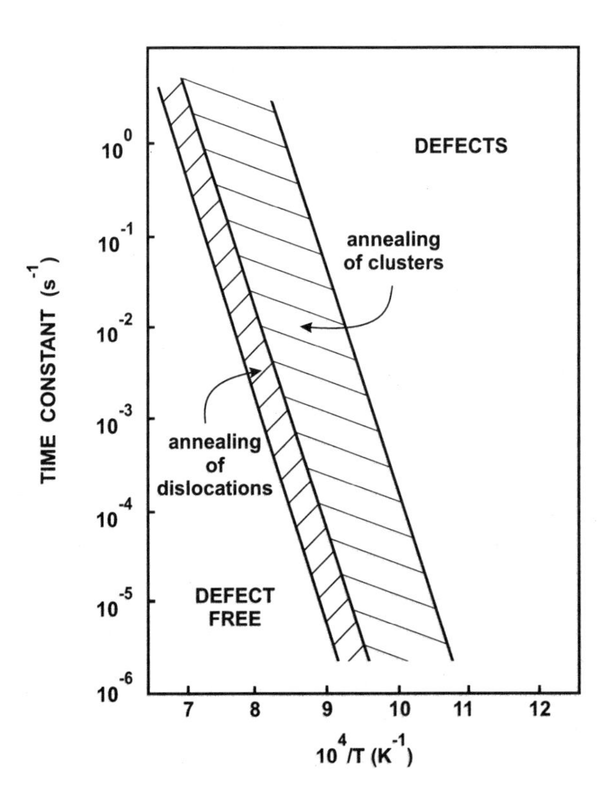

FIGURE 2.5. Temperature dependence of the time constant of defect annealing in ion-implanted silicon.

of secondary defects. The activation energy for category II defect annealing remains invariable even when the process is enhanced by supersaturation of implanted gallium, phosphorus, or arsenic. Incomplete annealing of secondary extended defects takes place in the implanted layers supersaturated with aluminum and antimony. There is always a dislocation network formed.

The annealing of category II defects and interstitial dislocation loops obviously requires vacancy generation or interstitial sinks. Supersaturated solutions of substitutional impurities diffusing partly via interstitials, i.e., boron, phosphorus, arsenic, and antimony [116–126], create the necessary conditions. Self-interstitials kick out substitutional impurities from the regular lattice positions, favoring the dissolving of interstitial dislocations. In this way recrystallization and secondary extended defect evolution in implanted layers followed by nonequilibrium point defect generation can influence impurity atom behavior during RTP.

2.3. BEHAVIOR OF IMPLANTED SUBSTITUTIONAL IMPURITIES

Electrical activation of implanted impurities without redistribution was considered to be one of the advantages of RTP. In fact, estimates for the known diffusivities of substitutional impurities in silicon at temperatures of 1100–1200 °C [127], even for seconds, give impurity atom diffusion lengths less than 10 nm. Charge and concentration effects can stimulate two- to threefold deeper impurity diffusion [128, 129]. The diffusion is negligible relative to implanted layer thicknesses usually employed (100–500 nm). Earlier investigations on thermal processing of implanted silicon by incoherent light within seconds [130–134] demonstrated a good agreement between experimental and theoretically calculated broadening of implanted profiles. However, consecutive investigations performed in the extended time and temperature range [85–89, 135–139] revealed unexpected deep redistribution of implanted substitutional impurities, commonly referred to as transient enhanced diffusion. It was observed only for certain regimes of implantation and subsequent RTP.

Comparative experiments carried out in oxygen-containing ambients as well as in inert gas or in vacuum show that this effect is not related to the substitutional impurity diffusion enhancement observed in silicon oxidation [140–146]. Furthermore, heating of implanted samples by incoherent light and by low-energy electrons from the side of the implanted layer, and from the back side of the wafer show that ionization in the energy-absorbing layer has practically no influence [44, 147]. Therefore, the enhanced impurity diffusion during RTP has a different origin. In this section the most representative experimental RTP data are considered in analyzing the enhanced diffusion of implanted boron, phosphorus, arsenic, and antimony in single-crystal silicon.

2.3.1. *Boron*

Boron, having the highest solubility in silicon of the elements in the third group of impurities, has the widest technological applications. Boron is a light ion and when implanted is not accompanied by surface layer amorphization. At traditionally employed implant energies, 10–100 keV, the relatively large projected range of the boron ion restricts the ability to fabricate thin doped layers. This difficulty is solved by the use of heavier, boron difluoride molecular ions, but it leads not only to surface amorphization, but also to fluorine accumulation in the implanted layer. Therefore, different behavior is observed for layers implant with boron versus boron difluoride ions during the subsequent thermal processing.

Implanted boron redistribution in silicon after RTP is both in agreement with [130, 133, 151] and enhanced [102, 152–176] when compared to results from conventional long-duration diffusion experiments. In the last case the factors that appear to be important are (1) the temperature range, (2) the amount of damage in the as-implanted layer (i.e., the presence of disorders or complete amorphization), and (3) the impurity concentration.

The enhanced diffusion of boron based on the experimental data for heating up to 700–1100 °C [152–162, 173] displayed anomalously rapid deepening of the impurity profile tail by 60–100 nm. This implies effective diffusivities at these temperatures of 5×10^{-13}–1×10^{-11} cm^2/s, which is several orders of magnitude greater than the equilibrium value [127–129]. The enhanced diffusion occurs within seconds at 1000–1100 °C, but it lasts for 45 min when the temperature is 800 °C [159, 161]. No enhancement was observed after annealing at temperatures below 600 °C [149] or above 1150 °C [130, 134, 148]. These temperature limits agree with those for radiation defect-enhanced diffusion observed in conventional annealing of implanted layers and during hot ion implantation in silicon crystals [177]. This shows the dominant role of nonequilibrium point defects in the diffusion process. The implantation-produced damage, supersaturated boron solution, and boron precipitates can be the source of the nonequilibrium defects. In comparable RTP experiments involving single-crystal and self-ion-amorphized silicon implanted with boron [155, 157, 161], it was found that enhanced diffusion is most pronounced in the nonamorphized samples. Transmission electron microscopy showed nonequilibrium interstitials to release from the disorder created by implantation and transformed during thermal processing into category I defects. As these defects are located around the maximum of implanted boron, impurity-enhanced diffusion zones directly border this area as illustrated in Fig. 2.6.

In amorphized silicon, damage annealing occurs via solid-state epitaxial recrystallization. The defects (category II) capable of releasing interstitials are located beneath the amorphous–crystalline interface where the impurity concentration is low. Therefore, any diffusion enhancement that takes place [156, 163, 164] during rapid thermal processing is insignificant [161].

Moreover, experiments with carbon-preamorphized layers [175, 176] have shown the importance of this impurity in boron behavior during RTP. Boron transient enhanced diffusion was completely eliminated and end-of-range dislocations were significantly reduced when the preamorphization had been performed with 90-keV carbon at 2×10^{15} $ions/cm^2$. Electrical activation of implanted boron is increased under such conditions [178, 179]. The effects are explained by gettering of silicon interstitials with carbon. SiC agglomerates are supposed to form and act as a sink for the interstitials.

Based on the assumption that the concentration of nonequilibrium point defects decreases exponentially in time, the experimental effective diffusivity of boron in the temperature range of 800–1000 °C for processing durations of 5–150 s is approximated by the expression [162]

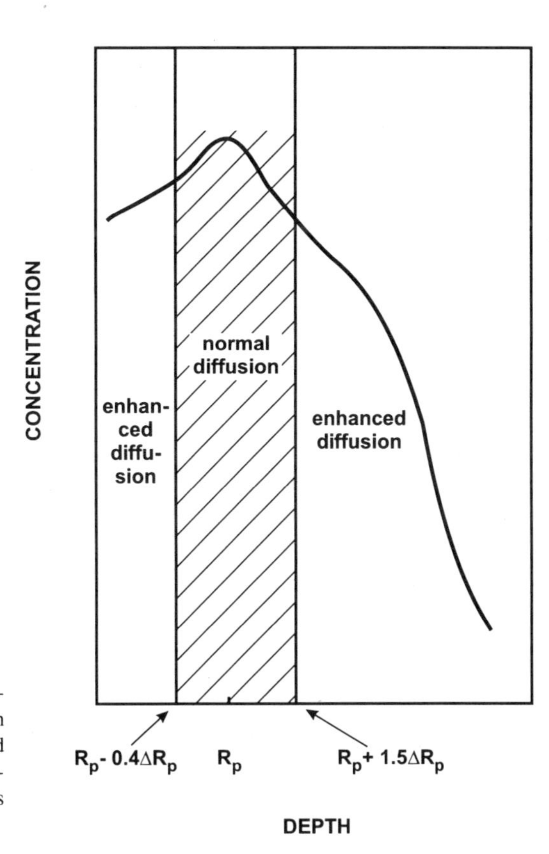

FIGURE 2.6. Schematic diagram of the transient point defect model for boron diffusion during rapid thermal processing. The hatched region corresponds to the implant-induced dislocation network where normal diffusion occurs [152].

$$D_{\text{eff}} = \frac{\tau D_{t0}}{t} \tag{2.13}$$

where D_{t0}, the diffusivity defined at the initial moment of processing ($t = 0$), is

$$D_{t0} = 1.4 \times 10^{-7} \exp(-1.1/kT) \quad \text{cm}^2/\text{s} \tag{2.14}$$

and the relaxation time is

$$\tau = 2.9 \times 10^{-6} \exp(1.57/kT) \quad \text{s} \tag{2.15}$$

The activation energy of 1.1 eV was attributed by Miyake and Aoyama [162] to boron diffusion as vacancy–boron complexes. However, any vacancy complex formation seems to have a low yield in the region saturated with the reconstructing defects of interstitial type. Therefore, the discussed diffusion parameters are more closely related to the boron interstitials or to boron complexes with interstitial silicon. Activation energies for transiently enhanced boron diffusion as low as 1.46 eV [152] and 2 eV [153] have also been reported. These were associated with the diffusion governed by nonequilibrium point defects generated during implantation damage annealing. The value 1.46 eV is calculated as a difference between the activation energy for intrinsic boron diffusivity (D_i =

$0.76\exp(-3.46\ \text{eV}/kT)\ \text{cm}^2/\text{s}$ [128]) and that of 2 eV used in the empirical formula describing boron effective diffusivity during RTP [152]

$$D_{\text{eff}} = D_{\text{i}}\,[1 + 5 \times 10^{-6} \exp(2/kT) \exp(-t/\tau)] \tag{2.16}$$

where $\tau = 4.4$ s. The above features are general for implanted boron redistribution when the impurity concentration is lower than the boron solubility (4×10^{20} atoms/cm^3, see Section 2.4). There are additional concentration effects observed, when this level is exceeded, including formation and dissociation of supersaturation, clustering, and precipitation. A considerable number of the impurity atoms and substitutional impurities are electrically inactive under these conditions.

Figure 2.7 shows the impurity and hole depth profiles in boron-implanted silicon after RTP. Prior to boron implantation the silicon crystals were predamaged by self-ion implantation (amorphization was not achieved) to reduce boron channeling. The concentration of holes does not exceed the boron solubility but forms a plateau in the region where the boron solution is supersaturated. Boron precipitates were also found in this region, and they are also observed in conventional furnace annealed samples [165]. The estimated effective diffusivity [157] appears to exceed the equilibrium one by up to four orders of magnitude. However, there is practically no difference for the low-dose-implanted samples, where the maximum boron concentration is below its steady-state solubility in silicon [130–134].

Boron behaves in another way when it has been introduced by implantation of boron difluoride molecular ions [112, 150, 151, 166–168]. An amorphous layer created by implantation with a dose above 10^{16} ions/cm^2 and fluorine contamination plays a role.

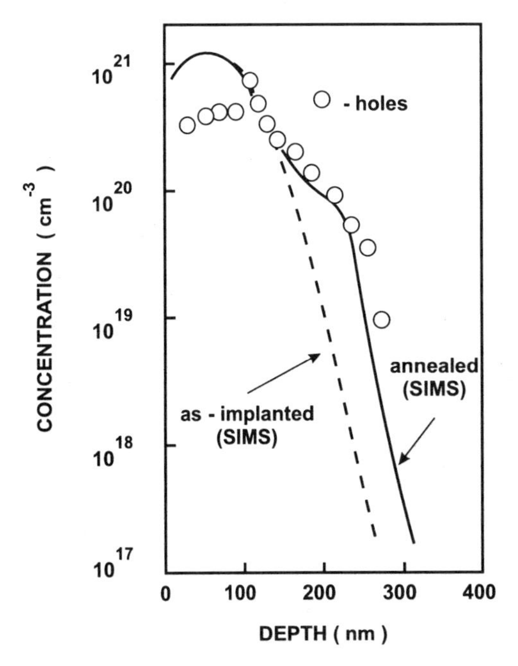

FIGURE 2.7. SIMS and carrier concentration profiles in predamaged boron-implanted silicon annealed at 1000 °C for 20 s [157].

Solid-state epitaxial recrystallization of the amorphized layer, as compared to annealing of implantation-induced disorder, causes more effective electrical activation of the impurity. Figure 2.8 shows this difference measured at temperatures up to 1200 °C. There have also been observations of boron-enhanced redistribution at least until its concentration returns below the solubility limit [112, 166–168].

Implanted fluorine diffuses to the surface and into the bulk accumulating in the region of category II defects, i.e., at the amorphous–crystalline interface. Surface concentration of fluorine reaches 10^{20} atoms/cm^3 after processing at 900–1000 °C for 10 s [107, 168]. There is boron gettered in both regions after annealing at temperatures as low as 925 °C, and two peak profiles are formed characterized by a difference in maximum and minimum concentrations of two- to threefold [107, 112, 168]. RTP at 1150 °C and above causes the fluorine to evaporate, lowering the concentration to 10^{17}–10^{18} atoms/cm^3 [168].

Impurity accumulation in gallium [169, 170] and indium [169, 171, 172] implanted silicon at the depths where category II defects are also observed after RTP within milliseconds and seconds. The impurity atoms rapidly diffuse from the region of their maximum concentration to the amorphous–crystalline interface, where a dislocation network is formed. The process is characterized by at least an order of magnitude increased diffusivity. However, there is no enhanced diffusion outside this region. Thus, enhanced impurity diffusion is obviously correlated to the secondary extended defect evolution in the implanted layers. The assumption that nonequilibrium interstitials are responsible for the enhanced diffusion adequately explains the experimental observation that the highest efficiency is observed for boron, whereas those for gallium and indium are much lower.

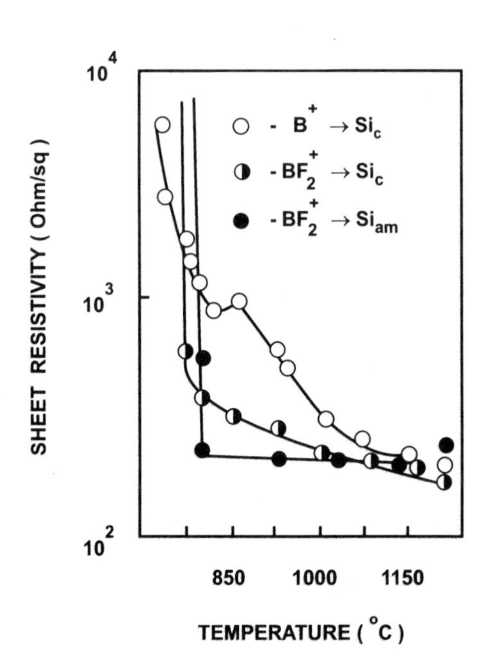

FIGURE 2.8. Sheet resistivity of *p*-type layers created by implantation of boron (150 keV/2 × 10^{16} ions/cm^2) or boron difluoride (50 keV/1 × 10^{16} ions/cm^2) in single-crystal silicon and of boron difluoride (50 keV/1 × 10^{16} ions/cm^2) in preamorphized silicon on 10-s annealing [168].

2.3.2. Phosphorus

Phosphorus is the fastest diffuser in silicon of the traditional group-V dopants. This property determines its unique behavior during RTP. Annealing of phosphorus-implanted silicon crystals for times less than a second [135, 180, 181] and within seconds at a temperature of 900 °C results in a complete electrical activation of the impurity without any noticeable redistribution. Measured sheet resistivity in this case agrees with that calculated for undisturbed as-implanted profiles [182]. An increase of the induced temperature to 950 °C and above, for a processing duration of seconds, produces significant impurity redistribution [137, 156, 183, 184].

Enhanced phosphorus diffusion has been noted at comparatively low implant impurity concentrations of 10^{18}–10^{19} atoms/cm^3 [183] as well as at the concentrations near the phosphorus steady-state solubility in silicon [86, 137]. In the low-concentration range it is in general identical to the boron diffusion discussed above. In fact, low-dose phosphorus ion implantation, at 10^{13}–10^{14} ions/cm^2, with energies higher than 50 keV does not amorphize the crystal. Thus, annealing of the implant-induced disorder produces category

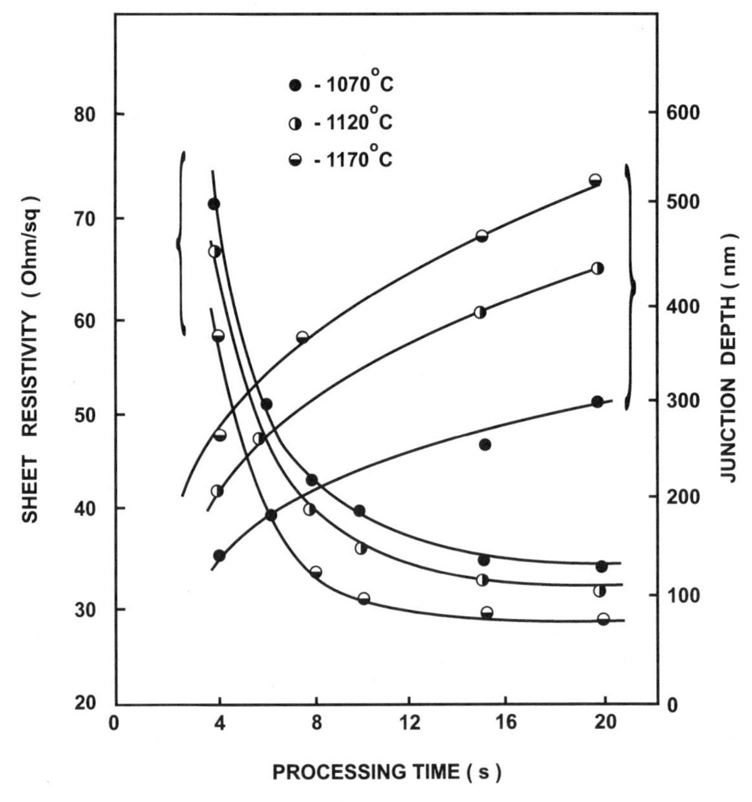

FIGURE 2.9. Sheet resistivity and *p–n* junction depth of phosphorus-implanted (30 keV/1.9 × 10^{16} ions/cm^2) silicon layers versus incoherent light rapid thermal processing parameters [137].

I defects and nonequilibrium interstitials which enhance phosphorus redistribution by diffusion.

There is another feature of enhanced diffusion in heavily implanted layers (at 10^{20}–10^{21} atoms/cm^3; implant dose of 1×10^{16} ions/cm^2 and higher). Figure 2.9 shows the sheet resistivity and *p–n* junction depth as a function of RTP parameters for these layers [137]. Incoherent light heating to 1070–1170 °C within 4–20 s was used. The most rapid decrease of the sheet resistivity was observed in the first 6–8 s of processing, when implanted layer recrystallization and phosphorus electrical activation were taking place. There is then a delay in the resistivity decrease closely correlated with the increase of the doped layer depth. The lowest sheet resistivity is 28–30 Ω/sq. After conventional furnace annealing at 900 °C for 20 min, 38–40 Ω/sq is produced. The calculated resistivity is 80 Ω/sq assuming undisturbed and complete activation of the as-implanted impurity profile [182].

Electron depth profiles in the above samples are presented in Fig. 2.10. Processing for 4 s is not enough for complete electrical activation of the implanted impurity. Steady-state induced temperature is actually not achieved for this duration of incoherent light heating. Processing for 10 s causes complete activation but impurity redistribution becomes valuable. The diffusivity estimated from changes in the *p–n* junction depth are two orders of magnitude higher than the equilibrium value, and ten times greater than the diffusivity estimate including concentration effects [128].

The phosphorus diffusivities obtained in Arrhenius coordinates are approximated well by the activation energy of 2.0 ± 0.1 eV and preexponential factor of $4.9 \pm 0.6 \times 10^{-4}$ cm^2/s. For comparison, the equilibrium parameters determined from conventional furnace

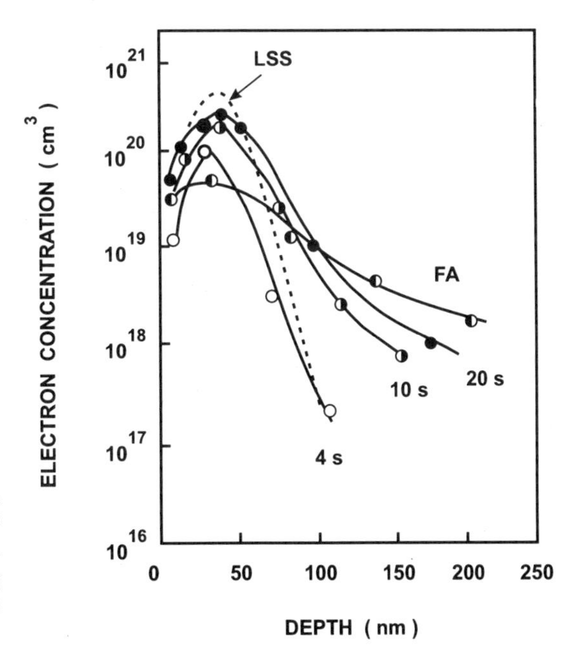

FIGURE 2.10. Electron depth profiles in phosphorus-implanted (30 keV/1.9×10^{16} ions/cm^2) silicon after rapid thermal processing at 1170 °C and furnace annealing (FA) at 900 °C for 20 min. Dashed line: calculated values from LSS theory [137].

experiments are 3.69 eV and 10.5 cm^2/s [127]. Therefore, the enhanced diffusion for implanted phosphorus is characterized by a lower activation energy.

The enhanced phosphorus redistribution was typically observed for implanted doses greater than 1×10^{16} ions/cm^2. It is independent of ion energy being strictly controlled by the maximum concentration of the impurity. Figure 2.11 displays RTP-induced sheet resistivity variation when the peak as-implanted phosphorus concentration was 2.6×10^{20}–2.0×10^{21} atoms/cm^3. Decrease of sheet resistivity with the processing temperature increase from 950 to 1100 °C is associated with phosphorus depth redistribution (see Fig. 2.11). This has been confirmed by neutron activation analysis. However, there is no redistribution in the layer at the maximum impurity concentration of 2.6×10^{20} atoms/cm^3.

Taking into account that amorphous layers were formed in all implantation regimes used, the observed differences cannot be attributed to different concentration of point defects, which are released during thermal processing. On the other hand, the critical

Calculated Peak Concentration of Phosphorus (at/cm^3) Implanted into Silicon

Implanted dose (ion/cm^2)	1×10^{15}	5×10^{15}	1×10^{16}
$E = 18$ keV $R_p = 25$ nm, $\Delta R_p = 11$ nm	○ 3.6×10^{20}	◑ 1.8×10^{21}	——
$E = 36$ keV $R_p = 50$ nm, $\Delta R_p = 20$ nm	● 2.6×10^{20}	◐ 1.0×10^{21}	◒ 2.0×10^{21}

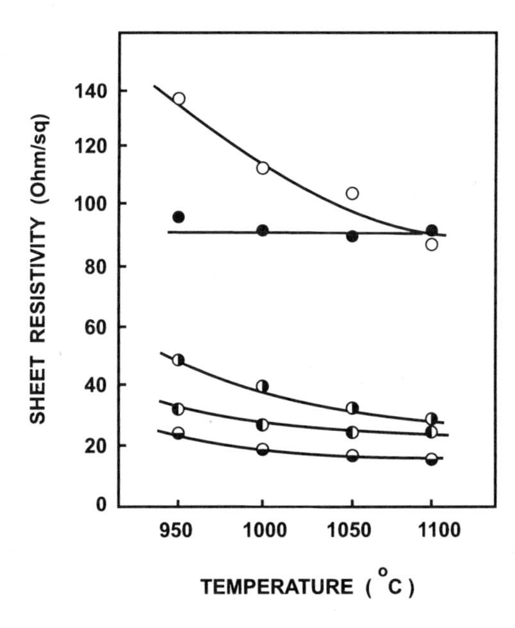

FIGURE 2.11. Sheet resistivity of phosphorus-implanted silicon after 10-s incoherent light annealing [86].

concentration, beyond which an enhanced diffusion occurs, correlates with the steady-state phosphorus solubility in silicon (see Section 2.4). These results suggest that the diffusion enhancement is by nonequilibrium point defects created by dissociation of supersaturated phosphorus solution.

2.3.3. Arsenic

Implanted arsenic behavior during RTP has received the widest interest because of its application for thin heavily doped *n*-type layers in silicon. The experimental studies can be divided into two groups: first, the work from the early 1980s where arsenic redistribution from an implanted layer was not detected [131–134, 185–187], and second, later investigations carried out using an extended range of implantation parameters and RTP regimes in which arsenic-enhanced diffusion was definitely observed [85, 87, 139, 149–153, 188–191].

The experiments that found no arsenic redistribution are limited by implants in the energy range of 50–100 keV with doses of 10^{14}–10^{15} ions/cm^2. This produces peak impurity concentrations less than 1×10^{20} atoms/cm^3. Rapid annealing performed at 1000–1100 °C for 10 s produced complete electrical activation of the implanted impurity. Sheet resistivity was found to agree well with values [182] calculated for electron profiles corresponding to as-implanted impurity distributions. The investigation included the combined free carrier and atomic profile analysis. An increase of implant doses to 10^{16} ions/cm^2 results in disagreement in the experimental and calculated sheet resistivities. The effect of different processing times and temperatures on sheet resistivity of arsenic-implanted silicon is illustrated in Fig. 2.12. Implantation energies of 10–70 keV and doses of 1×10^{15}–1×10^{16} ions/cm^2 were employed. These correspond to projected ranges and as-implanted peak concentrations from 11 to 50 nm and from 2.5×10^{20} to 5.0×10^{21} atoms/cm^3, respectively.

Two different correlations between sheet resistivity and processing conditions can be identified: For implants with peak impurity concentrations below 3.0×10^{20} atoms/cm^3 (50 and 70 keV/1 $\times$ 10^{15} ions/cm^2) no change in sheet resistivity as a function of the processing temperature is observed, whereas for concentrations above 4.5×10^{20} atoms/cm^3 (the remainder of the implants) there is a decrease in sheet resistivity with increasing temperature. The sheet resistivities which are independent of the processing temperature coincide satisfactorily with the calculated values. Qualitatively and quantitatively identical behavior was found in the temperature range up to 1300 °C for processing times up to 30 s [148, 166, 185, 187].

Analysis of arsenic depth profiles indicates that sheet resistivity variation is connected with impurity redistribution in the bulk of the semiconductor crystal from the surface. Impurity out-diffusion and evaporation takes place through unprotected surfaces during annealing at temperatures as high as 1150 °C [148, 185, 186]. In this case sheet resistivity remains the same, or even increases with increased processing duration. A protective silicon dioxide film, 50 nm thick, or a protective inert ambient at atmospheric pressure suppresses arsenic evaporation.

The consistency of sheet resistivity with processing time and temperature reflects the stability of as-implanted impurity distribution. A sheet resistivity decrease is inevitably

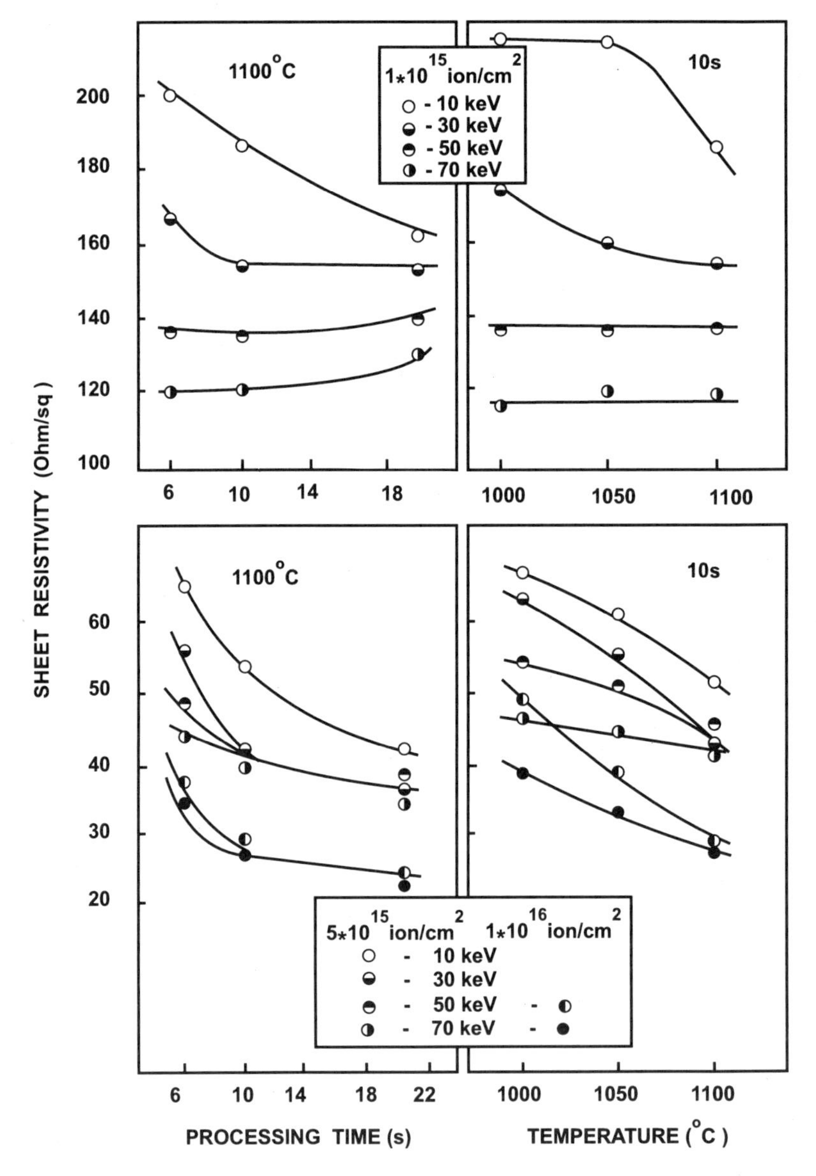

FIGURE 2.12. Sheet resistivity variation with processing time and induced temperature for arsenic-implanted (100) silicon crystals [87].

a result of impurity diffusion. Typical arsenic redistribution is illustrated by the profiles in Fig. 2.13.

The following features shall be stressed for implanted arsenic redistribution with RTP. First, the redistribution occurs in the already recrystallized layer. The process is extended to a depth 1.5–2 times greater than the depth of an implantation-amorphized

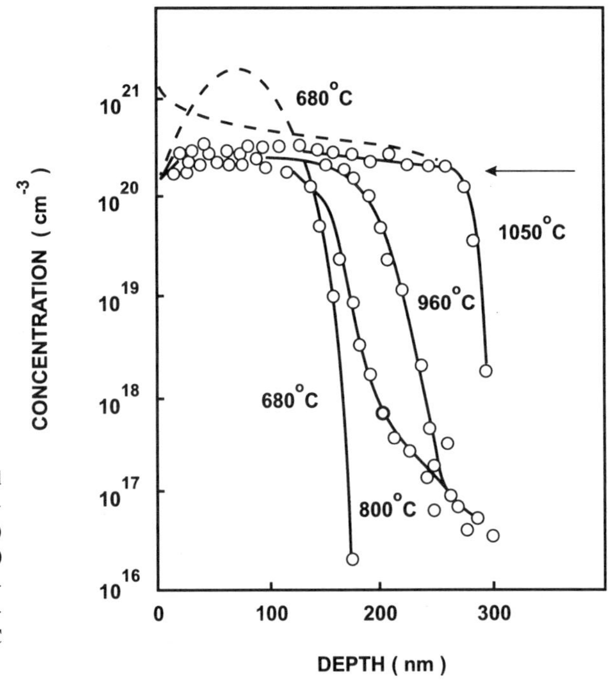

FIGURE 2.13. Electron (solid lines) and atom (dashed lines) concentration depth profiles in (100) silicon implanted with arsenic (100 keV/2 × 10^{16} ions/cm^2) and incoherent light annealed for 10 s. Arsenic steady-state solubility at 1050 °C is indicated by the arrow [189].

layer. Second, the redistributed depth was a function of the processing temperature and is time dependent, which confirms the diffusion mechanism for the impurity redistribution. Third, regardless of the implanted dose and energy, an enhanced arsenic diffusion is observed in the layers where the as-implanted impurity concentration was above 1×10^{20} atoms/cm^3. RTP of these layers forms flat profiles with maximum concentrations of electrically active impurities at a level corresponding to the arsenic steady-state solubility in silicon, at the given temperature. The depth of the redistribution appears to correspond to an arsenic diffusivity 2–3 orders larger than the values known from conventional furnace annealing experiments [127, 128].

Enhanced diffusion starts to be realized for processing at temperatures higher than 900 °C. It lasts 1–30 s until the maximum concentration of the arsenic is stabilized at its steady-state solubility level. Diffusion activation energy has been estimated to be in the range of 1.7–2.2 eV [87, 152, 153, 188], and 3.4–3.6 eV [139, 189], which is always less than the 4.05 eV [128] observed for a traditional thermal process.

It should be pointed out that after RTP at 600–800 °C for seconds, as well as at higher temperatures (~1300 °C) for milliseconds, when the processing duration limits the solid-state reaction to regrowth of an amorphized layer, a supersaturated arsenic solution may be formed in the silicon [192–195]. Impurity substitutional fraction in such solutions amounts to 80–90%, i.e., (2–3) × 10^{21} atoms/cm^3.

Increase of processing time to some minutes at 600–800 °C leads to a decrease of the substitutional fraction, and to clustering. The last process has an activation energy of

1.1 eV [194, 195]. Supposing the dissociation of the supersaturated solution to be controlled by diffusion of impurity interstitials, the activation energy may be attributed to interstitial arsenic diffusion.

Vacancies [152] or interstitials [149] associated with the implantation defect removal are considered to be responsible for the transient enhanced diffusion. There has been an attempt to describe enhanced diffusion at a high concentration of arsenic (up to 7×10^{20} atoms/cm^3) with a model that includes a double negatively charged vacancy diffusion-controlled process [191].

The approaches that assume the defects that stimulate diffusion are only vacancies or only interstitials have limited success in explaining the complete set of accumulated experimental data. This also contradicts the concept of a dual diffusion mechanism for substitutional impurities in silicon [123]. Assuming that both vacancies and interstitials participate in the process seems to be the best explanation. This approach is based on the diffusion enhancement by nonequilibrium interstitials, generated during rearrangement of dislocations, and by interstitials and vacancies, produced by dissociation of supersaturated impurity solutions in silicon.

2.3.4. Antimony

Antimony is an impurity in silicon of technological interest. In comparison with the other widely used group-V dopants, namely, phosphorus and arsenic, it has a lower solubility, $(6–8) \times 10^{19}$ atoms/cm^3, and much lower diffusivity. This suggests investigating the possibility of creating antimony supersaturated solutions, and its diffusion behavior in rapid thermally annealed implanted layers [89, 136, 149, 169, 192, 196–199].

Antimony supersaturated solutions in silicon have been shown to form at low temperatures, 700–900 °C, and with processing durations of 15–20 s [192, 196–199]. In these regimes the recrystallization front comes through the whole implanted layer reaching the surface. Because of low diffusivity there is practically no impurity redistribution, except for antimony concentrations higher than 5×10^{20} atoms/cm^3, when some antimony is pushed to the surface by the recrystallization front.

Figure 2.14 shows antimony and electron depth profiles in the implanted layers subjected to low-temperature RTP. For impurity peak concentrations above 5×10^{20} atoms/cm^3 one can find a maximum electron concentration of $4–5 \times 10^{20}$ /cm^3. This is a factor of 7 higher than the antimony solubility limit in silicon and more than 30 times its solubility at 700 °C. The increase of implant dose above 2×10^{16} ions/cm^2 at 80 keV, provides no increase of doped layer conductivity, but the fraction of substitutional impurity is reduced.

Nonsubstitutional antimony is located in the region spreading from the depth of peak in impurity concentration, to the surface [89, 198]. Impurity atoms form precipitates, defect complexes and are condensed at dislocation loops. Similar behavior was also observed for long-time annealing experiments [149]. Mössbauer spectroscopy data show the substitutional antimony fraction to be reduced with increased processing time because of precipitate and defect complex formation. Figure 2.15 presents the relationships between these fractions. Antimony complexes start growing at relatively low processing temperatures, when the length of their diffusion, estimated from the known data [128], is

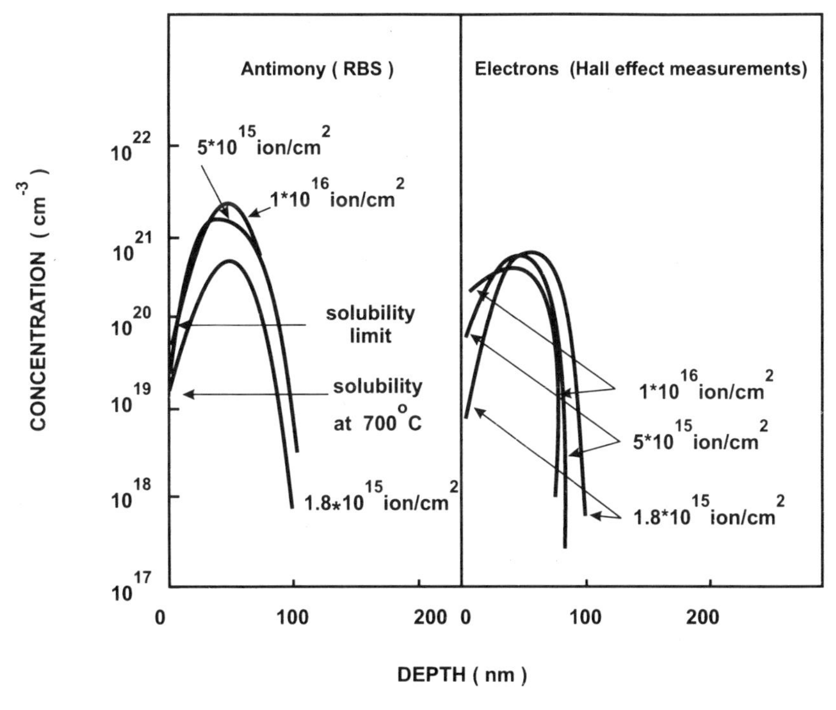

FIGURE 2.14. Antimony and electron concentration depth profiles for 80-keV antimony-implanted silicon after incoherent light annealing at 700 °C for 15 s [196].

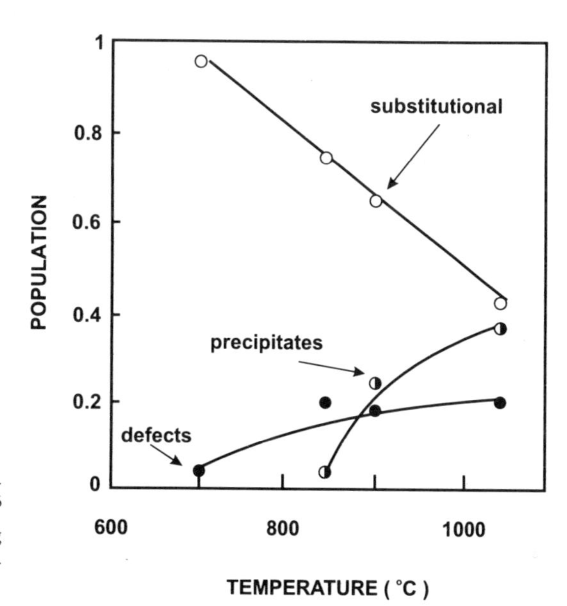

FIGURE 2.15. Population of different implanted antimony (80 keV/2 $\times 10^{16}$ ions/cm^2) sites as a function of processing temperature for 15-s incoherent light annealing [197].

limited by several interatomic distances. At 900 °C and higher the diffusion length becomes comparable with the distance between impurity atoms which leads to the primary growth of precipitates.

Antimony diffusion from implanted layers appears to be noticeable at processing temperatures higher than 1050 °C [89, 136, 198, 199]. Diffusion to the surface is followed by impurity accumulation there in the form of a continuous film [198] and by evaporation [89]. The substrate impurity content is reduced by 1.1–1.5 times for heavily implanted layers, typically disappearing for layers containing impurities below the solubility level.

The depth redistribution of antimony is a function of the impurity concentration. Figure 2.16 shows antimony profiles created by RTP of samples implanted at two doses

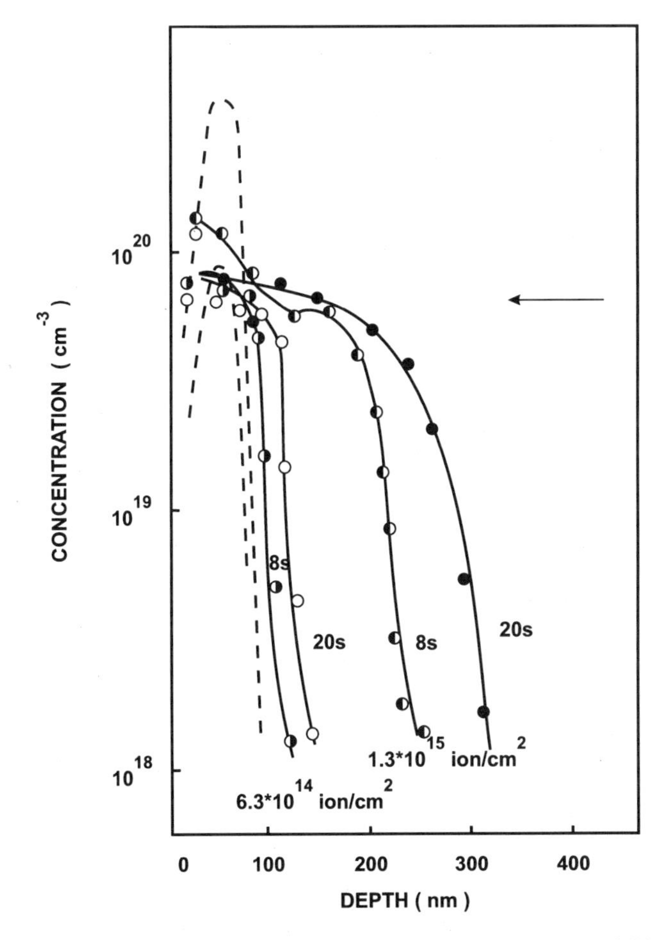

FIGURE 2.16. Antimony profiles created in (111) silicon by implantation with energy of 60 keV and transiently annealed at 1100 °C. As-implanted LSS calculated profiles are represented by the dashed lines. Antimony steady-state solubility is indicated by the arrow [89].

with peak impurity concentrations of 9.0×10^{19} and 4.7×10^{20} atoms/cm^3 (dashed lines) [89]. These were chosen to study the antimony behavior at concentrations near and above its solubility limit. In the heavily doped sample the impurity distribution is extremely deep compared to that in the lightly doped case.

The estimated diffusivity gives values that are one to three orders higher than the equilibrium ones. For an induced temperature of 1100 °C the diffusivities were 3×10^{-12} and 4×10^{-11} cm^2/s in the low- and high-dose-implanted samples, respectively. Antimony equilibrium diffusivity at this temperature is only 2.5×10^{-14} cm^2/s [128]. As for the other substitutional impurities, antimony supersaturated solutions are not formed on RTP of implanted layers at temperatures above 1000 °C and at concentrations higher than the steady-state solubility impurity diffusion is enhanced.

The enhanced antimony diffusion is interpreted according to several approaches. A vacancy mechanism was developed by Holland and Fathy [198] and Nylandsted Larsen *et al.* [199] within the concept of concentration-enhanced diffusion. A high electron concentration in the doped layer is supposed to produce single and double negatively charged vacancies and vacancy–impurity complexes, which may carry the impurity. This approach fails to explain the diffusion in heavily doped layers, where the electron concentration is stabilized at the solubility level of antimony but the impurity concentration exceeds this level. The models involving nonequilibrium point defects created during annealing of implantation damage [149], and during dissociation of the supersaturated solution [89, 136] seem to be more consistent. The low activation energy of the enhanced diffusion, 1.8 eV [149], allows one to suppose that nonequilibrium interstitials stimulate the process, though their role for antimony is not as striking as for phosphorus, arsenic, or boron.

2.3.5. *Elastic Stresses*

Elastic stresses act in a crystal as an inner source of stored energy. They create drift fields and generate point defects during a thermal process. A stressed state is usually formed during the stages of sample preparation, in particular by mechanical polishing of wafers [200], and during ion implantation [201–203]. Therefore, it is reasonable to include estimates of the elastic stresses for the study of implant behavior under RTP.

The existing models for the process of point defect generation, during relaxation of stresses in a semiconductor crystal, are not commonly accepted. One of the recognized points of view is based on the formation of Frenkel pairs (vacancy–interstitial) as a result of an elastic stress field interacting with lattice defects [203, 204]. The members of the Frenkel pairs thus become separated in the lattice. Interstitial atoms move in a direction of elastic pressure spreading, and vacancies toward it. Migration of these nonequilibrium point defects influences impurity diffusion [205–208], causing nucleation and growth of dislocations [209]. The relaxation time for elastic stresses, estimated from the experiments on the dynamics of high-temperature deformation in silicon [210, 211], and in single-crystal silicon structures with a silicon dioxide film [212] varies from one to tens of seconds in the temperature range of 800–1200 °C.

The influence of internal elastic stresses on the electrophysical parameters of phosphorus-, arsenic-, and antimony-implanted silicon subjected to RTP at 900–1100 °C was extensively studied [213]. Dislocation-free (111) silicon uniformly doped with boron (1.4 $\times 10^{14}$ atoms/cm^3) was used for the experiments. Phosphorus, arsenic, and antimony ions have been implanted with energies of 30, 50, and 60 keV, respectively, to produce doped layers at identical depths. Implant doses were chosen to provide peak impurity concentrations below (low dose) and above (high dose) steady-state solubility of these impurities at high temperatures. Phosphorus and arsenic doses were 2.5×10^{14} and 7.5×10^{15} ions/cm^2, antimony doses were 6.3×10^{13} and 1.3×10^{15} ions/cm^2. Sheet resistivity and electron mobility were measured by the Hall effect. Impurity atom profiles were determined using layer-by-layer neutron activation analysis. Internal stresses were measured by an acousto-optical resonance technique [214].

Both electrophysical parameters and elastic stresses were observed to have no variations, as a function of the processing regime, for the low-dose-implanted rapidly annealed samples. Table 2.3 presents the experimental results. The high stability of the created phosphorus, arsenic, and antimony substitutional solutions leads us to conclude that they are in thermodynamic equilibrium. Sheet resistivity and electron mobility values agree well with the theoretical ones predicted for complete electrical activation of as-implanted impurity profiles.

The situation changes dramatically in the high-dose-implanted samples. All of the parameters have become a function of the processing temperature, as displayed in Fig. 2.17. Internal stresses appear to be significantly reduced with a temperature increase to 950 °C and correlate to the maximum of the impurity concentration. The resistivity of phosphorus- and arsenic-doped layers is reduced monotonically with processing temperature rise. At the same time, electron mobility is increased after processing at temperatures above 1000 °C. Antimony-doped layers demonstrate some specific features. There is a minimum of sheet resistivity and almost invariable carrier mobility. This is connected with the antimony evaporation discussed above.

Supersaturated substitutional impurity solutions formed on rapid annealing at temperatures lower than 950 °C create elastic stresses. The processing temperature rise results in an almost twofold order-of-magnitude fall in the level to $2–5 \times 10^6$ N/m^2. The maximum impurity concentration approaches the steady-state solubility and a balance is achieved in less than 10 s at a processing temperature of 950 °C. This is followed by impurity redistribution and the electrically inactive fraction of the impurities is consis-

TABLE 2.3. Parameters of Low-Dose-Implanted Silicon after Rapid Thermal Processing at 900–1100 °C for 10 s

Impurity	Peak impurity concentration (atoms/cm^3)	Sheet resistivity (Ω/sq)	Integral carrier mobility (cm^2/V·s)	Elastic stresses ($\times 10^6$ N/m^2)
P	5.9×10^{19}	280 ± 10	90 ± 2	45 ± 15
As	5.9×10^{19}	370 ± 5	68 ± 1	25 ± 5
Sb	2.1×10^{19}	800 ± 30	175 ± 3	50 ± 15

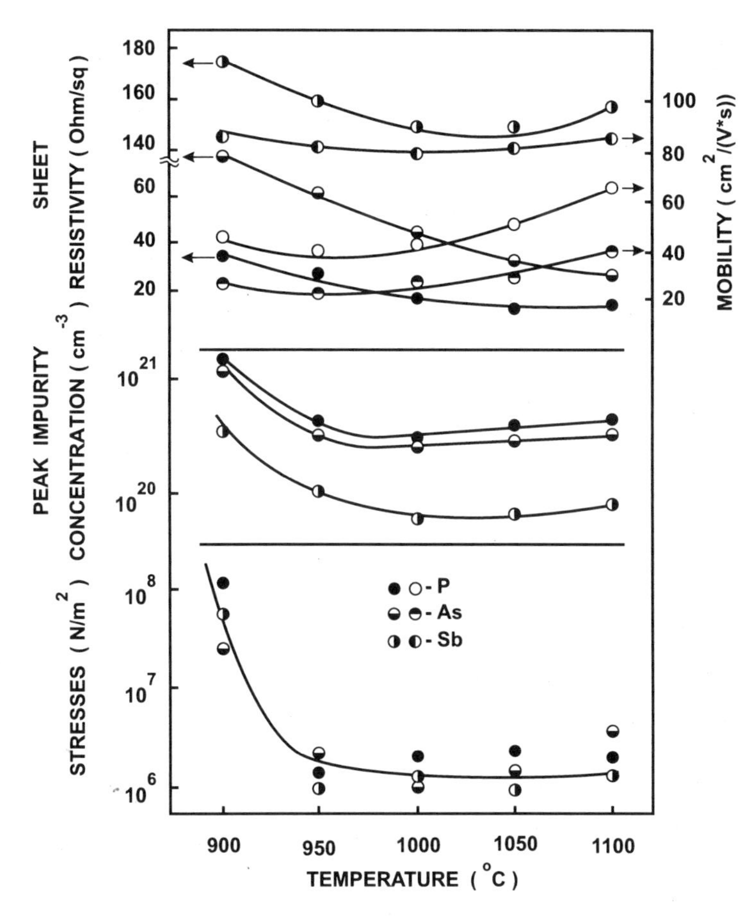

FIGURE 2.17. Sheet resistivity, free carrier mobility, maximum impurity concentration, and elastic stresses in phosphorus, arsenic, and antimony heavily implanted (111) silicon after 10-s incoherent light annealing [213].

tently reduced. The same features were observed by other researchers [215–217]. This actually shows the importance of the elastic stresses in impurity atom behavior in supersaturated substitutional impurity solid solutions created by ion implantation and RTP.

2.3.6. *General Features*

Summarizing the experimentally observed features of implanted substitutional impurity behavior in silicon under RTP, the following aspects are clear. Heating to 700–900 °C for 10–15 s causes complete solid-state recrystallization of the implanted layers accompanied by the formation of almost 100% of the substitutional fraction with concentrations up to 10^{21} atoms/cm^3. Practically no impurity redistribution takes place.

In the case of rapid annealing at 900–1300 °C two distinct situations have to be considered. They are connected with the peak implanted impurity concentration below and above steady-state impurity solubility. Impurity redistribution in the undersaturated layers is usually insignificant and is described well by the equilibrium diffusivities. Supersaturated impurity solutions have a transient enhanced effective diffusivity 1.5–2 orders greater than the equilibrium values and activation energies reduced to 2–3 eV. The maximum concentration of the impurities in electrically active substitutional positions is stabilized at the level corresponding to the steady-state solubility of the impurity in silicon at the processing temperature.

The enhanced diffusion takes place in the layer that is already recrystallized with the participation of mostly interstitial nonequilibrium point defects. These are generated during subsequent dissociation of supersaturated substitutional solution, created in the process of solid-state epitaxial regrowth, and the associated elastic stresses relax. Nonequilibrium point defects released from clusters may also enhance diffusion. For example, annealing of disorder produced by boron implantation stimulates diffusion at concentrations less than its solubility limit.

Implanted impurity redistribution occurs via their diffusion and drift in the field of elastic stresses related to nonequilibrium point defects generated during RTP. The mechanism for the diffusion includes vacancies, silicon and impurity interstitials for boron, phosphorus, arsenic, and antimony in silicon [116–126]. This has been used successfully for simulation of the impurity profiles and electrophysical parameters of the doped layers formed by implantation and subsequent RTP [218, 219].

2.4. DISSOCIATION OF SUPERSATURATED SOLID SOLUTIONS AND IMPURITY SOLUBILITY

In experiments on RTP of implanted silicon layers, a trend is observed, namely, that the maximum impurity concentration stabilizes at the level corresponding to the impurity solubility. This indicates that a study of the stability of supersaturated substitutional solutions in silicon crystals is an important consideration in rapid heating. There is a need for verification of available data on substitutional impurity steady-state solubility. It is known that supersaturated substitutional impurity solutions are unstable in semiconductor crystals [220]. Conventional longer-duration thermal treatment is followed by their dissociation, which means reduction of the number of impurity atoms in the positions typical for the thermodynamically stable undersaturated solid solution of the impurity. It is accompanied by abnormal impurity diffusion, clustering, precipitation, and impurity atom condensation at structure defects in electrically inactive states. Impurity interstitials, vacancy–impurity complexes, or both are supposed to take part in the processes. The leading role of one type of the defects is still in question. Nevertheless, dissociation can definitely be considered a process responsible for nonequilibrium point defect generation. The kinetics of supersaturated solid impurity solution dissociation controlled by the diffusion of excess impurity atoms (to sinks) is described by [221, 222]

$$N(t) = N_0 \exp[-(t/\tau)^m] \tag{2.17}$$

where $N(t)$ is the impurity concentration in the nondissociated solid solution, N_0 its initial concentration, and m the empirical factor determined by the geometry of the second phase produced in the dissociation process. Theoretical estimates show $m = 3/2$ for spherical precipitates, $m = 1$ for disk-shaped and cylindrical ones of small size [222]. Relaxation time is determined by the integral density of the precipitate nucleation centers N and by the impurity diffusivity D

$$\tau = 1/DN \tag{2.18}$$

The temperature dependence of the relaxation time is usually presented in the Arrhenius form

$$\tau = \tau_0 \exp (E_a/kT) \tag{2.19}$$

where E_a is the activation energy of dissociation.

Impurity solution dissociation in single-crystal silicon has mostly been investigated in conventional furnace experiments [223–231]. In these studies the thermal processing time was varied from tens of minutes to several hours at temperatures not higher than 900 °C. The data obtained agree well in general, but there is a significant scatter in the values of activation energy and preexponential factor. Rapid heating experiments extending the temperature range to 1100 °C provide more precise characteristics for phosphorus and arsenic supersaturated solid solution behavior [232–234]. Supersaturation is created by liquid-phase nanosecond laser annealing of ion-implanted samples in these experiments. The concentration of the impurities in electrically active substitutional positions was found to exceed their steady-state solubility by five- to eightfold. The dissociation process initiated by incoherent light heating for seconds was controlled by monitoring of integral free carrier concentration and sheet resistivity in the doped layers. The reference undersaturated layers exhibited no changes in the above parameters on processing at 300–1100 °C for seconds. The supersaturated solutions demonstrate variation of the parameters depending on the processing temperature. Two specific temperature ranges can be distinguished. There is a 70–85% drop of free carrier concentration after processing at 300–700 °C. However, at 1000 °C the free carrier concentration is increased to a value corresponding to the implant dose [232–234].

Among the substitutional impurities, phosphorus behavior in supersaturated solutions has been the most extensively investigated [228, 230, 231, 233, 234]. The effects observed seem to be general.

Figure 2.18 shows the kinetics of the integral electron concentration drop in the phosphorus supersaturated layers subjected to thermal processing at 300–700 °C. The experimental data are presented in semilogarithmic coordinates in order to evaluate the fit of Eq. (2.17) with $m = 1$. Two linear approximations are evident from the fit, which means there are two different relaxation times: a rapid dissociation stage that takes place within the first 10–40 s of thermal processing, and a slow one when the duration is greater than 60 s. The temperature dependence of relaxation time for both stages is presented in Fig. 2.19. Rapid dissociation is characterized by two activation energies, i.e., 1.10 eV at 300–400 °C and 0.33 eV at 400–700 °C. For the slow stage only one activation energy (0.58 eV) is apparent. The data obtained for other impurities are collected in Table 2.4. Most of the values for the activation energy occupy two bands, 0.3–0.4 and 0.9–1.3 eV.

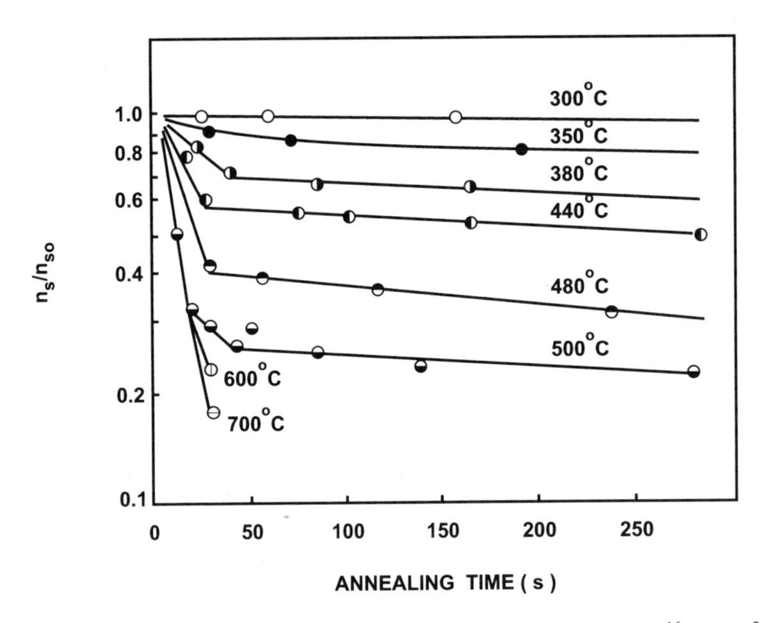

FIGURE 2.18. Relative electron concentration in phosphorus-implanted (40 keV/2 × 10^{16} ions/cm^2) liquid-phase laser-annealed (111) silicon as a function of the subsequent processing duration at different temperatures [234].

A reasonable explanation of the dissociation is as follows: Electrical deactivation of substitutional impurities obviously requires two elementary acts. First, an impurity atom is released from a substitutional position and, second, it diffuses to a sink. The rate at which these processes occur depends on the temperature, hence influencing the dissociation kinetics. The split in the activation energies supports the conclusion that dissociation is limited by the rate of impurity interstitial diffusion in the low-temperature range, and at higher temperatures the rate of impurity atom release from substitutional positions becomes the limiting factor, being directly connected with the vacancy removal process. In the latter case the activation energy obtained for phosphorus is 0.4 eV [228] and 0.33 eV [234], for arsenic 0.33 eV [232], and for antimony 0.33 eV [227]. These correspond well to the migration energy of neutral vacancies in silicon, 0.33 eV [235, 236]. An activation energy in the range of 0.9–1.3 eV is an indication of a process limited by impurity interstitial diffusion. It is normally associated with metals diffusing in silicon via dissociative mechanism. This conclusion is also supported by the available data on the diffusion of group-III and -V interstitials: boron 1.1 eV [162], phosphorus 1.1–1.4 eV [233, 237], arsenic 1.1 eV [193], and antimony 1.8 eV [92, 149].

The most representative extrapolation to 1000 °C of relaxation times for supersaturated impurity solutions in silicon, are in the range of seconds to hundreds of seconds (see Table 2.4). Experimental investigations of boron, phosphorus, arsenic, and antimony behavior in supersaturated solutions [233, 238] show that dissociation occurs within the first 15–20 s at temperatures of 900–1100 °C. The invariability of the sheet resistivity with the processing time duration confirms this. Also, there were no changes in impurity

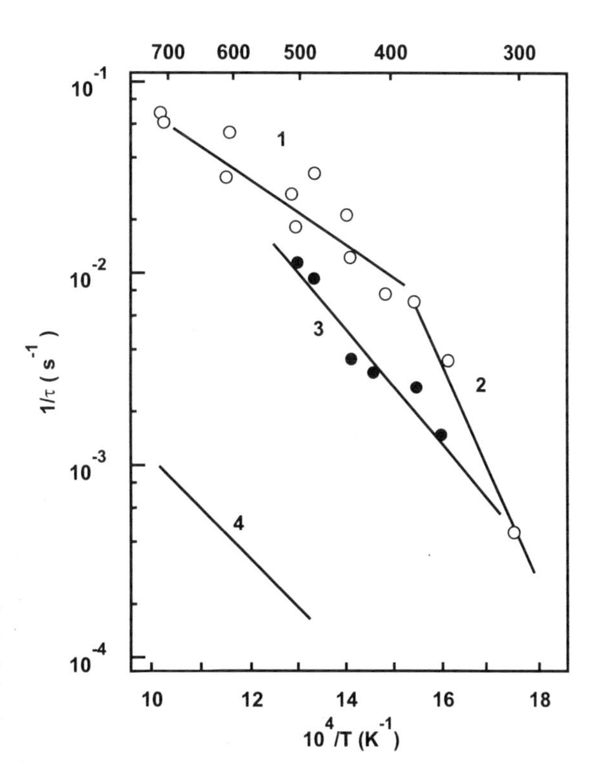

FIGURE 2.19. Relaxation time of supersaturated phosphorus solution in silicon. Duration of thermal processing: (1, 2) 10–30 s; (3, 4) 1–45 min. Data for (1–3) from Ref. 234; data is (4) and Ref. 230.

concentration profiles, which sets the free carrier distribution. Flat top-shaped profiles with a maximum atom and free carrier concentration corresponding to the impurity solubility are formed.

The theory discussed in Section 2.3 requires that the steady-state solubility be accurately determined to correctly describe the diffusion behavior of substitutional impurities. The steady-state impurity solubility is the maximum concentration of impurity atoms allowing thermodynamic balance between crystal and another neighboring phase, which can be dislocations, grain boundaries, precipitates, amorphous, or a liquid phase.

The first attempt to summarize experimental data for silicon was carried out by Trumbore in 1960 [239]. His data have been widely used for a long time. Subsequently, accurate experiments in the extended range of RTP show that Trumbore's values of substitutional impurity solubility should be corrected [240–260]. They were extensively reexamined by Borisenko and Yudin [238] as illustrated in Fig. 2.20. In particular, the solubility of boron, phosphorus, and arsenic are evidently lower than the values presented by Trumbore, although the data for antimony are the same.

Thermodynamic parameters related to impurity interaction in a silicon matrix are derived from the experimental data. The balance between the two phases supposes equality of temperature and chemical potential. The atomic diffusion in liquid and solid

TABLE 2.4. Parameters of Supersaturated Impurity Solution Dissociation in Single-Crystal Silicon

Impurity	Temperature range (°C)	τ_0 (s)	Activation energy, E_a (eV)	m	Ref.	τ (1000 °C)
Cr	170–230	2.7×10^{-9}	0.91	1	223	1.1×10^{-5}
Fe	100–500	—	0.58	1	224	—
Au	500–700	6.9×10^{-5} [a]	1.31[a]	1	225	11
Ir	500–900	2.4×10^{-3}	1.2	1	226	138
B	800–900	2.1×10^{-10} [a]	3.2	0.61	230	1039
	800–900	—	3.3	—	231	—
P	350–700	—	0.4	1	228	—
	350–500	2.8[a]	0.5	1	230	269
	300–400	4.3×10^{-7}	1.10	1	234	—
	400–700	0.3	0.33	1	234	6
	700–1000	—	1.9	—	231	—
As	350–410	1.6×10^{-13} [a]	2.0	1	229	1.4×10^{-5}
	400–600	0.67	0.5	1	230	64
	800–1100	—	0.3	1	232	—
	700–1100	—	3.4	—	231	—
Sb	600–900	0.44	0.33	1	227	9

[a] Calculated from the experimental data presented in the reference.

states differs considerably, and therefore a steady-state balance between them is not possible. Therefore, a short-time quasi-steady-state equilibrium is realized. Amorphous material resembles more closely the crystalline phase indicating that a steady-state thermodynamic balance is easily achieved at higher temperatures.

The analogy between amorphous and liquid phases was used to calculate the solid steady-state solubility of an impurity in a semiconductor crystal. According to the theory of regular solutions [261], the impurity solubility can be defined from the variation of enthalpy, ΔH, and entropy, ΔS, corresponding to the impurity atom exchange between the two phases. The solid steady-state mole fraction of the impurity can be written [261] as

$$X_2^{\mathrm{S}} = X_2^{\ell} \exp\left[\frac{\Delta H_2 - Q_{\mathrm{i}}^{\mathrm{S}}}{R^* T} - \frac{\Delta S_2^{\mathrm{tr}}}{R^*} + \frac{Q_{\mathrm{i}}^{\ell}(1 - X_2^{\ell})^2}{R^* T}\right] \tag{2.20}$$

where Q_{i} is the interaction parameter, R^* the universal gas constant, superscripts "s," "ℓ," "tr" correspond to the crystalline phase, the contacted phase, and to the atom exchange s $\leftrightarrow$ ℓ in the couple, respectively. Subscript "1" denotes the semiconductor and "2," the impurity. Equation (2.20) is valid for $X_2^{\mathrm{s}} \ll 1$.

The mole fraction of the impurity in the other phase is $X_2^{\ell} = 1 - X_1^{\ell}$, where

$$\ln X_1^{\ell} = -\frac{\Delta H_1^{\mathrm{tr}}}{R^*}\left(\frac{1}{T} - \frac{1}{T_1^{\mathrm{tr}}}\right) - \frac{Q_{\mathrm{i}}^{\ell}(1 - X_1^{\ell})^2}{R^* T} \tag{2.21}$$

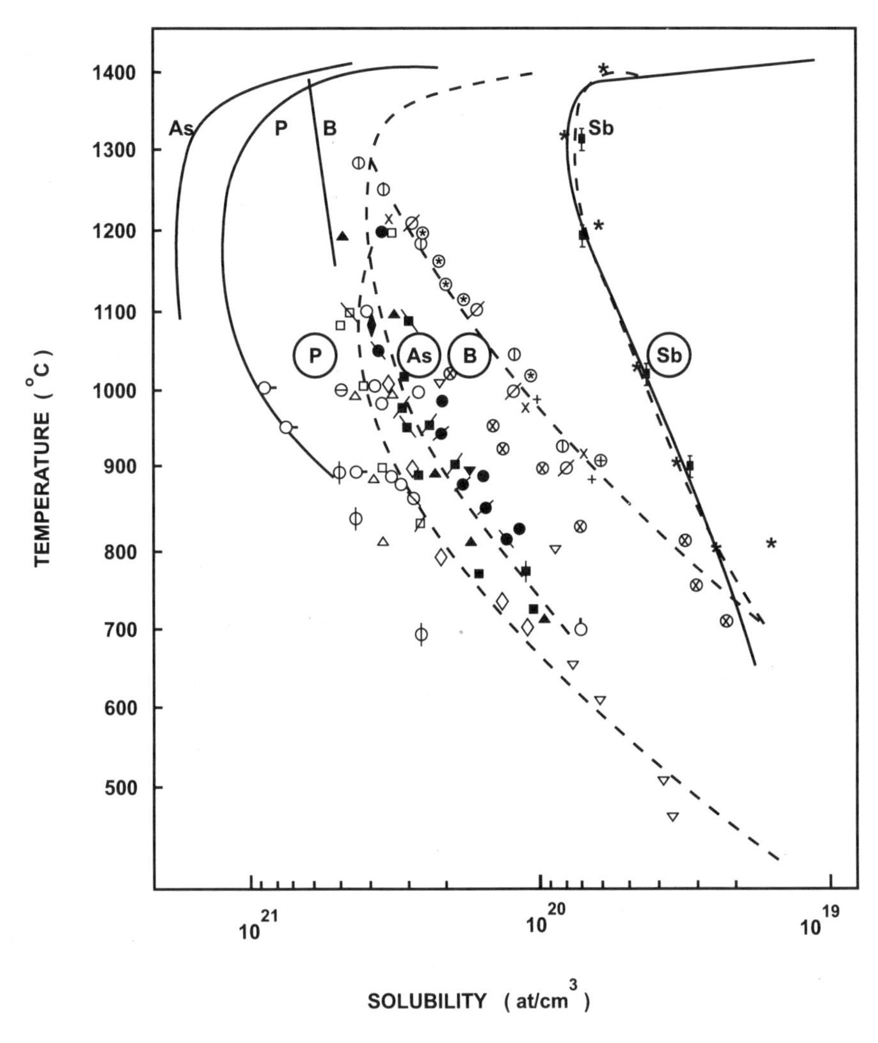

FIGURE 2.20. Steady-state solubility of substitutional impurities in silicon. Experimental values. Solid lines represent Trumbore's data. Dashed lines are the result of calculations with the corrected thermodynamic parameters. Experimental results:

P: □ [238], [240], ⊖ [243], ▽ [245], [247], ○- [248], ⃠ [249], △ [253], ○ [256], ◇ [257], [258], ɸ [260]
As: ● [238], [245], ◆ [250], [251], [252], [253], [254], ● [257], ▲ [259], ■ [260]
Sb: ■ [238], * [256],
B: ∅ [238], ⊛ [242], ⊗ [242], ⊕ [244], ⦶ [246, 247], + [255], × [256]

The experimental results were treated in Ref. 238 according to the above mathematics to obtain thermodynamic parameters related to the interaction of boron, phosphorus, arsenic, and antimony with a silicon matrix. The enthalpy of impurity exchange between the phases ΔH_2^{tr} was assumed to have a linear dependence on the temperature. The values of the parameters necessary for the calculations were obtained by extrapolation of the data for crystalline and liquid silicon to the temperature range investigated (see Table 2.5). The temperatures T_1^{tr} and T_2^{tr} correspond to the phase transitions of pure silicon and the impurity. Taking into consideration the temperature dependence of ΔH_2^{tr}, a more correct value for the parameter Q_i^S responsible for atomic interaction between impurity atoms and silicon is produced. The temperature-dependent part of this parameter was separated and added to the entropy of the phase transition. Such an approach provides a better agreement between the experimental data and calculated solubilities encompassing the ionic properties of the impurity atoms [220]. Positive values of Q_i^S indicate that the interaction between identical atoms is dominant. Relatively high values of this parameter are caused by the retrograde solubility of the substitutional impurities in silicon.

The values obtained for the thermodynamic parameters characterizing the interaction of boron, phosphorus, arsenic, and antimony with silicon allow accurate numerical simulation of the steady-state solubility of these impurities at temperatures from 700 to 1300 °C.

Except for the thermodynamically consistent approach discussed, the simplified empirical Arrhenius expression $N = N_0 \exp(-E_a/kT)$ may be employed to describe the experimentally observed impurity solubility in silicon [264, 265]. It is valid over a limited temperature range because of the monotonic character of the exponential function. Appropriate activation energies and preexponential factors are summarized in Table 2.6. They have been obtained by fitting the experimental data. Note that it is incorrect to extrapolate the Arrhenius exponential relation over the temperature range noted in Table 2.6.

TABLE 2.5. Thermodynamic Parameters Determining Interaction of Substitutional Impurities with Silicon Matrix[a]

	Introduced from Refs. 262, 263			Calculated in Ref. 238		
Impurity	ΔH_2 ($\times 10^3$ J/mol)	T_2^{tr} (K)	$Q_i^S = \Delta H_2 - \Delta H_2^{tr}$ (10^3 J/mol)	ΔS_2^{tr} (J/mol·K)	$\Delta S_2 = \Delta H_2/T_2^{tr}$ (J/mol·K)	X_2^S/X_2^l at $T = T_1^{tr}$
P	16.7	866	53.0 8.0 [220]	12.53	19.28	0.123 0.35 [239] 0.206 [220]
As	21.8	1090	61.7 24.9 [220]	12.48	20.0	0.095 0.3 [239]
Sb	20.1	904	54.0 58.1 [220] 53.0 [220]	32.08	22.24	0.014 0.023 [239] 0.017 [220]
B	50.2	2350	118.8		21.36	0.09 0.8 [239]

[a] $\Delta H_1^{tr} = 49.06 \times 10^3$ J/mol; $T_1^{tr} = 1685$ K; $R^* = 8.314$ J/mol·K; $Q^l = 28 \times 10^3$ J/mol [220].

TABLE 2.6. Parameters of Arrhenius Approximation for Solid Steady-State Solubility of Substitutional Impurities in Silicon [264]

		Solid-state solubility coefficients	
Impurity	Temperature range (°C)	N_0 (× 10^{22} atoms/cm^3)	Activation energy, E_a (eV)
B	900–1325	9.25	0.73
P	750–1050	1.8	0.40
As	800–1050	2.2	0.478
Sb	850–1150	0.38	0.56

2.5. IMPURITY DIFFUSION FROM A SURFACE SOURCE

Diffusion doping of semiconductors based on the use of a surface impurity source and RTP has been proposed by Borisenko and Nylandsted-Larsen [266] as a simpler, less expensive, and less time-consuming technique to ion implantation. The enhanced impurity diffusion observed in heavily implanted transiently annealed silicon layers suggests this method is worth pursuing. The first experiments on incoherent light heating-induced arsenic diffusion were in silicon from a spin-on source in the time range of seconds [266]. Later experiments include phosphorus and boron [267–269]. The results obtained were confirmed and extended by other researchers [270–277].

The proposed method includes two principal steps: impurity source deposition and RTP. Doped silicon dioxide is widely used as a surface impurity source. It is formed by depositing a few drops of a commercial dopant emulsion [278–282] onto a semiconductor wafer followed by spinning at 2000–4000 rpm for 10–20 s. The emulsion consists of liquid silicon organic (e.g., tetraethoxysilane) with dissolved impurity oxides or salts in it. Impurity concentration in the spin-on deposited film can be varied in the range of 10^{20}–10^{22} atoms/cm^3. Its thickness is about 100–300 nm. The wafer with the deposited film is subjected to RTP.

2.5.1. *Arsenic*

Figures 2.21–2.23 present the characteristics of arsenic-doped silicon from the Emulsitone source [266]. The doping process proceeds most effectively at temperatures above 1000 °C. This is displayed by a sufficient decrease of sheet resistivity to 60–70 Ω/sq after heating to 1200 °C. At the same temperatures, samples processed for 25 s had 1.3- to 2.0-fold lower sheet resistivity than those heated for 10 s. The doped layers are characterized by a perfect crystalline structure with more than 82% of the impurity atoms occupying substitutional sites. The exception is the thin surface layer where nonsubstitutional arsenic is located.

Arsenic concentration depth profiles are typical of diffusion, being dependent on the processing temperature and time (Fig. 2.23). Surface concentration of the impurity remains approximately constant at 8 × 10^{19} atoms/cm^3 for the temperature range of 1050–1150 °C, and somewhat increases at 1200 °C. Increasing the processing time from

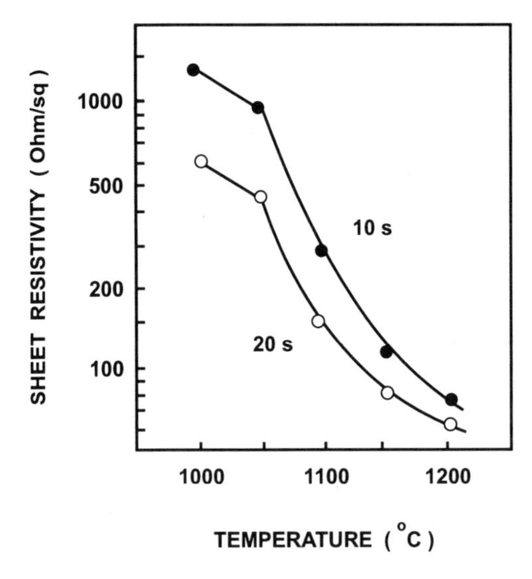

FIGURE 2.21. Sheet resistivity of arsenic-doped (100) silicon layers created by rapid diffusion from the spin-on deposited doped oxide film.

10 to 25 s is accompanied by an almost twofold increase in the doped layer depth to a maximum value of 400 nm.

The effective diffusivity of arsenic was estimated to be an order of magnitude greater than the equilibrium one. The diffusion activation energy is 4.6 ± 0.3 eV which correlates to the known isothermal values, although the preexponential factor is three times higher.

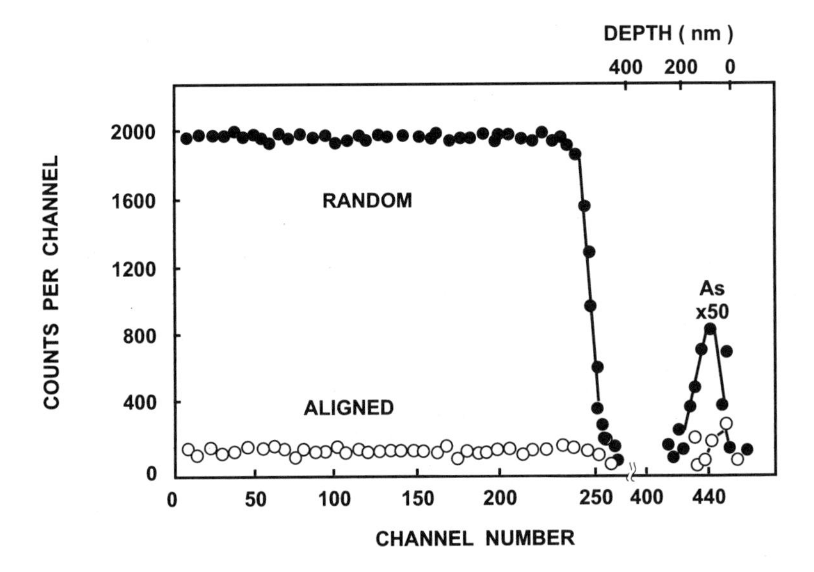

FIGURE 2.22. 2-MeV He^+ ion backscattering spectra for (100) silicon doped with arsenic from the spin-on source at 1150 °C for 10 s.

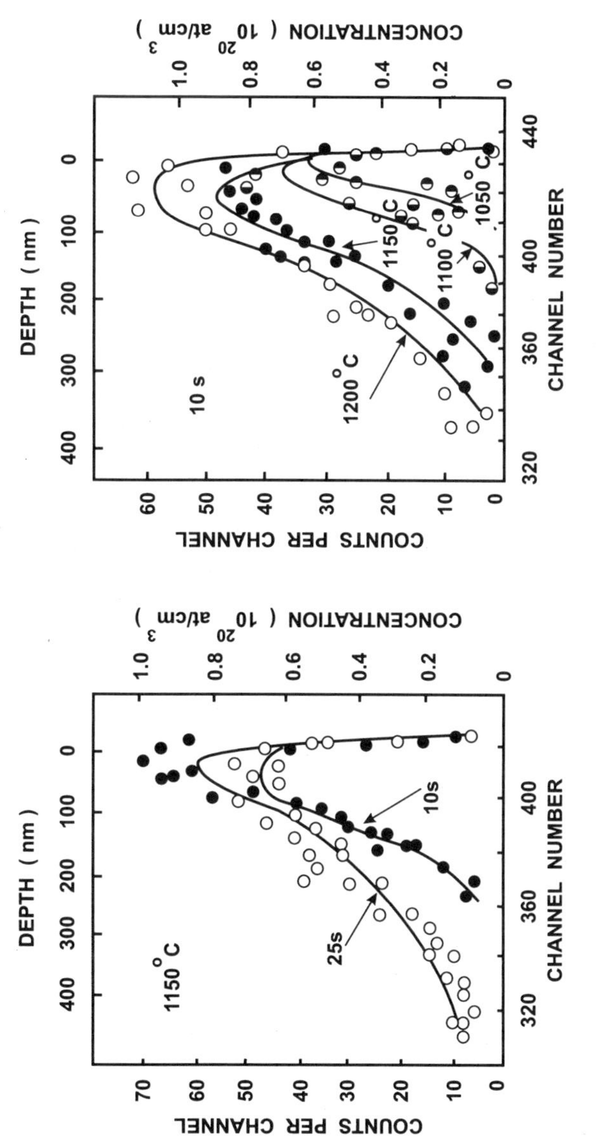

FIGURE 2.23. Arsenic depth profiles produced in (100) silicon by diffusion from the spin-on source.

2.5.2. Phosphorus

The influence of the impurity content for phosphorus spin-on sources on sheet resistivity and doped layer depth was investigated, as shown in Fig. 2.24 [270]. Sheet resistivity is rapidly reduced when the phosphorus concentration is increased in the range of 5 to 15 wt % content of P_2O_5 in the glass. The resistivity value saturates at 30 wt %. There is a proportional increase of the doped layer depth in this range of P_2O_5 content. The data are corroborated by experiments using a 20 wt % P_2O_5 source [271]. Sheet resistivity, *p–n* junction depth (in crystals uniformly doped by boron to 1.4×10^{15} atoms/cm^3), and electron concentration profiles versus thermal processing regimes are shown in Figs. 2.25 and 2.26 for concentrated emulsions (50 wt % P_2O_5) [267, 268]. Thermal processing at 900–1200 °C for 6–20 s is characterized by rapid lowering of sheet resistivity. The depth of the doped layer increases, and with longer processing the sheet resistivity and the doped layer depth tend to be stabilized at the levels defined by the processing temperature. The higher temperature produces a lower sheet resistivity, and a deeper doped layer. Minimum sheet resistivity of 10 Ω/sq and maximum *p–n* junction depth of about 550 nm were measured.

Sheet resistivity variation as a function of thermal processing regimes is mainly determined by the depth of the doped layer. It can be deduced from the electron depth profiles. The surface electron concentration is about $(1–3) \times 10^{20}$ /cm^3 and correlates well with steady-state phosphorus solubility at these temperatures. The electron concentration profiles have a plateau region near the surface and a kink in the tail, typical for phosphorus-doped layers [282]. Effective diffusivities were estimated to be 0.5–1.5 orders of magnitude higher than the values known from traditional furnace diffusion. Such an enhancement is also a function of the concentration of impurity atoms in the surface. Diffusion activation energies and preexponential factors were evaluated from the experimental results presented in Table 2.7 [270, 271]. The enhanced diffusion characterized by the reduced activation energy apparently occurs only at high impurity concentration in the surface source.

The phosphorus concentration is about 10^{21} atoms/cm^3 in the films containing 40–50 wt % P_2O_5. It is worth assuming that a metastable supersaturated impurity solution is

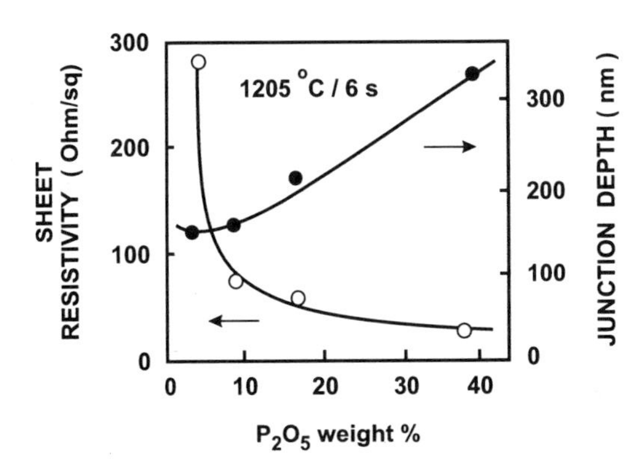

FIGURE 2.24. Sheet resistivity and *p–n* junction depth of phosphorus-doped layers as a function of phosphorus content in the spin-on source [270].

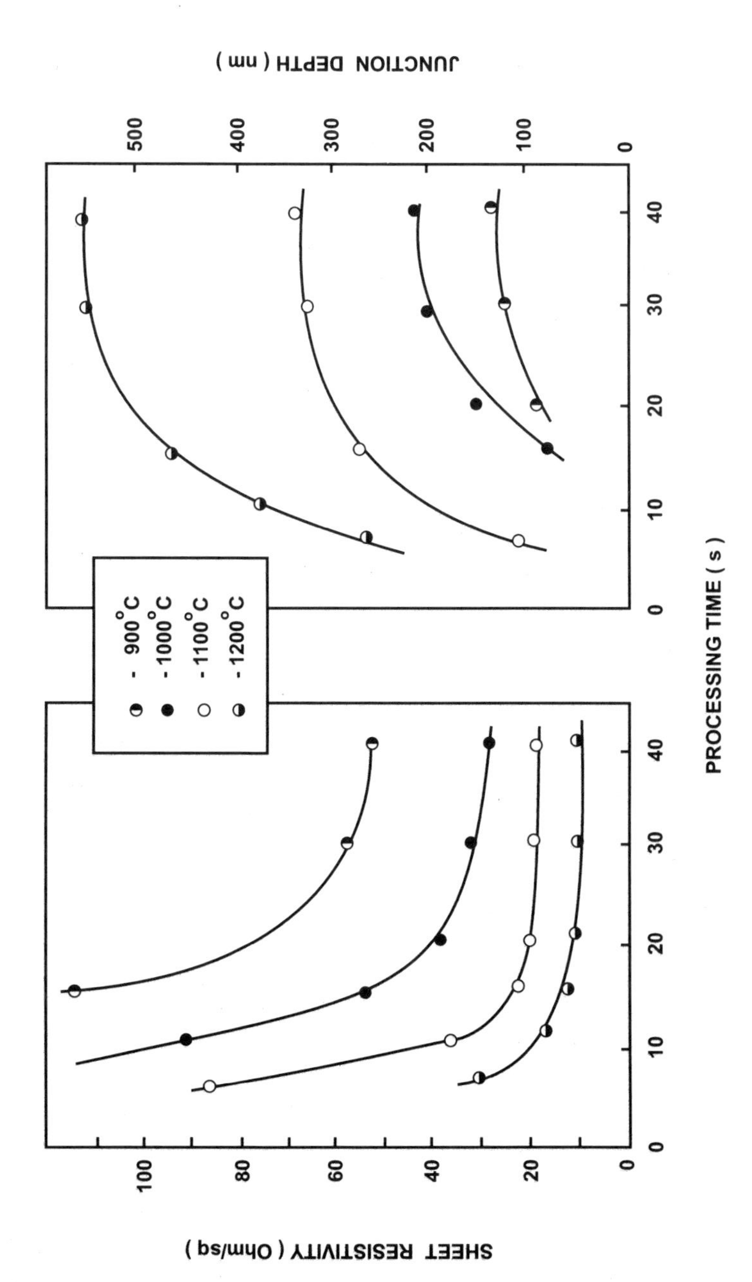

FIGURE 2.25. Sheet resistivity and *p–n* junction depth of spin-on phosphorus-doped (111) silicon layers as a function of rapid thermal processing regimes [268].

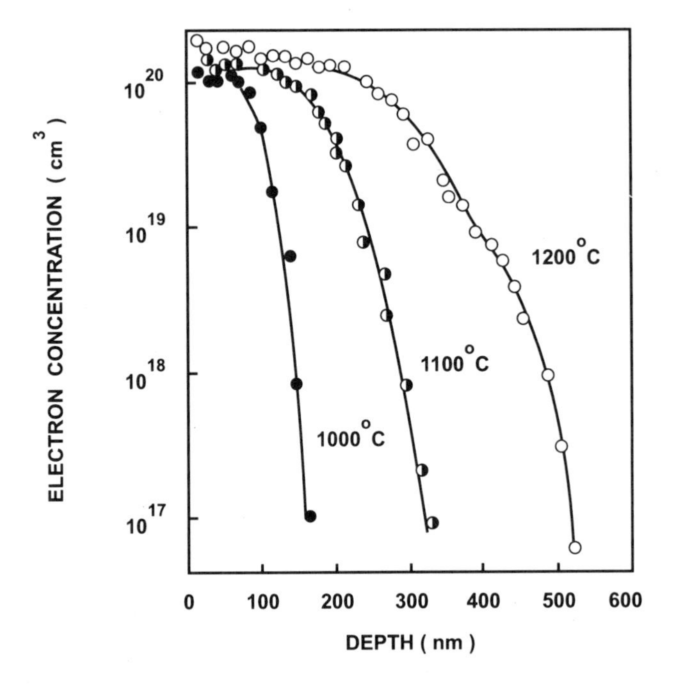

FIGURE 2.26. Electron concentration profiles in (111) silicon doped by phosphorus diffusion from the spin-on source for 20 s [268].

formed at the silicon surface during phosphorus in-diffusion from the spin-on source. Dissociation of the solution produces interstitials giving rise to phosphorus diffusion. Moreover, chemical reactions providing silicon oxidation at the film–substrate interface add silicon interstitials in the diffusion zone. The activation energy derived from the effective diffusivity is therefore reduced.

2.5.3. Boron

Similar features were found for boron diffused into silicon from the doped oxide films [272, 273, 277] at a concentration of 1×10^{22} atoms/cm^3. Hole depth profiles in the

TABLE 2.7. Phosphorus Diffusion from Spin-On Source into Single-Crystal Silicon Initiated by Rapid Thermal Processing at 1000–1200 °C

P_2O_5 content in the source (wt %)	Maximum electron concentration in the doped layer (cm^{-3})	Diffusion	Diffusivity coefficients D_0 (cm^2/s)	E_a (eV)	Ref.
~20	$(6–9) \times 10^{19}$	normal	—	—	271
37.5	$(1.1–1.5) \times 10^{20}$	enhanced	78	3.40	270
50	$(1.1–2.7) \times 10^{20}$	enhanced	3.4×10^{-6}	1.8	268
diffusion under intrinsic conditions			10.5	3.69	127

doped layers are shown in Fig. 2.27, where the surface hole concentration is always higher than 10^{20} p/cm^3 reaching 2×10^{21} p/cm^3 on processing at 1250 °C. There are plateaus in the hole distributions correlated to the boron solubility, $(2–3) \times 10^{20}$ atoms/cm^3, at these temperatures. Formation of supersaturated substitutional solution in the near-surface silicon region is obviously confirmed. The estimated effective diffusivity indicates a one to two order-of-magnitude enhancement. The process is characterized by a reduced activation energy of 2.1 eV. As in the case of phosphorus, this supports the conclusion that the diffusion process under consideration is mediated by nonequilibrium interstitials.

2.5.4. *Metal Impurities*

In contrast to substitutional impurities, there has been no enhancement observed for interstitial ones. Iron and gold diffusion from a surface metallic film was investigated for thermal processing within milliseconds [283] and for gold within seconds [284, 285]. 2-MeV He^+ ion channeling indicated that a considerable fraction of the nonsubstitutional gold (90–95%) had diffused into the silicon at 950–1100 °C for seconds [284]. The surface concentration reaches $(6–7) \times 10^{19}$ atoms/cm^3, which exceeds the equilibrium gold solubility in silicon [239].

Diffusion of impurity in the near-surface region is characterized by an activation energy of 1.19 eV and a preexponential factor of 3.5×10^{-4} cm^2/s [284]. This is in good agreement with the data for the dissociative diffusion of gold during traditional thermal processing, at 1.11–1.14 eV [137]. Different diffusion characteristics were obtained for low concentrations (10^{14}–10^{16} atoms/cm^3) at the impurity tail penetrating into silicon crystal for 200–500 μm [285]. The E_a and D_0 are 0.29 eV and 1.74×10^{-4} cm^2/s, respectively.

2.5.5. *General Features*

Thus, by examining the set of experimental results for implanted layers, the impurity concentration is critical, as well, in the behavior during RTP-activated diffusion from a

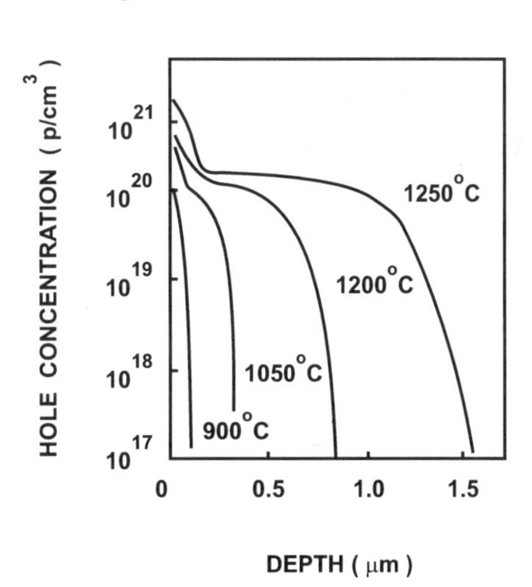

FIGURE 2.27. Hole concentration profiles in the doped layers formed by boron diffusion into silicon from the oxide film source for 120 s [272].

surface source. An enhanced diffusion takes place when the concentration of substitutional impurities at the surface is higher than the steady-state solubility. Activation energy for the process is usually reduced by 2–3 eV. There is no enhancement for concentrations lower than this level, and also for metal impurities, diffusing exclusively via a dissociative mechanism. This stresses the decisive role of nonequilibrium point defects generated during RTP. Such defects are readily produced during the dissociation of metastable supersaturated substitutional impurity solutions, created in silicon just beneath the surface-impurity source.

Oxidation processes in the silicon-based spin-on impurity sources may also contribute to the diffusion enhancement. Oxidation-related changes in the infrared spectra of the surface oxide film were observed [269, 277]. Silicon oxidation produces interstitials, which can enhance diffusion of substitutional impurities even at low concentration [286]. However, this enhancement factor seems to be not as effective when compared to the dissociation of supersaturated solutions.

2.6. IMPURITY GETTERING AND DIFFUSION IN POROUS SILICON STRUCTURES

Porous silicon formed by anodization of single-crystal silicon in hydrofluoric acid aqueous solutions is one of the promising materials of the modern solid-state electronics [287]. A wide range of novel technologies exploiting the unique properties of the material have been developed. Gettering of fast-diffusing impurities and deep doped layer formation are traditionally among them. Intense luminescence of porous silicon demonstrated in 1990 by Canham [288] opened a new era of silicon-based optoelectronics [289].

Anodic treatment of silicon in fluorine-containing electrolytes creates narrow-pore channels directed preferentially from the surface into the crystal. Crystalline structure is preserved in the porous layer, while an extremely developed inner surface appears. Pore size varies from a few to hundreds of nanometers and is dependent on the type of silicon conductivity and anodic current regime.

The investigations of single-crystal silicon with a surface porous layer subjected to RTP by incoherent light for seconds include gold and copper gettering [290, 291], arsenic [292], and erbium [293, 294] diffusion.

2.6.1. Gettering

Metal impurities are well known to create deep levels in the silicon band gap. They cause degradation of the material electrophysical parameters responsible for a semiconductor device performance. Porous silicon with high surface area is attractive for gettering of fast-diffusing metal impurities. The effectiveness of porous layer gettering of copper and gold was measured by monitoring of minority free charge carrier lifetime at the back of the silicon wafer [290, 291]. Neutron activation analysis and RBS were also employed. These experiments were performed on (111) silicon wafers uniformly doped with boron (4×10^{17} atoms/cm^3) or antimony (5×10^{18} atoms/cm^3) with a 2-μm, lightly phosphorus-doped epitaxial layer (3×10^{15} atoms/cm^3). The choice of a substrate with *p*- or *n*-type conductivity produces porous silicon layers with different structures and properties. The samples were saturated with gold or copper by long-time diffusion from the metal surface film, at 1125 °C. A metal-enriched surface layer was then chemically etched. Porous silicon layers 10 μm thick with a density of 1.2–1.8 g/cm^3 were formed at the back side of the wafers. Porous layers covered half of the wafer in order to have a reference region on the same sample.

The minority carrier lifetime in the epitaxial layer is illustrated in Fig. 2.28, as a function of RTP regime. The lifetime increases after the first 4–8 s of thermal processing, reaching its maximum value at 8 s. Further processing time of 20 s reduces the value to that obtained with a furnace anneal (FA). This change is typical for silicon contaminated with either copper or gold.

A significant increase in impurity concentration has been shown by neutron activation analysis and RBS. Gold distributions in the porous layer and nearby silicon crystal are presented in Fig. 2.29. Initially, impurity atoms were uniformly distributed over the wafer thickness with a concentration of 1×10^{17} atoms/cm^3. RTP of the samples produced an accumulation of gold in the porous layer, reaching 2×10^{14} atoms/cm^2 for n-type and 3×10^{14} atoms/cm^2 for p-type silicon after 10 s. The reference porous sample, annealed in a furnace for 10 min, produced a higher integral concentration of trapped impurity, 5×10^{14} and 7×10^{14} atoms/cm^2 in n- and p-type silicon, respectively.

There is a gold-depleted region beneath the boundary of the porous layer formed on a p-type wafer. In contrast, in n-type silicon the impurity atoms are accumulated in the nearby silicon substrate, as well as in the porous layer. With increased processing time, the depleted region is located deeper in the bulk of the crystal.

Gold is a fast diffuser in silicon and migrates over a distance of about 100 μm in 1 s at a temperature of 1100 °C. Therefore, impurity atoms accumulated in the porous layer may be collected from the whole bulk of the semiconductor wafer (typical thickness of 300–400 μm).

Trapping of metal impurities by the surface of crystalline silicon has been observed after both conventional furnace [295] and rapid thermal [296, 297] processing. Impurity atoms are fixed by broken lattice bonds at the surface. Gettering efficiency is considerably enhanced by the developed surface area in porous silicon.

The maximum impurity amount that can be trapped by the getter is the gettering capacity of the layer. During thermal processing of seconds this limit is not reached. Traditional isothermal annealing at the same temperature brings about a further increase in the accumulated impurity. Thus, a porous silicon layer acts as a constant sink for gold atoms causing an impurity gradient favorable for their diffusion into the porous region.

Gold diffusivity was estimated from the total amount of the impurity accumulated in the porous layer according to Boltaks [298]

$$W^* = 2\, N_0\, [D(t_p/\pi)]^{1/2} \tag{2.22}$$

where N_0 is the initial impurity concentration. For processing at 1100 °C within seconds the diffusivity is invariable with time. It is $(6–8) \times 10^{-7}$ cm^2/s in n-type silicon structures and $(1.1–1.4) \times 10^{-6}$ cm^2/s in p-type ones. Conventional furnace annealing gives 4×10^{-8} and 1×10^{-7} cm^2/s, respectively. The last value is close to the available equilibrium diffusivity of gold in silicon [127]. Copper demonstrates identical behavior in porous silicon structures.

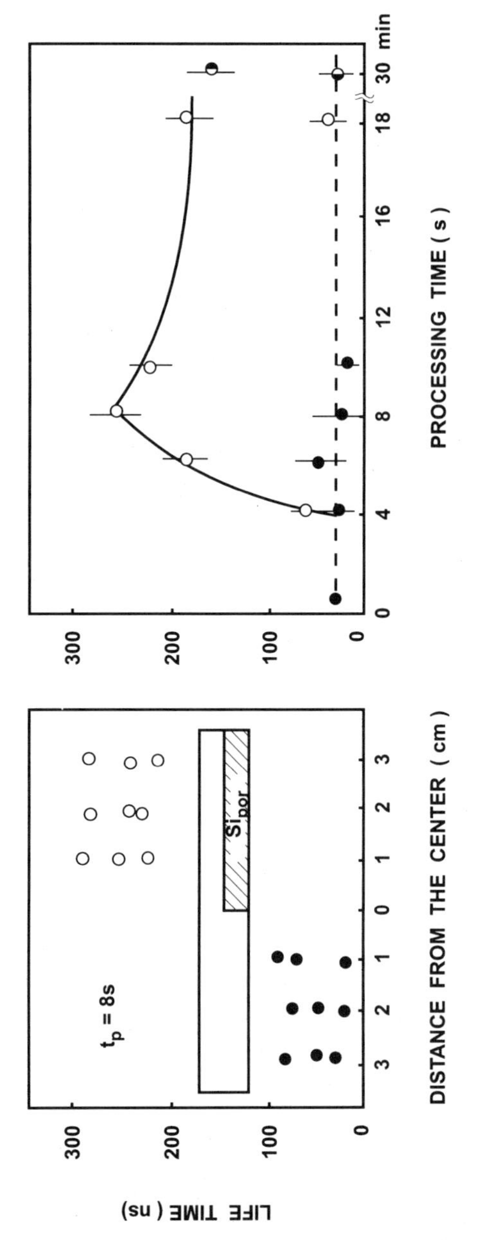

FIGURE 2.28. Minority carrier lifetime in reference and gettered regions of the silicon wafer subjected to rapid thermal processing at 950 °C [290].

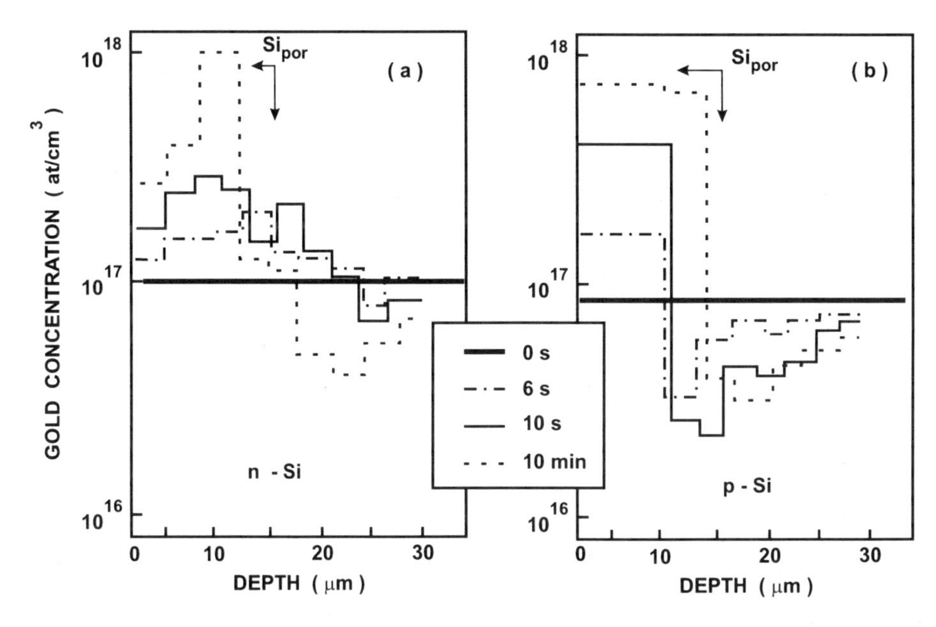

FIGURE 2.29. Neutron activation analysis deduced gold distributions in porous silicon structures formed on n- and p-type (111) silicon after rapid thermal processing at 1100 °C [291].

The enhanced diffusion mobility of gold and copper under RTP has to be related to the elastic dynamic stresses generated during heat-induced structural changes in the porous silicon [299, 300]. This would explain the differences between the impurity profiles at the boundary of porous layers formed on n- and p-type silicon wafers. These layers differ in the structure at the boundary with the substrate.

A porous layer on p-type silicon has a narrow transition region. The thickness of the layer, where porous material density reaches the crystalline silicon one, consists of not more than 10% of the total porous layer depth. It is extended to 40–50% in porous layers formed on n-type silicon. Hence, elastic stresses generated in such a structure should be reduced by their spreading in the intermediate region.

2.6.2. *Diffusion*

Porous silicon was noted to provide extremely deep boron diffusion under conventional furnace annealing [301]. Among substitutional impurities, only arsenic diffusion was investigated in porous silicon structures subjected to RTP of seconds duration [292]. In this study, porous layers 1.2–1.5 μm thick, with a density of 1.5–1.9 g/cm^3, were locally formed on p- and n-type silicon crystals. A spin-on-deposited arsenic-doped film was used as the impurity source. Figure 2.30 shows RBS spectra for p-type silicon samples. The depth of the arsenic-doped layer in the single-crystal silicon was 100–200 nm. That is in good agreement with the previously obtained data [266]. In the neighboring porous regions, arsenic diffused through the whole porous layer and penetrated into the underlying silicon substrate to a depth of about 100 nm.

Impurity atoms penetrate into the porous silicon just at the source deposition stage. They become uniformly distributed there after RTP, as displayed in Fig. 2.30. Arsenic concentration depends on the conductivity and type of the silicon wafer, being 5.5×10^{20} atoms/cm^3 for *p*-type and 2.5×10^{20} atoms/cm^3 for *n*-type material. Assuming impurity atoms are located not only in the bulk, but also at the pore surface, the difference observed is associated with the structure of pores formed by anodization of *p*- and *n*-type silicon. Arsenic diffusion into the silicon crystal beneath the porous layer is found to be independent of both the porous layer thickness and its structure.

An influence of the processing time on the arsenic diffusion depth was definitely observed, but no effect was observed at temperatures in the range of 950–1100 °C. Figure 2.31 presents arsenic depth profiles as a function of the processing time. The doped layer depth in the crystal beneath the porous layer increases from 200 to 300 nm on increasing the processing duration from 10 to 40 s. This time dependence is typical for the diffusion mechanism. The effective diffusivity of arsenic is estimated to be about 5×10^{-12} cm^2/s, which is more than two orders of magnitude greater than the equilibrium value at 1100 °C [128]. The independence of arsenic diffusion on processing temperature suggests that this process is governed by the effects in the porous layer and at its boundary. It is supported by the experiments on doping of porous silicon with erbium [293, 294].

Erbium demonstrating self-attributed luminescence at 1.54 μm in silicon and other electronic materials [302] is very attractive for silicon-based optoelectronics. Because of the low diffusivity in monocrystalline silicon it is usually introduced by ion implantation. Meanwhile, porous silicon has been shown to be effectively doped with this impurity by RTP-activated diffusion from surface electrochemically deposited erbium [293] or spin-on sources containing erbium [294]. In both cases *p*-type wafers were used. Porosity was

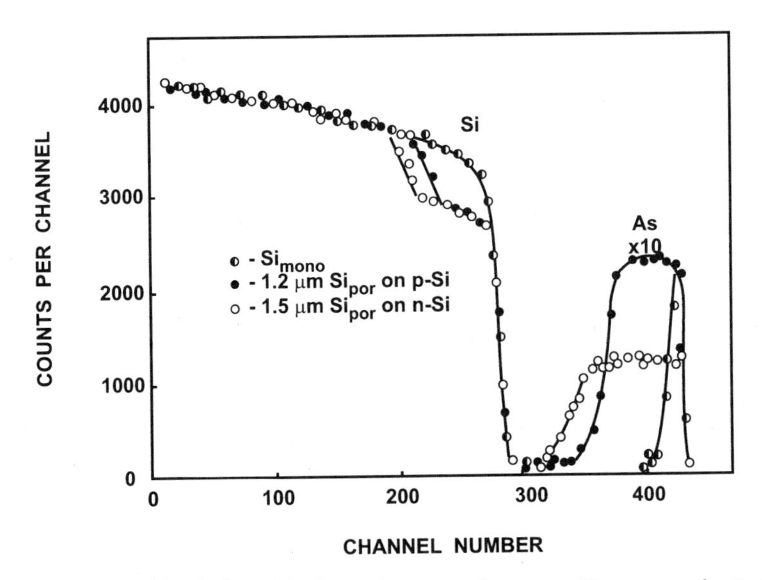

FIGURE 2.30. 2-MeV He^+ ion Rutherford backscattering spectra for porous silicon on *p*- and *n*-type silicon (111) crystals doped with arsenic from the spin-on source at 1100 °C for 10 s [292].

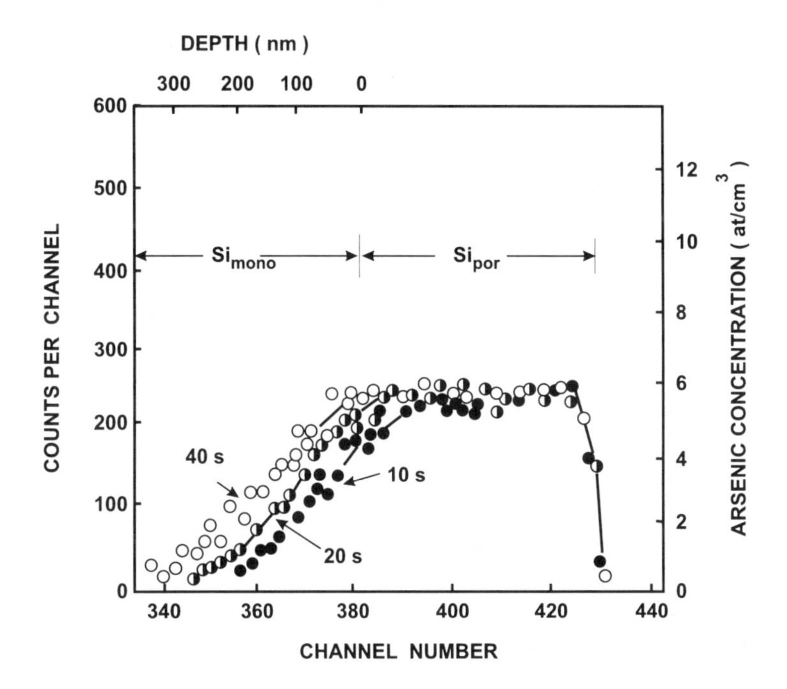

FIGURE 2.31. RBS arsenic concentration profiles in porous layers on *p*-type (111) silicon created by diffusion from the spin-on source at 1000 °C [292].

in the range of 40–60%. Processing at 900–1300 °C for 4–30 s in an oxidizing ambient provided practically uniform erbium concentration at a level of about $10^{19}/cm^3$ within the whole porous layer 5–10 mm thick. Etching of SiO_2 and additional high-temperature processing of the samples showed that the impurity diffused into the crystalline matrix in the porous layer.

In explaining the fast penetration of impurities into porous silicon, two main factors will be considered. First, pore walls provide channels for surface diffusion which is much faster than the bulk one. Second is the point defect generation during structural changes in a porous layer.

Heat-induced structural changes in porous layers have to be considered first. High-temperature processing has been established [287] to cause sintering of porous silicon accompanied by intense generation of nonequilibrium vacancies. Hence, the boundary between the porous layer and the silicon substrate acts as a sink of self-interstitials. Moreover, there are considerable elastic stresses relaxing in the porous layer [299]. These stresses and the associated point defect generation seem to have a much greater influence on the different structure of the porous layers than the temperature regime and doping type of the silicon wafers.

The observed behavior of fast-diffusing metals, arsenic, and erbium in wafers with porous silicon gives an indication of the valuable effects of the relaxation processes in porous layers on impurity diffusion. New experiments are needed to clarify these issues.

2.7. SIMULATION OF DIFFUSION PROCESSES INITIATED BY TRANSIENT HEATING

The main features that come to light from investigations of solid-state processes in silicon subjected to RTP may be summarized as follows:

1. Implant layer recrystallization and formation of secondary structural defects, impurity diffusion and segregation are thermally assisted processes. Recrystallization is completed as soon as the temperature reaches 750–800 °C, producing in heavily doped layers metastable supersaturation of substitutional impurities.
2. Annealing of implants produced disordered clusters, releasing nonequilibrium point defects (mainly interstitials), and causing enhanced diffusion of substitutional impurities.
3. Substitutional impurity behavior in the doped layer is determined by its concentration. Undersaturated impurity solutions are thermodynamically stable, whereas supersaturated ones dissociate at high temperatures (~1000 °C) within seconds, providing excess vacancies and interstitials which enhance impurity diffusion.
4. Internal elastic stresses dramatically influence impurity behavior in supersaturated layers and in structures with interface boundaries.

The above list of characteristics has been applied to simulation of the impurity behavior in transiently radiation-heated silicon [218, 219]. Impurity diffusion in implanted layers subject to RTP is a good representative test of the model.

First, numerical estimates of nonequilibrium point defect generation during annealing of implantation clusters and during dissociation of supersaturated solutions are of practical importance. Clusters of disorder are usually produced by light ions, e.g., boron. For heavy group-V ions that appear after a low-dose implant or after a hot implantation, clustering and its influence on the impurity behavior become detectable at impurity concentrations of 10^{19}–10^{20} atoms/cm^3. Effects related to supersaturated impurity solutions take place in the impurity concentration range of 10^{20}–10^{21} atoms/cm^3. The release of active point defects during annealing is somewhat proportional to the impurity concentration, and the dissociation of supersaturated solutions has a one- to twofold higher efficiency. Therefore, nonequilibrium point defect generation by cluster annealing may be ignored. However, there is no difficulty in introducing it in a simulation procedure, as both processes have first-order reaction kinetics described by Eq. (2.17), with $m = 1$ [48]. Therefore, the mathematics presented below are adequate for both cases while supersaturated impurity solutions are mainly involved.

Figure 2.32 schematically summarizes the features of implanted substitutional impurity redistribution and sheet resistivity variation on RTP at 800–1200 °C. Two distinct regions of implantation arise: (1) low dose (D_1) and (2) high dose (D_2). At low dose, one assumes the peak impurity concentration is below the steady-state solubility. At the high-dose implant it exceeds this level.

In the first case, RTP causes practically no redistribution. Concentration profile broadening is limited to the depth of equilibrium impurity diffusion predicted from

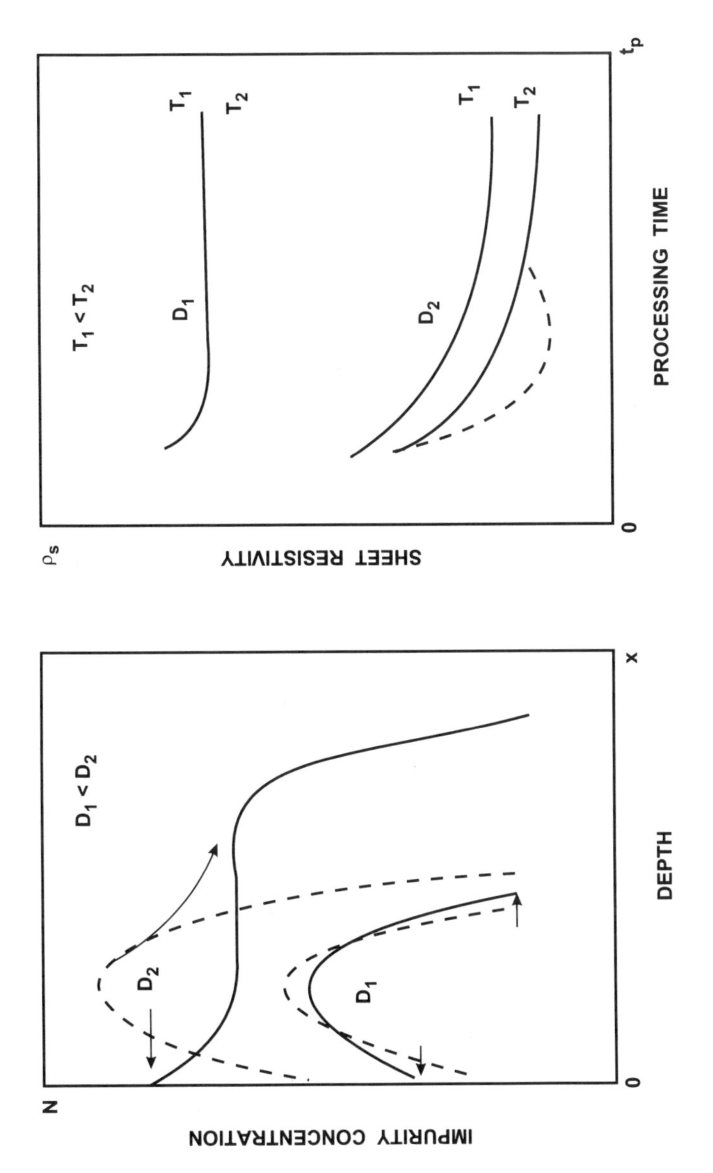

FIGURE 2.32. Redistribution of implanted substitutional impurities in silicon on rapid thermal processing. Dashed line shows as implanted condition.

conventional furnace experiments [127–129]. After recrystallization of the implanted layer, its sheet resistivity becomes processing time and temperature independent.

In the second case, an enhanced impurity redistribution takes place. It is characterized by an effective diffusivity several times greater than the equilibrium value. Maximum impurity concentration in the electrically active substitutional positions is stabilized at the level of its steady-state solubility, at the processing temperature. Sheet resistivity monotonically decreases with the processing time and temperature. There could be a minimum resistivity corresponding to maximum electrically active fraction just after a complete recrystallization.

Recrystallization in implanted layers starts first and proceeds at relatively low temperatures while rapid heating to 900 °C provides solid-state epitaxial regrowth completed within several milliseconds. Implanted impurity redistribution caused by evaporation and segregation may be calculated from the diffusion equation

$$\frac{\partial N}{\partial t} = \frac{\partial}{\partial x}\left(D_{\mathrm{a}} \frac{\partial N}{\partial x}\right) \tag{2.23}$$

where D_{a} is the impurity diffusivity in amorphous silicon. The boundary condition accounting for impurity evaporation with the rate β_{i} takes the form

$$-D_{\mathrm{a}} \left.\frac{\partial T}{\partial x}\right|_{x=0} = +\beta_{\mathrm{i}} \left. N\right|_{x=0} \tag{2.24}$$

The condition describing solid-state impurity segregation at the developing amorphous–crystalline interface is

$$D_{\mathrm{a}} \left.\frac{\partial N}{\partial x}\right|_{x=x_{\mathrm{a}}-\ell} = - [1 - k'(t)]\, v_{\mathrm{c}}(t) \left. N\right|_{x=x_{\mathrm{a}}-\ell} \tag{2.25}$$

here, the regrowth rate

$$v_{\mathrm{c}}(t) = v_{\mathrm{o}} \exp[- E_{\mathrm{a}}/kT(t)] \tag{2.26}$$

and x_{a} is the initial amorphous layer thickness, $k'(t)$ the segregation coefficient, and ℓ the thickness of the layer regrown within time t as defined from (2.10).

In the recrystallized layer, impurity atoms occupy substitutional positions. A metastable supersaturated solution is formed in heavily implanted layers. It inevitably dissociates leading to a reduction of the concentration of impurity atoms in substitutional positions to the steady-state solubility level. Nonequilibrium impurity interstitials and vacancies are produced. Fast-diffusing interstitials enhance impurity redistribution. Vacancies annihilate with self-interstitials at inner sinks and at the sample surface.

Solid-state recrystallization of implanted layers is not the only method to create supersaturated solutions of substitutional impurities. They also may be formed by liquid-phase laser annealing of implanted layers and by saturation diffusion from an external chemically active source of impurities. Nevertheless, the kinetics of dissociation do not depend on the supersaturation prehistory. This enables one to consider impurity diffusion in the supersaturated layers to take place with the generation of nonequilibrium point defects. Moreover, excess silicon interstitials mediating impurity diffusion can be generated during annealing of extended secondary defects related to ion implantation. It enhances diffusion at an undersaturated concentration, which is typical particularly for boron.

We assume that there are two components in the flux of the diffusing species. The first one is represented by equilibrium impurity atoms in substitutional positions. This fraction is limited by the steady-state impurity solubility. Fast-diffusing nonequilibrium impurity interstitials form the other component. Indicating the corresponding variables with labels V and I, respectively, the total concentration of impurity atoms is

$$N^{\mathrm{V}} + N^{\mathrm{I}} = N \tag{2.27}$$

The concentration of interstitials is controlled by the dissociation kinetics of the supersaturated solution

$$N^{\mathrm{I}} = (N - N_{\mathrm{s}})\,[1 - \exp(-t/\tau)]\,f(N) \tag{2.28}$$

where N_{s} is the steady-state impurity solubility, τ the supersaturation relaxation time, $f(N)$ the stepwise function represented by $f(N) = 0$ at $N < N_{\mathrm{s}}$ and $f(N) = 1$ at $N \geq N_{\mathrm{s}}$. The diffusion equation for the equilibrium and nonequilibrium components can be written

$$\frac{\partial N^{\mathrm{V}}}{\partial t} = \frac{\partial}{\partial x}\left[D^{\mathrm{V}}\frac{\partial N^{\mathrm{V}}}{\partial x} - V^{\mathrm{V}} N^{\mathrm{V}}\right] + \frac{N^{\mathrm{I}} - N_0^{\mathrm{I}}}{t_{\mathrm{i}}} \tag{2.29}$$

$$\frac{\partial N^{\mathrm{I}}}{\partial t} = \frac{\partial}{\partial x}\left[D^{\mathrm{I}}\frac{\partial N^{\mathrm{I}}}{\partial x} - V^{\mathrm{I}} N^{\mathrm{I}}\right] - \frac{N^{\mathrm{I}} - N_0^{\mathrm{I}}}{t_{\mathrm{i}}} \tag{2.30}$$

The term $(N^{\mathrm{I}} - N_0^{\mathrm{I}})/t_{\mathrm{i}}$ describes annihilation of nonequilibrium impurity interstitials, where N_0^{I} is their equilibrium concentration and t_{i} the lifetime of nonequilibrium interstitials.

The diffusivity D^{V} is the equilibrium one taken from the data obtained in conventional annealing experiments. Accounting for the charge state of the vacancies, involved in the diffusion process, it is [128]

$$D^{\mathrm{V}} = h_{\mathrm{e}}\,[D^0 + D^+(p/n_{\mathrm{i}}) + D^-(n/n_{\mathrm{i}}) + D^{=}(n/n_{\mathrm{i}})^2] \tag{2.31}$$

where h_{e} is the internal electric field enhancement factor

$$h_{\mathrm{e}} = 1 + z\,N\frac{\partial \ln(N/n_{\mathrm{i}})}{\partial N} \tag{2.32}$$

p, n are the local hole and electron concentrations, z the impurity charge state, and $n_{\mathrm{i}} = [1.5 \times 10^{33}\,T^3 \exp(-1.21/kT)]^{1/2}$ /cm^3 [203]. Arrhenius parameters for the diffusivities used in D^{V} calculation are presented in Table 2.8. Diffusivities of impurity interstitials $D_{\mathrm{i}}^{\mathrm{I}}$ are available from the data collected in Table 2.9. The parameters V^{V} and V^{I} are the drift velocities of substitutional and interstitial impurity atoms in a field of internal stresses.

The sum of (2.29) and (2.30) with substitution (2.27) and (2.28) for usually valid $D^{\mathrm{I}} >> D^{\mathrm{V}}$ and $V^{\mathrm{I}} >> V^{\mathrm{V}}$ gives the impurity diffusion equation in the general form

TABLE 2.8. Parameters of Substitutional Impurity Diffusion in Silicon [128]

	D^0		D^+		D^-		$D^=$	
Impurity	D_0 (cm^2/s)	E_a (eV)	D_0 (cm^2/s)	E_a (eV)	D_0 (cm^2/s)	E_a (eV)	D_0 (cm^2/s)	E_a (eV)
B	0.037	3.46	0.76	3.46	—	—	—	—
Al	1.385	3.41	2480	4.20	—	—	—	—
Ga	0.374	3.39	28.5	3.92	—	—	—	—
In	0.785	3.63	415	4.28	—	—	—	—
Tl	1.37	3.70	351	4.26	—	—	—	—
P	3.85	3.66	—	—	4.44	4.00	44.2	4.37
As	0.066	3.44	—	—	12.0	4.05	—	4.32
Sb	0.214	3.65	—	—	15.0	4.08	—	—
Bi	1.08	3.85	—	—	396.0	4.12	—	—

TABLE 2.9. Diffusion of Metastable Interstitials in Silicon

Interstitial	Temperature range (°C)	Diffusivity coefficients D_0 (cm^2/s)	E_a (eV)	Refs.
Si^-	~ – 90	—	0.35	304
Si^+	100 – 150	—	0.85	304
Si^0	270 – 330	—	1.5	304
Si	150 – 350	—	0.13	235
	700 – 800	3.75×10^{-9}	0.13	305
	1100 – 1200	8.6×10^5	4.0	306
	470 – 1200	0.14	1.68	307,308
	800 – 1200	1.3×10^6	3.22	285
	460 – 1200	0.335	1.86	309
B	300 – 900	—	0.62	310
	–98 – –78	—	0.6	311
	–73 – –23	0.04	0.58	312
	800 – 1000	1.4×10^{-7}	1.1	162
P	300 – 900	—	0.34	310
	950 – 1100	4.47×10^{-6}	1.39	233
	600 – 800	5.6×10^{-9} [a]	1.3	237
As	1100 – 1150	$0.0083bX$	1.8	149
	600 – 1100	—	1.1	194,195
	1000 – 1150	3.3×10^{-8} [a]	1.5	313
Sb	650 – 750	0.04	1.8	92
	700 – 850	$0.0083bX$	1.8	149
Bi	750 – 800	$0.0083bX$	1.8	149

[a] Calculated from the experimental data presented in the reference.
$b = 0.01$ for implantation at liquid-nitrogen temperature.
$b = 0.02$ for implantation at room temperature.
X = impurity atomic fraction.

$$\frac{\partial N}{\partial t} = \frac{\partial}{\partial x} \{D^{\mathrm{V}} + D^{\mathrm{I}} [1 - \exp(-t/\tau)] f(N)\} \frac{\partial N}{\partial x}$$
$$- V^{\mathrm{V}} N - V^{\mathrm{I}}(N - N_{\mathrm{s}}) [1 - \exp(-t/\tau)] f(N)\} \tag{2.33}$$

The dissociation of the supersaturated solution stops at $t >> \tau$ where the impurity concentration drops to its steady-state solubility. Nonequilibrium point defects are no longer generated and impurity diffusion is governed by the known equilibrium parameters.

The approach discussed supposes impurity drift migration in the field of elastic stresses. They are induced by the point defects generated during dissociation of supersaturated impurity solutions, chemical reactions at the surface, and structural changes in polycrystalline and porous layers.

The force driving all impurity atoms characterized by the dilation volume U^{V} in the substitutional position and U^{I} in the interstitial state, is [314]

$$F^{\mathrm{V}} = k^* \, U^{\mathrm{V}} \, \mathrm{grad}[E(r)] \tag{2.34}$$

$$F^{\mathrm{I}} = k^* \, U^{\mathrm{I}} \, \mathrm{grad}[E(r)] \tag{2.35}$$

$E(r)$ represents the field of lattice deformation and k^* the modulus of volume compression, which equals $E^*/[3(1 - \upsilon)]$.

Supposing that the gradient of nonequilibrium concentration of point defects and the temperature gradient cause identical lattice deformations, then the field of these deformations may be written in the form [315]

$$E(r) = \frac{(1 + \nu)}{(1 - \nu)} \frac{1}{N^*_{\mathrm{Si}} \, a} [(\Delta a_{\mathrm{i}} + \Delta a_{\mathrm{v}}) \, N^{\mathrm{I}} + \Delta a_{\mathrm{s}} \, N^{\mathrm{V}}] \tag{2.36}$$

where N^*_{Si} is the host atom concentration, υ Poisson's ratio, a the lattice constant, and Δa_{i}, Δa_{v}, Δa_{s} the lattice constant deviations related to the interstitial, vacancy, and impurity atom in the substitutional position, respectively. Substitution of (2.36) into (2.34) and (2.35) for the one-dimensional case gives

$$V^{\mathrm{V}} = \frac{D^{\mathrm{V}}}{kT} F^{\mathrm{V}} = \frac{E^*(1 + \nu) D^{\mathrm{V}} U^{\mathrm{V}}}{3(1 - 2\nu)(1 - \nu) a k T N^*_{\mathrm{Si}}} \frac{\partial}{\partial x} [(\Delta a_{\mathrm{i}} + \Delta a_{\mathrm{v}}) N^{\mathrm{I}} + \Delta a_{\mathrm{s}} N^{\mathrm{V}}] \tag{2.37}$$

$$V^{\mathrm{I}} = \frac{D^{\mathrm{I}}}{kT} F^{\mathrm{I}} = \frac{E^*(1 + \nu) D^{\mathrm{I}} U^{\mathrm{I}}}{3(1 - 2\nu)(1 - \nu) a k T N^*_{\mathrm{Si}}} \frac{\partial}{\partial x} [(\Delta a_{\mathrm{i}} + \Delta a_{\mathrm{v}}) N^{\mathrm{I}} + \Delta a_{\mathrm{s}} N^{\mathrm{V}}] \tag{2.38}$$

where E^* is Young's modulus. The parameters involved in the drift velocity calculation are summarized in Table 2.10. Coordinate dependence of the driving force always follows the impurity concentration profile.

Figure 2.33 shows a comparison of the calculated and experimental results for phosphorus-implanted silicon subjected to transient heating with incoherent light. The time dependence of the induced temperature was used in the calculations. Phosphorus atom profiles were obtained from layer-by-layer neutron activation analysis. The calculated profiles fit the experimental data when the interstitial phosphorus diffusivity is

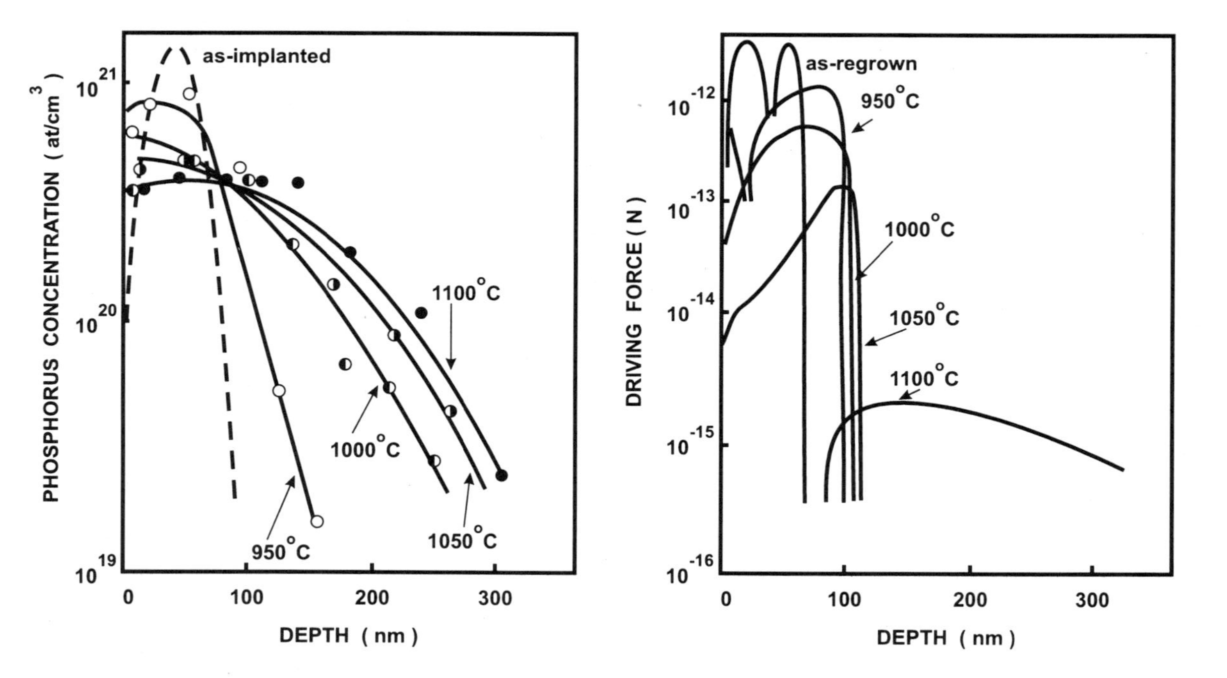

FIGURE 2.33. Calculated (lines) and experimental (circles) depth profiles of phosphorus and internal interstitial impurity driving force in phosphorus implanted (30 keV/7.5 × 10^{16} ions/cm^2) 10-s incoherent light annealed silicon [218].

TABLE 2.10. Silicon Crystal and Point Defect Parameters

$a = 5.43 \times 10^{-10}$ m	$E^{*} = 16.6 \times 10^{10} - 1.5 \times 10^{7} \times T$, N/m^2 where $T = 300$–900 K [316]
$\Delta a_i = 0.12a$ [200]	$U^{I} = 2.5a^3$ [317]
$\Delta a_v = -0.02a$ [200]	$U^{V} = 4\pi/3(r_i^3 - r_s^3)$
$\Delta a_s = 0.85 \times 10^{-3}a$ [219]	r_i, r_s — covalent radii of impurity and host atoms
$v = 0.26$ [200]	

approximated by $D^{I} = 9.0 \times 10^{-6} \exp(-1.43/kT)$ cm^2/s and the relaxation time is $\tau = 0.3 \exp(0.34/kT)$ s. The calculated driving force for interstitial phosphorus is closely related to the impurity profile characterized by the highest value corresponding to the maximum impurity gradient and maximum supersaturation.

The maximum internal driving force calculated for interstitial boron, phosphorus, arsenic, and antimony in silicon as a function of the supersaturation of these impurities is presented in Fig. 2.34. The force achieves a step rise when the impurity concentration

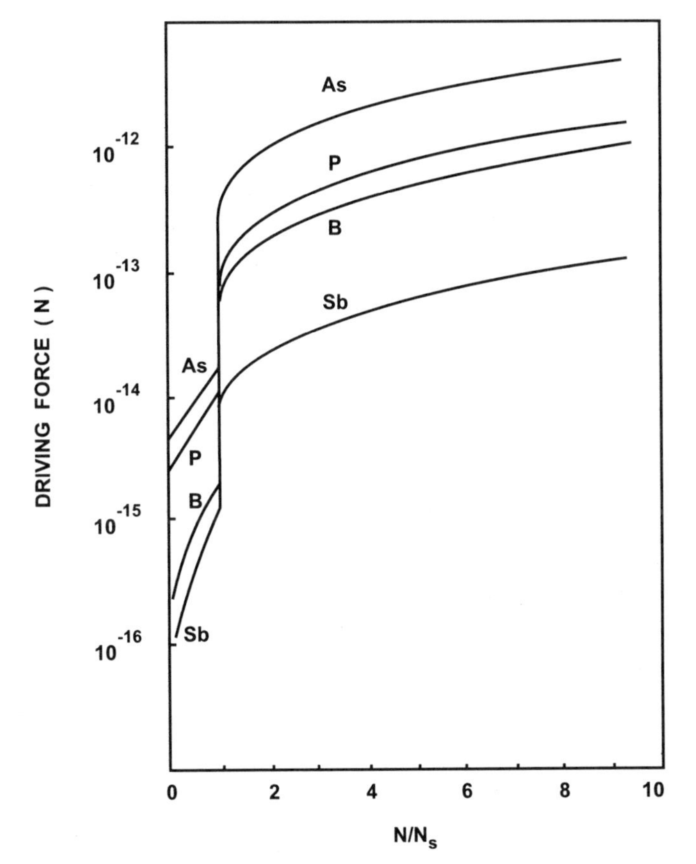

FIGURE 2.34. Calculated internal force driving interstitial boron, phosphorus, arsenic, and antimony in single-crystal silicon at different supersaturation at 950 °C for $t = \tau$ [218].

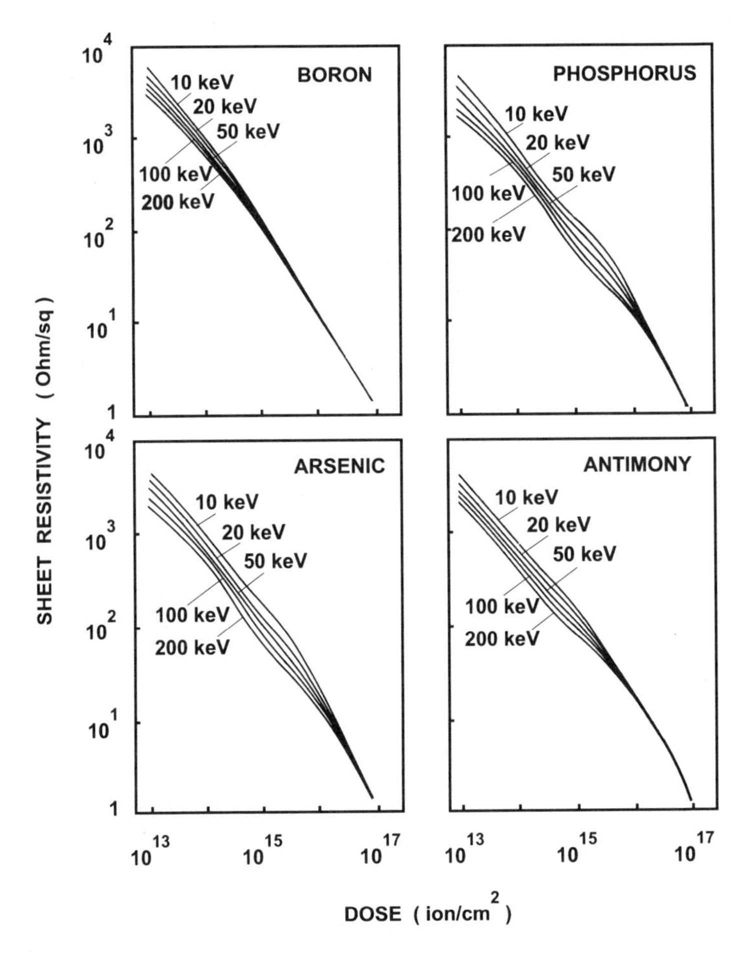

FIGURE 2.35. Sheet resistivity of ion-implanted transiently annealed silicon layers as a function of energy and dose of implants [218].

exceeds its solubility. The steady-state solid solubility, impurity dilation volume, and point defect-induced lattice deviations define the absolute value of the driving force for the substitutional impurities migrating via vacancies and interstitials. Among the impurities analyzed, antimony has the lowest solubility in silicon and produces the lowest driving force. The enhanced diffusion phenomenon is therefore less pronounced for this particular impurity.

The consistent agreement between the relaxation processes, yielding nonequilibrium point defects, supports the concepts presented above. Supersaturated impurity solutions and their dissociation during RTP have been considered to be responsible for enhanced substitutional impurity diffusion. The dissociation process results in the generation of nonequilibrium impurity interstitials and vacancies. Fast-migrating impurity interstitials and internal stresses connected with point defects have been accounted for in the general diffusion equation (2.33). The effective diffusivity D_{eff} based on this approach has the form

$$D_{eff} = D^{V} + D^{I}[1 - \exp(-t/\tau)]f(N) \tag{2.39}$$

It is quite obvious that diffusion activation energy defined in RTP experiments may vary from the values attributed to interstitial atom migration, 0.34–1.80 eV, to the equilibrium ones found in traditional isothermal experiments, 3.69–4.20 eV.

The above model has been used for calculation of sheet resistivity of layers created with boron, phosphorus, arsenic, and antimony implantation and subsequent RTP [218]. Impurity atoms implanted below their solubility were assumed not to diffuse significantly, conserving the as-implanted profile. In the heavily implanted layers, enhanced diffusion related to the dissociation of the supersaturated impurity solution resulted in an increase in the depth of the doped layer. Impurity evaporation was neglected, and all impurity atoms were believed to occupy electrically active substitutional positions. The impurity concentration dependence of free carrier mobility was taken from Masetti *et al.* [318] and Plunkett *et al.* [319].

Figure 2.35 presents the results of sheet resistivity calculations for implant energies from 10 to 200 keV in the dose range of 10^{13}–10^{17} ions/cm^2. The dose dependence of sheet resistivity has two specific regions. The first one covers the low-dose range (10^{13}–10^{15} ions/cm^2) where undersaturated layers are formed. At a fixed dose, in this region sheet resistivity decreases when the energy of the implanted impurity is increased. The second region covers the dose range of 10^{16} ions/cm^2 and higher where supersaturation is achieved. Sheet resistivity of heavily implanted layers becomes independent of the implant energy. The calculated sheet resistivities were shown to agree well with the experimental data obtained by different researchers in a wide range of implantation and RTA regimes.

Supersaturated substitutional impurity solutions are not the only source of nonequilibrium point defects. Self-interstitial cluster annealing, chemical reactions (e.g., oxidation, silicidation) are able to produce the same effect. The approach presented here is by no means limited to the simulation of impurity behavior in silicon, but can foreseeably be applied to other impurities and semiconductors.

3

Crystallization, Impurity Diffusion, and Segregation in Polycrystalline Silicon

Polycrystalline silicon is widely employed in semiconductor integrated circuits as a material for gates in MOS devices, emitters in bipolar transistors, and interconnections where low sheet resistivity is important [1, 2]. Heavily doped layers are also attractive as diffusion sources for doping the underlying substrate [3]. Finally, polysilicon has been recrystallized successfully into device-quality material for three-dimensional integrated circuit devices [1]. The successful technological application of polysilicon requires precise control of the material parameters through appropriate choice of the material deposition conditions, doping, and annealing conditions. The prospect that polysilicon will find greater utility is increased considerably with the use of RTP [4].

The sheet resistivity of polysilicon layers is defined by the number of free carriers and their mobility, which depends on the impurity atom diffusion behavior and structural changes initiated by heat treatment. The time and temperature dependence of the solid-state diffusion processes in polysilicon may be classified according to the length of atomic diffusion, $l_d = [D(T)t]^{1/2}$, where $D(T)$ is the impurity diffusivity and t the time. The general features presented here are taken from conventional furnace annealing experiments, where long-duration heating of the samples has been performed. When diffusion occurs for a distance of approximately the interatomic spacing, the result is impurity electrical activation [5–13] and impurity cluster formation [13]. In implanted layers defect annealing and recrystallization take place inside the crystal grains [6, 7]. As the impurity diffusion length increases to the same order as the grain size, one observes impurity atom segregation at grain boundaries [13–19], and diffusion from the boundaries into the grains, where substitutional impurities become electrically active [5, 9, 12]. For high temperature processing (~1000 °C) the diffusion length is comparable to the thickness of the polysilicon layer [20–33], and therefore impurity atoms either diffuse through the layer into the substrate [34–38] or evaporate from the surface [39]. In addition, self-diffusion leads to grain growth [40–46].

The particular features that arise during RTP [33, 47–72] are compared with conventional furnace-annealed polysilicon in this chapter.

3.1. ELECTRICAL ACTIVATION AND SEGREGATION OF IMPURITIES

Polysilicon doped by substitution impurities, both during deposition and by subsequent implantation, has few electrically active impurity atoms [5]. Electrical activation occurs when the impurity atoms migrate to a substitutional position inside the grains. Heat processing leads to point defect annealing and to intragrain recrystallization in implanted layers. A diffusion length of several interatomic spacings is sufficient for a fraction of impurity atoms to reach the nearest grain boundary and segregate there. These processes are the fastest, occurring at processing times as short as 10^{-2} s when the induced temperature is 600–800 °C. The electrical activation and segregation of the impurity atoms appear to be closely correlated.

3.1.1. Experimental Facts

Heavily phosphorus- or arsenic-implanted polysilicon layers flash annealed with a xenon lamp for 10^{-2} s at an energy density of about 50 J/cm^2 inducing a temperature of 650–700 °C, show electrical activation of the impurities occurring simultaneously with recrystallization [47, 48]. The fraction of electrically active impurities is 85%, in agreement with conventional furnace annealing in this temperature range [6, 7]. Further increase in energy density is accompanied by a decrease in the number of electrically active arsenic atoms as seen in Fig. 3.1. Their fraction is lowered to 50% at 75 J/cm^2 (T = 900 °C) and then starts growing again. The nonmonotonic behavior of the electrically active atom concentration results in a maximum of sheet resistivity correlated with the minimum free carrier concentration, and a slight growth in carrier mobility. The decrease of electrically active atom concentration is related to the segregation of impurity atoms at grain boundaries, where they are electrically inactive [47, 48].

The growth of carrier mobility in the phosphorus- and arsenic-doped layers is explained by a reduction of carrier scattering by ionized impurity atoms in the grains. From Fig. 3.1 it follows that the impurity segregation over a temperature range of 800–900 °C is provided by impurity diffusion from the bulk to the grain boundaries. The impurity diffusion length appears to be comparable with the average size of the grains. Impurity diffusivity estimated according to $\ell_d = (D\,t_{pr})^{1/2}$, where $\ell_d = l_g/2$ ($l_g = 0.2 \times 10^{-4}$ cm is the average grain size, $t_{pr} = 10^{-2}$ s), is about 10^{-8} cm^2/s. It is four to five orders of magnitude greater than grain boundary diffusivity of group-III and -V impurities in polysilicon [27, 29] and six to seven orders greater than that for single crystal silicon [73] at 900 °C.

There are at least two reasons for this. The first concerns an increased effective duration of thermal processing related to the relatively slow sample cooling by heat diffusion and radiative heat loss mechanisms. The second one may involve the formation of impurity clusters inside grains [13]. Segregation is of equal importance and leads to a decrease in the electrically active impurity concentration. However, these factors as well as nonequilibrium implantation-produced point defects can increase the estimated dif-

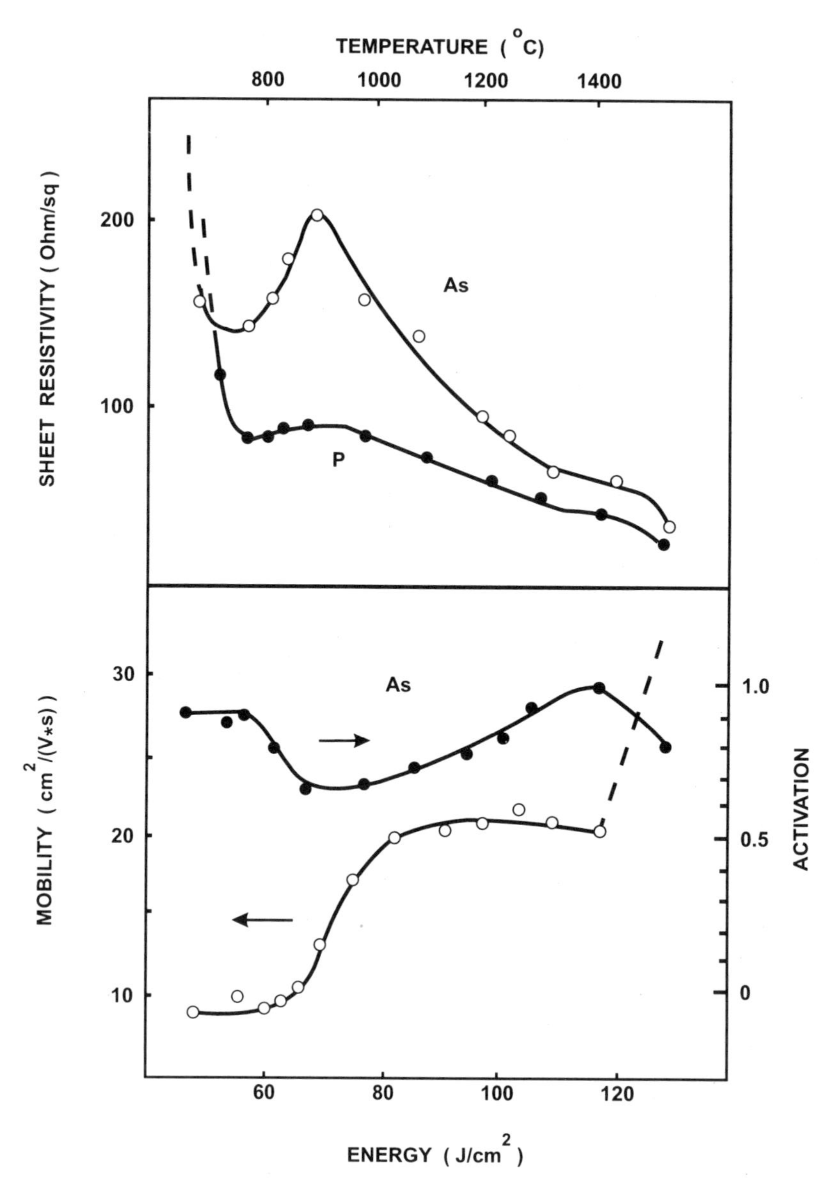

FIGURE 3.1. Sheet resistivity, Hall mobility, and carrier concentration versus energy density for arsenic- and phosphorus-implanted polycrystalline silicon [47, 48].

fusivity by only two orders of magnitude and, therefore, there must be another mechanism to account for the observed enhancement. Implantation and transient heating-induced stresses may be involved [70]. However, to analyze their effect in the millisecond time processing regime, experimental data with better time resolution are required.

The electrical activation and segregation of impurity atoms introduced into polysilicon during deposition has been investigated in detail for processing times of seconds [43, 50]. The sheet resistivity of heavily phosphorus-doped polysilicon, deposited on single-crystal silicon and oxidized silicon substrates was studied for periods of 4–40 s and temperatures of 900–1150 °C. Figures 3.2 and 3.3 show that the sheet resistivity decreased monotonically with both the processing time and the maximum induced temperature. Over the whole temperature range investigated, the sheet resistivity of polysilicon subjected to standard furnace annealing of 30 min is lower than that for RTP. The redistribution of impurities during the furnace annealing is responsible for this difference in the polysilicon layers on single-crystal silicon substrates. For the oxidized ones it may be explained by more complete activation of impurity atoms because of the slower crystallographic relaxation effects for oxidized substrates [41, 42].

Free carrier concentrations were measured in these samples by infrared absorption [49]. The maximum impurity activation was achieved on the single-crystal substrates by 20 s of thermal processing. Incomplete electrical activation occurs at shorter durations. Furnace annealing decreases the concentration of electrically active atoms in a polycrystalline layer as a result of diffusion of impurity atoms into the substrate. Electron mobility in this case appears to be almost constant in the range of 20–25 cm^2/V·s.

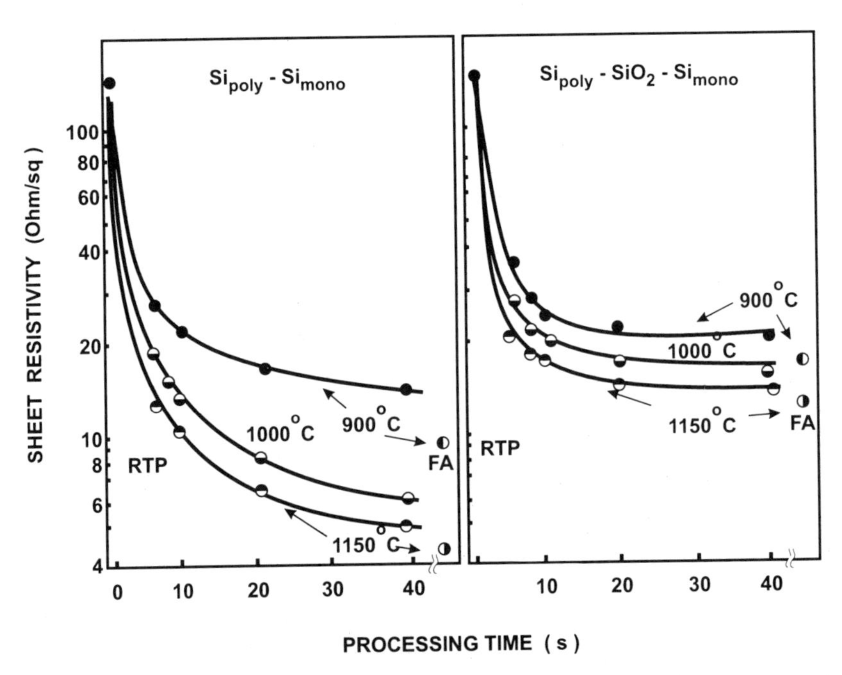

FIGURE 3.2. Sheet resistivity of phosphorus-doped polycrystalline silicon as a function of processing time for rapid incoherent light and furnace (30 min) annealing.

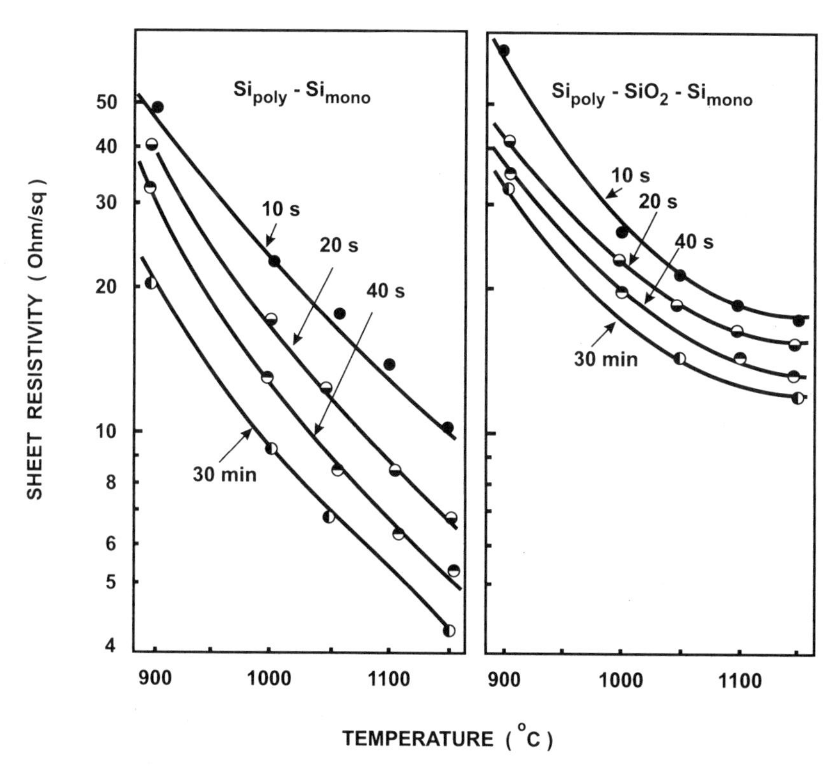

FIGURE 3.3. Sheet resistivity of phosphorus-doped polycrystalline silicon as a function of processing temperature.

In contrast, the preference for impurities to appear at grain boundaries with polysilicon doped during deposition, ion implantation provides a uniform impurity distribution inside grains (in planes parallel to the sample surface). Subsequent thermal processing leads to redistribution of impurity atoms which determines their electrical activity.

The sheet resistivity of polysilicon implanted with phosphorus and arsenic in a dose range of 5.0×10^{15}–2.5×10^{16} ions/cm^2 decreases monotonically with the anneal temperature, both for RTP and for furnace annealing [50, 52]. This is illustrated in Fig. 3.4, where the sheet resistivity after RTP at 1000–1100 °C is much higher than that for furnace annealing. Moreover, furnace annealing results in steady-state values of the resistivity, while further temperature increases of RTP cycles lead to a decrease in sheet resistivity. The exception is at high doses of arsenic where the sheet resistivity rises twice after processing at temperatures above 1200 °C. The lowest dose of implanted phosphorus gives sheet resistivities corresponding to single-crystal silicon. For doses of 1×10^{16} and 2.5×10^{16} ions/cm^2 the lowest sheet resistivities of the polysilicon are 34 and 18 Ω/sq for arsenic and 32 and 18 Ω/sq. for phosphorus. Electron concentration increases monotonically with the anneal temperature, up to 1200 °C, for both the conventional furnace and RTP.

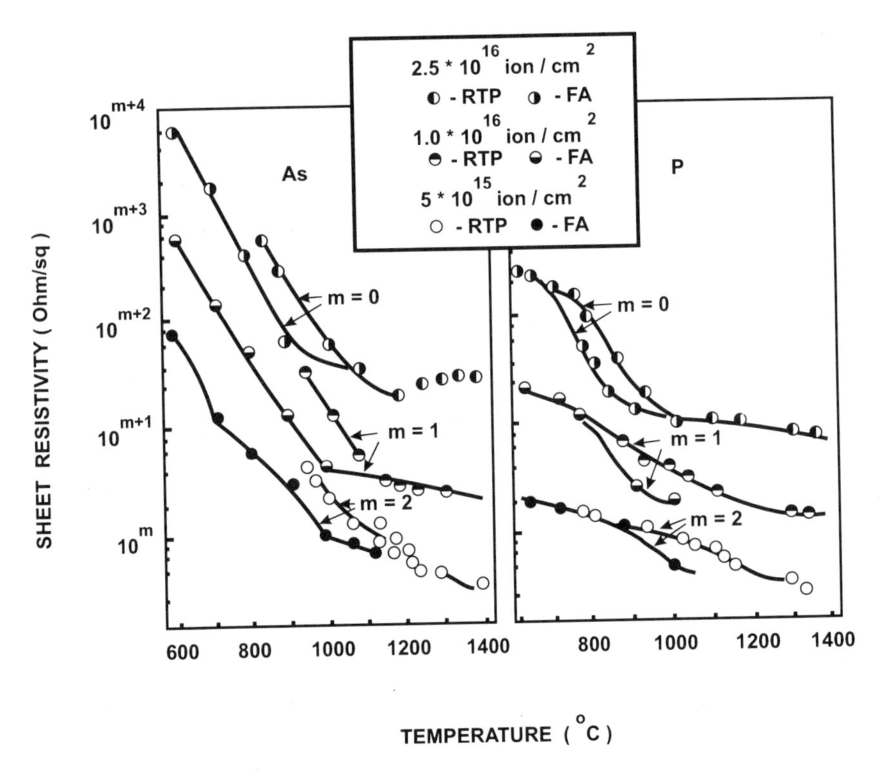

FIGURE 3.4. Sheet resistivity of arsenic- and phosphorus-implanted polycrystalline silicon versus processing temperature for 10 s (RTP) and 30 min (FA) annealing.

The influence of RTP duration on the electrical activation and segregation of impurity atoms is clearly observed at lower temperatures [63]. Figure 3.5 presents the variation of sheet resistivity and electron concentration in phosphorus-implanted polysilicon layers as a function of the processing time duration at 800 and 1000 °C. The symmetrical character of the sheet resistivity data shows that the major effect is the electrical activation of phosphorus atoms. Note that the electron mobility in the doped layer remains practically constant. The general tendency is for the sheet resistivity to decrease and the electron concentration to increase after the first 10 s of heating. At longer processing times the high-temperature (1000 °C) curves reach saturation. At the lower temperature (800 °C) extrema at 10 s are followed by a factor of two decrease in electron concentration and increase of sheet resistivity. Phosphorus and electron depth profiles presented in Fig. 3.6 show that these features are related to the temperature-dependent activation of impurity atoms. Thermal processing at 800 °C starts with an increase in phosphorus electrical activation, and its diffusion redistribution in the polycrystalline layer. The highest degree of activation is achieved after 10 s of heating and is characterized by a maximum electron concentration of $5.0 \times 10^{20}/cm^3$. Further increase in processing time lowers the electron concentration to $2.5 \times 10^{20}/cm^3$.

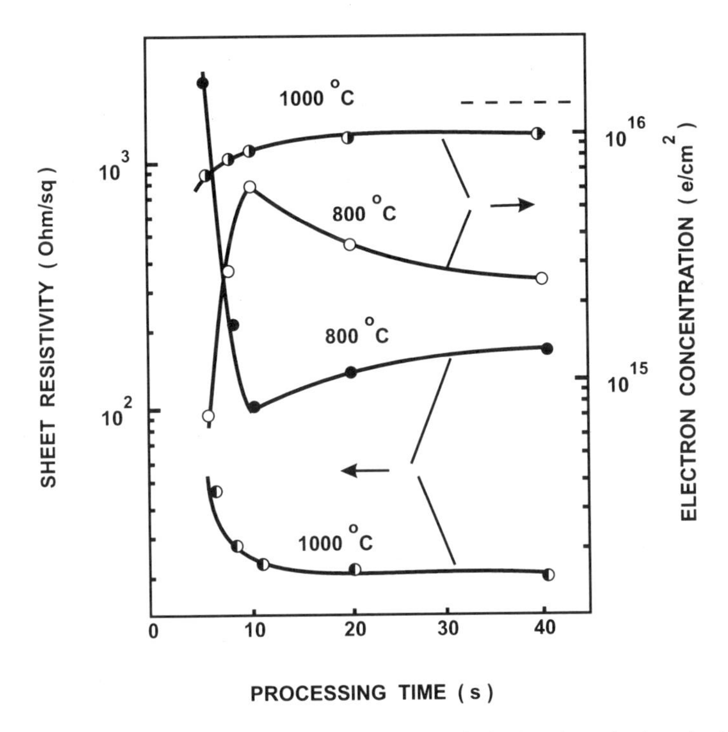

FIGURE 3.5. Sheet resistivity and integral electron concentration in the phosphorus-implanted polycrystalline silicon after rapid thermal processing at 800 and 1000 °C. The dashed line indicates the concentration level corresponding to full electrical activation of the implanted impurity.

A comparison of the depth profiles indicates that a considerable number of phosphorus atoms are electrically inactive after processing at 800 °C. Their maximum concentration is about 10^{21} atoms/cm^3 and they are mainly located in the implanted layer. During heating for 10 s and longer, phosphorus atoms diffuse through the 0.45-μm polycrystalline layer to the substrate interface, while electrical activity of the impurity is observed only in the 0.2-μm surface layer.

For implanted phosphorus annealed at 800 °C, intragrain recrystallization occurs demonstrating that a supersaturated impurity solution was formed. The maximum electron concentration of 5.0×10^{20}/cm^3 was detected after heating for 10 s to 800 °C, while the steady-state phosphorus solubility in single crystal silicon at this temperature is only 1.6×10^{20} atoms/cm^3 [74]. An increase in the processing time stabilizes the electrically active phosphorus concentration at its steady-state solubility level. Impurity diffusion and segregation are competing processes in this case.

For boron-implanted polysilicon subjected to RTP at 960–1150 °C, carrier mobility is higher than after furnace annealing at the same temperatures [69]. This finding was explained by a nonuniform impurity distribution inside grains and by the related decrease of carrier scattering at grain boundaries.

Implanted phosphorus behavior at 800 °C demonstrates that a supersaturated impurity solution was formed when intragrain recrystallization is completed. The maximum

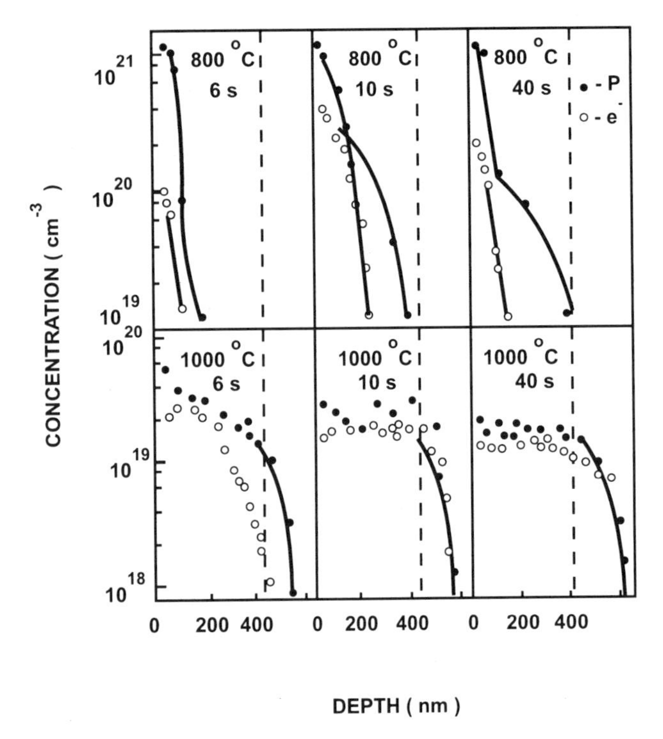

FIGURE 3.6. Phosphorus (●) and electron (○) depth profiles in phosphorus-implanted polycrystalline silicon after rapid thermal processing at 800 and 1000 °C for 6–40 s. The dashed lines indicate the interface between the polycrystalline silicon and single-crystal substrate.

electron concentration of $5.0 \times 10^{20}/cm^3$ was detected after heating for 10 s to 800 °C, while the steady-state phosphorus solubility on monocrystalline silicon at this temperature was only $1.6 \times 10^{20}/cm^3$ [74]. An increase in processing time stabilizes the electrically active phosphorus concentration at the level of its steady-state solubility.

For temperatures above 1200 °C, and for heavily arsenic-implanted polysilicon (dose of 2.5×10^{16} ions/cm^2) anomalous segregation takes place. The electron concentration is decreased as a result of impurity diffusion to electrically inactive positions at grain boundaries and at the surface. Electron mobility is reduced by the increase of the energy barrier at grain boundaries and by retarded grain growth. Both effects are related to the impurity atom accumulation at the boundaries. Phosphorus-doped polysilicon has no electron concentration or mobility reduction in the same dose and temperature range.

The analysis of arsenic depth profiles shows that impurity redistribution becomes detectable after 10 s of processing at temperatures above 840 °C. Figure 3.7 illustrates this effect in a polycrystalline layer, which is 0.35 μm thick [52]. At the lowest temperature (840 °C) the impurity peak does not change from its as-implanted position. At higher temperatures, impurity atom diffusion into polysilicon is initiated. At 1080 °C, impurity profiles become flat throughout the polycrystalline layer. The uniform doping level is retained up to 1200 °C. This is the temperature range where the experimental carrier

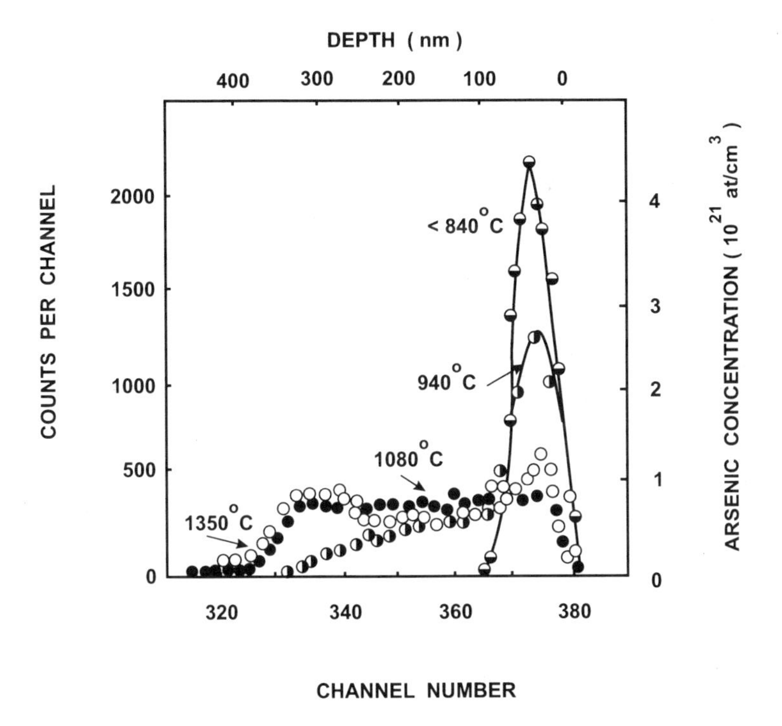

FIGURE 3.7. RBS (1.4-MeV He^{+}) arsenic depth profiles in polycrystalline silicon implanted with 50 keV, dose of 2.5×10^{16} ions/cm^2, and transiently annealed for 10 s.

concentration fit the clustering model. That also means the average grain size is large enough to minimize segregation at grain boundaries under these conditions.

The above experimental data enable one to conclude that some rapid-heating-induced solid-state processes affect the electrical activation of implanted boron, phosphorus, and arsenic differently. Initially in the first 10 s of heating, recrystallization inside the grains takes place. It is accompanied by the movement of impurity atoms to substitutional positions, where they display electrical activity. Then, impurity diffusion and segregation are the processes achieving the balance.

3.1.2. Model

Segregation under nonequilibrium thermodynamic conditions is controlled by impurity diffusion in grains and by capture/release of impurity atoms at grain boundaries [75]. Diffusion of the electrically active substitutional fraction of impurity atoms $N(r,t)$ inside a grain of radius r_g in spherical coordinates is described by

$$\frac{\partial N(r,t)}{\partial t} = D\frac{1}{r^2}\frac{\partial}{\partial r}\left[r^2\frac{\partial N(r,t)}{\partial r}\right] \tag{3.1}$$

where D is the bulk impurity diffusivity. Initial conditions for $N(r,t)$ and $N_{gb}(t)$, which is the concentration of electrically inactive impurity segregated at grain boundaries, are

$$N(r,0) = n_0, \qquad 0 \le r \le r_g \tag{3.2}$$

$$N(r_g,0) = N'_{gb}(0) = 0 \tag{3.3}$$

Boundary conditions for this case may be written in the linear form

$$\frac{dN'_{gb}(t)}{dt} = \gamma \frac{4}{3} \pi r_g^3 N(r_g, t) - \beta N'_{gb}(t) \tag{3.4}$$

where γ is the rate of impurity atom capture in electrically inactive positions at grain boundaries and β the injection rate of impurity atoms from grain boundaries into electrically active positions inside grains. The total amount of impurity atoms inside grains and at grain boundaries is

$$N'_{tg} = \frac{4}{3} \pi r_g^3 n_0 = N'_g(t) + N'_{gb}(t) \tag{3.5}$$

where

$$N'_g(t) = \int_0^{r_g} N(r,t)\, 4\, \pi\, r^2\, dr \tag{3.6}$$

Equations (3.1)–(3.6) have an analytical solution in the form [75]

$$N'_g(t) = N'_{tg} \left\{ 1 - \frac{\gamma}{\gamma + \beta} \left[1 - \exp\left(-6 \frac{(\gamma + \beta)}{\gamma} \frac{t}{t_i} \right) \right] \right\} \tag{3.7}$$

where $t_i = D/r_g^2$ is the mean time of impurity atom diffusion inside a grain. The parameters γ and β can be defined from the equality (3.7) as $t \to \infty$ to the segregation formula for equilibrium conditions [13]

$$N_g(t) = N_{tg} \left[1 + \frac{A^* Q_s}{l_g N_{Si}} \exp\left(\frac{E_a}{kT} \right) \right]^{-1} \tag{3.8}$$

where $A^* = \exp(-\Delta S/k)$, ΔS the difference between oscillation entropy of impurity atoms at the grain boundary and inside the grain, l_g the average grain size, Q_s the surface density of vacant sites for impurity atoms at the grain boundary, and N_{Si} the atom density of silicon. $Q_s / N_{Si} = 1.62 \times 10^{-7}/l_g$ for l_g in cm. The dimensionless factor A^* depends on the type of impurity and its concentration. It is 3.02 for arsenic and 2.46 for phosphorus [13] at a concentration of 2×10^{19} atoms/cm^3. The activation energy $E_a = 0.456$ eV is for both of these impurities. A comparison of (3.7) and (3.8) at $t \to \infty$ yields

$$\frac{\gamma}{\beta} = \frac{A^* Q_s}{l_g N_{Si}} \exp\left(\frac{E_a}{kT} \right) \tag{3.9}$$

The above formulas allow an appropriate calculation of the phosphorus and arsenic segregation in polysilicon subject to RTP. Impurity segregation is not significant for boron during conventional furnace annealing [76] and has not been observed after RTP [58, 65, 68].

The large concentration of electrically inactive impurity atoms is not adequately explained by impurity segregation. Impurity cluster-formation inside grains may also contribute to a decrease in the concentration of electrically active atoms in polysilicon. With arsenic over the temperature range of 1000–1200 °C the correlation between total impurity concentration, N_t, and electron concentration, n, is described very well by [52]

$$N_t = \frac{n(1 + k_{eq} n^3)}{1 - 2 k_{eq} n^3} \tag{3.10}$$

where the equilibrium clustering coefficient is

$$k_{eq} = 1.258 \times 10^{-70} \exp\left(\frac{2.062}{kT}\right) \quad \text{cm}^9 \tag{3.11}$$

There is good agreement between experimental results and theoretical predictions in the high-temperature range. A slight concentration dependence clearly indicates that the high concentration of electrically inactive arsenic can be quantitatively accounted for by impurity clustering in the grains. Annealing times of 10 s were found to be long enough for the clustering process to be completed at these temperatures. However, a discrepancy between experimental and theoretical results is observed at lower temperatures. This is believed to reflect a greater fraction of the impurity atoms residing at grain boundaries. Several factors can produce this aggregation including nonuniform impurity distribution, small grain size, and segregation [13].

Summarizing, electrical activation of substitutional impurity atoms in polysilicon layers subjected to RTP can be achieved within milliseconds at induced temperatures as high as 700–900 °C. Rapid heating for seconds is followed by impurity segregation at grain boundaries and clustering inside grains. At temperatures over 900 °C, impurity diffusion and related effects become dominant.

3.2. IMPURITY DIFFUSION

In an analysis of impurity behavior in polysilicon layered structures usually employed in semiconductor electronics, two aspects have to be considered. The first is diffusion in a polysilicon layer itself. The second relates to impurity diffusion from a polysilicon layer into the underlying single-crystal silicon substrate. These are the topics for discussion in this section.

3.2.1. Diffusion in Polysilicon Layers

There is considerable solid-state diffusion of substitution impurities in polysilicon under transient heating. The time scale for significant redistribution in the range of milliseconds remains obscure. The most favorable estimate, assuming the highest grain boundary diffusivity, suggests that the shorter processing times of 10^{-2} s are insufficient for impurity diffusion through a polysilicon layer with a thickness of 0.5 μm. However, experimental results [47, 48] contradict this statement.

Arsenic redistribution was detected by Rutherford backscattering (RBS) in 0.2-μm-thick polysilicon subjected to incoherent light heating for 10^{-2} s at an energy density of

80 J/cm^2 [47]. The induced temperature was estimated to be about 1000 °C. Processing at an increased energy density of 110 J/cm^2 ($T \sim 1250$ °C) results in a uniform impurity distribution throughout the polycrystalline layer. However, layer-by-layer Hall effect measurements performed in identical experiments provide no evidence of any redistribution at temperatures up to 1300 °C [48]. No impurity loss by evaporation was observed in these experiments. This contradiction appears to be related to the analysis technique employed, one study providing atomic profiles [47] and the other, electron profiles [48]. Moreover, measurement of the temperature profiles at a millisecond time resolution is extremely challenging. The temperature can be changed considerably, being dependent on the geometry of the light source, sample holder, and processing chamber as much as on the annealing environment [77].

The most reliable data characterizing substitutional impurity diffusion in polysilicon were obtained for RTP of seconds. The behavior of boron [56], gallium [60], phosphorus [58, 63], and arsenic [33, 54–57, 64, 66, 70] has been investigated. It has been experimentally observed that an enhanced impurity diffusion begins at temperatures above 800 °C, the same as with conventional furnace annealing.

In Fig. 3.8 the free carrier depth profiles clearly show redistribution of implanted boron in polysilicon [58]. Boron redistribution, as well as that of gallium [60], is characterized by fast impurity diffusion through the polycrystalline layer to the interface with the oxidized silicon substrate yielding practically uniform impurity concentration within the layer. Out-diffusion and evaporation at lower impurity concentration in the region near the surface, extended to a depth of some 50 nm. A silicon dioxide protective film with a thickness of 60 nm suppresses this phenomenon, which is important at temperatures above 1100 °C.

Phosphorus-implanted polysilicon impurity redistribution is notably manifested after processing at 800 °C. Figure 3.6 showed phosphorus and electron depth profile changes

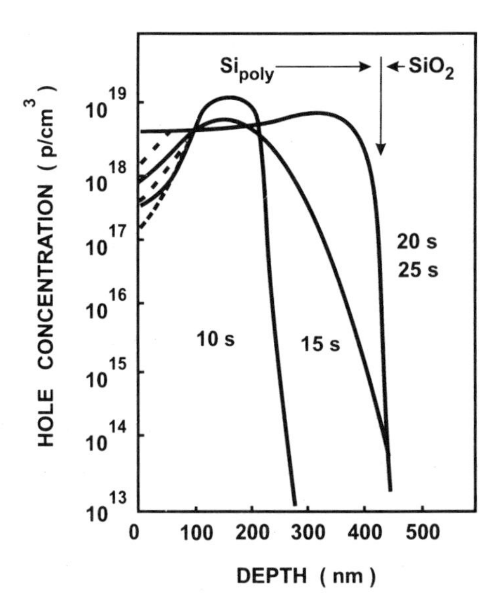

FIGURE 3.8. Hole concentration profiles of boron implanted into 0.45-μm thick polycrystalline silicon on silicon dioxide after rapid thermal processing at 1200 °C.

as a function of the processing time ranging from 6 to 40 s for induced temperatures of 800 and 1000 °C [63]. Only 6–10 s is needed for phosphorus to diffuse through the whole polycrystalline layer at 1000 °C. Further increase of the processing time causes impurity penetration into the single-crystal substrate. When polysilicon was deposited onto oxidized silicon substrates [58], implanted phosphorus redistribution resulted in a uniform concentration in the polycrystalline layer. Phosphorus redistribution in polysilicon is characterized by an anomalously high diffusivity. The estimates made based on the experimental results at 800 and 1000 °C gave values of 1×10^{-10} and 2×10^{-9} cm^2/s, respectively. Conventional furnace annealing experiments at the same temperatures predict diffusivities of 6×10^{-17} and 3×10^{-14} cm^2/s in single-crystal silicon [73] or 4×10^{-14} and 1×10^{-12} cm^2/s for grain boundary diffusion in polysilicon [26].

The activation energy of the enhanced phosphorus diffusion was 1.76 eV. This is comparable with the values obtained in the furnace annealing experiments for grain boundary diffusion as is evident from the data in Table 3.1. The existence of a deep tail of electrically inactive phosphorus (Fig. 3.6) supposes that fast-diffusing impurity atoms occupy grain boundaries, where their electrical activity is not displayed.

Similar features were also observed for arsenic. Figure 3.9 presents impurity depth profiles in the arsenic-implanted polysilicon subjected to RTP by radiation with a graphite heater for 10 and 12.5 s [54]. They have typical Gaussian impurity profiles [33, 55–57, 64, 70]. Accounting for the time of temperature rise and fall, the effective parameters for thermal processing were 1.25 and 2.0 s at 1050 and 1120 °C, respectively. Impurity

TABLE 3.1. Diffusion Parameters of Substitutional Impurities in Polycrystalline Silicon

Impurity	Temperature range (°C)	Preexponential factor D_0 (cm^2/s)	Activation energy (eV)	Diffusion process	Refs.
		Rapid thermal process			
P	800–1000	1.9×10^{-2}	1.76	grain boundary	63
As	700–1050	0.28	2.84	—	33
As	800–1000	8.4×10^{4}	3.8	tail region	70
As	800–1000	1700	3.8	peak region	70
		Furnace process			
B	1000–1200	6.6×10^{-4}	1.87	grain boundary	22
P	1000–1200	4.0×10^{-3}	1.71	grain boundary	20, 22
P	900–1100	5.3×10^{-5}	1.95	grain boundary	26
P	700–850	44.8	3.4	combined	31
P	700–1100	1.0	2.75	combined	30
P	700–1050	1.81	3.14	—	33
As	750–950	8.6×10^{4}	3.9	grain boundary	27
As	700–850	10	3.36	grain boundary	29
As	950–1100	0.63	3.22	combined	34
As	700–1000	1.66	3.22	combined	30
Sb	1050–1150	13.6	3.9	bulk	32
Sb	1050–1150	812	2.9	grain boundary	32

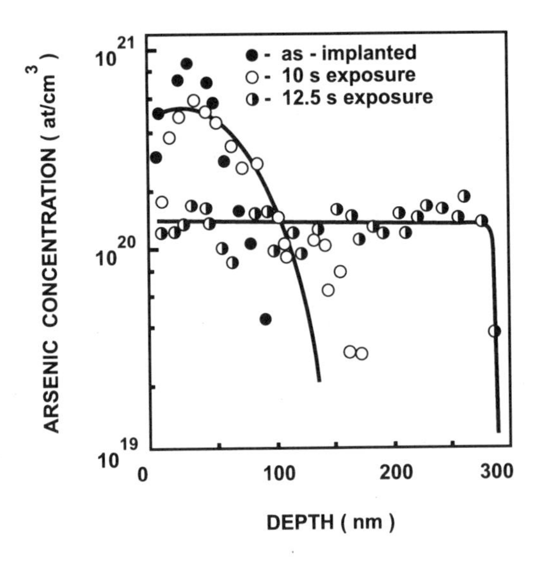

FIGURE 3.9. RBS arsenic depth profiles in polycrystalline silicon implanted with a dose of 5×10^{15} ions/cm^2 after rapid thermal processing by graphite heater. Solid lines show the profiles calculated for arsenic diffusivities of 4×10^{-12} and 8×10^{-11} cm^2/s.

diffusivities in these two regimes were estimated to be 4×10^{-12} and 8×10^{-11} cm^2/s. A comparison of arsenic diffusivity data displayed in Fig. 3.10 shows them to agree well with those obtained by Takai *et al.* [33].

In the temperature range investigated, arsenic diffusivity in polysilicon is 1.5–2 orders of magnitude greater than that in a silicon crystal. Its temperature dependence based on published results [27, 29, 30, 33, 34] differs both in preexponential factor D_0 and in activation energy E_a. One of the reasons for such a discrepancy is related to the inadequate choice of the numerical fitting procedure based on the analysis of the spread of a surface peak [34]. A tail region of the redistributed depth profile seems to supply correct information. It should be noted that a spread of the surface peak permits an estimate of the impurity diffusivity inside single-crystal silicon grains. Moreover, scattering in the diffusivity parameters can be explained by the difference in experimental conditions determined by way of the impurity introduction [29], or by the structure of polysilicon [20], or the silicon substrate, and by the geometry of the layer in which the diffusion is studied [28]. Even sample geometry can play a role via its influence on the heating rate and induced temperature [78]. Anomalously high arsenic diffusivity obtained by Swaminathan *et al.* [27] may be a result of impurity redistribution at the substrate interface, where stress relaxation plays a definite role in the impurity migration.

Arsenic diffusion activation energy values in RTP cycles are 2.84 eV, which is lower than those observed in conventional furnace annealing experiments, both for the diffusion via grain boundaries 3.36–3.9 eV, and for the combined grain boundary and bulk diffusion 3.22 eV (see Table 3.1). The enhanced arsenic diffusion that occurs during RTP is caused by the rapid release of impurities captured by defects as a result of the high heating rate [55]. In contrast, furnace annealing provides slow step-by-step impurity defect rearrange-

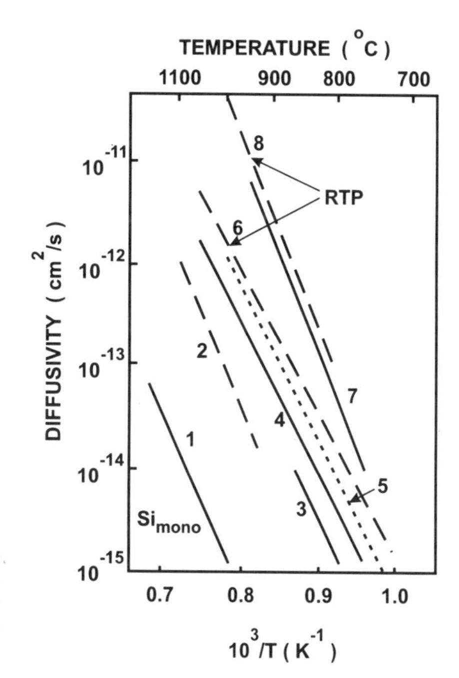

FIGURE 3.10. Arsenic diffusivity in single-crystal (1, [68]) and polycrystalline silicon: (2, [34]), (3, [29]), (4, [33]), (5, [30]), (7, [27]) from conventional furnace annealing experiments; (6, [33]), (8, [70]) from rapid thermal processing for seconds.

ment which yields stable high-temperature complexes. These complexes restrict diffusion redistribution of these impurities.

In the analysis of substitutional impurity behavior it is also necessary to account for their limited solubility in silicon, which leads to dissociation of supersaturated solid solutions inside grains under rapid heating conditions. Nonequilibrium point defects, i.e., vacancies and impurity interstitials, are generated, which enhance impurity diffusion, both inside grains and at grain boundaries.

3.2.2. Diffusion from Polysilicon into Crystalline Silicon

Impurities located in a polysilicon layer deposited onto a silicon crystal were shown to diffuse into the substrate under certain heating regimes. This is of scientific and practical interest, because heavily doped polysilicon layers are used in semiconductor technology as a surface impurity source to form emitters of bipolar transistors, source and drain regions of MOS devices, solar cells, and other devices.

RTP-induced impurity diffusion in such structures was investigated for boron [65], phosphorus [50], and arsenic [64–66]. Figure 3.11 shows typical phosphorus and electron depth profiles in polycrystalline silicon on single-crystal silicon substrates subjected to RTP of seconds' duration, and conventional furnace annealing of minutes' duration [50]. The atomic profiles were obtained by neutron activation analysis. The free carrier profiles were constructed from layer-by-layer Hall effect measurements. The maximum concentration of electrically active impurity atoms does not exceed 3.5×10^{20} atoms/cm^3, which corresponds approximately to the steady-state phosphorus solubility in silicon [74].

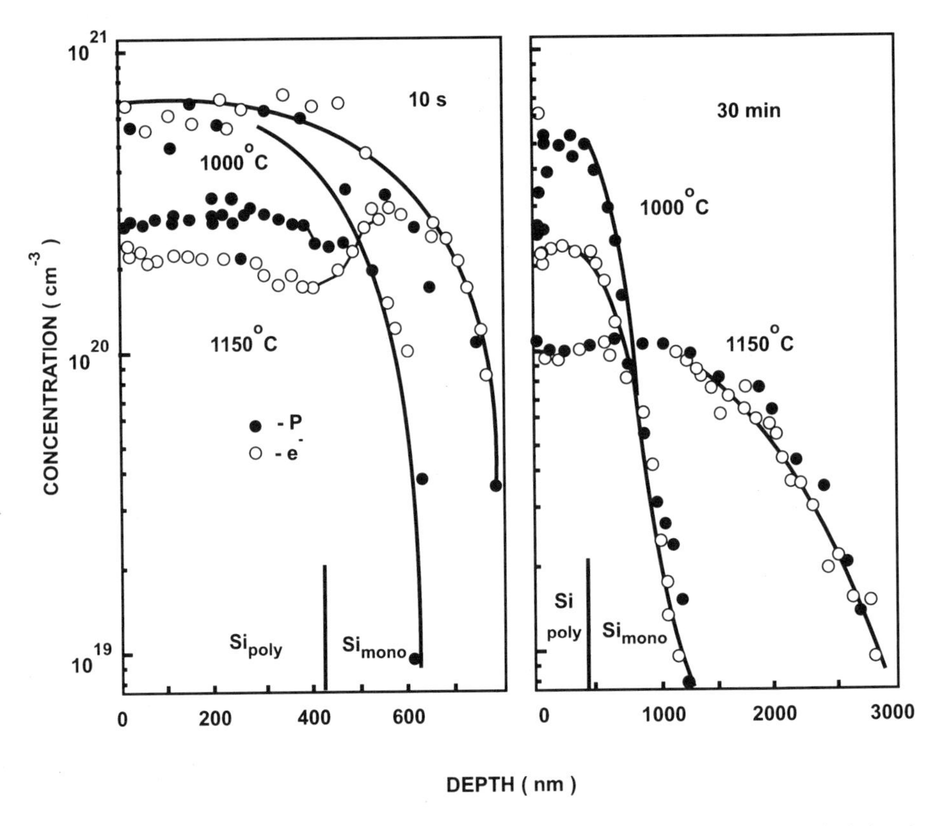

FIGURE 3.11. Phosphorus (●) and electron (○) concentration depth profiles in the as-deposited doped polycrystalline silicon on a single-crystal substrate after rapid thermal processing for 10 s and conventional furnace annealing for 30 min.

Diffusivity estimation shows that the rapid-heating-induced diffusion is enhanced 1.5- to 2-fold. Experimental diffusivity data for phosphorus as well as for boron and arsenic are summarized in Table 3.2.

Boron and phosphorus diffusion from a heavily doped polycrystalline layer is characterized by a decreased activation energy in comparison with the diffusion in a silicon crystal free of coating. The activation energies for arsenic, in the case of diffusion via negatively charged vacancies, are 4.05 and 4.32 eV, which are very similar [73]. However, absolute estimates of impurity diffusion remain enhanced under the rapid heating conditions. The enhancement is obviously related to the stresses induced by thin native oxide at the substrate interface and by segregated impurities [34, 38, 65]. These effects are enhanced by rapid heating and cooling of the samples [48, 65].

Impurity out-diffusion and evaporation are observed in heavily doped polysilicon heated for some seconds in a vacuum [53–55] and are pronounced for arsenic at temperatures of 1200 °C and above. Phosphorus has a higher diffusion mobility and demonstrates a detectable evaporation at lower temperatures. Surface protection by silicon dioxide or phosphosilicate glass films can be successfully employed to prevent

TABLE 3.2. Diffusion Parameters of Substitutional Impurities Introduced from Polycrystalline Silicon Layers into Silicon Crystals

Impurity	Crystal orientation	Processing mode	Temperature (°C)	Preexponential factor, D_0 (cm^2/s)	Activation energy, E_a (eV)	Ref.
B	(100)	RTP	1050–1150	0.21^a	2.76^a	65
P	(111)	RTP	900–1150	6.9×10^{-4}	1.78	50
P	(111)	FA	900–1150	5.4×10^{-2}	2.63	50
As	(100)	RTP	1050–1150	$2.1 \times 10^{5\,a}$	4.3	64
As	(100)	RTP	1050–1150	840^a	4.2	66
As	(111)	FA	850–1050	0.90^b	3.37	35

[a] Calculated from the experimental data presented in the reference.
[b] Calculated for arsenic concentration of 1×10^{20} atoms/cm^3.

this phenomenon during heating to high temperatures in vacuum [53, 58]. RTP in air or an inert ambient at temperatures of 1000 °C and lower also eliminates evaporation.

3.3. GRAIN GROWTH

High-temperature processing for a few seconds is sufficient to observe self-diffusion in polysilicon. Silicon atoms that migrate through grain boundaries mediate recrystallization. Grain growth recrystallization occurs because polysilicon has a higher free energy than single-crystal silicon. Heating provides the thermal activation energy to overcome the potential barrier and the transformation from the high-energy to the lower-energy form. Driving forces and mechanisms for the transformations are analyzed in detail elsewhere [46]. The main features are summarized here as follows for different temperature ranges.

As the temperature of a sample is raised in the range of 500–800 °C, the first process observed is usually the solid-state regrowth of amorphous regions in direct contact with grains. This occurs because of the higher mobility of the amorphous–crystalline boundary and the large driving force present. The regrown material reproduces the orientation of the underlying silicon seed grains, although competitive solid-state growth may also give rise to a slightly larger grain size. Tiny grains may be present in deposited layers, apparently entirely amorphous by X-ray analysis, and these can act as nuclei during subsequent annealing. The grain size and morphology will be determined by the average spacing and location (i.e., surface or bulk) of these nuclei.

At somewhat higher temperatures, 850 °C or above, grain boundary mobility becomes high enough for some grain boundary motion to occur. In polycrystalline layers with grain sizes smaller than the layer thickness, the main driving force comes from reduction in grain boundary length and curvature. It gives rise to a characteristic growth of most of the grains to an average limiting diameter approximately equal to the layer thickness. This process is generally known as primary recrystallization [40].

At still higher temperatures, 950 °C and above, a few grains may grow to a much greater size, eventually consuming all others. This process has been called secondary or even tertiary recrystallization, or secondary grain growth. When the driving force is the minimization of surface energy, the term surface-energy-driven secondary grain growth has been used to describe the process [44].

Conventional furnace annealing provides a parabolic dependence of average grain size on the processing time $l_g \sim t^{1/2}$. It corresponds to the model of grain growth controlled by the boundary energy [40]. In the case of RTP, time dependence of the sample temperature can be accounted for with an effective parameter approach. The model assumes that grain boundary motion is responsible for grain growth. As a consequence of this model, the grain size follows

$$l_g \sim (D\, t/T)^{1/2} \tag{3.12}$$

Here, effective values are taken for all of the parameters on the right side of the expression. The equation (3.12) is valid for average grain sizes smaller than the thickness of the polycrystalline layer. Grain growth slows down at larger grain sizes.

Recrystallization processes in polysilicon grains epitaxially aligned to a single-crystal silicon substrate [56, 64–66, 72] produce texturing of the layer deposited onto silicon dioxide [61, 62]. Figure 3.12 demonstrates He^+ channeling spectra for heavily

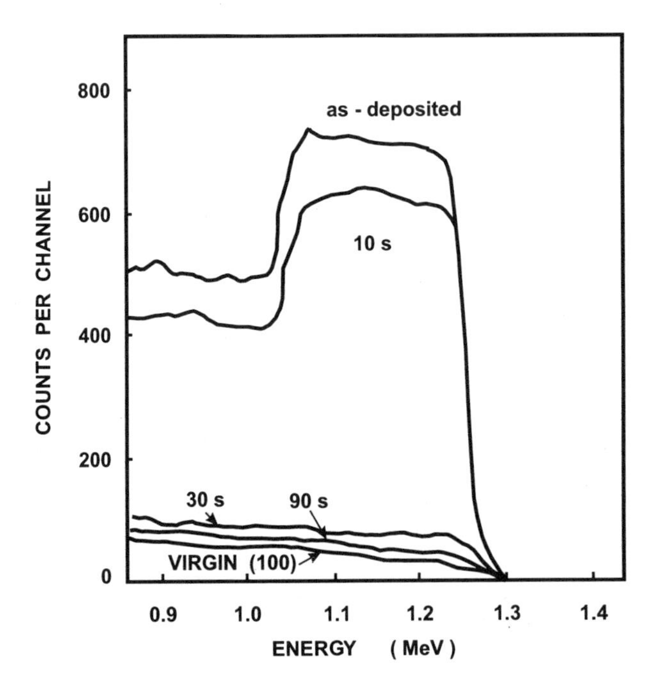

FIGURE 3.12. Channeling spectra of 2.2-MeV He^+ for 0.48-μm-thick arsenic-implanted polycrystalline silicon on (100) silicon after rapid thermal processing at 1100 °C.

arsenic-implanted polycrystalline layer on a (100) single-crystal silicon substrate subjected to RTP at 1100 °C [64]. Good epitaxy is evident after 90 s of annealing. It is characterized by χ_{min} = 5.5% (typical χ_{min} = 3.5% for virgin silicon). A temperature increase to 1150 °C results in the same perfection achieved after 10 s of annealing. The activation energy for solid-state epitaxial recrystallization has been evaluated at 4.5 eV [72], 7–9 eV [66] in unimplanted and 5–6 eV [71] in implanted layers.

Impurities in polysilicon considerably influence the recrystallization, secondary grain growth, and epitaxial perfection [45, 56, 64, 71, 79]. For substitutional impurity concentrations at or above 5×10^{20} atoms/cm^3, phosphorus and arsenic markedly enhanced the grain growth while boron had little effect. The influence of substitutional impurities on recrystallization is determined by elastic stresses created by impurity supersaturation inside grains and by impurity segregation at grain boundaries. Both of these phenomena cause a decrease in the threshold for plastic deformation in silicon, making it easier for grain boundary movement.

Grain growth is restricted by chemically active dopants such as oxygen [67, 80] and carbon [81]. Even additional implantation of boron, phosphorus, or arsenic produces no effect in polysilicon containing oxygen at 10–30%, and transiently annealed at 1150–1200 °C for 10 s [67]. This is a consequence of oxide precipitate formation at grain boundaries. Silicon and impurity atom diffusion through the grain boundaries is limited; however, the maximum electrical activation of implanted substitutional impurities is achieved [68].

Grain growth in substitutional impurity doped polysilicon is followed by an increase of carrier mobility and hence material with a lower resistivity. RTP permits good control of the material parameters.

4

Component Evaporation, Defect Annealing, and Impurity Diffusion in the III–V Semiconductors

The III–V semiconductors GaAs, InP, and related ternary compounds are important for making sensors, solid-state lasers, high-frequency devices, and integrated circuits [1, 2]. They have unique and promising optical and electrophysical properties compared to silicon and germanium. However, their utilization is limited by the dramatic influence that thermal processes employed during device fabrication have on these properties based on primarily their decomposition and incongruent component evaporation at high temperatures. RTP has become one of the effective ways to resolve this problem of III–V semiconductor decomposition [3]. The most acceptable processing conditions are in the heat balance regime, because of minimal temperature gradients and processing simplicity. Overall, a short processing time, typically seconds, has been found optimal for the best properties in doped layers of III–V semiconductor compounds [4–6].

4.1. DECOMPOSITION AND COMPONENT EVAPORATION

The main factor leading to the decomposition of semiconductor compounds under thermal processing is loss of stoichiometry by evaporation. The surface layer structure changes significantly when the layer is depleted by several atomic percent of one of the components, producing a solution of the III–V compound in the majority component [7–11]. Crystalline structure is preserved in the less depleted layers, although defects can be created. For example, Fig. 4.1 schematically shows the surface layer structure produced by heating of GaAs. Atom evaporation and condensation keep the stoichiometric balance at the surface. The competition between evaporation and condensation rate determines the overall evaporation rate. If they are equal, the process is called equilibrium (or Knudsen). If they are not equal, the process is nonequilibrium. The conditions of heating the material in a vacuum are referred as free surface (or Langmuir) evaporation.

Changes in stoichiometry are initiated by decomposition of the semiconductor compound and by the loss of one of the components, usually a group-V element at high

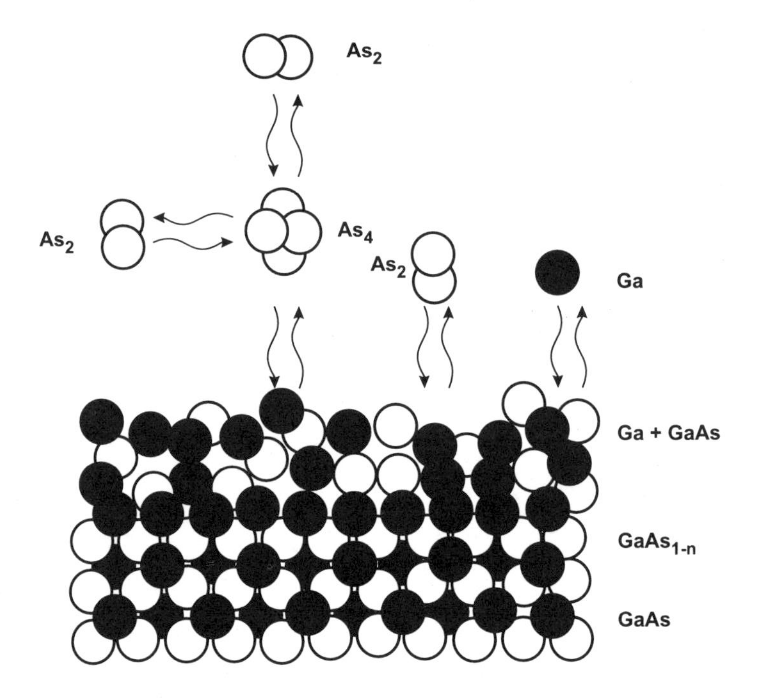

FIGURE 4.1. Processes at the surface of heated gallium arsenide [12].

temperatures. Surface evaporation causes diffusion of these atoms from the bulk to the surface. An increase in processing temperature, producing higher evaporation and diffusion rates, as well as a processing duration increase, leads to greater variation in stoichiometry from the bulk of the semiconductor crystal.

Solid–vapor equilibrium for III–V semiconductor compounds is characterized by a wide variety of reactions at the surface. The major ones are shown in Table 4.1. Group-III elements evaporate atom by atom, while two- and four-atom molecular fragments of group-V elements take part in the process. The concentration of two-atom molecules at temperatures below the melting point has been found to be much higher than that of four-atom ones [7–11].

Thermodynamic calculations, taking into account all of the reactions, are quite complicated and time consuming. However, the theory developed for individual evaporation can be successfully applied.

Evaporation Theory

Vapor pressure is the main parameter that determines equilibrium evaporation rates. Experimental vapor pressures of the components for GaAs, GaP, InAs, and InP as a function of temperature are shown in Fig. 4.2. In the limited temperature range shown, the vapor pressure is accurately described by the Arrhenius-type relation $P = P_0 \exp(-E_a/kT)$.

TABLE 4.1. Reactions at the Surface of Heated III–V Compound Semiconductors [7–9]

Compound	Reactions[a]	Enthalpy of the reaction ($\times 10^3$ J/mol)
GaAs	$GaAs(s) \rightarrow Ga(s) + As(v)$	314
	$GaAs(s) \rightarrow Ga(s) + \frac{1}{2} As_2(v)$	188
	$GaAs(s) \rightarrow Ga(s) + \frac{1}{4} As_4(v)$	123
	$2\ As_2(v) \rightarrow As_4(v)$	–262
	$GaAs(s) \rightarrow Ga(v) + As(v)$	649
	$4\ As(s) \rightarrow As_4(v)$	144
	$2\ As(s) \rightarrow As_2(v)$	224
	$As(s) \rightarrow As(v)$	300
	$\frac{1}{2} As_2(v) \rightarrow As(v)$	188
	$Ga(s) \rightarrow Ga(v)$	275
	$Ga(s) + As(s) \rightarrow GaAs(s)$	–74
InP	$InP(s) \rightarrow In(s) + \frac{1}{2} P_2(v)$	157
	$2\ P_2(v) \rightarrow P_4(v)$	–243
	$In(s) \rightarrow In(v)$	243
GaP	$GaP(s) \rightarrow Ga(s) + \frac{1}{2} P_2(v)$	191
	$GaP(s) \rightarrow Ga(s) + \frac{1}{2} P_4(v)$	134
	$2\ P_2(v) \rightarrow P_4(v)$	–288
	$4\ P(s) \rightarrow P_4(v)$	129
	$2\ P(s) \rightarrow P_2(v)$	179
	$2\ P(l) \rightarrow P_2(v)$	142
	$\frac{1}{2} P_2(v) \rightarrow P(v)$	245
	$P(s) \rightarrow P(v)$	334
	$GaP(s) \rightarrow Ga(v) + P(v)$	711
	$Ga(s) + P(s) \rightarrow GaP(s)$	–101
InAs	$2\ InAs(s) \rightarrow 2\ In(s) + As_2(v)$	325
	$As_4(v) \rightarrow 2\ As_2(v)$	228

[a]Phases: s, solid; l, liquid; v, vapor.

The values of preexponential factor and activation energy calculated from the experimental data are summarized in Table 4.2. According to the vapor pressure of volatile components (group-V elements), the thermal stability of the semiconductor compounds is ranked as follows: GaP > GaAs > InAs > InP.

Equilibrium evaporation may be characterized by equal concentrations of the group-III and -V elements in the vapor phase. The temperature at which this occurs is determined from the congruent evaporation condition [6]

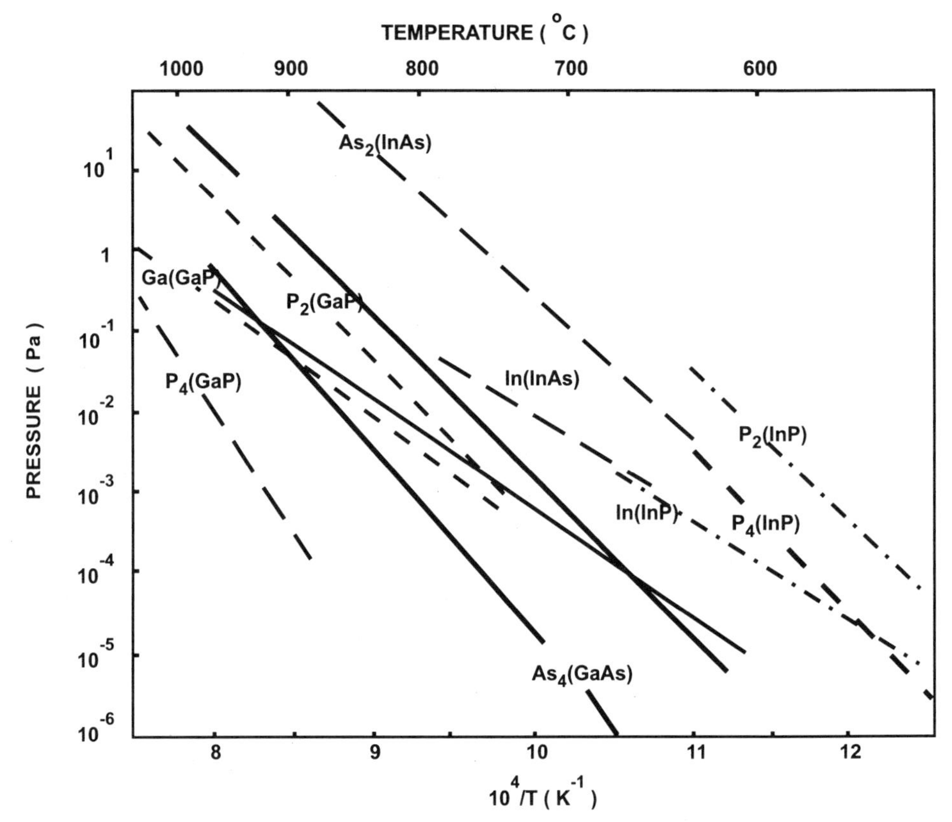

FIGURE 4.2. Equilibrium vapor pressure of the components at the heated surfaces of III–V semiconductors [7–11].

TABLE 4.2. Equilibrium Vapor Pressure Parameters for III–V Compound Semiconductors

Compound	Component	Preexponential factor, P_0 (Pa)	Activation energy, E_a (eV)	Temperature range (°C)	Ref.
GaP	Ga	3.94×10^{10}	2.78	800–1400	7
	P_2	6.38×10^{16}	3.00	800–1400	7
	P_4	1.4×10^{21}	5.73	950–1400	7
GaAs	Ga	1.14×10^{10}	2.63	600–800	8
	Ga	3.60×10^{10}	2.69	765–950	10
	As_2	1.58×10^{17}	3.96	600–1150	8
	As_2	3.43×10^{16}	3.83	750–965	10
	As_4	2.72×10^{20}	5.12	750–900	8
InP	In	3.83×10^{10}	2.50	550–660	11
	P_2	3.28×10^{16}	3.29	550–620	11
	P_4	6.93×10^{19}	4.01	550–620	11
InAs	In	4.45×10^{9}	2.31	680–760	10
	As_2	1.30×10^{17}	3.48	650–810	10

$$P(\mathrm{A}) = 2\,P(\mathrm{B}_2) + 4\,P(\mathrm{B}_4) \tag{4.1}$$

where P represents the partial pressure of evaporating particles of the A (group III) or B (group V) element shown in parentheses. The experimental values of the congruent evaporation temperatures for III–V semiconductors are presented in Table 4.3. The finding that III–V semiconductors have congruent evaporation point provides several important recommendations. In particular, processing at this temperature is the most desirable to preserve compound stoichiometry. The choice of a processing temperature above or below that point provides the opportunity to vary the concentration of point defects produced by dominant evaporation of the host atoms from the group-III or -V element sublattice. Equilibrium evaporation conditions are difficult to realize for compound semiconductor technology. In practice, the conditions of thermal processing are closer to free surface evaporation.

Kinetic gas theory can be employed to calculate free evaporation rate [13]. At a given temperature, T, the rate is constant and defined by the equilibrium pressure, P_{eq}, of evaporating particles in the Hertz–Knudsen equation:

$$\omega_{\mathrm{e}} = \frac{a_{\mathrm{v}} P_{\mathrm{eq}}}{(2\,\pi\,M\,kT)^{1/2}} \tag{4.2}$$

where a_{v} is the evaporation coefficient and M the molecular mass of the evaporating particles.

Generally, the rate of loss of the most volatile component is equal to the difference between the rates of evaporation and condensation. For the free evaporation condition the pressure in the system is believed to be equal or close to zero. Hence, the rate of loss is described by (4.2).

Free evaporation also has a congruent evaporation point. It is determined by the equality of the evaporation rates from the equation [6]

$$\omega_{\mathrm{e}}(\mathrm{A}) = 2\omega_{\mathrm{e}}(\mathrm{B}_2) + 4\omega_{\mathrm{e}}(\mathrm{B}_4) \tag{4.3}$$

The substitution of (4.2) into (4.3) gives a relation between evaporation coefficients, equilibrium vapor pressures, and masses of evaporating particles

$$\frac{a_{\mathrm{v}}(\mathrm{A})P(\mathrm{A})}{[M(\mathrm{A})]^{1/2}} = \frac{2a_{\mathrm{v}}(\mathrm{B}_2)P(\mathrm{B}_2)}{[M(\mathrm{B}_2)]^{1/2}} + \frac{4a_{\mathrm{v}}(\mathrm{B}_4)P(\mathrm{B}_4)}{[M(\mathrm{B}_4)]^{1/2}} \tag{4.4}$$

Congruent free evaporation takes place at higher temperatures than does equilibrium evaporation (see Table 4.3). At temperatures above the free congruent evaporation point, group-V elements evaporate faster. The ratio of the evaporated group-V and -III atoms

TABLE 4.3. Temperature (°C) of Congruent Evaporation of III–V Compound Semiconductors

Evaporation conditions	GaAs	InP	GaP
Langmuir (free surface evaporation)	660 [7, 9]	365 [11]	680 [7]
Knudsen (equilibrium evaporation)	625 [9]	356 [11]	672 [11]

exceeds unity, and increases rapidly as the temperature rises. The stoichiometry correlation is normally thought to be less than unity at temperatures below this point, but this has not been observed in conventional long-duration experiments [9, 10]. The fluxes of evaporating atoms remain equal over a rather wide temperature range. The reason is that the evaporation rate of the group-III atoms cannot exceed the evaporation rate of the group-V ones; otherwise, the former would form a continuous layer. At higher temperatures, group-V atoms have a zero sticking coefficient in the absence of free group-III atoms [9] based on the lower sublimation temperature. Therefore, it is more correct to discuss the upper limit of congruent evaporation instead of the lower limit. Nevertheless, an increased rate of gallium evaporation compared with arsenic has been observed during RTP of GaAs [14–16]. Figure 4.3 shows the experimental and calculated values of evaporation rates, which are in satisfactory agreement, both above and below the congruent evaporation limit. This phenomenon is observed in short-time thermal processing, when the amount of evaporated atoms is small, and there is enough gallium at the surface to hold arsenic atoms.

Thermal decomposition and evaporation of one of the components changes the stoichiometry and structure of the surface layer and leads to degradation of electrophysical and optical characteristics. For example, because arsenic loss in the undoped semi-insulating GaAs results in a decrease of its surface concentration from 52.5% to 47.5%, *GaAs is usually grown arsenic rich to create a semi-insulating substrate* with a 300-fold reduction of resistivity [17]. Furthermore, there was a conversion of the conductivity type from electron to hole. Such undesirable changes can be prevented by a suitable choice of the thermal processing regimes and by surface protection layers.

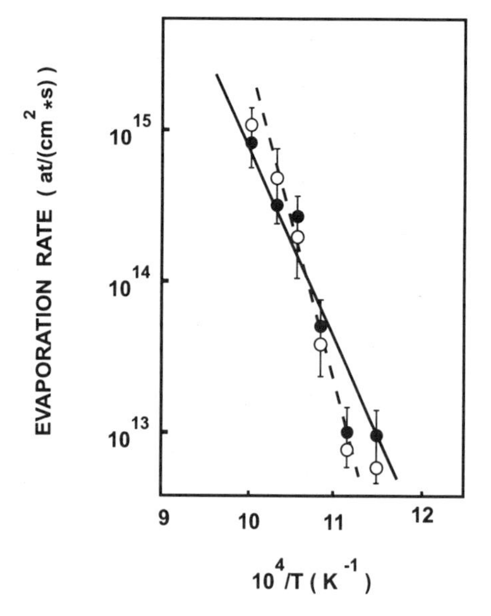

FIGURE 4.3. Arsenic (○ and dashed line) and gallium (● and solid line) evaporation rates from gallium arsenide transiently heated for 30 s. Points represent experimental values. Lines are the data calculated according to the Hertz–Knudsen equation [16].

4.2. HEATING REGIMES AND SURFACE PROTECTION

Thermally induced decomposition of semiconductor compounds and component evaporation can be reduced by reducing the process duration or by decreasing the evaporation rate with protecting ambients, custom sample holders, or surface layers.

4.2.1. Sample Holders

Solid-phase annealing of GaAs without surfaces encapsulants illustrates the possibility of finding a combination of processing temperature and time that results in minimal changes in the surface layer [17–31]. Typical sample holders are shown Fig. 4.4, in which the GaAs wafer is placed in a local arsenic overpressure to restrict incongruent evaporation.

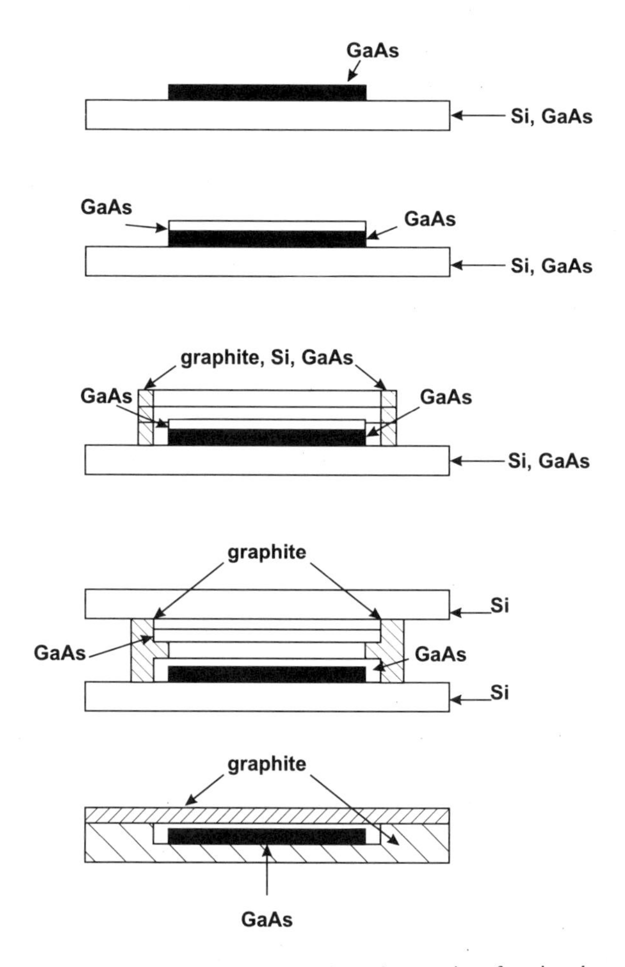

FIGURE 4.4. Sample holders employed for rapid thermal processing of semiconductor compounds.

The simplest holder based on the so-called proximity technique consists of a silicon, or semiconductor compound, substrate larger than the wafer size. The wafer processed is placed face to face with the holder and an equilibrium vapor pressure is set up in the submicron clearance between the wafer and substrate. Within 10^{-6} s of commencing high-temperature processing, the vapor is established which suppresses evaporation of the semiconductor compound components [16]. An analogous effect can be achieved by placing additional protective substrates on top of the wafer. Increased heating and cooling rates at the wafer edge are the limiting factors for the simple proximity holders. Radial temperature gradients induce elastic stresses resulting in slip-line formation. Minimizing the temperature gradient is very important for the III–V semiconductors because they are more plastic than silicon. Reduced temperature gradients are achieved simply by applying guard rings surrounding the wafer [32–35]. The rings are made of silicon, graphite, or a semiconductor compound.

The temperature gradient can be reduced and an overpressure of the volatile component produced at the same time effectively with an inner closed cavity design shown in Fig. 4.4 [36–40]. The overpressure is achieved by saturation of the inner surfaces with it or by placing an additional substrate of the same semiconductor material.

The effectiveness of the above sample holders is demonstrated by the following examples. Decomposition of unprotected InP surface appears just after processing at 350 °C for 4 s [23]. In the holder, where the wafer is covered by another InP substrate, phosphorus evaporation remains negligible after 4 s of processing at 850 °C. Further increase in processing temperature and/or time results in a considerable phosphorus loss. A GaAs cover permits an increase of the temperature to 900 °C because it forms an effective diffusion barrier on the InP surface [30, 31]. InAs is synthesized at the surface of the InP wafer when arsenic occupies the sites of evaporated phosphorus. At high temperatures, InAs is more stable than InP. This results in a superthin layer of InAs which forms a barrier layer suppressing phosphorus evaporation. The GaAs substrate can be repeatedly used as a cover holder for the InP substrate. The additional superthin layer does not introduce any degradation in high-quality MIS structures [31]. Decomposition of GaAs is slower than InP because of the lower evaporation rate of arsenic.

4.2.2. Processing Regimes

Figure 4.5 shows time–temperature limits over which little change occurs in the surface layer stoichiometry. The acceptable regimes are given below the experimental curves. Some change does occur at an unprotected free surface. Placing the wafer face to face with a silicon or GaAs holder increases the range of temperature and processing time available. The best results have been obtained with holders in which a closed inner cavity maintains a local arsenic overpressure inside. Defect-free GaAs processing can be achieved at temperatures up to 1100 °C for up to 10 s processing time. Excellent results are also obtained in an arsine ambient [41].

4.2.3. Encapsulating Films

Protective surface layers considerably decrease component evaporation from the semiconductor wafer. For example, GaAs powder, covering a GaAs wafer, has been used

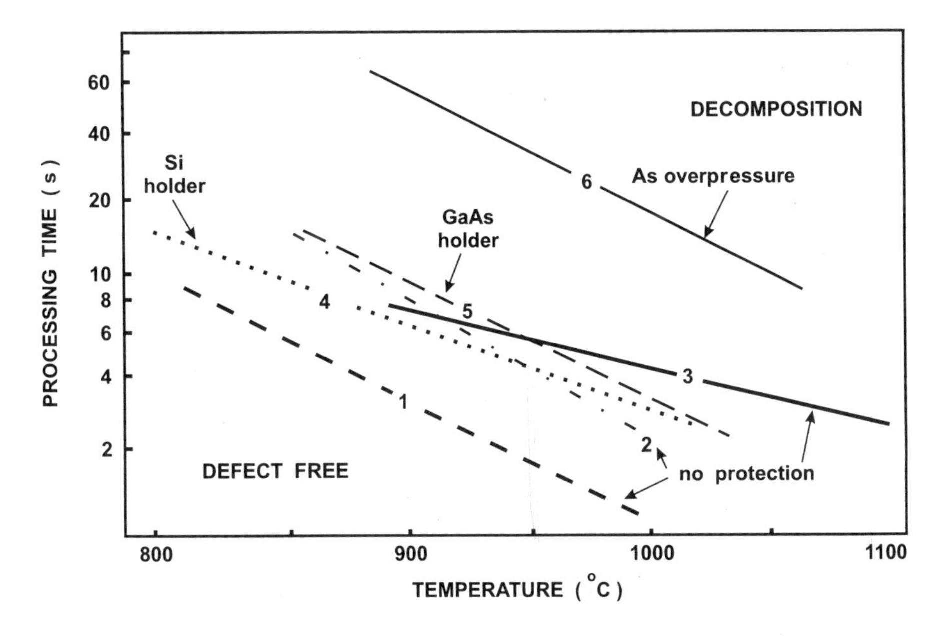

FIGURE 4.5. Temperature–time range providing rapid thermal processing of nonencapsulated gallium arsenide: 1, 2, 3, without any protection [20, 24, 26]; 4, face to face with the silicon holder [25]; 5, face to face with the gallium arsenide holder [25]; 6, local arsenic overpressure [36].

successfully to suppress arsenic evaporation [42]. In spite of the simplicity of the technique, the sample surface is safely protected for up to 30 min at 800 °C. However, the developed powder surface is an effective getter for uncontrolled impurities, which can diffuse into the wafer.

The application of encapsulating films has obvious technological advantages. Silicon dioxide, nitride, and oxynitride, silicate glasses, and aluminum nitride are widely employed for this purpose [43–72]. For an appropriate choice of encapsulant one must consider the influence of the film thickness, structure, and composition, on the quality and stoichiometry of the semiconductor surface layer. The decrease of arsenic loss from GaAs capped with different films is shown in Fig. 4.6. Arsenic evaporation is completely suppressed by a 20-nm-thick silicon layer, 16-nm-thick silicon dioxide or silicon nitride film at processing temperatures up to 800, 850, and 900 °C, respectively. At higher temperatures, these films reduce arsenic loss by three to four orders of magnitude compared to free evaporation estimated from the Hertz–Knudsen equation. Calculated arsenic evaporation rates based on known arsenic diffusivity in the encapsulating materials were found to be about nine orders of magnitude lower than experimental values [43]. This indicates that evaporation takes place through pores and defects in the film, whose size and concentration are determined by the film composition and by the deposition technique.

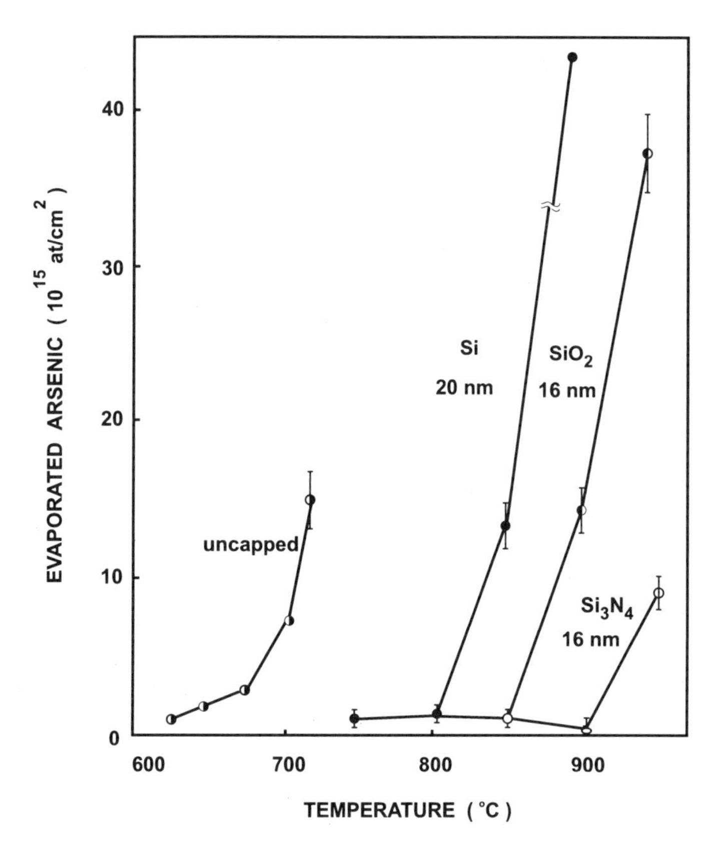

FIGURE 4.6. Temperature dependence of arsenic evaporation through various caps on gallium arsenide during thermal processing for 10 s [43].

Silicon nitride films are mainly used for GaAs surface protection during activation of group-VI implants [44, 45]. The films deposited by evaporation usually have poor reproduction of the protective action, especially at high temperatures (950 °C and above). Ion-plasma and pyrolytically deposited films are much better. It is worth mentioning that film deposition is a process requiring increased temperature. Hence, it can result in partial annealing of implantation-induced defects. This is illustrated by InP implanted with selenium [46]. The crystalline structure of material with implant doses of 1×10^{13} and 1×10^{15} ions/cm^2 was characterized by a relative yield of backscattered helium ions of 46 and 82%, respectively. After pyrolytic deposition of silicon nitride film at 550 °C, the value of this parameter was reduced to 9 and 73%, respectively, for the above implanted doses. Some restrictions can appear when a silicon nitride cap is used for annealing of silicon-implanted GaAs. The cap-related tensile stresses [47] as well as diffusion of remaining hydrogen from the cap into the implanted region [48] reduce the concentration of donor states in the doped layer. Silicon nitride films are not recommended for processing at temperatures higher than 900 °C because of their degradation [49, 50]. At high temperatures a better choice is aluminum nitride, silicon dioxide, or oxynitride encapsulants.

Aluminum nitride films provide rather good results for activation of implanted group-IV impurities as well as of group-II impurities. The films are usually deposited by reactive sputtering of aluminum in an ammonia ambient. They have satisfactory protective properties at temperatures of 1000 °C and higher [50–52]. The thickness of the film necessary for effective protection is about 30–60 nm smaller than for other materials. Aluminum nitride films have rather good adhesion to GaAs. However, there is a risk of aluminum-enriched intermetallic compound formation at the wafer surface resulting in difficulties in encapsulant removal. A simple test was proposed by Bensalem *et al.* [50] to examine films deposited by different techniques. If an aluminum nitride film deposited onto a reference glass substrate dissolves in hot water, it can be used for protection; otherwise, the film may not be easily dissolved after high temperature processing.

Silicon dioxide films are mainly employed for activation of group-II and -VI implants. They are formed by chemical vapor or plasma deposition. A thickness of 100–150 nm prevents decomposition of the GaAs surface at processing temperatures up to about 1000 °C. At higher temperatures, cracks may occur in the film even after short processing times (seconds) [53], and therefore silicon oxynitride [54] or silicate glasses [55] are a better choice.

Silicon-based emulsion films have also been used as encapsulants. They are deposited by spinning-on at 3000–5000 rpm [56–59]. However, they may contain doping impurities, in the form of oxides or salts which can diffuse into the substrate. Therefore, an additional film to encapsulate the substrate wafer is used. It was experimentally determined that the thickness of the double film sandwich must not exceed 400 nm, in order to avoid peeling [56]. Spin-on caps for GaAs withstand heating up to 900 °C for 60 min. On InP they withstand heating up to 700 °C for 30 min. For short durations (seconds), the temperature limits may be increased by 100–150 °C.

Tantalum-based films were found [60] to be appropriate encapsulants for GaAs. Amorphous $Ta_{74}Si_{26}$ is a good cap up to 800 °C and amorphous $Ta_{36}Si_{14}N_{50}$ up to 850 °C. While the effectiveness of tantalum-based caps is improved significantly by addition of either nitrogen or silicon, the best result is provided by a cap containing both elements.

4.2.4. Diffusion into Encapsulating Films

Diffusion of semiconductor compound components into encapsulating films influences the electrical activation of implanted impurities. Among the components of III–V semiconductors, gallium is the fastest diffuser in silicon dioxide [61, 62]. Gallium diffusion in the film, in the presence of water molecules in the annealing ambient, is followed by gallium oxide formation on the film surface and at the interface.

The diffusion process can be modified by doping the film with group-III and -V impurities. Spin-on films produced with emulsion compositions based on tetraethoxysilane or silanol are a suitable choice [56, 59]. Impurity concentration in these films may reach several weight percent which is easily monitored by IR spectroscopy. Gallium diffusion from the wafer into the film is enhanced by the presence of phosphorus or arsenic in the film. The process slows down for gallium-doped films, and it stops when the concentration in the film reaches 5.3%. Out-diffusion of gallium leaves vacancies in GaAs, which increase electrical activation of group-II and -IV impurities [64].

When choosing encapsulants for InP, one should take into account indium accumulation at the interface, particularly with silicon nitride and phosphosilicate glass films [65, 66]. This effect is, however, not observed with a silicon dioxide cap, although phosphorus fast diffuses in this material.

4.2.5. Thermal Stresses in Films

Thermal expansion of the film and the wafer also needs to be considered. The necessary data are shown in Table 4.4. Differences in expansion coefficient lead to thermal stress which induces generation of dislocations in the wafer and macrodefects in the film. Thermal processing of GaAs covered with 100 nm of silicon dioxide produces stresses of about 10^7 N/m^2 [52]. They increase in proportion with the film thickness. Thermal expansion of GaAs is close to that of aluminum nitride and this encapsulant has been used up to 2 μm thick. The proper thermal expansion coefficients are achieved for spin-on caps by varying the dopants [56, 72]. Thermal stresses induced in the capped wafer affect the electrical activation of implanted impurities. For example, implanted selenium and tin are activated more effectively in GaAs protected by a binary silicon nitride/aluminum nitride film, as compared to the separate use of these encapsulants [26]. Activation of zinc implanted into GaAs is higher under a silicon nitride cap and lower under an aluminum nitride one [51].

The appearance of macrodefects in the film is also the consequence of the film and wafer thermal expansion. Binary thin-film structures were found to be the most effective. Layers composed of silicon nitride and silicon dioxide, or of silicon nitride and phospho-

TABLE 4.4. Thermal Linear Expansion Coefficient of III–V Compound Semiconductors and Encapsulant Materials

Material	Expansion coefficient ($\times 10^{-6}$ °C^{-1})	Refs.
GaAs	6.4	67, 68
$Ga_{0.26}In_{0.74}As_{0.60}P_{0.40}$	5.4	70
$Ga_{0.47}In_{0.53}As$	5.7	70
GaP	5.9	69
GaSb	6.7	69
AlAs	5.2	68
$Al_nGa_{1-n}As$	$6.4 - 1.2\,n$	67, 68
InAs	5.2	55, 70
InP	4.6	70
AlN	6.3	45, 69
Si_3N_4	3.6	71
SiO_2	0.6	71
Phosphosilicate glass	3.4	55
Borosilicate glass	3.5	55

silicate glass allowed long processing times for InP at 800 °C and above without cracking [65].

Finally, for an appropriate choice of the encapsulant for III–V semiconductors, consideration must be given to both the encapsulating properties and thermal expansion of the materials employed.

The differences in film and wafer thermal expansion coefficients impose restrictions on the film thickness. The bigger this difference, the thinner the protective film that must be used. Heating-induced stresses at the film/wafer interface increase with the film thickness. On the other hand, a thin film becomes less effective in suppressing evaporation. Here the main factor is not the film material but its quality defined by the presence of pinholes, pores, and cracks. The ideal encapsulating film should have the following features [7, 73]: (1) low deposition temperature, (2) low oxygen content, (3) prevent erosion of the wafer surface, (4) not allow in-diffusion of impurities from either the film or external sources, (5) not induce stresses at the surface region, and (6) be removed by benign means.

4.3. DEFECT GENERATION AND ANNEALING

Generation and annealing of point defects in III–V semiconductors is an inherent result of RTP. They have been extensively investigated in GaAs and InP because of the practical application of these materials. There are many deep-level defects in as-grown GaAs [74–85], some of which are removed by RTP, while new ones inevitably appear [54, 86–94]. The electronic properties of the traps generated in GaAs are summarized in Table 4.5.

The electron traps M1, M3, and M4 were reported to exist in GaAs grown by molecular beam epitaxy [74, 76–78]. Figure 4.7 shows variation of their concentration with the temperature induced during RTP. M1 and M4 are stable at 700 °C, but they disappear with processing at 900 °C. It was noted [27] that the decreased concentration of M1 and M4 is correlated with increased concentration of N2 and N3. M1 and M4 are assumed to be related to vacancy–impurity complexes involving an arsenic vacancy [77]. Hence, the decrease of their concentration on RTP can be associated with the production of N2 and N3 traps via creation of more stable arsenic divacancy–impurity complexes.

The traps EL3, EL5, and EL6 usually observed in the bulk of as-grown GaAs crystals disappear after RTP at 700–900 °C [88]. They are dominantly associated with native point defects and vacancy–impurity complexes. For example, EL5 is assumed to be a complex involving a gallium vacancy, such as V_{Ga}–As_{Ga} and V_{Ga}–Si_{Ga}. RTP at 800–900 °C produces M2, N1, and N3 traps in epitaxial layers and ED1, ED2, and ED3 traps in GaAs crystals [89, 91]. In spite of the identity of activation energy for M2 and N1, they belong to different defects. These are either defects with several stable states or amorphouslike defects. N3 and ED2 are evidently associated with one defect, which is present both in epitaxial layers and in the bulk. ED1 is related to interstitial arsenic complexes and ED3 to its vacancies [88].

Analyzing the trap family EG6, EG7, and EG8 of GaAs epitaxially grown on silicon generated by 20 s of processing at 800 °C [92], one could find identical traps, i.e., N2, ED1, and EN1, respectively, reported in earlier investigations (see Table 4.5).

TABLE 4.5. Electron Traps Generated in n-GaAs by Rapid Thermal Processing within Seconds

Trap name	Level (eV)	Capture cross section (cm^2)	Refs.
M1	$E_c - 0.18$	3.2×10^{-15}	87, 91
M2	$E_c - 0.64$	—	89
M3	$E_c - 0.33$	7.6×10^{-14}	87, 91
M4	$E_c - 0.51$	5.8×10^{-13}	87, 91
N1	$E_c - (0.50–0.72)$	$(2–100) \times 10^{-15}$	87, 91
N2	$E_c - 0.36$	3.0×10^{-15}	91
N3	$E_c - 0.49$	1.1×10^{-14}	91
ED1	$E_c - 0.26$	2.4×10^{-14}	88
ED2	$E_c - 0.49$	1.7×10^{-15}	88
ED3	$E_c - 0.55$	5.1×10^{-13}	88
EN1	$E_c - 0.20$	5.4×10^{-16}	86
EL2	$E_c - 0.82$	1.3×10^{-13}	87
	$E_c - 0.81$	1.5×10^{-13}	88, 92
	$E_c - 0.82$	1.3×10^{-14}	91
EL3	$E_c - 0.61$	9.5×10^{-13}	88
EL5	$E_c - 0.40$	4.1×10^{-14}	88
EL6	$E_c - 0.35$	1.3×10^{-13}	88
EG6	$E_c - 0.36$	1.7×10^{-13}	92
EG7	$E_c - 0.27$	6.8×10^{-15}	92
EG8	$E_c - 0.20$	1.8×10^{-15}	92

There is a correlation between the defects forming the N1, EN1, and EL2 traps. Concentration of these traps reaches a maximum of $10^{16}–10^{17}/cm^3$ after processing at 800–850 °C for seconds. They are annealed at 950 °C. One of the peculiarities of these traps is their nonuniform distribution through the wafer surface. Trap concentration at the wafer edge is two to three orders of magnitude higher than that in the central part of the wafer [91].

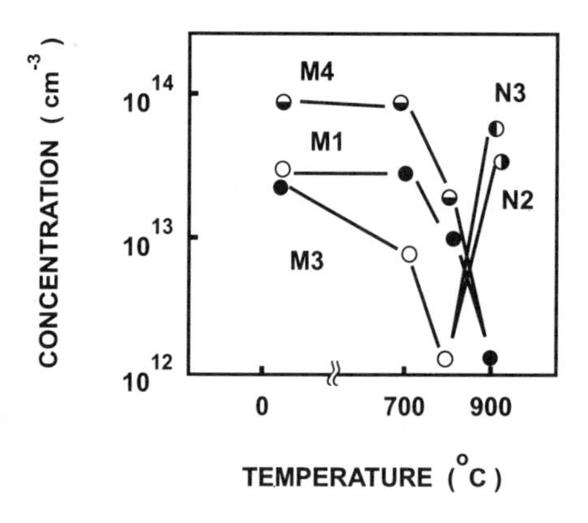

FIGURE 4.7. Concentration of electron traps in epitaxial n-GaAs versus temperature of processing within 6 s [91].

The EL2 trap is known [79, 80, 82] to originate in stressed GaAs crystals. It is generally believed to involve As_{Ga} [83] while discussion remains as to whether or not some other constituents are present, e.g., interstitial arsenic resulting in As_{Ga}–As_i complex. Formation of EL2 is directly associated with the native point defects and plastic deformation. RTP by radiation induces maximum elastic stresses and plastic deformation at the wafer edge, hence resulting in a maximum concentration of EL2 there. N1 is also connected with dislocations appearing in the plastically deformed regions [81].

There are several models describing the electrical active defect formation under the influence of elastic stresses. They are mostly related to the EL2 trap. Interstitial arsenic generated by stresses is postulated to create complexes with arsenic substituting gallium in its sublattice $(As_i)n$–As_{Ga} [96, 97]. Stresses also cause gallium vacancies, which interact with arsenic to form arsenic antisite defect V_{As}–As_{Ga} [98] and divacancy–As_{Ga} complex [99]. Arsenic precipitation stimulated by stresses may also take place at dislocations [100]. The above presentation shows that the concentration of electrically active defects introduced by RTP may be decreased by minimization of radial temperature gradients. Therefore, the sample holders discussed in Section 4.2 are frequently employed. In GaAs layers doped by implantation, additional electron traps associated with defect–impurity states appear [88, 93, 94, 95].

4.4. RECRYSTALLIZATION OF IMPLANTED LAYERS

The III–V semiconductor compounds are damaged more than silicon by ion implantation. Amorphization of the GaAs and InP crystals occurs during implantation at room temperature at an energy density of about 2.5×10^{20} keV/cm^2 [101, 102]. This is an order of magnitude lower than for silicon. Moreover, stoichiometric changes produced by forward recoil implantation and sputtering in the implanted layer play an important role in the damage formation [103]. Surface layers of the implanted semiconductor compound are enriched by the atoms of heavier components, while light atoms are moved beneath the peak of the implanted impurity concentration profile. The imbalance becomes more pronounced with an increase in the difference in atomic masses of the compound components, and for implantation of heavy ions.

The following stages in rapid solid-state annealing of implantation amorphized GaAs can be discerned [73, 104, 105]:

1. At 100–350 °C epitaxial regrowth of the amorphous layer takes place. The regrown region contains a lot of defects, which are microtwins, polycrystallites, dislocation loops, and stacking faults. The set of the residual defects is defined by the implantation regime.
2. At 350–600 °C microtwins and stacking faults are removed, while dislocation loops still exist in the doped layer. Implanted acceptor impurities start to become electrically active.
3. At 600–950 °C dislocation loops are partly removed. There is an increase in impurity activation.
4. At 950–1000 °C implanted donor impurities are being activated.

The recrystallization rate of GaAs decreases with increasing implant dose [106]. It is lowest for crystal direction ⟨111⟩ and highest for ⟨100⟩. This is a result of arsenic cluster formation at (111) grain boundaries retarding recrystallization. Layers regrown at 150–250 °C improve faster when implanted at liquid-nitrogen versus room temperature [107], because defect complexes formed at room temperature hinder the epitaxial regrowth. Regrowth of amorphized GaAs and InP takes place at high rates, illustrated by the data in Table 4.6. Recrystallization is completed within the first 3–5 s of heating to temperatures of 200–400 °C. However, secondary defect annealing demands higher processing temperatures.

The degree of perfection in the regrown layer depends on both the induced temperature and the heating rate. Figure 4.8 shows for GaAs implanted by zinc an amorphous layer recrystallized at 600 °C irrespective of the heating rate. The crystalline perfection is far from that of the virgin crystal because of secondary defects, for the most part, microtwins and dislocation loops. There annealing requires a temperature of 850 °C. Processing at the higher heating rates is more effective. Restoration of GaAs lattice implanted with heavy ions (selenium, tellurium) requires an increase of processing temperature to 1000 °C at a processing time of 3–5 s [49, 53, 113, 114].

Generally, the temperature range of 800–1000 °C may be recommended for recrystallization and secondary defect annealing in implanted GaAs. There is a greater degree of perfection in rapid thermal annealing (RTA) layers, compared to the case when conventional furnace annealing is used. It is characterized by χ_{min} = 4–5% for light impurity atoms and 5–7% for heavy ones.

RTP in the temperature range of 800–1000 °C is also available to anneal implanted InP [115–117]. The annealing temperature may be reduced to 700 °C [46], but this ensures an increase of the processing time up to several dozens of seconds. However, complete restoration of the layers, implanted at doses of 1×10^{14} ions/cm^2 and greater, cannot be achieved. Experimental data illustrating this finding are shown in Fig. 4.9. The best crystalline perfection of the layer implanted by 200-keV selenium ions at a dose of 1×10^{14} ions/cm^2 was characterized by χ_{min} = 6.5% on annealing at 700 °C for 80 s. The

TABLE 4.6. Solid Phase Recrystallization Parameters for (100) Ion-Implanted Amorphized Gallium Arsenide and Indium Phosphide

Compound	Implant	Temperature range (°C)	Preexponential factor, v_o (μm/s)	Activation energy, E_a (eV)	Recrystallization rate (400 °C)[a] (μm/s)	Refs.
GaAs	Be	180–250	2.9×10^{10} [a]	1.45	0.38	108
GaAs	Te	250–400	6.0×10^{11}	1.6	0.59	109
GaAs	As	230–330	2.5×10^{11}	1.6	0.25	111
InP	As	220–340	1.5×10^{11} [a]	1.55	0.35	110, 111
InAs	As	130–200	4.0×10^{12}	1.3	700	111
$In_{0.55}Ga_{0.45}As$	As	230–300	7.0×10^{10}	1.4	2.2	111
GaP	P	370–500	1.0×10^{9}	1.7	1.7×10^{-4}	111
GaSb	Si	280–370	3.0×10^{6}	1.25	1.3×10^{-3}	111

[a]Calculated from the experimental data presented in the reference.

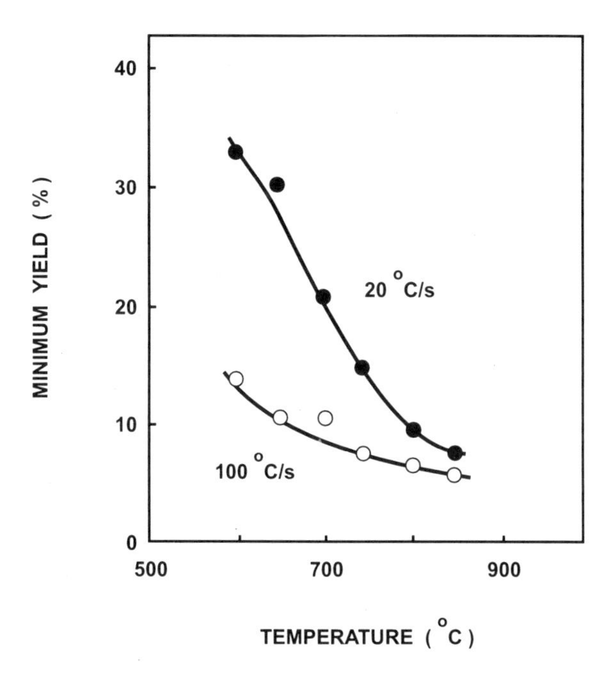

FIGURE 4.8. Minimum yield of channeled 380-keV He^+ ions for gallium arsenide implanted by zinc (150 keV/5 × 10^{14} ions/cm^2) as a function of the annealing temperature achieved with different heating rates: ●, 20 °C/s; ○, 100 ° C/s [112].

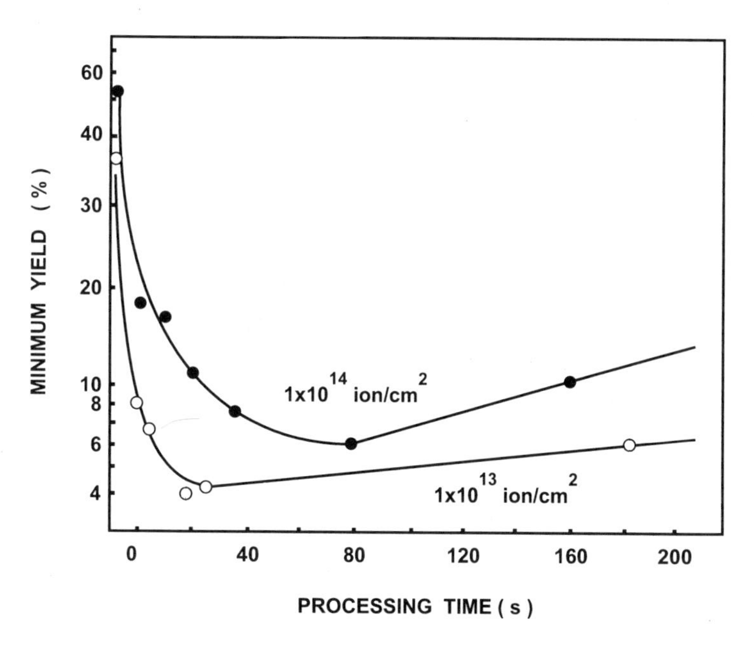

FIGURE 4.9. Minimum yield of channeled 1.5-MeV He^+ ions for indium phosphide implanted by 200-keV selenium ions as a function of processing time at 700 °C. Origin of the time axis corresponds to the as-implanted sample at (●) 1 × 10^{14} ions/cm^2 and (○) 1 × 10^{13} ions/cm^2. The zero point presents the result just after silicon nitride cap deposition [46].

layers implanted with doses of 5×10^{14} ions/cm^2 and greater can only be restored to a polycrystalline state at this temperature.

InP heavily implanted by silicon (180–200 keV at $(2–3) \times 10^{14}$ ions/cm^2) was found [115, 117, 118] to be recrystallized at 800 °C for 10 s, at 850 °C for 5 s, and at 950 °C for 1 s. The best crystalline perfection was achieved in the samples implanted at elevated temperature (175–200 °C). The same effect was observed for GaAs implanted by zinc, silicon, sulfur, and selenium at 170–300 °C [119, 120]. It is important to note that implantation at elevated temperatures does not amorphize III–V semiconductor crystals [121, 122]. Amorphous layers are usually formed after implantation at liquid nitrogen or room temperature, and also depend on the implant mass, energy, dose, and dose rate. Contrary to expectation from the regrowth of implanted silicon, the most complete restoration of the semiconductor compound lattice is achieved during annealing of disorders left by low-dose or hot implantation. That is not surprising because of the determining role of stoichiometry, which is more pronounced for amorphized semiconductor compounds.

4.5. IMPLANTED IMPURITY BEHAVIOR

The electrical activity and diffusion behavior of impurity atoms are determined by the position they occupy in the III–V semiconductor lattice [123, 124]. Group-II dopants (beryllium, magnesium, zinc, cadmium, mercury) substituting the host group-III atoms demonstrate acceptor properties. Group-VI impurities (sulfur, selenium, tellurium) are donors in the substitutional positions of the group-V component.

Impurities from group IV (silicon, germanium, tin, carbon) are amphoteric. They may substitute atoms of both components in semiconductor compounds, while exhibiting donor properties in the group-III component sublattice and acceptor properties in the sublattice of the group-V component. Specific point defects and related impurity behavior in both sublattices require separate analysis of electrical activation and diffusion of implanted acceptor, donor, and amphoteric impurities.

4.5.1. Acceptor Impurities

The general features of implanted acceptor impurity behavior during conventional furnace annealing are electrical activation at comparatively low temperatures and considerable diffusion redistribution.

4.5.1.1. Beryllium. Because of its light mass and consequent high electrical activation, beryllium is the most attractive acceptor dopant. Activation of beryllium implanted in GaAs may be achieved at relatively low temperature [125]. For doses of $10^{13}–10^{14}$ ions/cm^2 and an implantation energy of 200 keV, about 60% of the implanted impurity is activated at 550 °C for 20 s. The activation is worse for shorter durations. The values of carrier mobility obtained are 175 and 280 cm^2/V·s for the higher and lower doses, respectively. They were found to be independent of the processing time. The same results may be obtained with conventional furnace annealing at 850 °C for several dozens of minutes.

A higher degree of activation of beryllium implanted in GaAs is achieved by RTA at 800–1000 °C [40, 113, 126–131]. It is associated with a more complete removal of secondary defects. The extended temperature range may also be used for the simultaneous activation of implanted acceptor and donor impurities, when structures with p–n junctions are built.

Figure 4.10 illustrates the typical variation of sheet carrier concentration in GaAs implanted with different impurities as a function of temperature for 3 s annealing. In the case of beryllium, 90–100% electrical activation for implanted doses of 10^{13}–10^{14} ions/cm^2 is achieved after annealing at 750 °C. High implant doses, 10^{15} ions/cm^2 and above, require a temperature increase up to 800 °C; however, the degree of activation does not exceed 70%. Hole mobility is within 100–200 cm^2/V·s. Maximum hole concentration amounts to about 3×10^{19} p/cm^3.

The same features are observed for beryllium implanted in InP [132, 133], indium antimonide [134], and ternary compounds $Ga_{0.47}In_{0.53}As$ [135–137], $In_{0.52}Al_{0.48}As$ [138],

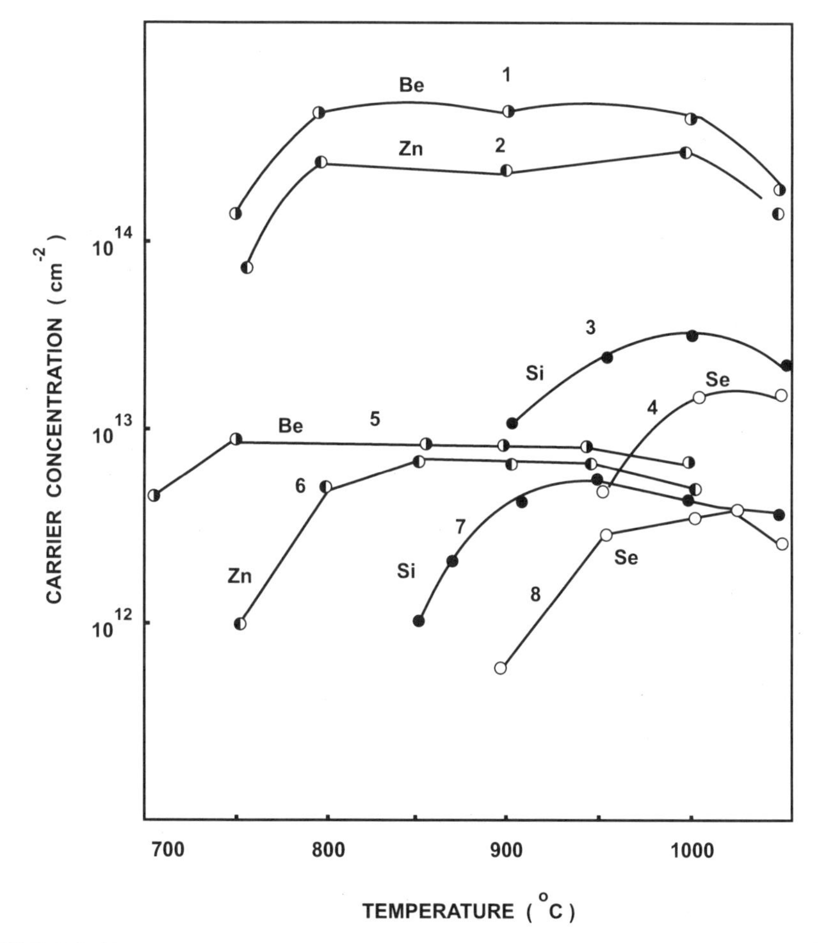

FIGURE 4.10. Sheet carrier concentration obtained from Van der Pauw measurements on gallium arsenide implanted by beryllium (1, 5), zinc (2, 6), silicon (3, 7), and selenium (4, 8) with energies of 40, 100, 60, and 100 keV, respectively, at doses of 1×10^{15} (1–4) and 1×10^{13} ions/cm^2 (5–8) on rapid annealing at various temperatures for 3 s [113].

and $Al_nGa_{1-n}As$ [139], where poor activation has been found for implanted doses higher than 10^{14} ions/cm^2. It is the result of beryllium out-diffusion and evaporation, which has a stronger effect in shallow layers created by low-energy ion implantation. The contribution of this effect is significantly reduced with RTP of implanted layers up to 500 nm thick. Besides, BeO_n formation occurs when out-diffused beryllium reacts with native gallium or arsenic oxides [131]. There is no evidence for pairing of beryllium and oxygen in AlGaAs [139] in contrast to the situation in GaAs. The unique feature of RTP of beryllium-implanted InSb is the reduced processing temperature of 400–450 °C, while durations of 20 s are employed [134]. Compared to conventional furnace annealing, a higher hole concentration has been achieved with insignificant impurity redistribution.

4.5.1.2. Magnesium. Electrical activation of magnesium implanted in GaAs is achieved by RTP at 700–950 °C [41, 64, 130, 140–144]. The optimum annealing temperature is about 800 °C. There is impurity out-diffusion into the nitride encapsulating film above this temperature. As a result, one can observe a gradual decline in the electrical activity with increasing temperature. Nevertheless, annealing in an arsine ambient allows processing temperature to be increased to 930 °C (plateau time of 6 s) at which the minimum sheet resistivity is produced [41].

About 100% of the impurity atoms become electrically active at an implantation dose of 5×10^{13} ions/cm^2 and lower. The degree of activation drops with increasing implant dose, constituting about 25% for 1×10^{15} ions/cm^2. Sheet carrier concentration and hole mobility (100–180 cm^2/V·s) show a slight dependence on annealing temperature in this range, although changes of carrier concentration profiles are rather important. It should be noted that these changes are mostly seen for the heavily doped layers created by implantation with doses of 5×10^{14} ions/cm^2 and higher.

Hole concentration–depth profiles in GaAs implanted with magnesium and subsequently annealed at 700–800 °C are shown in Fig. 4.11. Attention is drawn to the minimum in the experimental profiles at a depth corresponding to the projected range of the ions. It is smoothed out with increase of the annealing temperature at the expense of impurity redistribution. The lower magnesium activation in this region was related to the arsenic deficit, which arose as a result of implantation [140]. It was proposed to implant arsenic together with magnesium to compensate for arsenic loss [141, 143], choosing implantation energies in such a way that depth profiles of both implants would be as close as possible to each other. Moreover, RTP was carried out with a local arsenic overpressure and a silicon nitride cap.

Arsenic co-implantation was found to cause a slight decrease in hole mobility because of defects. The percentage activation of magnesium implanted with a dose of 1×10^{15} ions/cm^2 increased to 40%, from 27% without arsenic implantation. Co-implantation suppresses impurity diffusion to the surface and into the bulk of the semiconductor crystal, and carrier profiles become very close to as-implanted distributions. Maximum hole concentration was increased up to $4 \times 10^{19} p/\text{cm}^3$. The result is not dependent on the sequence of arsenic and magnesium implantation.

The effect of stoichiometric imbalance created by magnesium implantation-induced phosphorus knock-on in InP was investigated [144]. The upper limit for substitutional incorporation of the impurity was found to be of the order of 1×10^{19} atoms/cm^3 on annealing at 800–850 °C for 5 s. Although co-implantation of phosphorus and magnesium

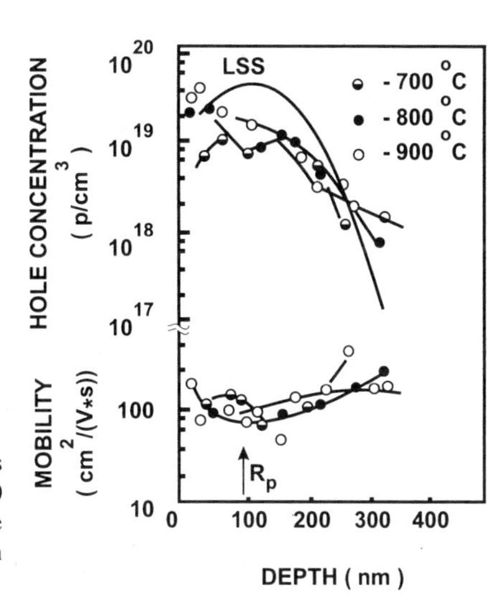

FIGURE 4.11. Hole concentration and mobility versus depth in gallium arsenide implanted by magnesium (100 keV/1 × 10^{15} ions/cm^2) after annealing for 10 s with the silicon nitride cap. Solid line: calculated values based on LSS theory [140].

has a positive influence on the impurity activation and redistribution, it is not as pronounced as in the case of GaAs.

4.5.1.3. Zinc. Behavior of zinc implanted in GaAs has been studied most extensively. Conventional furnace annealing results in low activation of the implant and in considerable diffusion broadening of its depth distribution [145]. It occurs either on processing with an encapsulating film or without it in an arsine ambient. Without encapsulation, more than 5% of the implanted zinc is evaporated. RTP-induced electrical activation of zinc implanted into GaAs takes place in the temperature range of 700–900 °C [30, 112, 113, 130, 146]. At low doses, up to 5 × 10^{13} ions/cm^2, practically all of the implanted impurity occupies electrically active substitutional positions after several seconds of processing at these temperatures and no impurity redistribution is detected. In heavily doped layers, greater than 1 × 10^{14} ions/cm^2, the degree of activation changes nonmonotonically with increasing annealing temperature, as shown in Fig. 4.12. Heating rate is also important. At 100 °C/s 95% activation is achieved at 800 °C, and at 20 °C/s no more than 85% activation after annealing at 850 °C is produced. The concentration of holes does not exceed (7–8) × 10^{19} p/cm^3, with a depth distribution profile characterized by slight impurity diffusion into the surface. The decrease of electrical activation with increasing processing temperature (Fig. 4.12), as well as with time, at 800 °C, leads us to conclude that there is no equilibrium in the heavily doped layers.

At higher annealing temperatures, 950–1100 °C, there is a higher degree of activation reaching 60–90%, suggesting zinc redistribution [51, 114, 128, 147]. In order to avoid surface decomposition at these temperatures, GaAs wafers have to be encapsulated. The choice of protective material has an effect on implanted zinc activation [51, 147]. Better results are obtained for silicon dioxide or silicon nitride caps, compared to aluminum nitride ones. This can be explained by vacancy formation in the gallium sublattice at the expense of its diffusion into the protective film (see Section 4.2). In fact, silicon dioxide favors gallium loss. Zinc redistribution in high-temperature RTP is mostly exhibited for

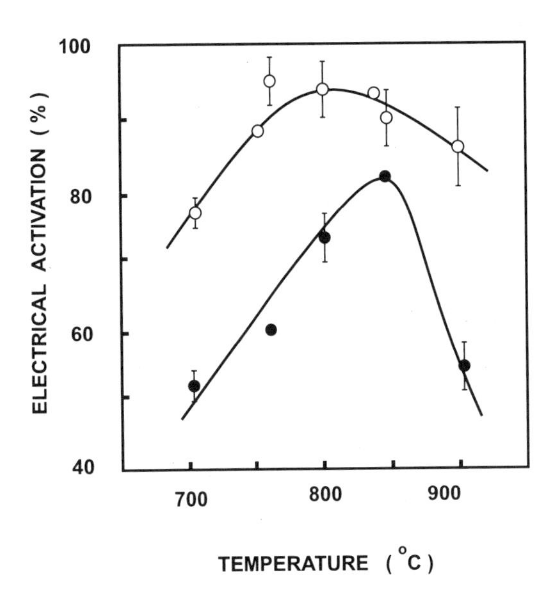

FIGURE 4.12. Electrical activation of zinc implanted into gallium arsenide versus induced temperature for different heating rates: ○, 100 °C/s; ●, 20 °C/s [112].

heavily doped layers, ~10^{20} atoms/cm^3. An increase in the depth of the doped layer of 100–200 nm is produced after processing at 1000–1150 °C for 3–10 s. Arsenic coimplantation reduces zinc redistribution [51, 147] and increases electrical activation of the impurity. The hole depth distribution also appears sharper, close to the as-implanted impurity distribution. Maximum carrier concentration reaches 9×10^{19} p/cm^3.

4.5.1.4. Cadmium. Cadmium is an acceptor impurity in III–V semiconductors but being a heavier atom is less attractive than beryllium, magnesium, and zinc. Implanted cadmium creates significant damage in the lattice. RTP experiments with GaAs [30, 148] and $Al_nGa_{1-n}As$ [149] implanted with cadmium showed poor electrical activation and low hole mobility in the doped layers.

4.5.1.5. Mercury. The behavior of implanted mercury in GaAs and InP subjected to RTP for seconds established that acceptor properties could only be realized when the samples were implanted at elevated temperatures (200 °C) [150–153]. No more than 20–30% of the implanted mercury becomes electrically active. Multilayer capping with silicon nitride and aluminum nitride films was used for annealing of InP at 800–850 °C for 60 s and GaAs at 900 °C for 3 s. Enhanced impurity diffusion was observed. For the samples implanted at room temperature the doped layers were semi-insulating, independent of the annealing regime.

4.5.2. Donor Impurities

Donor group-VI impurities in III–V semiconductors have a lower diffusion mobility than acceptors. Their electrical activation in implanted layers is achieved at higher temperatures, i.e., at 900 °C and above, which requires the use of capping layers. Impurity behavior is therefore controlled not only by the temperature–time regime, but also by the processes in the encapsulating film.

4.5.2.1. Sulfur. Sulfur displays the highest diffusion mobility among donor impurities in GaAs. Its redistribution in implanted layers annealed in a conventional furnace at 800–900 °C is considerable [64, 154, 155]. RTP of seconds allows substantial reduction of the redistribution, even at higher temperatures, which are necessary for the highest electrical activation of the impurity [154–157]. Figure 4.13 shows experimentally determined electron concentrations in sulfur-implanted GaAs after rapid and conventional furnace annealing. The much higher efficiency of rapid annealing is evident. It ensures that 50–70% of the implanted impurities become electrically active. A silicon nitride cap produced better activation than did a silicon dioxide one [155]. There is no sulfur redistribution in lightly doped layers at a concentration below 2×10^{18} atoms/cm^3, even after annealing at 1000 °C [156], but it does occur at higher concentrations, as demonstrated by the profiles in Fig. 4.14. Peak electron concentrations of 6×10^{18}/cm^3 were obtained, three times higher than after conventional furnace annealing. More impurity redistribution was observed for the furnace anneal, in spite of the fact that RTP was performed at a temperature 150 °C higher. Sulfur-enhanced diffusion in implanted layers is a function of the number of implant defects and increased by two orders of magnitude in GaAs [158]. The effect was more pronounced as the surface acts as a powerful sink for point defects. Short-time high-temperature regimes rather than long-time lower-temperature annealing have been stressed to prevent carrier mobility degradation [158].

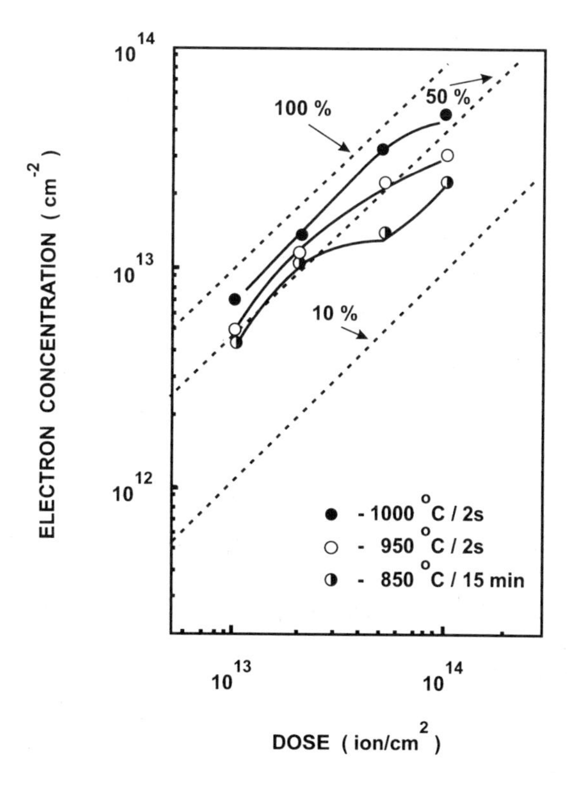

FIGURE 4.13. Sheet concentration of electrons in gallium arsenide implanted by sulfur with an energy of 150 keV and subsequently annealed with silicon nitride cap. Dashed lines and values represent the percentage of electrical activation [156].

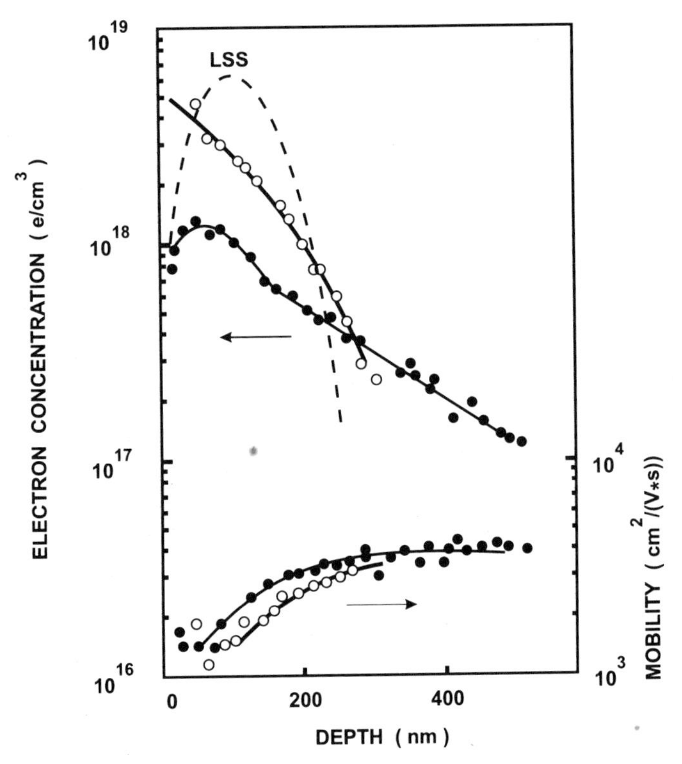

FIGURE 4.14. Electron concentration and mobility in gallium arsenide implanted by sulfur (150 keV/1 × 10^{14} ions/cm^2) on rapid thermal processing at 1000 °C for 2 s (○) and conventional furnace annealing at 850 °C for 15 min (●) Dashed line, calculated values based on LSS theory [156].

4.5.2.2. Selenium. Having essential mass, selenium rapidly amorphizes semiconductor compounds when implanted. Regrowth of the implanted layers and electrical activation of the impurity require relatively high temperatures, 1000–1150 °C for GaAs [49, 50, 53, 114, 159, 160] and 700–950 °C for InP [46, 121, 132]. The electron concentration in GaAs implanted with selenium depends on the temperature of RTP, as shown in Fig. 4.10. For the dose of 1 × 10^{13} ions/cm^2, good activation (~ 40%) is achieved after annealing at 950 °C. Higher doses, such as 10^{14}–10^{15} ions/cm^2, require increased annealing temperatures of 1050–1150 °C. Electrical activation of selenium depends considerably on the depth, which is determined by the implantation energy [49]. Experimental data illustrating this dependence are presented in Fig. 4.15. The best activation in the completely amorphized layer takes place at the interface with a single-crystal substrate. An increase in implantation energy is accompanied by expansion of the damaged region in the crystal, and a decrease in the peak concentration of defects, although an amorphous layer is not formed. Higher-efficiency annealing of implantation-induced disorders in III–V semiconductors is clearly demonstrated by the electron profiles in Fig. 4.15.

Implantation at elevated temperature can also prevent amorphization. Hot implantation was found to increase selenium activation after RTP in GaAs [53], InP [132], and

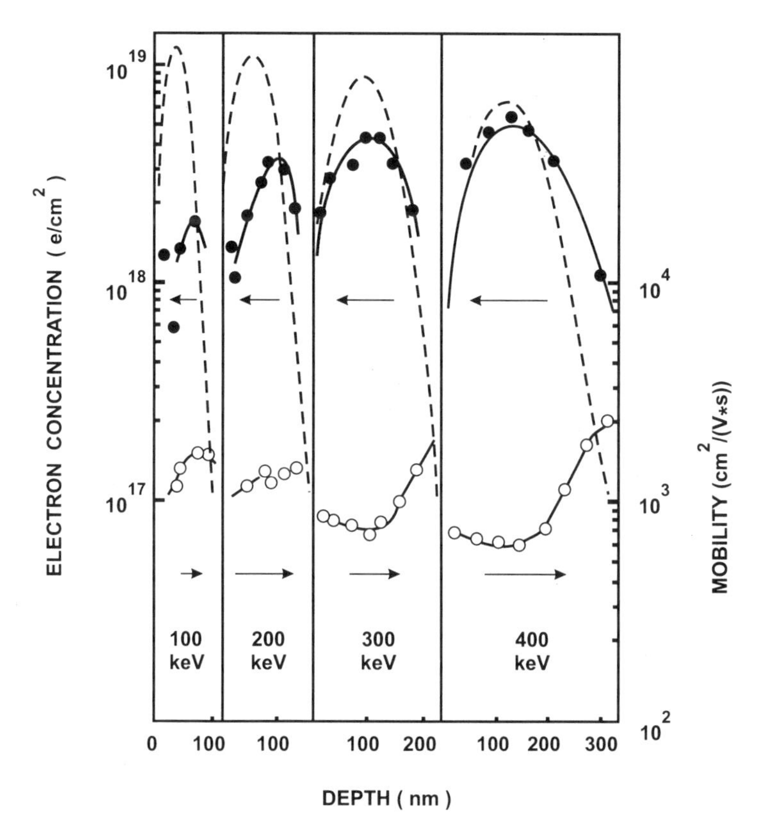

FIGURE 4.15. Electron concentration (●) and mobility (○) profiles in gallium arsenide implanted by selenium (1×10^{14} ions/cm^2) and annealed at 1000 °C for 2 s. Dashed line, calculated values based on LSS theory [49].

ternary semiconductor compound InGaAs [161]. However, even in these cases the electrically active fraction of the impurity in heavily doped layers does not exceed 50%, and maximum electron concentration is limited to 1×10^{19}/cm^3. Finally, selenium diffusion and redistribution in implanted layers is insignificant.

4.5.2.3. Tellurium. This element is the heaviest among donor impurities, and thus is not attractive for implantation in semiconductor compounds. The damage produced is greater and therefore harder to anneal. RTP of tellurium-implanted GaAs was carried out at 800–1000 °C for 5–60 s [162, 163]. Electrical activation of 10% was observed.

4.5.3. Amphoteric Impurities

Group-IV impurities display amphoteric properties in III–V semiconductor compounds. They may substitute host atoms in both sublattices, revealing a preference for the larger ones [123]. Localization of the impurity atom in one or another sublattice is determined by the processing temperature and by the vapor pressure of the volatile component at the surface. High temperature as well as overpressure of group-V components reduce vacancy concentration in the sublattice [164]. Under these conditions, amphoteric impurities preferentially occupy the substitutional positions of the group-III component, where they act as donors. Replacing the group-V component, initiated by this component evaporation, allows transition to the acceptor state. The donor states of silicon, germanium, and tin and the acceptor state of carbon are rather stable.

4.5.3.1. Silicon. Silicon is the lightest impurity that has been used to create electron-conducting layers in III–V semiconductors. The low level of radiation damage associated with silicon implantation promotes defect annealing and high impurity electrical activation with subsequent thermal processing. The temperature range employed for RTP of GaAs implanted with silicon is rather wide, from 800 to 1100 °C. The particular choice is determined primarily by the implanted dose, as presented in Fig. 4.16. In lightly implanted layers, doses less than 5×10^{13} ions/cm^2, a maximum activation of about 70% is achieved with processing of several seconds at 900–950 °C. Defect annealing and activation of silicon implanted with higher doses requires higher temperatures, up to 1050–1100 °C. Amphoteric properties of silicon in GaAs become appreciable after either conventional furnace or RTP at 950 °C. This causes a decrease in electron concentration of the doped layers (see Fig. 4.10) with a rise in annealing temperature [19, 25, 42, 114]. In InP such behavior begins to be observed after annealing at 850 °C for 10 s [117].

Luminescence analysis [165] and Raman spectroscopy [166] indicate that the decreased electron concentration in silicon-implanted GaAs layers is associated with

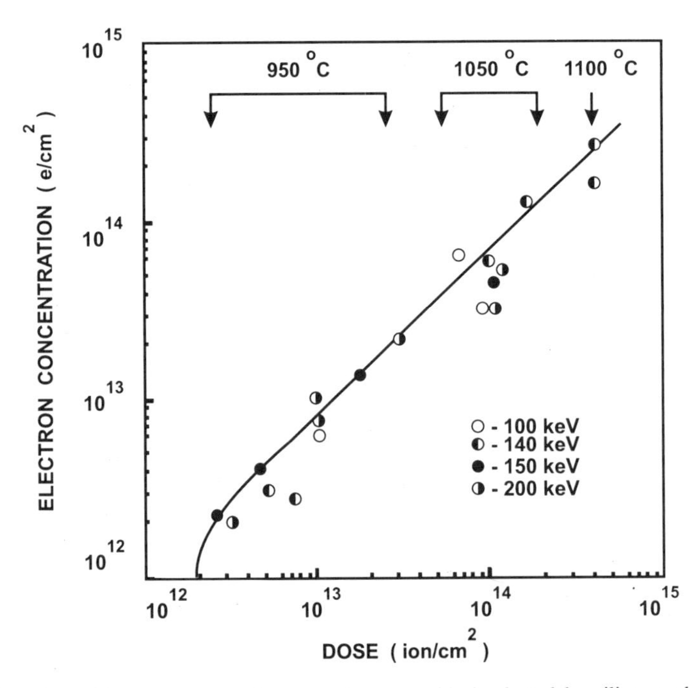

FIGURE 4.16. Sheet electron concentration in gallium arsenide implanted by silicon and subsequently annealed within 0–5 s at 950–1100 °C. Summary of published data from Gill and Sealy [5].

transition of the impurity atoms from donor to acceptor states. Next-neighbor Si_{Ga}–Si_{As} electrically inactive pairs are thought to be formed. Moreover, silicon containing so-called Si–X and Si–Y complexes may appear in heavily doped GaAs. Analysis by Wagner *et al.* [166] showed that use of any silicon dioxide cap during RTP led to formation of the Si–X defect. This defect is known to act as an acceptor [167, 168], which together with the Si_{As} acceptor limits electron concentration in the doped layers. The Si–X defect is thought to include Si_{Ga}–V_{Ga} and Si_{As}–V_{Ga} complexes [167,168], while Ono and Newman [169] identified the Si–Y defect as a Si_{Ga}–V_{Ga} complex. That makes the Si_{As}–V_{Ga} complex a likely candidate for the Si–X defect. It is evident that further investigations are necessary in order to understand the details of silicon electrical activation in GaAs. Considering the amphoteric behavior of silicon, the optimal temperatures for annealing implanted layers are 900–1000 °C for GaAs and 800–900 °C for InP. Vacancy formation in the sublattice of the group-V component is a factor limiting the maximum temperatures. Overpressure of this component at the wafer surface suppresses the process, hence permitting an increase of the annealing temperature, which increases activation by 10–20%. The same effect may be produced by coimplantation of group-V ions [17, 170].

Activation of implanted silicon as donors is influenced to a great extent by the encapsulant material. A silicon dioxide cap on GaAs has already been mentioned for extracting gallium from the wafer, more effectively than a silicon nitride one [64, 155]. A Ga-doped spin-on silica glass cap offered higher silicon activation than conventional SiO_2 films [59].

For InP rather high silicon activation (up to 98%) is obtained with silicon dioxide, silicon nitride, and silicate glass caps [107, 152, 171–173]. Electron depth profiles for silicon implanted and transiently annealed in GaAs and InP indicate insignificant impurity redistribution for processing within seconds. Typical electron concentration distributions in GaAs are shown in Fig. 4.17. In the near-surface region, from the surface to a depth corresponding to the projected range (closer to the surface at higher implant doses), the silicon does not display donor properties [25, 128]. Here the electron conductivity must be compensated by holes. Actually, arsenic evaporation enriches the region beneath the surface with arsenic vacancies and silicon becomes an acceptor occupying these vacancies. Another origin of the near-surface compensated layer may be associated with the accumulation of chromium atoms there, when semi-insulating chromium-doped GaAs wafers are used [174]. The solid solubility limit of chromium in GaAs is 1.2×10^{17} atoms/cm^3 [175]. In fact, its concentration at the surface of rapid thermal annealed GaAs is usually an order of magnitude higher [174] and electron conductivity is detected when the donor concentration exceeds this level. Chromium redistribution also explains enhanced electrical activation and diffusion of silicon implanted in semi-insulating GaAs processed at 850–950 °C for 15 s [176]. The effect is clearly observed at a depth of 50–300 nm, where according to the data of Kanber and Whelan [174] a chromium-depleted layer appears as a result of diffusion into the surface. A maximum electron concentration of 7–8 $\times 10^{18}$/cm^3 is achieved in silicon-implanted GaAs and InP with RTP within seconds, 1.5–2 times higher than after conventional furnace processing [105, 116].

RTP-induced diffusion behavior of silicon-implanted into GaAs has been shown [59, 122, 177] to be very sensitive to the defects in the near-surface region. Excess gallium vacancies are created there as a result of sputtering at the surface during implantation,

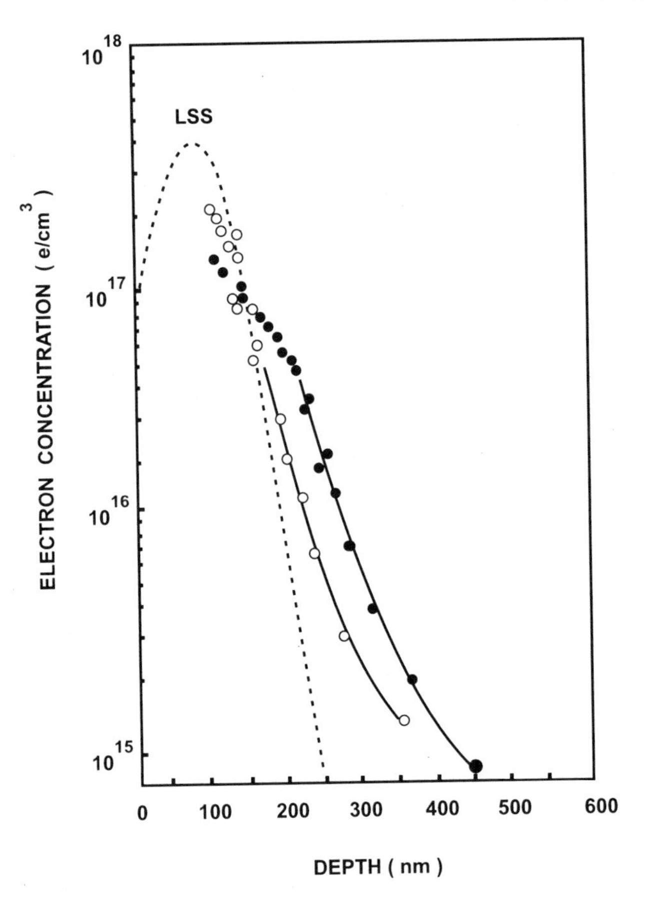

FIGURE 4.17. Electron concentration profiles in gallium arsenide implanted by silicon (100 keV/5 × 10^{12} ions/cm^2) on annealing at 950 °C for 2 s without a cap (●) and at 850 °C for 15 min with the silicon nitride cap (○) [25]. The dashed line is calculated by LSS theory.

particularly with low-energy ions. Excess gallium interstitials penetrate deeper into the crystal, as shown by theoretical simulation [177, 178] and confirmed in experiments with the gallium-doped cap [59], surface etching of the implanted layer prior to annealing [177], and implantation at elevated (about 40 °C) temperatures [122]. Silicon-enhanced diffusion was apparently demonstrated to occur only in cases where the implantation-produced gallium vacancies saturate the near-surface region.

4.5.3.2. Germanium. Experimental results have shown that germanium implantation into InP combined with RTP produces an n-type layer [121, 179]. The electrical characteristics of these layers are comparable to those obtained with silicon implantation. The highest concentration of electrically active germanium was 37% at a maximum electron concentration of 1 × 10^{19}/cm^3 obtained for implantation at 200 °C and subsequent annealing at 830 °C for 1 s.

4.5.3.3. Tin. RTP of tin-implanted GaAs was investigated to obtain heavily doped n-layers [50, 160]. The maximum electron concentration of (6–8) × 10^{19}/cm^3 was reached after annealing at 1050–1090 °C for 5 s. A double film cap consisting of aluminum nitride and silicon nitride was used. Donor impurity activation was 30% and there was significant impurity redistribution in the implanted layers. RTP of tin-implanted InP has shown [180] more promising results, in particular with phosphorus coimplantation. Processing at 800°C for 10 s resulted in about 70% donor electrical activation in the coimplanted samples for tin doses up to 10^{14} ions/cm^2. Maximum electron concentration of 1.5×10^{15}/cm^3 and mobilities in the range of 700–1000 cm^2/V·s were achieved.

4.5.3.4. Carbon. In contrast to amphoteric group-IV impurities, implanted carbon behaves predominantly as an acceptor in GaAs and related ternary semiconductor compounds subjected to RTP [181–185]. Relatively poor activation in GaAs, less than 3% [181], and no measurable effect in InGaAs and AlInAs [182] were observed in early experiments. There is a possibility of improved activation efficiency by hot implantation of 200 °C or by using gallium coimplantation in order to force carbon into the arsenic sublattice [183, 184]. In this case the activation yield becomes as high as 65–100% in low-dose (about 10^{14} ions/cm^2)-implanted GaAs annealed at 950–1000 °C for 5–10 s and 20% in both InGaAs and AlInAs annealed at 700 and 900 °C, respectively, for 10 s. Peak hole concentration of about 5×10^{19} p/cm^3 has been achieved. Excess carbon precipitates in clusters 3–5 nm wide [185].

4.5.4. General Features

The following trends in the behavior of implanted impurities in III–V semiconductors subjected to RTP can be noted. Acceptor group-II impurities are electrically activated in the temperature range of 700–900 °C. Their activation essentially depends on the damage in the as-implanted layer, being reduced with increasing implant dose. Increasing the annealing temperature to 1000–1100 °C slightly improves activation in heavily doped layers (10^{19}–10^{20} atoms/cm^3), but is accompanied by considerable impurity redistribution. Diffusion of the acceptors in GaAs is enhanced by arsenic deficiency in the implanted layers. It can be suppressed by arsenic coimplantation, which restores the stoichiometry and produces an increase in the acceptor activation and decrease in the impurity redistribution. Implanted group-VI donors and amphoteric group-IV impurities, with donor behavior, are activated in GaAs, InP, and related ternary semiconductors at 950–1100 °C. Damage in the as-implanted layers, particularly vacancies at the surface and interstitials below projected range, play a dominant role in the impurity behavior. In amorphous layers created by heavy ion implantation at room or liquid-nitrogen temperature, activation is low and impurity redistribution occurs. Implantation at elevated temperatures results in disordered regions being formed, a higher electrical activation, and reduced diffusion mobility of the impurities during RTP. A distinguishing feature of implanted impurity behavior in III–V semiconductors relative to silicon is the dopant electrical activation via rearrangement of secondary defects in the layer. The latter determine the fraction of substitutional impurities and the enhanced diffusion via defect–impurity complexes. The substitutional–interstitial mechanisms of diffusion originally proposed for silicon (see Chapter 2), is now becoming accepted as the explanation for impurity behavior in III–V semiconductors [122, 177, 178]. A substitutional impurity is thought to diffuse via jumping from a substitutional site, where it has a high solubility, to an interstitial site, where it has a high diffusivity. The discussed excess vacancies and interstitials present in implanted III–V crystals control the exchange between the sites.

4.6. IMPURITY DIFFUSION FROM A SURFACE SOURCE

Doping of III–V semiconductors from a surface source with RTP is an alternative technique to ion implantation for the formation of shallow, submicron, doped layers. Experimental investigations of zinc diffusion into GaAs [186–191], InP [192–194], GaInAs [195], and GaAsP [196]; and silicon [197], germanium [198], and tin [199] diffusion into GaAs utilize elemental, compound, and mixed surface films as diffusion sources.

4.6.1. Zinc Doping of GaAs and Related Compounds

A zinc film chemically deposited onto the GaAs substrate was used for doping at 700–800 °C for 30 s [186, 187]. The *p–n* junction depth was 120–250 nm after 10 s of processing and linearly increased with processing temperature. Impurity distribution in the doped layer is characterized by an invariable concentration of about 1×10^{20} atoms/cm^3 in the near-surface region, followed by a stepwise drop at the *p–n* junction. Much interest was devoted to the use of mixed oxide sources SiO_2:ZnO [188, 189, 196] and ZrO_2:ZnO [191]. They were formed by chemical vapor deposition or by the spin-on method. A wide range of impurity concentrations in the source and the possibility of introducing group-III or -V components governing the diffusion process are the main advantages of these sources.

Diffusion doping of GaAs and related ternary compounds (GaInAs, GaAsP) from a surface oxide source was performed in two temperature ranges: at 600–700 °C for 10–100 s and at 800–1000 °C for 3–10 s [191]. Parameters of the doped layers created in both regimes are in very close agreement. The doped layers were 100–300 nm thick, becoming deeper with increasing processing temperature and time.

Figure 4.18 shows zinc-doped layer depth versus processing parameters and typical impurity concentration profiles in GaAs. There is an impurity-enriched surface layer about 20 nm thick, where the zinc concentration is $(1–7) \times 10^{20}$ atoms/cm^3 independent of diffusion temperature and time. This layer may incorporate Zn_3As_2 formed during diffusion at the film–wafer interface. The concentration of zinc in the deeply diffused region is temperature dependent and in the range of 10^{19}–10^{20} atoms/cm^3. Not all impurity atoms demonstrate electrical activity in the doped layer, and therefore the concentration of free charge carriers is lower. Maximum hole concentration of 1.2×10^{19}/cm^3 was achieved after zinc in-diffusion at 700 °C for 30 s [192]. It increased to 1×10^{20}/cm^3 with high-temperature annealing, such as 950 °C for 5 s [163]. The corresponding carrier concentration is in the range of 1×10^{14}–5×10^{15}/cm^3, which is much lower than that produced by ion implantation.

Zinc diffusivity in GaAs calculated based on the RTP conditions [191] satisfies the experimental data [200, 201] obtained in conventional furnace diffusion experiments. However, a twofold increase in the diffusivity of zinc is observed at 800 °C with a surface oxide source and a GaAsP substrate [196]. It was also found that, in contrast to gallium,

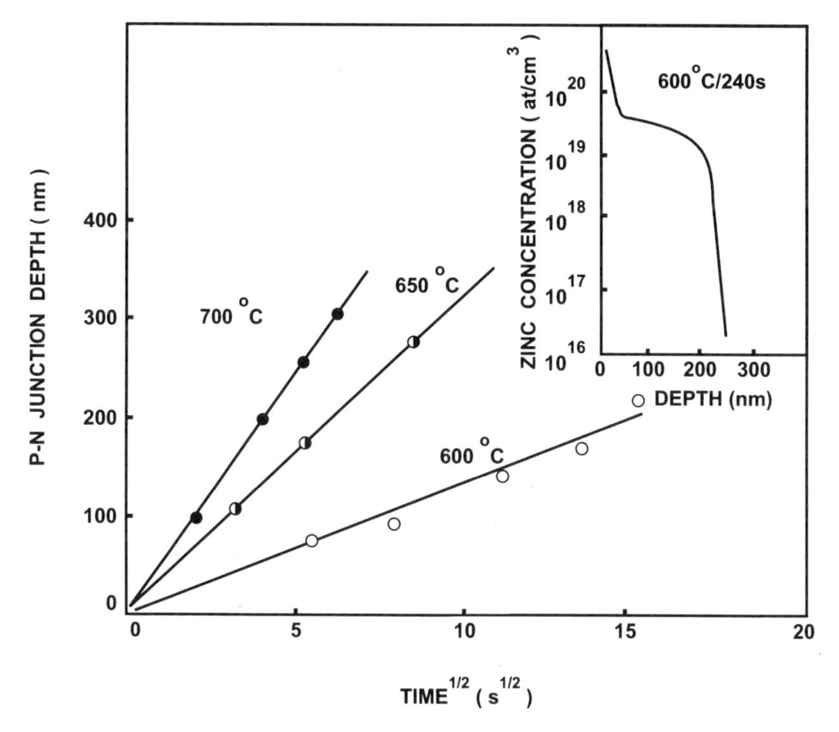

FIGURE 4.18. *p–n* junction depth and zinc distribution diffused in gallium arsenide doped from ZrO_2:ZnO surface source [191].

phosphorus introduced into the film source retarded zinc diffusion. The enhanced impurity diffusion seems to be associated with elastic stresses at the film–wafer interface, which are present because of the difference in the thermal expansion between the film and wafer materials (in addition to vertical temperature gradients).

Tungsten silicide films containing 5–10% zinc were proposed by Plumton [190] for use as an impurity source for GaAs. RTP-induced diffusion of zinc from the source has been found to be controlled by the ratio between silicon and tungsten atoms in the film and by the impurity concentration there. The films enriched by silicon (W/Si = 0.9) ensure marked zinc diffusion within 10–30 s at relatively low temperatures (600–700 °C). The depth of the doped layers increases in the range of 100–400 nm with increasing impurity content in the source. For a zinc concentration of about 10%, its diffusivity in GaAs is a factor of five larger than the equilibrium value.

Diffusion from the silicide source enriched by tungsten (W/Si = 3.7) has shown neither concentration dependence nor enhancement. It produced 300- to 900-nm-thick doped layers with processing at 850–950 °C for 10–30 s. Stresses at the film–wafer interface and difference in the mechanism of passing through this interface by the impurity atoms at low and high temperatures are thought to be responsible for the observed zinc diffusion behavior.

4.6.2. *Zinc Doping of InP*

Zinc diffusion doping of InP was performed with the use of ZnF_2 [192] and Zn_3P_2 [193, 194] surface sources. In order to prevent impurity evaporation, the source was protected by 100 nm of silicon dioxide. The Zn_3P_2 source is preferable in applications because the phosphorus suppresses loss from the wafer. Doped layers as thick as 1.5 μm with the hole concentration profile shown in Fig. 4.19 have been made at 400–600 °C in seconds. They are typical for diffusion from a limited-capacity surface source. The maximum hole concentration of $(1–2) \times 10^{18}$ p/cm^3 in semi-insulating InP is reached after diffusion at 520–540 °C for 15 s. Values of zinc diffusivity estimated from the concentration profiles are close to the values from conventional furnace experiments [202, 203].

Anomalous behavior for zinc diffusion from a surface source was observed with iron- and chromium-doped semi-insulating InP, previously processed at 850 °C for 15 s [194]. Zinc atoms penetrate surprisingly deep into the semiconductor crystal. The depth distribution has a double step shape, as shown in Fig. 4.20. There are three specific regions to it: (1) Zinc atoms accumulate in the surface region to a thickness of about 200 nm, reaching 2×10^{20} atoms/cm^3, although the maximum hole concentration is only about 1.5×10^{18} p/cm^3. (2) At a depth of 200–600 nm there is a "normal" diffusion region. (3) Finally, the region of zinc-enhanced diffusion occurs to a depth of 2.5–3.0 μm. These are consistent with the model [204] in which zinc diffuses in InP as an interstitial donor and is incorporated into the lattice both as the donor and as an immobile substitutional acceptor.

The profiles in Fig. 4.20 show that enhanced diffusion of zinc atoms takes place in a region of the semiconductor crystal, depleted of iron and chromium, that was created during preliminary RTP, which caused these impurities to diffuse to the surface, accumu-

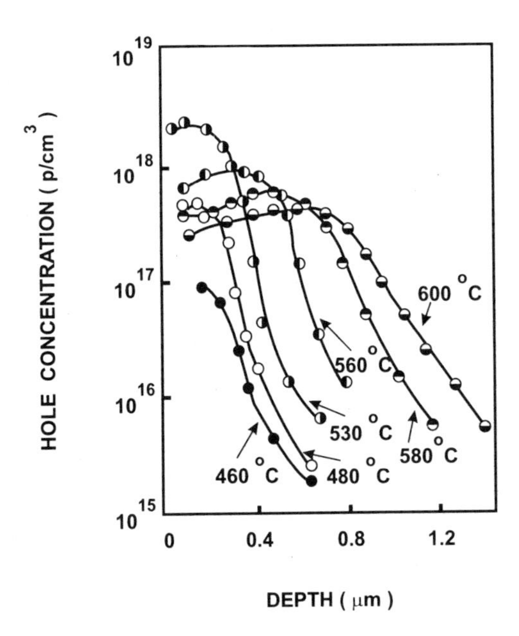

FIGURE 4.19. Hole concentration profiles in indium phosphide diffusion doped by zinc from the Zn_3P_2 surface source for 15 s [194].

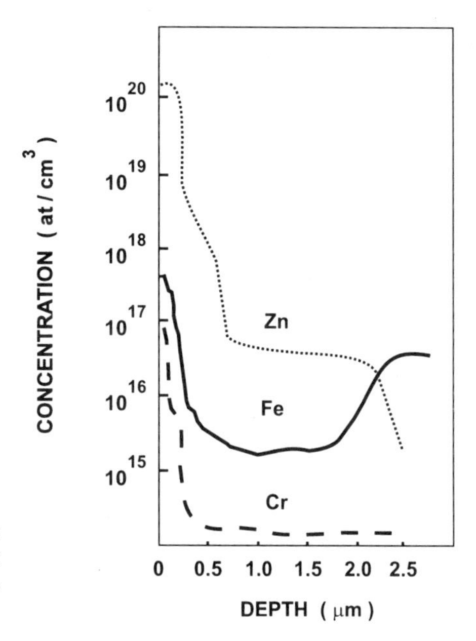

FIGURE 4.20. Impurity concentration profiles in semi-insulating indium phosphide annealed at 850 °C for 15 s and subsequently doped with zinc from the Zn_3P_2 surface source at 520 °C for 15 s [194].

lating there. The mechanism for the enhanced zinc diffusion is not clear. Nevertheless, preliminary high-temperature processing of semi-insulating InP crystals is definitely not advised when submicron zinc-doped layers are required.

4.6.3. *Silicon and Germanium Doping of GaAs*

Pure silicon and germanium films were used for diffusion doping of GaAs over the temperature range of 850–1050 °C and time duration of 3–300 s [197, 198]. Silicon films, 10 nm thick, were electron beam evaporated onto the substrate, and protected by a chemical vapor-deposited silicon nitride or silicon dioxide cap 500 nm thick. The samples capped with silicon nitride had a high resistivity, whereas the silicon dioxide capped samples produced doped layers, up to 300 nm thick, with a maximum electron concentration of $\sim 6 \times 10^{18}/cm^3$. This corresponds to ion implantation with doses of 10^{14} ions/cm^2. Silicon and electron depth profiles in the doped layer are shown in Fig. 4.21. Impurity atom concentration is an order of magnitude greater than the electron concentration because of the amphoteric behavior of silicon in GaAs [197].

Silicon is known to substitute gallium and arsenic in the GaAs lattice. A direct jump from one substitutional position to the nearest vacancy in the other sublattice is accompanied by a change of vacancy type and charge state of the silicon atom. In order to preserve charge neutrality, impurity migration is favored when two neighboring impurity complexes, Si_{Ga}–Si_{As}, take part in the process [197]. But this seems to be rare, particularly when excess gallium vacancies are generated by gallium out-diffusion into the silicon dioxide surface film. Combined substitutional–interstitial diffusion, mentioned in Section 4.5 for implanted impurities, may be a general mechanism for impurity behavior in semiconductors.

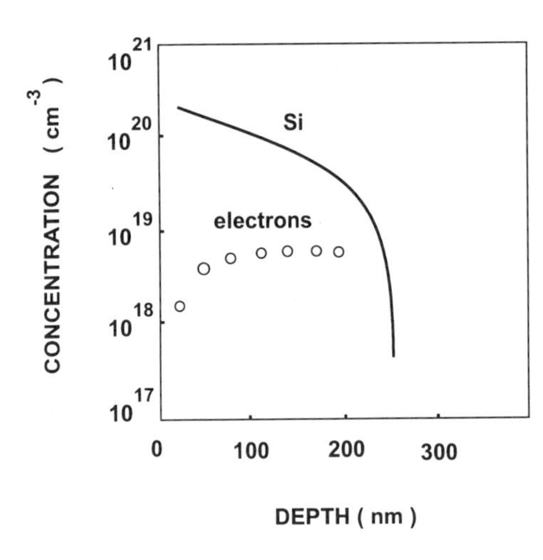

FIGURE 4.21. Impurity atom and electron concentration profiles in gallium arsenide diffusion doped by silicon at 1050 °C for 3 s from the elemental surface source protected by silicon dioxide cap [197].

Similar features were observed for germanium diffusion into GaAs [198]. An elemental surface source, protected with a silicon dioxide or phosphosilicate glass cap, was heated to 800–1050 °C for seconds. The thickness of the doped layer reached 100 nm with a maximum germanium concentration of ~10^{21} atoms/cm^3, although the electron concentration was no more than 5×10^{13}/cm^3. Low effective electron concentration in the doped GaAs layers is caused by germanium amphoteric behavior leading to conductivity compensation in the same way as for silicon. Compared to silicon dioxide, phosphosilicate glass more effectively suppresses arsenic evaporation and better enables gallium out-diffusion from the wafer, helping germanium display donor properties.

4.6.4. *Tin Doping of GaAs*

A spin-on glass source was used by Hernandes *et al.* [199] to dope GaAs with tin under RTP. Diffusion peculiarities were studied both with and without arsenic overpressure annealing ambients. The influence of gallium addition to the source was also analyzed.

Tin as $Sn(OC_3H_7)_4$ and gallium as $Ga(NO_3){\cdot}3H_2O$ were introduced into the tetraethoxysilane emulsion composition used in the experiments. After processing at 900–1000 °C for 60 s, tin-doped layers up to 500 nm thick were obtained. They have maximum electron concentration in the range of $(1–3) \times 10^{18}$/cm^3 and mobilities higher than 1000 cm^2/V·s.

Tin diffusivity has been observed to depend on arsenic overpressure. The higher the overpressure, the higher the diffusivity is. This was explained by the influence of point defects produced in the interface reactions between SiO_2 and GaAs. These reactions and related emission of point defects are reduced by increasing arsenic overpressure or by adding gallium into the spin-on surface. Activation energy of tin diffusion without arsenic overpressure was 2.3 eV, indicating that impurity atoms migrate predominantly via interstitials. Electrical activation of 70% of in-diffused impurity atoms was measured. Tin diffusion under arsenic overpressure is characterized by an activation energy of 3.2 eV

TABLE 4.7. Impurity Diffusion into III–V Compound Semiconductors

Compound	Impurity	Temperature range (°C)	Processing mode	Preexponential factor, D_0 (eV)	Activation energy, E_a (eV)	Impurity source	Ref.
GaAs	Zn	600–700	RTP	10.3^a	2.2	ZrO_2: ZnO film	191
GaAs	Zn	400–700	FA	26.2^a	2.3^a	SiO_2:ZnO film	201
GaAs	Zn	800–1000	FA	15	2.49	Zn vapor	200
GaAs	Zn	600–750	FA	$7.5 \times 10^{-4\,a}$	1.6	Evaporation of GaAs:Zn_2As_3 mixture	205
GaAs	Zn	600–750	FA	$2.5 \times 10^{-3\,a}$	1.7	Evaporation of GaAs:Zn_2As_3 Diffusion through the YSZ capping layer	205
GaAs	Zn	600–750	FA	$2.9 \times 10^{-3\,a}$	1.8	Zn vapor, diffusion through the YSZ capping layer	205
GaAs	Si	850–1050	RTP	0.11	2.5	Si film with the SiO_2 capping layer	197
GaAs	Sn	850–1000	RTP	4.3×10^{-2}	2.3	SiO_2:Sn film	199
GaAs	Sn	850–1000	RTP	49	3.2	SiO_2:Sn film, As overpressure ambient	199
GaAs	S	950–1050	RTP	4.0×10^{-3}	2.6	Epitaxial layer without a capping layer	158
GaAs	S	950–1050	RTP	4.7×10^{-3}	2.2	Epitaxial layer with the SiO_2 capping layer	158
GaAs	S	850–900	FA	4.0×10^{-4}	2.5	Epitaxial layer with the SiO_2 capping layer	158
GaAs	S	950–1200	FA	1.9×10^{-2}	2.6	S vapor	206
GaAs	S	900–980	FA	6.0×10^{-6}	2.6	S vapor, diffusion through SiO_2 capping layer	207
GaAs	Fe	700–900	FA	1.0×10^{3}	2.7	$Fe(NO_3)_3 9H_2O$ containing film	208
InP	Zn	520–600	RTP	0.36	1.46	Zn_3P_2 film	202
InP	Zn	400–600	FA	0.013^a	1.32^a	Zn vapor	202
InP	Zn	500–600	FA	1.65	1.35	Zn_3P_2 film with the Al_2O_3 capping layer	203

[a] Calculated from the experimental data presented in the reference.

and dopant electrical activation of 80%. The predominant diffusion mechanism was assumed to be substitutional.

4.6.5. General Remarks

The experimental results for RTP diffusion doping of III–V semiconductors from a surface source are far from complete. There is an urgent need to extend both impurity and semiconductor nomenclature in this area. Nevertheless, some general remarks which will need reexamination in the future are as follows: One of the most important questions concerns the mechanism of impurity diffusion under nonequilibrium rapid heating conditions. Table 4.7 collects the data available for impurity diffusion in GaAs and InP from a surface source. For comparison the results for sulfur redistribution in GaAs epitaxial layers are tabulated. Within the experimental uncertainty, activation energies for impurity diffusion initiated by RTP and conventional furnace annealing are in agreement. This suggests that identical impurity migration mechanisms occur under both heating conditions. However, the mechanism may change at high impurity concentrations ($>10^{20}$ atoms/cm^3), during joint diffusion of several impurities especially in the case of amphoteric impurities. Extended investigations should be carried out so as to clear up these points. Impurity diffusion under rapid heating conditions was experimentally observed to be either "normal" or enhanced. There are two factors considered responsible for the enhancement. The first one is associated with a predominant out-diffusion of one of the components into the encapsulating surface film, resulting in excess vacancy generation in the sublattice of this component. Elastic stresses at the film–wafer interface are another enhancement factor. There can be no doubt about the influence of these factors on impurity diffusion, but their role and mechanism under the nonequilibrium conditions of RTP need to be more accurately defined.

5

Diffusion Synthesis of Silicides in Thin-Film Metal–Silicon Structures

Metal silicides have received widespread attention as a result of their advantages in very large scale integrated microelectronics [1–10]. The low resistivity, high temperature, and chemical stability of these materials promote their application for interconnects, ohmic contacts, Schottky barriers, and diffusion barriers between aluminum metallization and silicon. Among the variety of approaches for silicide formation, heat-induced diffusion synthesis of thin-film metal–silicon structures appears to be the most practical. It is a simple and effective technique for control of thin film silicide characteristics.

The first experiments on solid-phase silicide formation by rapid thermal processing indicated that a processing time of seconds was necessary to provide accurate control of the processes involved [10]. However, transient heating influences the growth, phase composition, structure, and impurity behavior in the silicide produced. In this chapter the key features of rapid diffusion silicide formation are discussed and are compared with the features observed in conventional long-time furnace experiments. The most promising and already characterized microelectronic silicides are discussed.

The number of elements that can form silicides is rather imposing. There are usually more than three compounds in a metal–silicon phase diagram. Nevertheless, experiments show that not all are possible in films subjected to thermal processing. This does not mean that some of the equilibrium phases fail to originate, but that under conventional prolonged thermal processing they do not grow enough for their detection and study.

There are three main groups of silicides employed in thin-film structures [3]: (1) metal-rich silicides Me_xSi, (2) monosilicides MeSi, and (3) disilicides $MeSi_2$. Diffusion of metal atoms plays the primary role in the formation of the first group of silicides. Silicon diffusion dominates in the two other groups. We will analyze the metal–silicon systems (Table 5.1), in which silicides of the above-mentioned groups are formed. The metals used are representatives of refractory, transition, and noble metals.

TABLE 5.1. Silicides[a]

Metal	Eutectic temperature (°C)	Silicides	Formation temperature (°C)	Melting point (°C)	Silicide properties
Ti	1330	TiSi	500	1760	
		$TiSi_2$	600	1540	Lowest-resistivity silicide - 13–16 μΩ-cm (polysilicon substrate), 9–11 μΩ-cm (monocrystalline silicon substrate)
Ta	1385	$TaSi_2$	650	2200	Low resistivity, stable in oxidizing ambient
W	1140	WSi_2	650	2165	Low resistivity, thermally stable
Mo	1410	$MoSi_2$	525	1980	Low resistivity, thermally stable
Ni	966	Ni_2Si	200	1318	
		NiSi	350	992	
		$NiSi_2$	750	993	Low resistivity, epitaxial growth on silicon with the lowest lattice mismatch of 0.4%
Co	1195	Co_2Si	350	1332	
		CoSi	375	1460	
		$CoSi_2$	550	—	Low resistivity, thermally stable, epitaxial growth on Si with the lowest lattice mismatch of 1.2%
Cr	1300	$CrSi_2$	450	1550	Narrow-band-gap semiconductor (E_g = 0.35 eV)
Zr	1355	Zr_5Si_3	—	2325	
		$ZrSi_2$	700	1650	Thermally stable, Schottky barrier to silicon of 0.55 eV
Pd	720	Pd_2Si	100	1398	Low resistivity, Schottky barrier to silicon of 0.74 eV, epitaxial growth on (111) Si
		PdSi	800	—	
Pt	830	Pt_2Si	200	1100	
		PtSi	300	1229	Low resistivity, thermally stable, Schottky barrier to silicon of 0.87 eV, epitaxial growth on (111) Si

[a]Refs. 1–7.

5.1. CHARACTERISTICS OF THIN-FILM SILICIDE FORMATION DURING TRANSIENT HEATING

Transient heating of thin-film metal–silicon structures by radiation is controlled by two main factors. The first is the difference in the optical and thermophysical parameters of silicon and metals. Because of this, the heating dynamics of the films depend on the characteristics of the radiation used, and surface reflectivity, including which side of the substrate is subjected to irradiation. Another important factor is the high rate of oxidation and nitridation of most silicide-forming metals. Therefore, an inert ambient or protective encapsulating layer is needed. Radiation heating conditions can be optimized on the basis of numerical estimates of temperature profiles using the mathematics presented in Section 1.4. As far as the second factor is concerned, some practical recommendations can be gleaned from experiments on RTP of thin-film structures in vacuum, oxygen- and nitrogen-containing ambients [11–24], and when amorphous silicon, silicon dioxide or nitride caps and sublayers [19, 26, 27] are used.

5.1.1. Heating Efficiency

Let us evaluate numerically the heating efficiency of metal film/silicon structures when one side is uniformly irradiated by incoherent light, or electrons, in the heat balance regime. For incoherent light, incident energy distribution depends on the reflectivity of the irradiated surface. Metal reflectivity is typically between 0.6 and 0.9 (see Table 1.3), and therefore a major fraction of the incident energy is reflected. The reflectivity of electrons with energies from hundreds of electron volts to tens of kilo-electron-volts may be considered to be equal to zero for practical purposes. Calculated temperature–time profiles are presented in Fig. 5.1.

Any metal film on the irradiated surface produces a decrease in the induced temperature, whereas more effective sample heating occurs when the silicon surface ($R = 0.33$) is irradiated. For the same incident power density the maximum temperature is produced when the sample is heated by electrons ($R = 0$) and the induced temperature becomes independent of the sample surface exposed (metal film or silicon wafer). Therefore, low-energy electrons must be used for the most effective energy deposition. The limitations imposed by charge accumulation appear when substrates with dielectrics such as silicon dioxide or nitride are processed. Incoherent light heating does not have this problem. The best reproducibility of heating regimes and the highest efficiency of energy

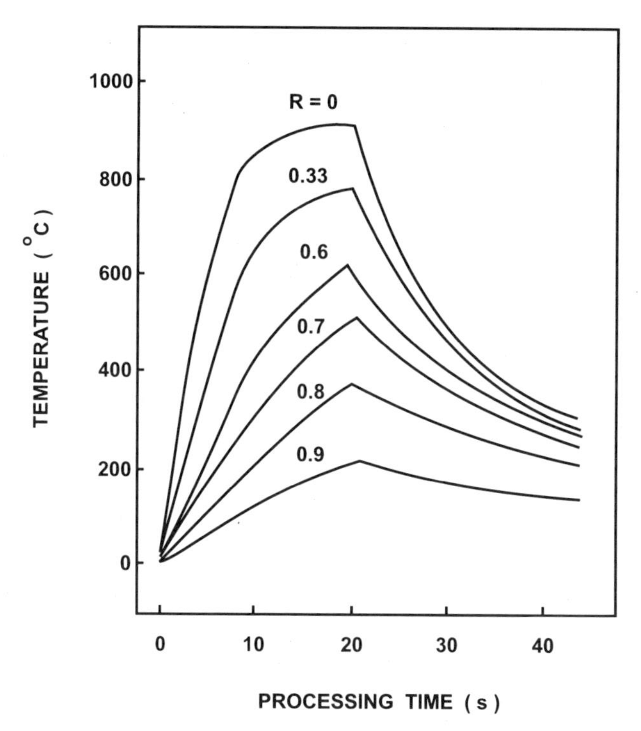

FIGURE 5.1. Calculated temperature versus time for silicon 380-μm-thick wafer with different reflecting films at the surface subjected to one-side incoherent light heating at $W_I = 8$ W/cm^2 for 20 s.

deposition are achieved on samples with differing surface films when the film free substrate surfaces are both irradiated (i.e., both sides of the wafer).

5.1.2. Influence of Ambient

The influence of processing ambient on silicide formation has been experimentally investigated for titanium [12, 13, 16–21], tantalum [14], tungsten [11, 23], and platinum [24, 25]. Tantalum and titanium films seem to be the most interesting as they have the highest oxidation rate among the silicide-forming metals. Titanium also has the highest nitridation rate, and therefore it is important to find out which of the two competing processes, i.e., oxidation (nitridation) or silicide growth, will dominate in an ambient containing oxygen (nitrogen) during RTP. On the other hand, films of titanium and tantalum oxide and nitride are of practical interest for passivation and for diffusion barriers. Titanium and tantalum are good oxygen getters. Conventional furnace thermal processing of titanium films on single-crystal silicon in a 50–100% oxygen ambient at atmospheric pressure produces complete oxidation of the metal at 600 °C so that silicidation was prevented [28]. Identical results have been found for titanium and tantalum films with RTP of seconds' duration [12, 14].

Backscattering spectra are presented in Fig. 5.2 for tantalum–silicon film structures transiently heated in air. An oxide layer appears on the metal film about 80 nm thick after processing at 500 °C. Its stoichiometry indicates Ta_2O_5, and the oxide thickness increases with increasing processing temperature, until complete oxidation, corresponding to a thickness of 240 nm, occurs at 700 °C. No silicide phases at the metal–silicon interface

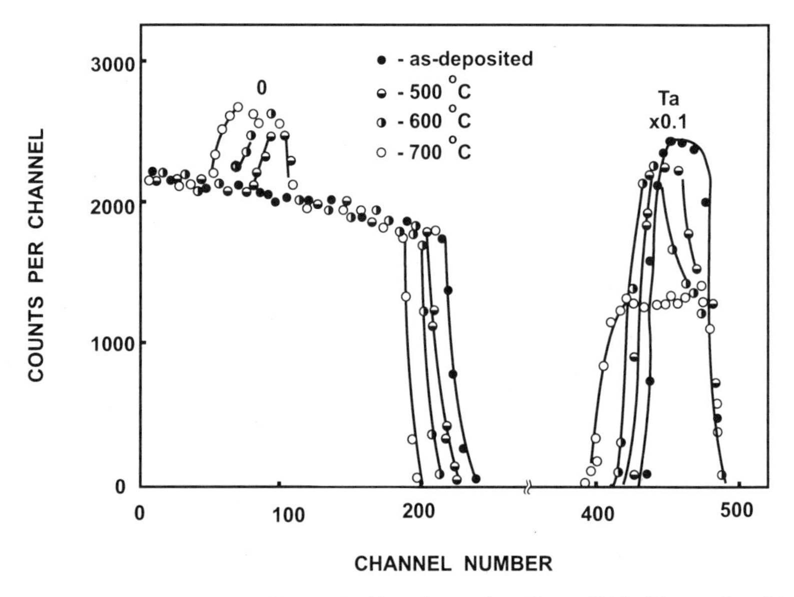

FIGURE 5.2. Backscattering spectra of 2-MeV He^+ ions for tantalum film on (111) silicon wafer subjected to incoherent light heating in air for 10 s [14].

were detected. Therefore, metal oxidation evidently proceeds considerably faster than silicide formation. Titanium and tantalum oxide films 100–300 nm thick allow oxygen atoms to diffuse through and do not restrict the oxidation rate of the underlying metal. These features are typical of other metal–silicon film structures; however, platinum has the lowest oxidation sensitivity.

The dominant role of oxidation can be suppressed when metal atoms are implanted into the silicon, such as tungsten [11]. In general, metal silicides with the best characteristics are formed during thermal processing in a neutral/reducing ambient (such as $Ar–H_2$) at atmospheric pressure and in vacuum, which is at least 10^{-5}–10^{-6} Torr, although conditions of high vacuum, 10^{-9} Torr, and lower are preferable. Experiments with tungsten–silicon structures indicate that residual oxygen at 10^{-9} Torr influenced the silicidation kinetics and properties of the tungsten silicide formed [22]. In spite of the high reactivity of most metals to oxygen, the silicides formed were stable against oxidation. Oxide growth rates for silicides are a little larger than for silicon [4, 29].

Nitrogen is less aggressive than oxygen in its interaction with silicide-forming metals, in particular with titanium which is the most active [28]. RTP of titanium films on silicon in nitrogen [15–17, 19–21] or in ammonia [17, 18] at 600–1100 °C ensures titanium nitride or oxynitride growth on the metal surface, while competitive silicidation at the metal–silicon interface occurs. The structures formed are schematically shown in Fig. 5.3a. Formation of titanium oxynitride on the metal surface occurs with processing in nitrogen and involves oxygen gettered from the as-deposited titanium film. The titanium is contaminated with oxygen from the residual atmosphere during its vacuum deposition [20, 30, 31] and from the annealing ambient before nitridation [19]. At temperatures up to 500 °C, oxygen diffusivity in titanium ($D = 1 \times 10^{-7} \exp(-1.2/kT)$ cm^2/s [32]) exceeds the diffusivity of nitrogen ($D = 0.2\exp(-2.32/kT)$ cm^2/s [33]) by more than an order of magnitude. Thus, the metal film is saturated with oxygen before a nitride or silicide layer starts to grow.

Oxygen solid solubility in titanium is 32 at. % [34]. Oxygen atoms are eventually expelled from the growing silicide layer at the surface. They react there with titanium nitride producing oxynitride. After processing at temperatures below 700 °C, the dominating phase in the silicide layer is a metastable C49 modification of $TiSi_2$. It also contains a small amount of monosilicide or other metal-rich phase (Ti_5Si_3). Higher processing temperatures lead to a double layer sandwich comprising oxynitride and C54-modified $TiSi_2$ developing. At 1100 °C and a processing time of up to seconds, oxygen atoms in the surface layer and the silicon in the disilicide layer are replaced by nitrogen. Uniform δ-TiN layer is formed. The rejected silicon is epitaxially regrown at the interface of the single-crystal substrate.

The Ti–Si–N phase diagram at 700–1000 °C [35] shows that $TiN–TiSi_2–Si$ is a three-phase region at equilibrium. The presence of the $TiN–TiSi_2$ tie line demonstrates that TiN rather than Si_3N_4 is the stable nitride on $TiSi_2$. Thus, during silicide nitridation on silicon substrates, TiN is formed on top of $TiSi_2$ and the liberated silicon atoms diffuse through the remaining silicide layer to become bonded to the silicon substrate [19]. When $TiSi_2$ is rapidly thermally nitrided in ammonia a small amount of silicon remains on the top surface mostly in the form of Si_3N_4 [22].

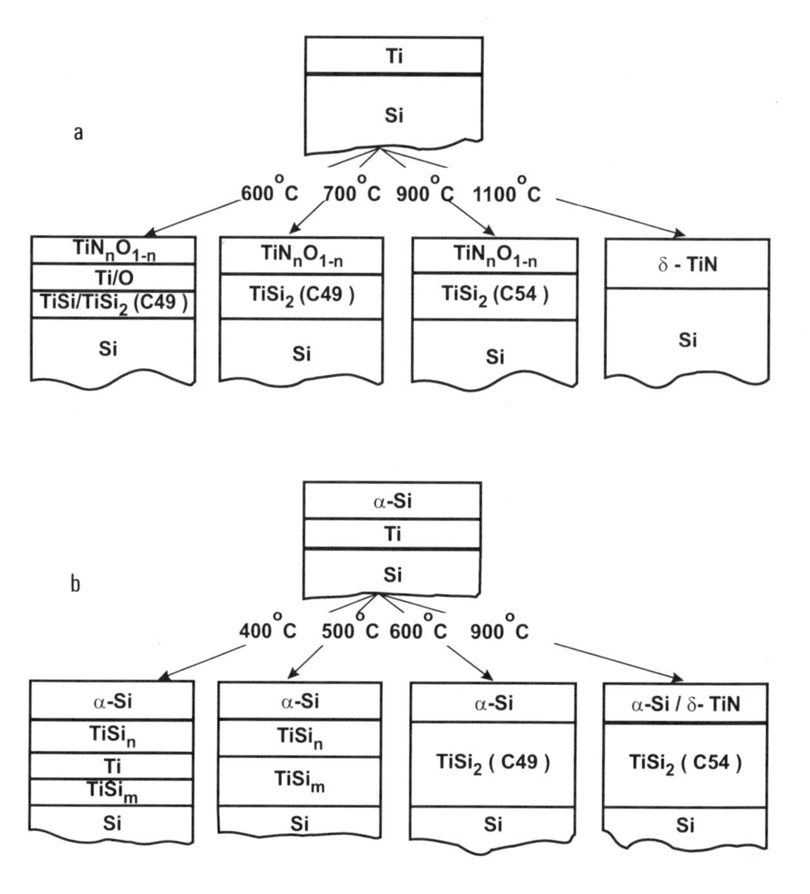

FIGURE 5.3. Schematics of titanium–silicon reaction products at various rapid processing temperatures in nitrogen [19].

In the case of platinum [24, 25], a partial pressure of oxygen at 10^{-7} Torr in the annealing ambient is sufficient to slow down silicide formation. Oxygen atoms incorporated in the metal film reduce the number of sites available for diffusing jumps of platinum, and thus decrease interaction diffusivity.

5.1.3. Influence of Capping and Sublayers

The metal surface can be protected by an amorphous silicon film [19, 27]. As-deposited amorphous silicon contains no more than a few atomic percent of oxygen. Films 10–30 nm thick provide the necessary protection for titanium during vacuum processing, but the thickness must be increased to 50–80 nm for processing in nitrogen at atmospheric pressure. Silicidation in such structures proceeds at both interfaces and contamination of titanium by oxygen or nitrogen from the annealing ambient is prevented. The reaction products at various temperatures are schematically shown in Fig. 5.3b. The high diffusivity of silicon atoms in the amorphous layer results in the silicide layer formed there containing more silicon than the layer adjacent to the silicon crystal substrate. Typical phases are $TiSi_2$, TiSi, and Ti_5Si_3. A uniform $TiSi_2$ layer is produced with processing at 900 °C for 10 s. Some patches of TiN can also be found at the surface. They are created by the diffusion of titanium atoms from the silicide and their interaction with nitrogen.

The process of silicide formation is critically dependent not only on the processing ambient and protective film composition, but also on the substrate material. Silicon dioxide or nitride films are usually employed to passivate the silicon wafer influencing the composition and structure of the silicides formed on the surfaces [29, 36]. This effect under RTP is more pronounced for titanium [19, 26].

Two types of reactions are possible during the interaction of silicides with silicon dioxide [29]. The first one yields silicon dioxide through the reaction of silicon atoms from the silicide with oxygen from the silicon dioxide. The silicide is reduced to a metal-rich phase and sometimes can result in elemental metal. Reactions of the second type promote oxidation of metal from the silicide, the silicon released being precipitated. The predominance of one of the reactions is determined by the difference in formation entropy for silicon dioxide and metal oxides. The more negative difference ensures that the silicide is more resistant to metal oxidation. Calculations indicate that almost all of the disilicides used, except $ZrSi_2$ and $HfSi_2$, result in silicon dioxide growth when interacting with oxygen [37]. $TiSi_2$ is an exception which has been studied in conventional furnace experiments [38, 39] and RTP [19, 26]. Titanium dioxide and nitride have been found when titanium silicides are formed on silicon dioxide or nitride sublayers.

The role of radiation heating conditions, annealing ambient, protective covers, and sublayers should be considered in experimental work. For the remainder of this chapter we have selected basic results obtained under processing conditions that minimize the influence of these factors. In particular, processing of thin-film metal–silicon structures transiently heated in a vacuum or inert gas ambient is reviewed.

5.2. PHASE CONTENT AND GROWTH KINETICS

Silicide formation initiated by thermal processing of metal films on a silicon wafer takes place over a wide temperature range (see Table 5.1). At low temperatures, a metal-rich silicide is formed [3, 4], and under the proper kinetic conditions the silicidation proceeds until the whole metal film is consumed. Then a silicide phase with a larger silicon content is formed. Such a transition usually occurs at a higher processing temperature. The process continues until a phase with the maximum allowable silicon content is reached. As a rule, this appears to be a disilicide or occasionally a monosilicide. The advantage of silicide synthesis by RTP is fast sample heating to temperatures of 500–1050 °C, which are much higher than the 300–700 °C typically employed for silicide formation in a conventional furnace. Therefore, the processes involved in silicide formation are governed by nonequilibrium thermodynamics, which affects the growth kinetics and phase content of the silicide layer.

Two kinds of silicide growth kinetics can be distinguished: Parabolic kinetics appear when the processing rate is controlled by component diffusion, but if the rate of chemical reaction is the limiting factor, linear kinetics occur. The thickness of the layer being formed is correspondingly

$$(x_j)^2 = Ut \tag{5.1}$$

$$x_j = Vt \tag{5.2}$$

where U is the parabolic rate constant and V the linear rate. They are usually presented by Arrhenius-type approximation with the preexponential factors U_0 or V_0 and activation energy E_a. Taking into account temperature variation, $T(t)$, during RTP the thickness of the layers, x_j, being formed can be calculated as follows:

$$(x_j)^2 = U_0 \int_0^\infty \exp[-E_a/kT(t)]\, dt \tag{5.3}$$

$$x_j = V_0 \int_0^\infty \exp[-E_a/kT(t)]\, dt \tag{5.4}$$

These phases and kinetics of silicide formation during RTP, as well as during conventional furnace annealing, have specific features for each metal–silicon system considered.

5.2.1. Titanium–Silicon

Silicide synthesis in thin-film titanium–silicon structures subjected to conventional long-time thermal processing is initiated at 335–350 °C [40–42]. It actively proceeds in the temperature range of 450–600 °C [9, 43–49] and the growth follows parabolic kinetics. The layer formed consists of Ti_5Si_3, Ti_5Si_4, TiSi, and $TiSi_2$. The richest silicon phase, which is the disilicide, becomes the only one after processing at 800 °C.

Titanium silicide formation during RTP has qualitatively analogous features, but the temperature range at which several phases are simultaneously observed is extended to higher values, and silicide layer growth is enhanced. Multiphase layers are formed at up to 700–750 °C [19, 50–54], but a uniform disilicide is produced at 800 °C and higher [19, 50, 55–62]. The above temperature range boundaries are quite mobile and they depend on the initial titanium film thickness and the processing time.

Figure 5.4 illustrates phase transitions in thin-film (30–40 nm) titanium-on-silicon structures subjected to RTP for seconds in vacuum. Metal-rich phases, i.e., Ti_5Si_4 and TiSi, start growing first. Disilicide appears after 10 s of processing at 600 °C, and growth occurs until a uniform film is produced with a processing time of 20 s.

In contrast to conventional furnace annealing, three or even more phases can coexist in the silicide layer synthesized by RTP. This indicates that nonequilibrium heating produces nucleation and growth of the silicon-rich phase before all of the metal has reacted to form the intermediate metal-enriched silicide phase. For example, titanium metal at a thickness of several hundreds of nanometers results in a phase transition that is incomplete even at 700–750 °C for seconds. Uniform disilicide is produced at processing temperatures of 800 °C [51]. Higher temperatures (as well as preliminary interface atomic mixing induced by ion implantation) result in $TiSi_2$ formation independent of the initial thickness of the metal film [55, 60–63].

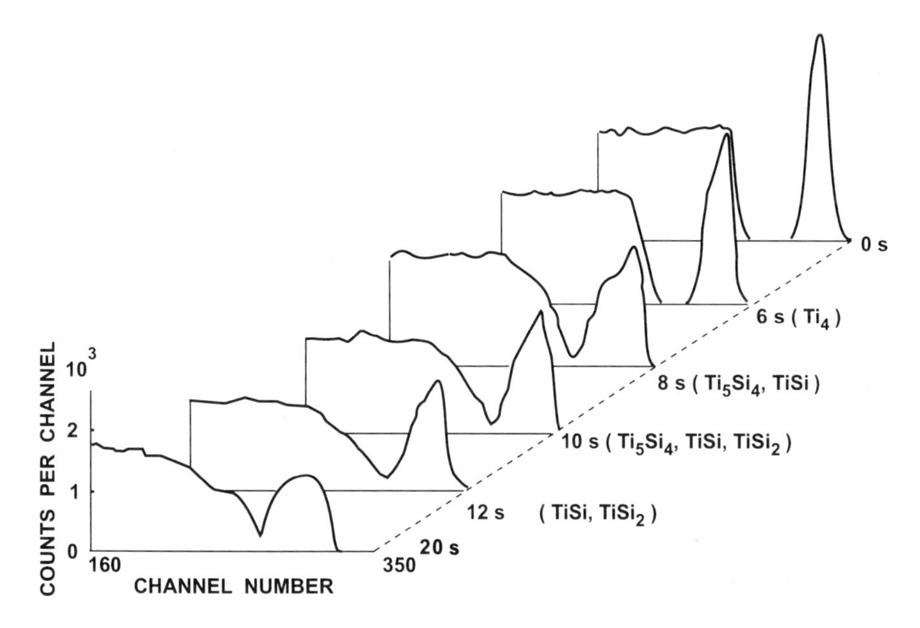

FIGURE 5.4. Backscattering spectra taken for titanium–silicon structure processed at 600 °C for various times [52].

$TiSi_2$ surfaces, even after processing in a vacuum or an inert gas ambient, are covered by titanium oxide TiO_n ($n = 1–2$) that is several tens of nanometers thick. Residual oxygen in the annealing ambient, or oxygen dissolved in the metal (pushed to the surface by the silicidation front) occurs in agreement with conventional furnace experiments [64]. RTP-induced growth of silicides is enhanced relative to conventional furnace annealing. The data in Table 5.2 show the kinetic parameters for these two processing regimes. Parabolic kinetics are evident for titanium silicide layer growth in both cases suggesting a diffusion-limited process. Lateral growth of the disilicide layer on silicon dioxide surfaces [52] indicates that silicon atoms are the most mobile species.

Extrapolation of the representative kinetic data from the conventional furnace experiments [46, 47] into a temperature range of 700–800 °C shows that RTP growth of titanium silicide proceeds at 1–1.5 orders of magnitude faster than a conventional furnace anneal. The process is characterized by two activation energies, namely, 2.6 eV for 700–750 °C and 1.5 eV for 750–800 °C. The latter agrees satisfactorily with values obtained for conventional furnace annealing (1.6–1.9 eV).

The activation energy change at 750 °C is systematic and is likely to be associated with a change in proportion of TiSi and $TiSi_2$ in the growing layer [51]. The above temperature is the upper limit for the existence of TiSi because only disilicide growth takes place at higher temperatures [43]. Moreover, C49-to-C54 structure transition, which will be discussed in Section 5.4, is initiated in the disilicide of this temperature, thus influencing the activation energy.

TABLE 5.2. Growth Parameters of Titanium Silicides

Phase	Substrate	Temperature range (°C)	Kinetics	Preexponential factor, U_0 (cm^2/s), V_0 (cm/s)	Activation energy, E_a (eV)	Refs.
		Rapid thermal processing				
$TiSi_2 + TiSi$	Si (111)	700–750	$t^{1/2}$	1.4×10^{3a}	2.6	51
$TiSi_2$	Si (111)	750–800	$t^{1/2}$	4.8×10^{-3a}	1.5	51
	Poly-Si	600–750	$t^{1/2}$		1.86	62
	Si on sapphire	600–750	$t^{1/2}$		1.65	62
		Furnace processing				
TiSi	Si (111)	275–340	$t^{1/2}$	2.8×10^{-7}	1.0	41
$TiSi + Ti_5Si_3$	Si (100)	350–425	t	760^a	1.6	42
$TiSi_2 + Ti_5Si_3$	Si (100), amorphous Si	475–650	$t^{1/2}$	4.0×10^{-3}	1.8	46, 47
$TiSi_2$	Si (100),	550–650	$t^{1/2}$	29^a	2.5	49
	poly-Si,	500–850	$t^{1/2}$		1.8	43
	lateral growth, from (100) Si	800–950	$t^{1/2}$	0.048^a	1.89	44

[a]Calculated from the experimental data presented in the reference.

The enhancement of disilicide growth may be explained by an increase in radiant energy absorption at the metal–silicon interface leading to additional heating. The temperature in this region appears to be higher than that measured for the whole sample. Moreover, the temperature gradient present enhances diffusion of silicon atoms. High heating rates limit oxygen and nitrogen penetration into the reaction zone from the surface, thus allowing the silicon to react with titanium.

5.2.2. *Tantalum–Silicon*

Information about phase and kinetic properties of silicide formed in the tantalum–silicon system is limited. The high oxidation rate of tantalum and low atomic transparency of its oxide and silicon dioxide for both tantalum and silicon atoms make it difficult to produce tantalum silicide by diffusion synthesis in thin-film metal–silicon structures. Conventional furnace annealing experiments [5] show that a disilicide phase is produced in this structure at 650 °C. The growth kinetics are linear with activation energies of 3.7 and 2.5 eV.

Rapid-heating-induced tantalum silicide formation has been predominantly investigated in mixed films deposited by cosputtering or coevaporation onto single-crystal silicon and oxidized silicon substrates and polycrystalline silicon layers [65–71]. These films were prepared with an excess of silicon. The atomic ratio between silicon and tantalum amounted to 2.2–3.0. The aim of thermal processing in this case was only to create conditions for local diffusion mixing and setting of the stoichiometry corresponding to the disilicide. The disilicide phase in such structures begins to be detected with heating to 700–800 °C for seconds. Its formation becomes completed within processing times of 3–30 s at 1000–1100 °C. Investigations of rapid diffusion synthesis of tantalum silicides with metal-on-silicon film structures are limited [72–75]. Disilicide was noted to be the first nucleated phase at 750 °C within seconds [72], but at 900 °C the processing time can be shortened to 0.1 s [74]. In oxygen-contaminated films tantalum disilicide formation starts at 850 °C [73], and it is completed within 0.1–10 s at 1100 °C. Symmetrical growth of disilicide grains with respect to the tantalum–silicon interface indicates that diffusion of both components is responsible for disilicide formation [73], i.e., reaction-limited growth kinetics [5, 75]. It can be attributed to the influence of oxygen on the silicidation. In the temperature range of 875–950 °C, the growth rate of tantalum disilicide follows [75]

$$V_{\mathrm{TaSi_2}} = 1300\exp(-3\mathrm{eV}/kT) \quad \mathrm{cm/s} \tag{5.5}$$

5.2.3. Tungsten–Silicon

Silicide formation data for the tungsten–silicon system are contradictory. The silicide layer formed in conventional furnace experiments starts growing at 650 °C [5], consisting of tungsten disilicide. Its growth kinetics are linear with an activation energy of 3.0 eV in the temperature range of 700–850 °C [76]. However, tungsten disilicide growth was observed to follow parabolic kinetics with activation energies of 3.4 and 2.2 eV, respectively, at 650–800 and 850–1100 °C [77, 78]. Metal-rich silicide W_5Si_3 was found in sufficient quantities for analysis only at the tungsten saturated interface in W–WSi_2 structures [79].

Investigations of RTP-induced synthesis of tungsten silicides [23, 77, 80–87] partly resolve these contradictions, specifying the process of silicide formation. Figure 5.5 shows the results of an X-ray phase analysis of the thin-film tungsten-on-silicon structure

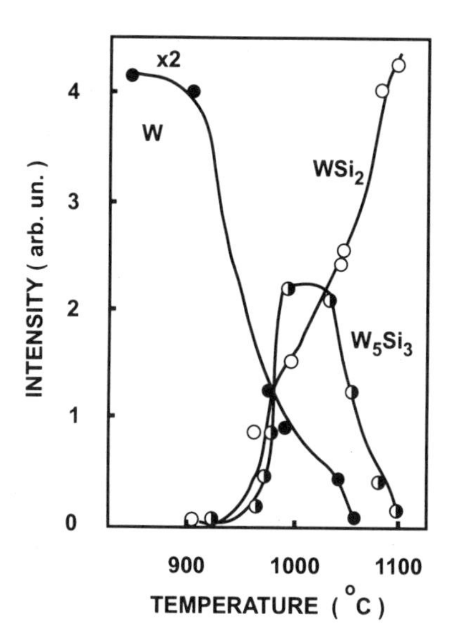

FIGURE 5.5. Integral X-ray diffraction peak intensities [W-metal: (110), (200), and (211); W_5Si_3: (211), (002), and (202); WSi_2: (002), (101), (202), and (213)] versus processing temperature of tungsten–silicon structure for 30 s [23].

subjected to thermal processing for 30 s at different temperatures. There are no phase changes up to 900 °C. The silicides W_5Si_3 and WSi_2 are formed with processing at 950 °C. Nonreacted metal decreases with an increase in processing temperature and it completely disappears above 1050 °C. The disilicide grows monotonically over the entire temperature range, while the phase W_5Si_3 reaches a maximum at 1000 °C, and is rapidly reduced at higher temperatures, disappearing above 1100 °C.

The stoichiometric WSi_2 is formed within seconds on processing at 1100 °C, and above, in both metal-on-silicon [23, 77, 87] and codeposited mixed metal–silicon [80–82, 85, 86] thin-film structures. Hexagonal WSi_2 nucleates at about 600 °C and starts to transform into the tetragonal phase above 900 °C [86]. The amorphous-to-hexagonal and hexagonal-to-tetragonal transitions are kinetically controlled with activation energies of 1.4 and 2.2 eV, respectively. The best uniformity of WSi_2 films is achieved by processing at 1200 °C with durations up to 10 s. High-energy ion implantation in the film destroys interface oxide layers and introduces mixing of the atomic components, resulting in a 50–100 °C decrease in temperature for uniform disilicide formation [82–84].

WSi_2 growth kinetics under RTP follow a parabolic law [23, 77, 87] indicating the process is controlled by component diffusion into an interphase boundary. The activation energy is 2.2 eV in good agreement with data from conventional furnace experiments (see Table 5.3). Silicon atoms are the most mobile in the growing layer. RTP promotes faster silicide layer growth. The enhancement factor varies from 2.5 to 10–15 times in comparison with the data from high-temperature (850–1100 °C [78]) and low-temperature (650–800 °C [77]) conventional furnace experiments, extrapolated to 1100–1200 °C. Elastic stresses generated in the film structure may be responsible for the observed enhancement [77].

5.2.4. Molybdenum–Silicon

Silicide formation in the molybdenum–silicon system has much in common with the processes occurring in the tungsten–silicon one. The experimental data are quite limited. In conventional furnace experiments [88] Mo_3Si was identified to be the first nucleated crystalline phase which is correlated to the stable structure of the amorphous Mo–Si

TABLE 5.3. Growth Parameters of Tungsten Silicides

Phase	Substrate	Temperature range (°C)	Kinetics	Preexponential factor, U_0 (cm^2/s), V_0 (cm/s)	Activation energy, E_a (eV)	Ref.
			Rapid thermal processing			
WSi_2	Si (100)	1150–1350	$t^{1/2}$	1.46^a	2.2	77
			Furnace processing			
W_5Si_3	WSi_2	1350–1870	$t^{1/2}$	13.2^a	3.58	79
WSi_2	Si (100)	700–850	t	5.8×10^{6a}	3.0	76
	Si (100)	650–800	$t^{1/2}$	6.6×10^{3a}	3.4	77
	Si-poly	850–1100	$t^{1/2}$	0.6	2.2	78

[a] Calculated from the experimental data presented in the reference.

TABLE 5.4. Growth Parameters of Molybdenum Silicides under Conventional Furnace Processing

Phase	Substrate	Temperature range (°C)	Kinetics	Preexponential factor, U_0 (cm^2/s), V_0 (cm/s)	Activation energy, E_a (eV)	Ref.
Mo_3Si	$MoSi_2$		$t^{1/2}$	0.26[a]	3.25	79
Mo_5Si_3	$MoSi_2$	1190–1715	$t^{1/2}$	38.4[a]	3.58	79
$MoSi_2 + Mo_3Si$	Si (111)	475–550	t^2	—	2.4	9
$MoSi_2$	Si-poly	850–1100	$t^{1/2}$	0.88[a]	2.2	78

[a]Calculated from the experimental data presented in the reference.

interlayer which occurs at 400 °C. At 440 °C and above Mo_5Si_3 is added. Molybdenum disilicide ($MoSi_2$) is known to start growing at 525 °C [5]. Metal-rich phases, in particular Mo_3Si and Mo_5Si_5, can also be present in the silicide layer [89]. The growth kinetics are parabolic, as indicated by the data summarized in Table 5.4. Activation energies for the metal-rich phases (Mo_3Si, Mo_5Si_3) are 3.25 and 3.58 eV, values 1–1.4 eV higher than for the disilicide.

Solid-phase molybdenum silicide formation during transient heating was mainly investigated in codeposited of mixed films [89–93]. A uniform stoichiometric disilicide with tetragonal structure is produced in the temperature range of 1050–1100 °C, with an optimal process duration of 10 s. There were no phase changes observed at higher processing temperatures.

Molybdenum-on-silicon film structures, mixed by ion implantation, begin to form a silicide layer at 600 °C, as indicated by the sharp decrease in sheet resistivity [84, 94]. A uniform disilicide layer forms within seconds when the sample is heated to 1100 °C. Data on the phase content of the silicide layer in the interval between these two temperatures are not available. Nevertheless, we would expect to see metal-rich phases in common with conventional furnace annealing.

Among the general features observed in silicide formation on molybdenum–silicon and tungsten–silicon films there are two ones to be stressed. First, in spite of the fact that silicides are nucleated at relatively low temperatures, i.e., at 525–650 °C, the silicidation becomes completed by disilicide formation only at 1100 °C, and within 1–10 s of processing. Second, atomic mixing by ion implantation decreases the temperature for formation of a uniform disilicide layer. Both of these features along with the linear kinetics of WSi_2 growth during prolonged furnace annealing [76] suggest that a barrier oxide at metal–silicon interface influences the silicidation process. The oxide layer restricts the rate of silicide production, so that either ion beam mixing or processing at higher temperatures is required. Metal-rich silicides for molybdenum and tungsten grow slower than their disilicides and hence little is formed.

5.2.5. *Nickel–Silicon*

Nickel silicide formation in a conventional furnace usually starts with the appearance of a Ni_2Si phase at temperatures as low as 200 °C. Growth of this phase is controlled by nickel diffusion [5, 95–103]. Parabolic kinetics result with activation energies of 2.48 eV for lattice diffusion, and 1.71 eV for grain boundary diffusion [103]. Development of the silicidation process causes new phases to be synthesized and as a rule there are two phases close in silicon content: (1) Ni_2Si/NiSi at temperatures up to 275–300 °C and (2) NiSi/$NiSi_2$ at higher temperatures [97, 98, 101–108]. An increase in the processing time displaces the balance of the growth of silicon-enriched phases resulting in the formation of a single-phase silicide layer. The process remains diffusion controlled; however, contrary to Ni_2Si, silicon atoms are the most mobile species in NiSi and $NiSi_2$ [96, 109, 110].

Specific features associated with RTP arise at early stages of nickel silicide formation. Four phases, namely, NiSi, δ-Ni_2Si, θ-Ni_2Si, and $Ni_{31}Si_{12}$, were found at processing temperatures of 175–300 °C for 1 s [111]. Mixing of nickel and silicon atoms occurs in the intermediate amorphouslike layer preceding silicide nucleation. Depending on the size and composition of this layer, silicide phases may nucleate both inside the layer and at its boundaries with the nickel film and silicon substrate. Theoretically, any of the equilibrium phases can originate from the mixed amorphous layer. In fact, the location of nucleation and the presence of elastic stresses in the structure influence the growth. Considering experimental data [112, 114, 115], for the RTP of nickel films on silicon up to 450 °C the initial growth favors δ-Ni_2Si, as illustrated in Fig. 5.6a. After RTP at

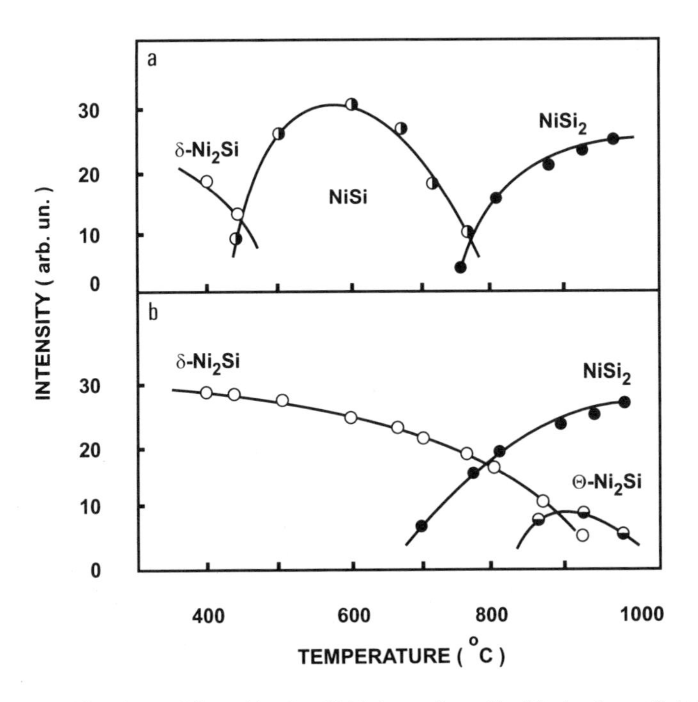

FIGURE 5.6. X-ray diffraction peak intensities for silicide layers formed by 30 s incoherent light heating (a) without implantation and (b) implanted with boron (30 keV/5 × 10^{14} ions/cm^2) nickel film on (100) silicon [114].

500–600 °C monosilicide (NiSi) becomes the dominant phase in the silicide layer. Processing at temperatures higher than 700 °C promotes $NiSi_2$ growth [116, 117]. The observed silicide phase transitions are identical to those that occur during conventional furnace annealing, although the temperature range and coexistence of some phases are shifted to higher temperatures.

Implantation of high-energy ions influences the sequence of phases during nickel silicide formation [114, 118–120]. The greatest effect is observed when the ion range is equal to the thickness of the metal film. Arsenic or argon implantation enhances growth of the disilicide during subsequent RTP, as it does in conventional furnace annealing [121]. Atomic mixing of the components promotes the preferential synthesis of $NiSi_2$ at lower temperatures, 600–650 °C, for processing times of 30 s.

Boron implantation also changes the consistency and thermal limits for phase formation [114, 115]. In Fig. 5.6b the intermediate phase NiSi disappears from the silicide layer and θ-Ni_2Si appears over the temperature interval 850–970 °C. At 700–750 °C this nickel-rich phase starts being directly transformed into disilicide at the expense of silicon diffusion from the substrate. The upper temperature limit for δ-Ni_2Si and the lower one for the disilicide are shifted to higher temperatures with the implantation of boron in the dose range of 5×10^{14}–5×10^{15} ions/cm^2 (ion energy 60 keV). In addition, in the samples implanted with doses higher than 2×10^{15} ions/cm^2 Ni_2Si (θ-phase) are formed with processing at temperatures above 800 °C, and boron compounds were not detected in the silicide layer.

The increased temperature stability of δ-Ni_2Si in the implanted layers indicates that boron passivates grain boundaries, producing a diffusion barrier. The barrier restricts nickel out-diffusion from the grains and silicon diffusion into the grains. It acts effectively up to 700–750 °C. At higher temperatures diffusion of silicon along grain boundaries and into the grains becomes the controlling process [3]. This results in disilicide formation by direct diffusion synthesis or through a Ni_3Si_2 phase which decomposes during sample cooling producing θ-Ni_2Si and $NiSi_2$ [1].

The growth of nickel silicide layers initiated by RTP follows parabolic kinetics, in agreement with conventional furnace annealing kinetics. The diffusion of nickel atoms influences the growth of the Ni_2Si phase [112, 113]. Silicon atoms are the most mobile species in the mono- and disilicide. Other data characterizing kinetics of nickel silicide growth are summarized in Table 5.5. Activation energies and growth rates of Ni_2Si are practically identical for RTA and conventional furnace processing. However, $NiSi_2$ grows an order of magnitude faster under RTP, although the activation energy is decreased insignificantly (from 1.65 eV to 1.4 eV). This difference is related to the higher probability of direct transition from Ni_2Si to $NiSi_2$ during rapid heating and cooling of the samples.

5.2.6. Cobalt–Silicon

Silicidation of cobalt–silicon proceeds in a similar manner to nickel–silicon. Simultaneous parabolic growth of two phases, Co_2Si and CoSi, occurs, as observed in conventional furnace annealing. The Co_2Si grows almost an order of magnitude faster than the CoSi [122–127]. Simultaneous growth occurs at 400–500 °C, where cobalt atoms are the most mobile species, until the metal is reacted to form Co_2Si. At 500–545 °C Co_2Si is

TABLE 5.5. Growth Parameters of Nickel Silicides

Phase	Substrate	Temperature range (°C)	Preexponential factor, U_0 (cm^2/s)	Activation energy, E_a (eV)	Refs.
		Rapid thermal processing			
Ni_2Si	Amorphous Si		0.017	1.3	112, 113
$NiSi_2$ + NiSi	Si (100)	700–900	0.029[a]	1.4[a]	114, 119
		Furnace processing			
Ni_2Si	Si (100)	200–325	2.0[a]	1.5	95–98
	Amorphous Si	210–335	0.35[a]	1.4	98
	Amorphous Si	240–300	6.3×10^{-4}[a]	1.1	99
	Amorphous Si	270–430	6.0	1.5	102
	Poly-Si	225–325	0.028[a]	1.3	100
	Lateral growth on Al_2O_3	400–600	0.14[a]	1.4	101
	Lateral growth from Si (100)	400–700	0.11[a]	1.5	104
	Lateral growth from Si (111)	400–700	0.34[a]	1.8	104
NiSi	Si (100), amorphous Si	250–400	0.85[a]	1.55	98, 105, 106
	Si (100)	275–400	2.0[a]	1.65	97
	Si (111)	275–400	0.29[a]	1.65	97
	Si (100)	300–360	9.2×10^{-4}[a]	1.23	107
NiSi + Ni_2Si	Si (111)	320–370	27.8[a]	1.83	107
$NiSi_2$	Amorphous Si	350–425	0.16[a]	1.65	108

[a] Calculated from the experimental data presented in the reference.

reduced by a growing CoSi phase. Further enrichment of the silicide layer by silicon produces $CoSi_2$. In the disilicide, silicon atoms are thought to be the most mobile species [134, 135]. However, a relatively high activation energy on its growth spread in the range of 2.3–3.7 eV [128–132] supports an argument that $CoSi_2$ growth is predominantly by a slow cobalt diffusion in this phase, which is characterized by the diffusivity $D = 0.21 \exp(-2.83\text{eV}/kT)\ cm^2/s$ [133]. Simultaneous existence of the metal and all of the mentioned cobalt silicides was observed [127] when an extremely thick (2 mm) cobalt source was used in conventional furnace experiments.

RTP of cobalt–silicon thin structures [112, 136–142] produces detectable Co_2Si at 300–370 °C. At temperatures around 500 °C two silicide phases, namely, Co_2Si and CoSi, are simultaneously growing within 10–50 s for 10–100 nm of as-deposited metal films until cobalt from the film is consumed by these silicides. Processing at higher temperatures leads to $CoSi_2$ formation. It becomes the only phase in the silicide layer above 700 °C. However, RTP at 800–1150 °C is used to obtain uniform polycrystalline and good-quality epitaxial disilicide layers.

The RTP-produced phase sequence Co–Co_2Si–CoSi–$CoSi_2$ is the same as for furnace processing. Diffusion-limited kinetics characterize the silicidation process. The growth parameters are shown in Table 5.6. Activation energies for silicide growth are very close for both processing modes but preexponential factors are higher for RTP. It can be inferred that the annealing ambient has less influence on the impurities for the reduced thermal budget processing.

5.2.7. *Chromium–Silicon*

The first phase nucleated for chromium–silicon subjected to conventional furnace annealing is $CrSi_2$ appearing after processing at 400 °C [3, 98, 101, 143–146]. Data about the growth kinetics are contradictory; both linear [98, 143, 144, 147], and parabolic kinetics have been observed [101, 148].

With RTP multiphase silicide layers of Cr_3Si, Cr_5Si_3, and $CrSi_2$ were produced by processing at 450–1100 °C for seconds [149–151]. The proportion of these phases varied with the processing temperature and time. $CrSi_2$ and Cr_3Si are detected in the layers synthesized at 1100 °C for 10 s. An increase in the processing time to 30 s results in the

TABLE 5.6. Growth Parameters of Cobalt Silicides

Phase	Substrate	Temperature range (°C)	Preexponential factor, U_0 (cm^2/s)	Activation energy, E_a (eV)	Refs.
		Rapid thermal processing			
Co_2Si	Si (100)	370–680	1.7	2.1	112, 136
CoSi	Poly-Si	470–545		2.09	141
	Si on sapphire	455–660		2.03	141
$CoSi_2$	CoSi	550–700	2×10^{3a}	2.6	141
	Poly-Si	530–590		2.91	141
	Si on sapphire	560–605		2.81	141
		Furnace processing			
Co_2Si	Si (111)	800–1050	1.45×10^{-5a}	1.16^a	127
	Co + Si	340–400		2.0	132
Co_2Si + CoSi	Si (100), Si (111)	400–545		1.75	122
	Si (100)	350–486	4.1×10^{-3a}	1.5	124
	Si (100)	385–490	0.11^a	1.85	126
CoSi	Si (100)	375–486	0.58^a	1.9	124
	Si (100)	385–490	0.030^a	1.8	126
	Si (111)	800–1050	3.5×10^{-3a}	1.62^a	127
	Amorphous Si	385–490	0.12^a	1.9	126
	Co_2Si	400–440		1.9	132
$CoSi_2$	Si (100)	550–575		2.6	130
	Si (111)	900–1050	2.3×10^{-2a}	2.3^a	127
	Amorphous Si	405–500	230^a	2.3	128
	CoSi	500–600	1.1×10^{4a}	2.8	129
	$Co_{60}W_{40}$	500–575	1.6×10^8	3.7	131
	CoSi	500–575		2.5	132

^{a}Calculated from the experimental data presented in the reference.

Cr_3Si phase being reduced, and Cr_5Si_3 inevitably appears with monotonically increased disilicide. The final uniform phase produced is the disilicide, which takes 90 s at 1100 °C, and longer at lower temperatures. Formation of metal-rich phases provides evidence of a limited silicon concentration at the metal–silicide interface. Silicon atoms become bonded with oxygen when they reduce chromium oxide in the metal film [31]:

$$2Cr_2O_3 + 3Si \rightarrow 3SiO_2 + 4Cr \tag{5.6}$$

Thus, the disilicide is "diluted" by free chromium atoms.

Rapid-heating-induced chromium silicide layer growth is governed by linear kinetics at temperatures up to 650 °C, and is parabolic at temperatures higher than 800 °C. Table 5.7 shows that the activation energy for parabolic growth by RTP is 1.5 eV, practically identical to the value (1.4 eV) obtained by Zheng *et al.* [101] where the same kinetics were observed in lateral growth of the disilicide. The absolute rate of silicide layer growth during RTP at 1000 °C is one to two orders of magnitude lower than that for conventional furnace annealing because of silicon oxidation according to Eq. (5.5). Chromium recovered from its oxide forms a nonequilibrium metal-rich phase that retards the disilicide growth.

5.2.8. *Zirconia–Silicon*

Little work has been done with thin-film zirconium silicides. The phase diagram has eight equilibrium compounds [152]. Furnace processing produces only zirconium disilicide at 600–800 °C [5, 145, 153], and zirconium monosilicide is expected [154] to precede it. The silicide layer starts forming at 500 °C when RTP of seconds is employed [155]. It consists of the disilicide and Zr_5Si_3 phase. After processing at 700 °C and above, it cannot be detected by X-ray analysis; however, the disilicide remains the only phase in the synthesized layer. Conventional furnace annealing performed at the same temperature

TABLE 5.7. Growth Parameters of Chromium Silicides

Phase	Substrate	Temperature range (°C)	Kinetics	Preexponential factor, U_0 (cm^2/s) V_0 (cm/s)	Activation energy, E_a (eV)	Refs.
		Rapid thermal processing				
$Cr_5Si_3 + CrSi_2$	Si (100)	450–650	t		0.7[a]	149–151
$CrSi_2$	Si (100)	800–1050	$t^{1/2}$	0.45[a]	1.5[a]	149, 151
		Furnace processing				
$CrSi_2$	Si (100)	450–525	t	5.2×10^{3}[a]	1.7	144
	Si (100)	500–700	t		1.7	143
	Si (111)	400–585	t	1.8×10^{2}[a]	1.5	146
	Si (100), Amorphous Si	400–505	t	8.0×10^{5}[a]	2.0	98
	Si (100)	400–475	$t^{1/2}$	5×10^3	2.3	148
	Lateral growth on Al_2O_3	500–650	$t^{1/2}$	2.2[a]	1.4	101

[a] Calculated from the experimental data presented in the reference.

produces only $ZrSi_2$. This confirms that Zr_5Si_3 is a nonequilibrium phase at these temperatures. The transition to disilicide suggests a diffusion mechanism for the silicide formation in thin-film structures.

5.2.9. Palladium–Silicon

Palladium in contact with silicon forms Pd_2Si at 100 °C and it is the only growing phase up to 800 °C [96]. The growth kinetics with conventional furnace annealing are exceptionally parabolic and controlled by diffusion of Pd atoms [100, 156–162], with an activation energy in the range of 1.0–1.5 eV (see Table 5.8). Silicide formation is usually performed at 200–300 °C where the growth rate varies over 1–1.5 orders of magnitude, and is a function of silicon wafer orientation and impurity concentration.

Parabolic growth of Pd_2Si occurs at higher temperatures, 360–500 °C, in seconds [163–167]. The rate is several orders of magnitude higher than values extrapolated from conventional furnace experiments, but the activation energy is in agreement at 1.1 eV. The reason for the enhanced growth has not been determined. A single phase of Pd_2Si is preserved up to 900 °C.

5.2.10. Platinum–Silicon

Silicide layers in thin-film platinum–silicon structures appear with furnace annealing at 200 °C [4]. Pt_2Si grows first and is transformed into PtSi with an increase in processing temperature and time. The monosilicide is the final phase in this system. The kinetics and phases are influenced by oxygen and carbon content in the deposited platinum film, and

TABLE 5.8. Growth Parameters of Pd_2Si

Substrate	Temperature range (°C)	Preexponential factor, U_0 (cm^2/s)	Activation energy, E_a (eV)	Refs.
	Rapid thermal processing			
Si (100)	350–450	1.48	1.1	163
	Furnace processing			
Si (100), (111), (110), amorphous Si	200–275	7.9[a]	1.5	156, 157
Si (111)	175–250	1.3[a]	1.4	158
Si (100)	215–290	2.7×10^{-4}[a]	0.95	159
Amorphous Si (evaporated)	215–290	2.4×10^{-4}[a]	0.9	159
Amorphous Si (sputtered)	240–300	2.8×10^{-2}[a]	1.2	159
Si (100) P doped at $5 \times 10^{13}/cm^3$	200–275	0.48	1.4	160
Si (100) As doped at $5 \times 10^{20}/cm^3$	200–275	1.2×10^{-4}	1.05	160
Si (100) P doped at $5 \times 10^{14}/cm^3$	200–250	0.18	1.35	160
Si (100) As doped at $5 \times 10^{20}/cm^3$	200–250	1.6×10^{-4}	1.05	160
Poly-Si	200–275	0.084[a]	1.2	100
Si (111)	170–230	0.49	1.32	161
Si (111)	160–250	1.1[a]	1.34	161

[a]Calculated from the experimental data presented in the reference.

by impurities in the silicon substrate. Formation of Pt_2Si and then PtSi takes place in films deposited onto lightly doped silicon and processed in high vacuum, when residual gas pressure is no higher than 10^{-7} Torr [168–170]. Monosilicide starts growing after a transition of platinum into Pt_2Si. Growth of Pt_2Si is governed by platinum diffusion and proceeds an order of magnitude faster than monosilicide growth in which diffusion of silicon atoms plays a major role. Oxygen and carbon contaminants in platinum films, as

TABLE 5.9. Growth Parameters of Platinum Silicides

Phase	Substrate	Temperature range (°C)	Preexponential factor, U_0 (cm^2/s)	Activation energy, E_a (eV)	U (400°C) (cm^2/s)[a]	Refs.
		Rapid thermal processing				
Pt_2Si	Si (100)	300–375	0.67	1.38	3.0×10^{-11}	25
Pt_2Si + PtSi	Si (111)	400–450	0.021	1.0	6.6×10^{-10}	176
PtSi	Si (100)	375–450	12.27	1.67	3.6×10^{-12}	25
PtSi	Si (111)	400–600	0.014	1.1	7.8×10^{-10}	176
	Poly-Si	375–450	60[a]	1.8	1.9×10^{-12}	177
	Si (100)	400	—	—	1.2×10^{-11}	179
		Furnace processing				
Pt_2Si	Si (100)	220–275	1.55[a]	1.39	5.8×10^{-11}	174, 175
	Si (100)	250–300	0.7	1.5	3.9×10^{-12}	173
	Si (100)	200–300	0.08	1.3	1.4×10^{-11}	170
	Si (100)	194–305	1.4[a]	1.4	4.4×10^{-11}	98
	Si (111)	200–500	1.5×10^{-3}	1.0	4.7×10^{-11}	171
	Si (100), (111)	260–350	8.9	1.6	8.8×10^{-12}	168
	Si (100), (111), poly-Si	250–310	4.5[a]	1.5	2.5×10^{-11}	169
	Amorphous Si	194–305	2.1[a]	1.4	6.6×10^{-11}	98
	Lateral growth from PtSi	500–800	0.4×10^{-3}	1.5	2.2×10^{-15}	173
PtSi	Si (100)	225–300	10.1	1.5	5.6×10^{-11}	170
	Si (100)	250–370	0.46[a]	1.5	2.6×10^{-12}	98
	Si (100)	295–350	21.3[a]	1.69	4.4×10^{-12}	174, 175
	Si (100) P $5 \times 10^{13}/cm^3$	300–400	7.8	1.45	1.0×10^{-10}	172
	Si (100) As $5 \times 10^{20}/cm^3$	300–400	3.3	1.45	4.3×10^{-11}	172
	Si (111)	300–500	0.25	1.41	6.6×10^{-12}	171
	Si (111) P $4.3 \times 10^{14}/cm^3$	300–400	130	1.6	1.3×10^{-10}	172
	Si (111) As $5 \times 10^{20}/cm^3$	300–400	54	1.6	5.3×10^{-11}	172
	Si (100), (111)	350–420	0.63	1.6	6.2×10^{-13}	168
	Si (100), (111), poly-Si	300–370	1.9[a]	1.5	1.1×10^{-11}	169
	Amorphous Si	250–370	0.65[a]	1.5	3.6×10^{-12}	98

[a]Calculated from the experimental data presented in the references.

well as substitutional impurities in silicon at concentrations close to their solubility limit, retard growth of Pt_2Si [98, 169–172] and lead to competition with the growth of PtSi. Unreacted platinum remains at the wafer surface. The growth kinetics of the silicide layer are always parabolic with activation energies of 1.3–1.5 eV for Pt_2Si and 1.4–1.6 eV for PtSi (see Table 5.9). The growth rates are practically identical for (100) and (111) silicon [168, 169].

RTP synthesis of platinum silicide has many similarities to conventional furnace processing [24, 25, 166, 176–181]. First Pt_2Si appears with processing at 250–300 °C for seconds; it is a metastable phase at these temperatures, and is therefore rapidly exhausted by the growing monosilicide. Typical evolution of the phases is shown in Fig. 5.7. A principal feature of platinum silicide formed in seconds is that two silicide phases coexist. These phases persist up to 400–500 °C for "pure" metal films [177, 178], and up to 800 °C when the as-deposited films are saturated with oxygen and carbon [24, 25]. Retarded silicide formation leaves unreacted platinum at the surface. A uniform monosilicide layer is formed after processing at 800 °C and above for 10–15 s [166, 179]. The growth kinetics of PtSi are parabolic [177–179] and there is no enhancement of

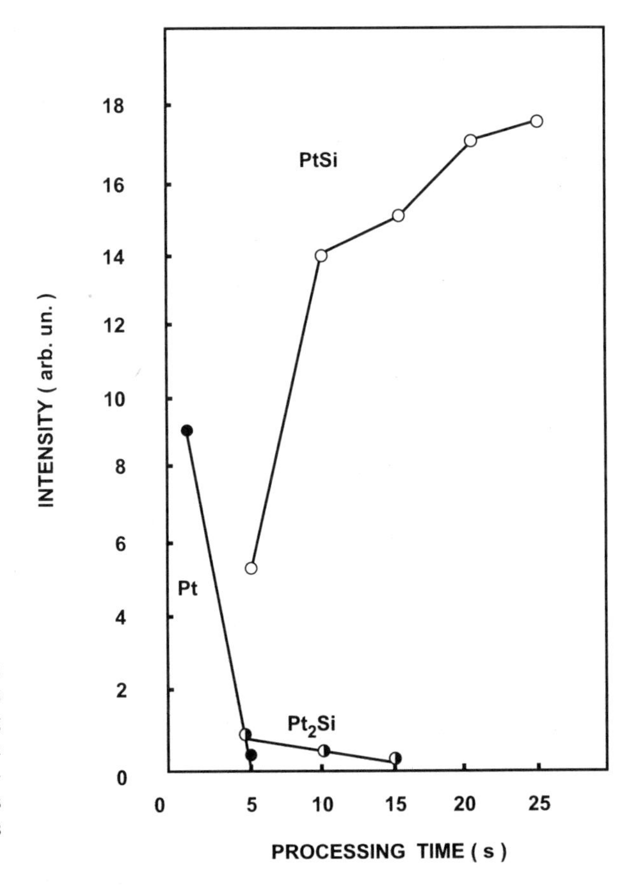

FIGURE 5.7. X-ray diffraction peak intensities for silicide layers formed by rapid thermal processing of 42-nm-thick platinum film on (100) silicon. The maximum temperature ranged from 250 °C for 5-s processed samples to 550 °C for 20-s processed ones [178].

growth with RTP (see Table 5.9). At the same time, comparative experiments on RTP of argon-implanted thin-film platinum–silicon structures [180, 181] showed that doses up to 5×10^{15} ions/cm^2 accelerate the phase formation. At doses of 5×10^{16} ions/cm^2 and higher the phase formation is reduced. Ion implantation at low doses produces atomic mixing at the metal–silicon interface, but at high doses the implanted species condense at the silicide grains restricting grain boundary diffusion of silicon and platinum.

5.2.11. *General Features*

The general features of silicidation in thin-film metal–silicon structures subjected to RTP by radiation may be summarized as follows:

1. The silicide formation processes are exclusively thermal in nature, being controlled by the induced temperature and processing time.
2. The silicide layer nucleates at the metal–silicon interface. Subsequent phase transitions occur via solid-phase reactions tending to produce a layer enriched with silicon.
3. The parabolic growth kinetics are dominant, indicating that atomic diffusion is the rate-limiting process. Enhanced growth of silicide layers is a consequence of parallel phase formation in contrast to sequential ones, in agreement with conventional long-time furnace processing.
4. Implantation and impurity effects are equally pronounced in silicide formation during RTP and conventional furnace annealing.

The above features are useful in constructing models for solid-phase diffusion synthesis of silicides.

5.3. SIMULATION OF SILICIDE SYNTHESIS

Silicide nucleates in a thin metastable amorphouslike layer at the metal–silicon interface where the components are mechanically mixed. This layer is formed during the metal deposition onto the silicon substrate [3] and grows at the beginning of silicidation [182]. Activation energy for growth ranges from 0.9 to 1.6 eV, indicating that diffusion of metal atoms is predominantly responsible for this process. Thickness of the layer reaches 4.5–27.0 nm before crystalline silicide phases nucleate and start growing in it. Subsequent thermal processing at higher temperatures and/or for longer times initiates qualitative development and growth of silicide phases.

The initial silicide nucleation has been studied for conventional prolonged isothermal processing. Walser and Bene [183] suggest that a silicide with its highest melting point near the lowest eutectic temperature on the binary phase diagram is the first to nucleate. In Pretorius's study [184] the compound heat of formation is employed as a criterion. He postulates that the first nucleating phase is a congruently melting silicide, possessing the highest negative effective heat of formation for component concentrations close to the lowest eutectic temperature. The kinetic approach developed by Bene [185] supposes the phase having the highest product $D\Delta G$ to be the first to nucleate and grow, where D is the effective diffusivity of the most mobile atoms in the growing silicide and ΔG is the difference in free energies for the nucleating phase and for the initial state of the system. There are a number of exceptions to the above rules. Moreover, their application to RTP, where phase diagrams can change significantly, is not quite correct. In any case, the dominant growing phase, rather than the first nucleating one, is of practical interest.

The model we will consider assumes that a metal-rich silicide starts growing first. Injection of metal atoms into the silicon is limited by their diffusivity, while silicon penetration into the metal is restricted by the slower and energy-consuming process of silicon bond breaking at the metal–silicon interface [3]. Therefore, primary diffusion of metal into silicon brings about formation of the metal-enriched silicide at the initial heating stage, for temperatures up to 200–300 °C. Further temperature increase enhances diffusion of silicon atoms in the silicide layer. It causes the growth of a silicon-rich silicide phase at the silicon substrate.

Analysis of the metal–silicon binary diagrams [1] indicates that three or four equilibrium silicides can be formed when the growth proceeds from the most metal-rich phase and evolves into a silicon-rich one (see Table 5.10). The general progress of the phase transformations comprises the following transitions:

$$Me_3Si \rightarrow Me_2Si \rightarrow Me_5Si_3 \rightarrow MeSi \rightarrow MeSi_2 \tag{5.7}$$

(with bypass arrows from Me_2Si to $MeSi$ above, and from Me_3Si to Me_5Si_3 below)

For a particular metal–silicon system some phases from (5.7) may be absent or substituted by other phases close in composition. Therefore, we suppose three phases to be involved in the synthesis of a silicide layer: disilicide $MeSi_2$ as the final silicon-rich phase and two intermediate phases represented by monosilicide MeSi and the metal-rich Me_2Si phase.

TABLE 5.10. Equilibrium Silicides

Metal	Me_3Si	Me_2Si	Me_5Si_3	MeSi	$MeSi_2$
Ti	Ti_3Si	—	Ti_5Si_3	TiSi	$TiSi_2$
Ta	Ta_3Si	Ta_2Si	Ta_5Si_3	—	$TaSi_2$
W	W_3Si	—	W_5Si_3	—	WSi_2
Mo	Mo_3Si	—	Mo_5Si_3	—	$MoSi_2$
Cr	Cr_3Si	—	Cr_5Si_3	CrSi	$CrSi_2$
Mn	Mn_3Si	—	Mn_5Si_3	MnSi	$MnSi_2$
Fe	Fe_3Si	—	Fe_5Si_3	FeSi	$FeSi_2$
Co	Co_3Si	Co_2Si	—	CoSi	$CoSi_2$
Ni	Ni_3Si	Ni_2Si	Ni_5Si_3	NiSi	$NiSi_2$
Zr	Zr_3Si	Zr_2Si	Zr_5Si_3	ZrSi	$ZrSi_2$
Nb	Nb_3Si	—	Nb_5Si_3	NbSi	$NbSi_2$
Pd	Pd_3Si	Pd_2Si	—	PdSi	—
Ir	Ir_3Si	Ir_2Si	—	IrSi	$IrSi_2$
Pt	Pt_3Si	Pt_2Si	—	PtSi	—

RTP provides formation of multiphase silicide layers when the induced temperature allows the growth of several silicide phases, shown schematically in Fig. 5.8. The multiphase composition is a result of the short processing time and limited diffusivity of the components. Silicon atoms are the most mobile species in disilicides and in most of the monosilicides [3, 4]. That is why the growth of silicide layers is connected with diffusion of silicon atoms through disilicide, followed by a disilicide-producing reaction at the $MeSi_2$–MeSi interface. Then surplus silicon diffuses farther through the monosilicide and reacts with metal at the MeSi–Me_2Si interface. Reactions at both interfaces cause the growth of silicon-enriched silicide phases:

$$Me_2Si + Si \rightarrow 2\,MeSi \tag{5.8}$$

$$MeSi + Si \rightarrow MeSi_2 \tag{5.9}$$

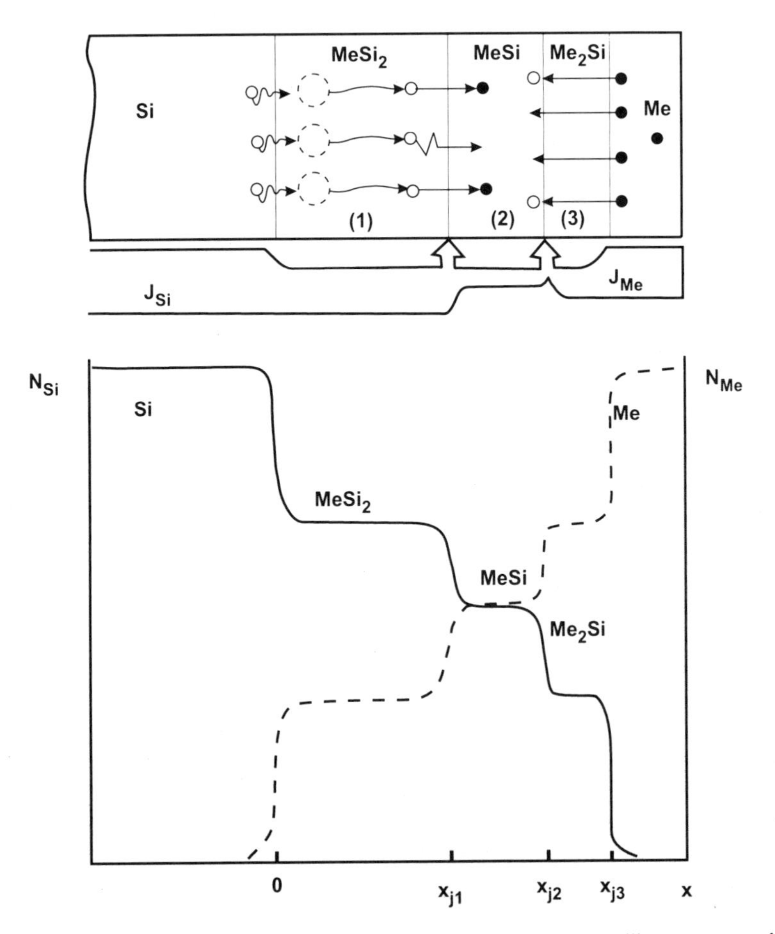

FIGURE 5.8. Diffusion processes and component distributions in a thin-film metal– silicon structure subjected to rapid thermal processing.

The counterdiffusion of metal atoms from the surface film provides the growth of the metal-rich phase. The diffusion process and interface reactions result in the redistribution of the components in the synthesized silicide layer. A stepwise distribution of the steady-state concentrations corresponds to the equilibrium silicide phases. An increase in processing time moves the metal–silicide interface toward the surface and the growing silicide layer consumes metal. Silicidation is completed when a uniform disilicide is formed. Disilicide can also be the only growing phase under RTP when its growth rate is much higher than that of the other phases. As a result, other phases may not grow to more than several monomolecular layers. Disilicide is considered to be the final silicon-richest phase. However, silicidation of platinum–silicon and palladium–silicon systems is complete with the formation of the monosilicide. In such cases the simulation procedure is limited to this phase.

The presented model summarizes the experimental results considered and allows numerical calculations. We place the zero point of the coordinate axis at the silicide–silicon interface. Positions of the corresponding phase boundaries are marked x_{j1}, x_{j2}, x_{j3} as shown in Fig. 5.8. Assuming diffusion-limited growth kinetics of the silicide layer (this is the most common case) the boundary positions after RTP are

$$(x_{j1})^2 = U_{01} \int_0^\infty \exp[-E_{a1}/kT(t)]\, dt \tag{5.10}$$

$$(x_{j2})^2 = U_{02} \int_0^\infty \exp[-E_{a2}/kT(t)]\, dt \tag{5.11}$$

$$(x_{j3})^2 = U_{03} \int_0^\infty \exp[-E_{a3}/kT(t)]\, dt \tag{5.12}$$

where the U's are the preexponential factors and E_a's are the activation energies in the parabolic rate constant Arrhenius expression and subscripts 1, 2, and 3 correspond to the parameters for the $MeSi_2$, MeSi, and Me_2Si phases, respectively.

The metal-rich phase Me_2Si grows until all of the metal is consumed. The boundary condition are therefore

$$\int_0^{x_{j3}} N_{Me} dx \leq \frac{\rho_{Me} n_A}{A_{Me}} \ell \tag{5.13}$$

where $N_{Me} = f(x)$ is the concentration profile of metal atoms in the silicide layer, n_A is Avogadro's number, and metal-related parameters are: ρ_{Me} volume density, A_{Me} atomic weight, and ℓ thickness of as-deposited film.

The integral condition (5.13) may be replaced by the sum with a precision sufficient for practical purposes:

$$\frac{\rho_1}{A_1} x_{j1} + \frac{\rho_2}{A_2}(x_{j2} - x_{j1}) + \frac{\rho_3}{A_3}(x_{j3} - x_{j2}) \leq \frac{\rho_{Me}}{A_{Me}} \ell \tag{5.14}$$

When the metal-rich phase Me_2Si and the intermediate phase MeSi are consumed, the third and second summands on the left side of (5.14) must be removed. Concentration profiles of the most mobile atoms in the silicide layer are obtained from numerical integration of the reaction diffusion equations with the corresponding initial and boundary conditions in each of the different regions. Coexistence of two phases in quantities sufficient for analysis is observed in experiments, and implies $x_{j2} = x_{j3}$. Accounting for this fact we can use the analytical solution of the reaction diffusion equation [186]. Then the distribution of silicon atoms in $MeSi_2$ is described by

$$N(x) = N(0-\delta) - [N(0-\delta) - N(x_{j1}+\delta)]\ \mathrm{erf}(x\cdot a_1/x_{j1})\ \mathrm{erf}(a_1) \tag{5.15}$$

and in the neighboring intermediate MeSi phase by

$$N(x) = \frac{[N(x_{j1}+\delta)\mathrm{erf}(a_2) - N(x_{j2}-\delta)\mathrm{erf}(a_1\, b) - N(x_{j1}+\delta) + N(x_{j2}-\delta)\mathrm{erf}(x\cdot a_2/x_{j2})]}{[\mathrm{erf}(a_2) - \mathrm{erf}(a_1\, b)]}$$

$$a_1 = \frac{x_{j1}}{2\,(D_1\, t)^{1/2}}, \quad a_2 = \frac{x_{j2}}{2\,(D_2\, t)^{1/2}}, \quad b = (D_1/D_2)^{1/2} \tag{5.16}$$

D_1, D_2 are the silicon diffusivities in $MeSi_2$ and MeSi, respectively. Regions near the interface are marked with x coordinate "$-\delta$" for the left side and "$+\delta$" for the right one.

The mathematics presented, and the experimental parameters in Tables 5.2–5.9 permit us to predict the phase composition and silicide layer thickness as a function of the heating regimes employed [187]. While basic equations (5.10)–(5.12) have been written in a parabolic form, one can substitute all or some of them with linear equations of type (5.4) in order to describe a reaction-limited process for any particular silicide phase formation. This seems to be more flexible and reliable than use of fitting and semiempirical parameters in classical reactive diffusion models [140, 186, 188–190].

5.4. SILICIDE LAYER STRUCTURE

The crystalline structure of bulk silicides has been studied extensively [1, 4]. Nevertheless, general features of the structural transformations in thin-film silicides have not been reported. The only exception is the group of epitaxially grown silicides including Pd_2Si, PtSi, $NiSi_2$, and $CoSi_2$ investigated for conventional furnace annealing [110, 131, 191–205] and RTP [116–120, 138, 164–166].

The experimental data for polycrystalline and epitaxial silicide layers produced by RTP are discussed in this section. A polycrystalline silicide layer is the most common product of solid-phase silicidation of a heated thin-film metal–silicon structure. Silicide crystallites nucleate in an amorphouslike intermediate layer or directly at the metal–silicon interface [27, 53]. The extension of the dominant silicide phase is accompanied by corresponding growth of grains as the processing temperature and time are increased.

5.4.1. *Grain Growth*

Figure 5.9 shows the kinetics of grain growth for WSi_2 [87] and PtSi [179] during their synthesis by RTP. Linearity in $t^{1/2}$ coordinates shows the parabolic law of the

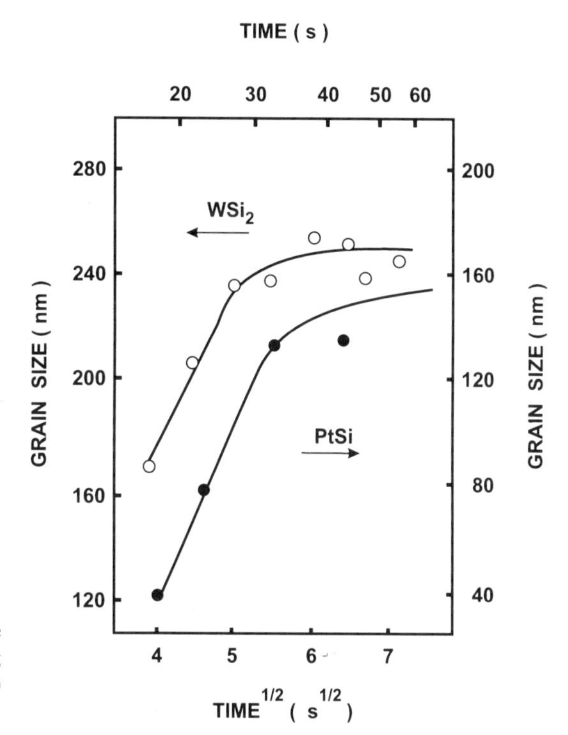

FIGURE 5.9. Grain growth kinetics of the silicides formed by rapid thermal processing of tungsten and platinum films on (100) silicon at 1100 and 400 °C, respectively.

crystallite growth, indicating the major role of the diffusion process in the synthesis of these silicide phases. Qualitatively identical results were obtained for $TaSi_2$ [72]. Thus, structural changes in the growing silicide layers are controlled by diffusion of metal and silicon atoms. With increasing time and temperature the silicide layer forms, and structural changes are stabilized. Recrystallization becomes a dominant process for which the processing temperature is a decisive factor. Coalescence of small grains of close crystallographic orientations is followed by growth of large crystallites as has been observed for PtSi [24, 180, 181].

Grain growth in the already synthesized silicide layer was observed for PtSi [24, 180, 181]. The experimental data are presented in Fig. 5.10. In the temperature range of 660–1050 °C the average grain size appears to be independent of the processing temperature. The main effects are associated with the induced temperature. Recrystallization of PtSi starts at 660 °C which corresponds to the correlation between the temperatures of recrystallization, T_r, and melting, T_m, for chemical compounds supposing $T_r = (0.5–0.6)T_m$ [206].

There are two stages in the PtSi recrystallization process, the first in the temperature range of 660–760 °C and the second at 850–1050 °C. An activation energy of 0.75 eV is typical for both stages. This value agrees well with the data for recrystallization of tungsten, molybdenum, and tantalum disilicides (0.65–0.70 eV) under conventional furnace annealing [207]. It is also not far from the activation energies (1.09 and 1.17 eV) for epitaxial regrowth of amorphous nickel and cobalt disilicides [205]. Identical activa-

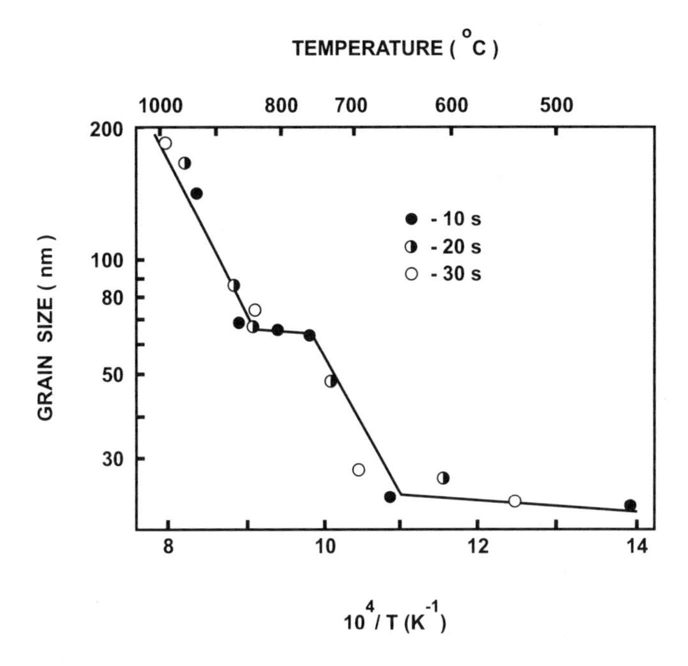

FIGURE 5.10. Average grain size of PtSi formed by rapid thermal processing of platinum film on (111) silicon as a function of the processing temperature [24].

tion energies for both stages suggest identical reaction mechanisms involving the coalescence of small grains superimposed with grain boundary migration at higher temperatures. At 800 °C and above, on (111) silicon, large grains appear whose size is defined by oxygen contamination in the as-deposited platinum film and by the impurities in the silicon substrate.

5.4.2. Grooving and Agglomeration

Heat-assisted grooving of grain boundaries and associated agglomeration in polycrystalline silicide films lead to their islanding during high-temperature processing [208]. The grooving process occurs as a result of a local equilibrium at the intersection of the grain boundary and the film surface and matter diffusion away from the high-curvature region formed. The equilibrium is defined by a balance of forces representing surface and grain boundary energies. Grooves develop at the silicide surface and silicide–substrate interface. Diffusion should continue until a constant curvature surface is reached. The resultant equilibrium morphology of a grain would be spherical caps on top and bottom, with the perimeter held at the equilibrium grooving angles. Based on this approach one can calculate critical grain size below which no agglomeration takes place.

Analysis of the agglomeration process performed by Maex [9] shows that the critical grain size increases for thicker films, lower grain boundary energy of higher interfacial and surface energies. Typical values of the critical grain size are on the order of 10 times the silicide film thickness, which implies 500-nm grain size for a 50-nm silicide film.

Agglomeration-related thermal stability of $TiSi_2$ and $CoSi_2$ is worse on polycrystalline than on monocrystalline silicon. Impurities at the interface or at the surface improve it. These are the only regularities reported. Systematic studies are in progress. Investigation of lateral confinement effects seems to be of particular practical importance.

5.4.3. Crystalline Structure Transitions

Some of the silicides can have two stable crystalline structures [1]. Consequent transitions between them are initiated by thermal processing. Among the silicides discussed, such structure transitions have been observed for $TiSi_2$, WSi_2, and $MoSi_2$. Two crystal modifications of $TiSi_2$ are characteristic of this material. In the thin-film disilicide orthorhombic base-centered lattice (C49 phase) is formed first at about 350 °C. A temperature increase to 600–800 °C initiates a transformation into the more stable orthorhombic face-centered configuration (C54 phase) [36, 209]. This transition takes place at 600–650 °C for as-deposited titanium films, thickness of 100–200 nm, and at 700–800 °C for thinner films (30–80 nm) [209] during conventional furnace annealing. Under conditions of RTP the C49-to-C54 transition takes place at 700–900 °C [19, 62, 210–216]. Moreover, the transition temperature depends on the width of the $TiSi_2$ lines when patterning of the film has been performed [211]. It increases by as much as 180 °C with a decreasing linewidth from 1000 to 100 nm.

Nucleation of the C54 phase predominantly take place at triple junctions of C49 grains with less interphase coherency [210]. No orientation relationships were found between the phases. The density of triple grain junction nucleation is higher than that at a grain boundary one. The final C54 microstructure is considered [211] to be a result of nucleation and growth of grains with different orientations. In the one-dimensional geometry imposed by narrow lines, elongated (040) grains oriented with the fast growth parallel to the lines are dominant. In two-dimensional blanket films, the isotropic (311) grains are more equally represented. Activation energy of the C49-to-C54 transition was reported to be 4.45 eV [36], 5.6–5.7 eV [212].

RTP-processed titanium disilicide has the same basic features as discussed for conventional furnace annealing, while the activation energy is reduced to 3.3–3.8 eV [62, 215, 216]. Impurities from the annealing ambient are thought to be responsible for the effect. They have enough time to diffuse into $TiSi_2$ during furnace annealing and form precipitates retarding the transition, whereas short RTP minimizes such possibilities. The dramatic role of impurities was shown in experiments [215] demonstrating an increase in the transition activation energy to 7.8 eV for the $TiSi_2$ formed on heavily arsenic-doped silicon substrates.

The importance of the stable low-resistivity C54 structure of $TiSi_2$ in practical applications has stimulated development and use of two-step RTP for its formation. It includes [215]: first annealing of a titanium–silicon thin-film structure at 700 °C for 30 s in order to get the C49 phase, etching of unreacted metal and TiN, and second annealing in the temperature range of 750–1000 °C for 10 s and longer in order to form the C54 phase. Particular temperature and time duration of the second step are mainly dependent on the thickness of the film. Higher temperatures are required for thinner films. Formation of homogeneous C54 $TiSi_2$ films thinner than 40 nm becomes difficult [9].

Lattice changes are also possible in WSi_2 [215–219] and $MoSi_2$ [220]. These compounds may have hexagonal and tetragonal lattices. The transition from the first to the second structure type occurs at 550–675 °C. The final phase is a tetragonal lattice disilicide which predominates in layers formed over 30–40 s at 1000–1100 °C, and is the most thermally stable [23, 87]. Structural changes were complete, confirmed by the invariability of the grain size with increases of processing time (see Fig. 5.9).

5.4.4. Epitaxial Growth

Formation of epitaxial silicides by RTP of metal films on single-crystal silicon was investigated for the typical epitaxially grown silicides: Pd_2Si [164–166], PtSi [166], $NiSi_2$ [116–120], $CoSi_2$ [117, 138, 221], as well as for $TiSi_2$ [13, 27, 222] and $TaSi_2$ [68]. Data are also available for conventional furnace-induced local epitaxy of disilicides of titanium [223–225], tungsten [217–219], molybdenum [217, 220], vanadium [226], chromium [146, 227, 228], iron [229–234], calcium [235], and erbium [236, 237]. These strengthen the belief that the list of RTP-produced epitaxial silicides is not limited to these examples.

The known epitaxial relationships between silicides and silicon are summarized in Table 5.11. $NiSi_2$ and $CoSi_2$ are not included because they have a cubic lattice identical to silicon. Their main planes and directions are marked with indexes parallel in the epitaxial structures being formed [191–205].

Data from Table 5.11 indicate that the (111) silicon surface is the most suitable for epitaxial growth. As a rule, epitaxial structures are formed at high temperatures, 1100–1300 °C, for both rapid thermal and conventional furnace annealing. The average size of epitaxial blocks varies from hundreds of nanometers to hundreds of micrometers.

Continuous epitaxial layers of palladium, platinum, nickel, and cobalt silicides were obtained with RTP. Epitaxial Pd_2Si was formed by processing palladium films 35–70 nm thick on (111) silicon with a scanning electron beam [164, 165]. The induced temperature was 500–700 °C and effective heating duration about 7 ms. The epitaxial layer grew from the metal–silicon interface. It had a perfection characterized by χ_{min} = 36–40% (for 2-MeV He^+ ion channeling) which is a little worse than χ_{min} = 20% for structures created by furnace annealing at 715 °C for 30 min. The considerable temperature gradients at the edge of electron beam scanning lines are responsible for this. However, epitaxial Pd_2Si with χ_{min} = 13% was already synthesized at 500 °C/15 s heating for analogous structures heated with incoherent light [166]. Epitaxial PtSi was formed by RTP of a "pure" platinum thin film on silicon structures [166]. It was characterized by χ_{min}= 19% (for 2-MeV He^+ ion channeling) equal to that for conventional furnace processing [191].

$NiSi_2$ and $CoSi_2$ synthesized on single-crystal silicon by RTP have the highest epitaxial perfection [116–120, 138]. For (111) silicon substrates, χ_{min} = 4.6 and 5.7% for nickel and cobalt disilicides, respectively [117]. This is better than that produced by conventional furnace annealing, but similar to the best formed by molecular beam epitaxy [194], with χ_{min} = 3–5% for $NiSi_2$ and χ_{min} = 2–4% for $CoSi_2$.

The induced temperature has an influence on the perfection of the synthesized epitaxial disilicides, as well as the sample cooling rate, illustrated by the results shown in Fig. 5.11. The induced temperature strictly controls the phase and structural transformations in the silicide layers. RTP of nickel–silicon structures at temperatures below 700 °C

TABLE 5.11. Epitaxial Relationships for Silicides on Single Crystal Silicon

Silicide	Silicon	Parallel planes: silicide//silicon	Parallel directions: silicide//silicon	Processing conditions temperature/time (°C/s)	Size of epitaxial blocks (μm)	Refs.
$TiSi_2$	(111)	$(\bar{3}13)//(2\bar{2}0)$[a]	$[\bar{1}01]//[111]$[a]	1000/20	~ 20	13
orthorhombic		$(004)//(02\bar{2})$	$[100]//[111]$			
	(111)	$(004)//(02\bar{2})$[a]	$[100]//[111]$[a]	500/60 min +	2–56	223
		$(400)//(02\bar{2})$	$[001]//[111]$	1100/60 min		
	$(1\bar{1}\,\bar{1})$	$(1\bar{1}\,\bar{1})//(\bar{1}01)$[a]	$[110]//[010]$[a]	700, 900/	0.2–2.0	222
		$(1\bar{1}\,\bar{1})//(\bar{1}01)$	$[110]//[121]$	few seconds		
		$(1\bar{1}\,\bar{1})//(\bar{1}01)$	$[110]//[1\bar{3}1]$			
		$(1\bar{1}\,\bar{1})//(\bar{1}01)$	$[110]//[161]$			
		$(1\bar{1}\,\bar{1})//(\bar{1}01)$	$[110]//[1\bar{4}1]$			
		$(1\bar{1}\,\bar{1})//(\bar{1}01)$	$[110]//[101]$			
		$(1\bar{1}\,\bar{1})//(\bar{1}01)$	$[110]//[141]$			
		$(1\bar{1}\,\bar{1})//(\bar{1}01)$	$[110]//[121]$			
		$(1\bar{1}\,\bar{1})//(010)$	$[110]//[001]$			
		$(1\bar{1}\,\bar{1})//(\bar{2}\,\bar{1}1)$	$[110]//[1\bar{1}1]$			
		$(1\bar{1}\,\bar{1})//(001)$	$[110]//[100]$			
		$(1\bar{1}\,\bar{1})//(100)$	$[110]//[001]$			
	(001)	$(113)//(040)$[a]	$[301]//[001]$[a]	1000/20 +	9–36	27
		$(111)//(220)$	$[110]//[001]$	1080/15		
		$(111)//(220)$	$[101]//[001]$			
	(001)	$(100)//(110)$	$[011]//[001]$	850/900	~ 5	224
		$(001)//(110)$	$[110]//[001]$			
		$(100)//(110)$	$[010]//[001]$			
$TaSi_2$	(111)	$(0003)//(22\bar{4})$	$[1\bar{1}00]//[111]$	1300/300	0.3–1.2	71
hexagonal		$(0003)//(2\bar{2}0)$	$[1\bar{1}00]//[111]$			
		$(0003)//(22\bar{4})$	$[1\bar{2}10]//[111]$			
	(001)	$(11\bar{2}2)//(2\bar{2}0)$[a]	$[1\bar{1}00]//[001]$[a]			
		$(1\bar{2}12)//(220)$	$[1\bar{2}13]//[001]$			
		$(1\bar{2}14)//(400)$	$[\bar{2}4\bar{2}3]//[001]$			
VSi_2	(111)	$(22\bar{4}0))//(22\bar{4})$	$[0001]//[111]$	400/60 min +	1–4	226
hexagonal		$(20\bar{2}0//(20\bar{2})$	$[0001]//[111]$	1000/60 min		
		$(20\bar{2}0)//(20\bar{2})$	$[1\bar{2}13]//[101]$			
		$(2\bar{2}0\bar{2})//(1\bar{3}\,\bar{1})$	$[1\bar{2}13]//[101]$			
		$(2\bar{2}02)//(3\bar{1}1)$	$[1\bar{2}1\bar{3}]//[141]$			
		$(20\bar{2}0)//(20\bar{2})$	$[1\bar{2}1\bar{3}]//[141]$			
WSi_2	(111)	$(20\bar{2}0))//(20\bar{2})$	$[0001]//[111]$	600/60 min +	0.5–1.0	218–219
hexagonal	(001)	$(20\bar{2}0)//(2\bar{2}0)$	$[0001]//[001]$	1100/60 min		
		$(2\bar{1}\,\bar{1}2)//(2\bar{2}0)$	$[\bar{2}4\bar{2}3]//[001]$			

(*continued*)

TABLE 5.11. (*Continued*)

Silicide	Silicon	Parallel planes: silicide//silicon	Parallel directions: silicide//silicon	Processing conditions temperature/time (°C/s)	Size of epitaxial blocks (μm)	Refs.
WSi_2	(111)	$(00\bar{4})//(20\bar{2})$	[110]//[111]	600/60 min +	0.5–1.0	218–219
tetragonal		$(11\bar{2})//(20\bar{2})$	[111]//[111]	1100/60 min		
	(001)	$(00\bar{4})//(2\bar{2}0)$	[110]//[001][a]			
		$(11\bar{2})//(2\bar{2}0)$	[110]//[001][a]			
		(004)//(220)	[100]//[001]			
$MoSi_2$	(111)	$(20\bar{2}0)//(20\bar{2})$	[0001]//[111]	450/60 min +	0.5–1.0	217, 220
hexagonal	(001)	$(20\bar{2}0)//(2\bar{2}0)$	[0001]//[001]	1100/60 min		
		$(2\bar{1}0\bar{1}2)//(2\bar{2}0)$	$[\bar{2}4\bar{2}3]//[001]$			
$MoSi_2$	(111)	$(00\bar{4})//(20\bar{2})$	[110]//[111]	450/60 min +	0.5–1.0	217, 220
tetragonal		$(11\bar{2})//(20\bar{2})$	[111]//[111]	1100/60 min		
	(001)	$(00\bar{4})//(2\bar{2}0)$[a]	[110]//[001][a]			
		$(11\bar{2})//(2\bar{2}0)$[a]	[110]//[001][a]			
		(004)//(220)	[100]//[001]			
$CrSi_2$	(111)	$(22\bar{4}0)//(22\bar{4})$	[0001]//[111]	400/60 min +	1–5	227
hexagonal		$(20\bar{2}0)//(20\bar{2})$	[0001]//[111]	1100/60 min		
		$(20\bar{2}0)//(20\bar{2})$	$[1\bar{2}13]//[101]$			
		$(022\bar{2})//(13\bar{1})$	$[1\bar{2}13]//[101]$			
	(111)	(001)//(111)	[100]//[011]	420–1250 °C	—	146
	(001)	$(2\bar{2}00)//(2\bar{2}0)$	$[\bar{2}\bar{2}43]//[001]$	1100/60 min	0.5–1.0	228
$FeSi_2$	(111)	(220)//(111)	$[0\bar{1}1]//[1\bar{1}1]$	300/60 min +	< 40	229
orthorhombic		(422)//(022)	$[0\bar{1}1]//[1\bar{1}1]$	1000/60 min		
		(022)//(004)	[100]//[100]			
	(001)	(001)//(100)	[110]//[010]	Molecular beam epitaxy at 400–450 °C	—	230
$FeSi_2$	(111)	(010)//(100)	[100]//[001]	300/60 min +	< 40	229
tetragonal		(001)//(010)	[100]//[001]	1000/60 min		
		(100)//(010)	[010]//[001]			
		(001)//(100)	[010]//[001]			
Pd_2Si hexagonal	(111)	(0001)//(111)	$[2\bar{1}\bar{1}0]//[11\bar{2}]$	250/120 min	—	238
PtSi	(111)	$(0\bar{1}0)//(111)$[a]	[001]//[011][a]	600/30 min	30–100	239
orthorhombic			[100]//[112]			
	(100)	(110)//(100)	—			
$ErSi_2$	(111)	(0001)//(111)	—	400 °C + 800 °C	~ 0.2	236
hexagonal		$(10\bar{1}0)//(4\bar{2}\bar{2})$				

[a]Dominant mode.

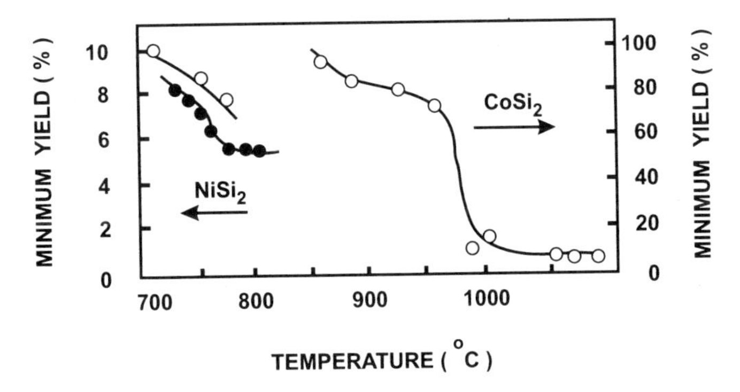

FIGURE 5.11. Minimum backscattering yield of 2-MeV He^+ ions channeling along ⟨111⟩ axis for $NiSi_2$ (125 nm) and $CoSi_2$ (90 nm) formed within 15 s from the metal films on (111) silicon via "fast" (○) and "slow" (●) cooling [117].

produces a layer containing mono- and disilicide. If the temperature is increased to 700–800 °C, a single-phase epitaxial disilicide is formed. The highest perfection is reached when the disilicide becomes the only phase at 770 °C (see Fig. 5.11). The quality of $NiSi_2$ was critically dependent on the cooling rate through a quenching effect. "Slow" cooling (to 200 °C within 60 s) improves χ_{min} from 7 to 5%, compared with "fast" cooling (to 200 °C within 20 s).

$NiSi_2$ epitaxial layers were also synthesized on (100) and (110) silicon [116, 117, 119]. A higher epitaxial perfection was observed for the (110) substrate. The thickness of the epitaxial layer influenced the quality as well, particularly for layers thicker than 100 nm, characterized by a two- to threefold increase of χ_{min} for 400-nm silicide layers. Preliminary ion beam mixing of the components at the metal–silicon interface with argon implantation stimulates phase formation and reduces this influence [119, 120].

Similar transformations occur for cobalt silicide layers, but at higher temperatures. In the conventional metal-on-silicon structures, epitaxial layers on (111) silicon begin at 1000 °C, at which temperature the disilicide phase is completely formed. In contrast to nickel, for which the disilicide grows simultaneously with the synthesis of the phase, epitaxial $CoSi_2$ is thought to appear as a result of recrystallization of the phase already formed. This may explain why the $CoSi_2$ has slightly lower crystalline perfection than $NiSi_2$.

Epitaxial growth of $CoSi_2$ on (100) silicon was shown to be difficult because of the competition between different orientations with similar lattice matching [240]. That is why preferential growth conditions at the silicon–silicide interface and temperature–time program of processing are important in achieving good epitaxy. Implantation of cobalt ions into monocrystalline silicon substrates [8], use of Ti–Co multilayer films [9] and their combination with two-step RTP [138, 221] resolve the problem mainly by eliminating the influence of interfacial impurities on the epitaxial nucleation and growth. It is achieved via atomic mixing by ion implantation or via gettering of the impurities by the titanium layer.

The two-step processing implies that the first step produces CoSi. It is achieved at 550 °C for 60 s in the samples which were cobalt implanted prior to the metal film deposition onto silicon [221] and at 650 °C for 60 s in the Ti–Co film on silicon substrates [138]. The second step is processing at 900 °C for 10–60 s after etching off unreacted metal and oxides from the surface. Epitaxial $CoSi_2$ nucleates and starts growing in the monosilicide at the interface with the monocrystalline silicon substrate. Its final crystalline quality on (100) silicon was characterized by $\chi_{min} = 5\%$ in the case of ion implantation and by $\chi_{min} = 15\%$ for Ti–Co structures.

In conclusion, it should be pointed out that structural transformations of the silicide layers synthesized by RTP are determined by thermal processes. They are governed by the same basic mechanisms as those produced by conventional furnace annealing. The specific features that are different are associated with the enhanced phase formation at the higher temperatures usually produced during RTP, and less influence of impurities from annealing ambient as a result of the reduced thermal budget.

5.5. IMPURITY BEHAVIOR

Silicides in microelectronics are usually formed on doped polycrystalline or single-crystal silicon. Impurity behavior in these structures is of practical importance in order to estimate possible changes in contact resistivity, and the depth of the underlying *p–n* junctions. Substitutional impurities in silicon, especially boron, phosphorus, arsenic, and antimony, are of particular interest.

For short time intervals in solid-phase synthesis of silicide, we can assume that there is no detectable diffusion of impurity atoms in doped layers. However, silicidation introduces point defects at the silicon surface, thus enhancing diffusion of both metal and silicon atoms. The doped silicon is consumed in the growing silicide and therefore will influence the impurity distribution even at low temperatures. The impurity redistribution observed for RTP has been confirmed by results of conventional furnace annealing [241–243]. We will distinguish the two cases practically important for impurity redistribution in: (1) metal–silicon structures and (2) silicide–silicon structures.

5.5.1. Metal–Silicon Structures

This type of structure is produced when a metal film is deposited onto a doped silicon substrate, or when an impurity is implanted through the metal into the silicon substrate. Impurity redistribution in such structures after RTP silicide formation has been investigated for titanium (boron, arsenic, antimony) [54, 55, 59, 244–246], tantalum (arsenic) [247], tungsten (boron, arsenic) [83, 84, 248], molybdenum (arsenic) [83, 84], cobalt (arsenic) [244], nickel and chromium (arsenic) [247], and palladium (arsenic) [167]. The results show that in general the impurity redistribution is characterized by: (1) segregation of impurity atoms to silicon by the growing silicide phase, (2) fast impurity diffusion through the silicide layer to the sample surface, and (3) impurity accumulation and evaporation from the surface.

For example, Fig. 5.12 shows the redistribution of arsenic in silicon in a growing titanium silicide for processing of seconds on (100) silicon. The major fraction of the

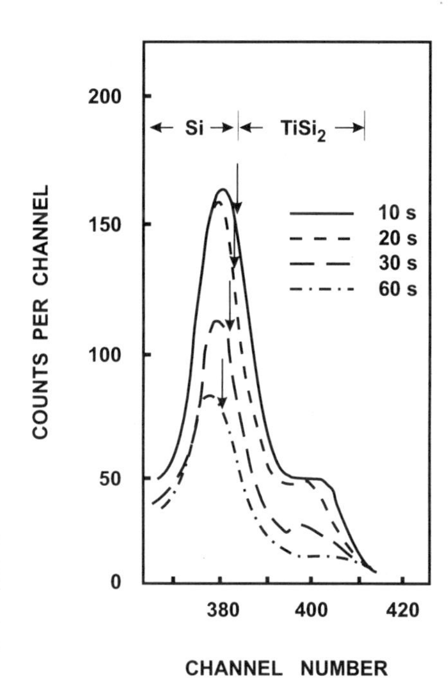

FIGURE 5.12. RBS arsenic profiles in Si–$TiSi_2$ structure formed by rapid thermal processing of titanium film on (100) silicon doped by arsenic (150 keV/5 × 10^{15} ions/cm^2 + 600 °C/30 min) at 850 °C for various processing times. The arrows indicate the position of the silicide–silicon interface [245].

impurity is localized in the silicon at the interface with the growing silicide, just as for conventional long-time furnace annealing [249–253]. The maximum arsenic concentration moves into the silicon as the silicidation front proceeds, although its magnitude decreases with diffusion of impurity atoms into the silicide. Impurity segregation into silicon during formation of $TiSi_2$ is most pronounced at 850–900 °C [54, 59, 245]. This effect is also present for cobalt [244], chromium [247], and palladium [167] silicides formed at temperatures below 900 °C. It is less marked for higher-temperature (1000–1100 °C) tantalum [249], tungsten, and molybdenum [83, 84] disilicide synthesis in which case impurity diffusion into the silicide layer becomes dominant.

Redistribution of substitutional impurities in silicon at the moving silicide front is determined by two groups of competing processes: (1) impurity segregation at the silicide–silicon interface and generation of nonequilibrium point defects in the silicidation reaction and (2) the capture of impurity atoms by the growing silicide layer and their fast diffusion in it.

The nonequilibrium defects influencing the impurity diffusion in silicon can be either vacancies or interstitials. When silicon atoms in a silicide are the most mobile species, the growth of the silicide layer is controlled by their diffusion, and nonequilibrium vacancies are generated at the silicide–silicon interface [246, 254]. In the case of metal-rich silicide formation, it is governed by diffusion of the more mobile metal atoms, as demonstrated for nickel, cobalt, palladium, and platinum silicides [3]. The reaction zone is saturated with interstitials which are produced by forcing silicon and impurity atoms from their regular lattice positions by the metal. This has been confirmed by observations in conventional furnace experiments of the enhanced redistribution of phosphorus, arsenic, and antimony at nickel, palladium, and platinum silicide fronts, and by the absence of such an effect during titanium, tantalum, molybdenum, and vanadium silicide growth [241, 243]. Arsenic has been noted to occupy nonsubstitutional electrically inactive positions in the region of enhanced redistribution.

The shift of the silicidation front into the silicon and diffusion of impurities into the silicide restricts their redistribution. Typical parabolic rates of silicide layer growth are 10^{-10}–10^{-12} cm^2/s. These are several orders of magnitude faster than the fastest interstitial impurity diffusion in silicon at temperatures of 800–900 °C. Thus, enhanced diffusion by itself is evidently unable to provide impurity redistribution in silicon in front of the growing silicide. However, the solubility of impurities in a silicide is a more important factor. The data for phosphorus and arsenic in WSi_2 [255], and $TiSi_2$ [257], indicate that the equilibrium solubility of group-III and -V impurities in silicide is in the range of 10^{19}–10^{20} atoms/cm^3 at 800–1000 °C. This is an order of magnitude lower than that in single-crystal silicon. Segregation of the impurities occurs and results in a diffusion barrier at the silicidation front. The barrier can be dissolved by impurity diffusion into the growing silicide, when the processing temperature is increased to 800–900 °C.

Impurity diffusion into the silicide becomes an important factor at higher temperatures. The limited experimental data available for phosphorus, arsenic, and antimony in titanium, tantalum, and tungsten disilicides are shown in Table 5.12. At 800 °C and above, the diffusion mobility is two to three orders of magnitude higher than that in single-crystal silicon. The diffusion activation energy not only depends on the impurity, but also to a greater extent varies from one silicide to another. Fast impurity redistribution in silicides under RTP conditions agrees well with calculated estimates, although the mechanism needs further study.

TABLE 5.12. Impurity Diffusion in Silicides

Silicide	Impurity	Temperature range (°C)	Preexponential factor, D_0 (cm^2/s)	Activation energy, E_a (eV)	Diffusivity, D (800 °C) (cm^2/s)[a]	Ref.
$TiSi_2$ (+ $TiSi$, Ti_5Si_3)	Sb	600–850	—	1.9	—	249
$TiSi_2$	As	675–900	—	0.95	—	253
	As	550–800	5.1×10^{-6a}	1.8	1.7×10^{-14}	258, 259
	P	550–800	5.4×10^{-5a}	2.0	2.1×10^{-14}	258
$TaSi_2$ crystal	P	660–910	6.44×10^{-10}	0.61	8.7×10^{-13}	260
$TaSi_2$ amorphous	P	660–910	2.0×10^{-8}	0.76	5.3×10^{-12}	260
WSi_2	P, As	800	—	—	3.3×10^{-12}	255
$MoSi_2$ polycrystalline	As	550–700	3.2×10^{-6}	1.4	8.2×10^{-13}	256
$CoSi_2$ polycrystalline	P	600–800	0.02	2.6	1.2×10^{-14}	261
	B	400–800	0.004	2.0	1.5×10^{-12}	261
	Ge	600–800	0.05–200	3.1	—	261

[a]Calculated from the experimental data presented in the reference.

5.5.2. *Silicide–Silicon Structures*

Silicide–silicon structures are formed by deposition of a silicide on polycrystalline or single-crystal silicon. Magnetron sputtering is traditionally used for this purpose as well as coevaporation and chemical vapor deposition (CVD). These techniques produce only a mixture of silicide forming components. Thermal processing is necessary to synthesize silicide phases (usually disilicides) with the appropriate structure and electrophysical parameters. Therefore, impurity redistribution is dependent on the conditions of conventional furnace annealing [242] and RTP [262]. Properties produced by RTP within seconds were investigated for silicides of titanium (phosphorus [242], arsenic [263, 264]), tantalum (phosphorus [265]), tungsten (boron [81], phosphorus [85], arsenic [82]), and molybdenum (boron [266], phosphorus [92]). Doped layers in a silicide–silicon structure can be in the silicon beneath the silicide interface, in the silicide, or on the silicide surface. The particular location determines the amount of impurity diffusion from the doped layer under thermal processing.

Silicide films deposited onto heavily doped polycrystalline silicon are most typical. They are widely used for so-called polycide interconnects in integrated circuits. RTP of such structures produces diffusion of impurities into the silicide and through it to the

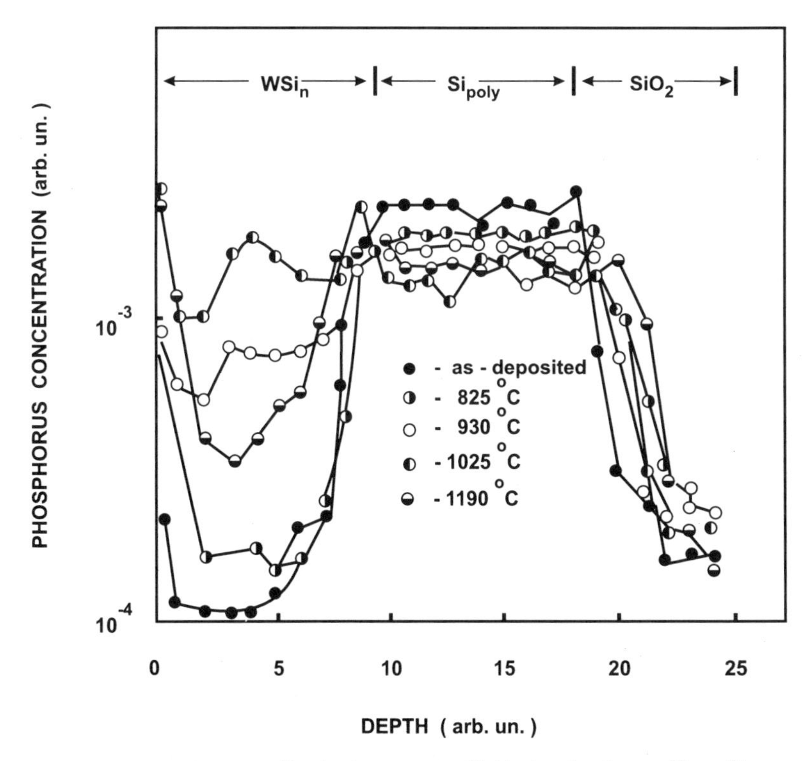

FIGURE 5.13. SIMS phosphorus profiles in the tungsten silicide-doped polycrystalline silicon structure subjected to rapid thermal processing for 6 s [85].

surface where their evaporation takes place [81, 85, 92, 242, 265]. Figure 5.13 illustrates typical impurity behavior for phosphorus redistribution in tungsten silicide on polycrystalline silicon. A silicide layer of $WSi_{2.6}$ was formed by CVD growth at 890 °C on the polycrystalline silicon saturated with phosphorus. Phosphorus redistribution in this structure already appears at 825 °C as the impurity atoms migrate from the polycrystalline silicon to the silicide, where they accumulate because of the limited rate of their evaporation from the surface. The higher the processing temperature, the faster the impurity concentration decreases in the silicon and increases in the silicide. A balance in concentration for both areas occurs with processing at 1025 °C for 6 s. A further temperature increase to 1190 °C changes the balance in favor of impurity evaporation from the surface, and results in a decrease in the impurity concentration in the silicide.

Boron, phosphorus, and arsenic evaporative losses from polycrystalline silicon covered by titanium, tantalum, tungsten, and molybdenum silicides reach 50–70% with processing for seconds at 1000–1200 °C. This loss is probably the result of rapid diffusion of the impurities along grain boundaries in the polycrystalline silicon, because it is not observed with single-crystal silicon [263]. The high diffusion mobility of the impurities in silicides results in a rapid redistribution in the implanted silicide layers [82, 264, 266] and diffusion from the surface through the silicide into the underlying silicon [85, 266].

Figure 5.14 shows implanted arsenic redistribution in CVD $WSi_{2.6}$ on single-crystal silicon. During subsequent annealing the impurity atoms diffuse both to the surface, where they are evaporated, and into the silicide–silicon interface and silicon. Arsenic depth profiles in the silicide are identical for rapid and conventional furnace annealing. However, in silicon there is a remarkable difference seen with RTP. There is evidence of enhanced impurity diffusion in the silicide, but no enhancement in single-crystal silicon.

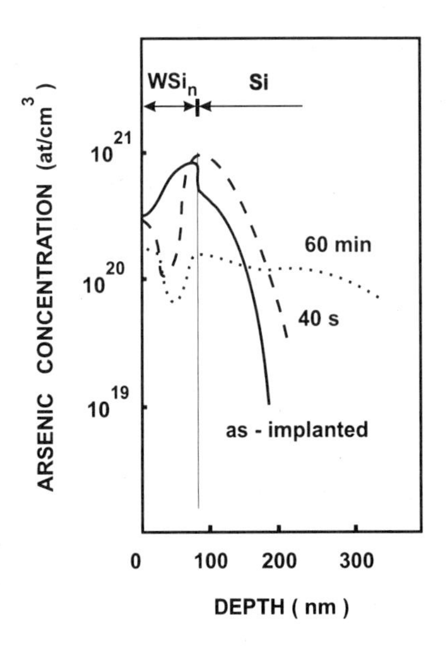

FIGURE 5.14. SIMS concentration profiles of arsenic implanted (100 keV/1 × 10^{18} ions/cm^2) in tungsten silicide film on (100) silicon and annealed at 1000°C [82].

The preliminary high-temperature annealing of as-deposited silicide for seconds has limited impurity redistribution for boron implanted into molybdenum silicide [266].

Similar features are seen for phosphorus diffused into tungsten silicide [85], and boron in molybdenum silicide [266] using an emulsion source for dopant. Diffusion concentration profiles of boron in as-deposited molybdenum silicide on single-crystal silicon substrates or preannealed for 30 s are shown in Fig. 5.15. The $MoSi_{2.3}$ layer was formed by magnetron sputtering. For the as-deposited samples impurity diffusion produced uniform boron distributions over the silicide layer. One can observe a clearly resolved impurity tail in the underlying silicon. Diffusion is slower in the preannealed samples and an erfc impurity depth profile is produced. Impurity penetration into silicon is always less for preannealed samples.

The high diffusion mobility of impurities in as-deposited silicide layers is the consequence of the stoichiometry and structural imperfection of the silicides. There is usually some silicon surplus in such layers to guarantee disilicide formation after annealing. The redundant silicon is pushed to grain boundaries when the disilicide phase is formed, and hence the channels for enhanced impurity diffusion are created. With an increase in processing time the silicon atoms leave the silicide layer or they form crystallites at the silicide grain boundaries. The flux of grain boundary fast-diffusing impurities is reduced and bulk grain diffusion is favored. Even in this case group-III and -V impurities still diffuse faster in silicide than in silicon. This allows silicides to be used as bulk impurity diffusion sources for shallow (< 100 nm) doped layer formation in monocrystalline silicon.

Effectiveness of the RTP-assisted doping processing with the use of silicides as a substitutional impurity (B, P, As) diffusion source was first investigated for $TiSi_2$ [267–269] and $CoSi_2$ [270–274], which are considered as prime candidates for contacts to shallow *p–n* junctions in silicon. $TaSi_2$ [269] and WSi_2 [271] have also been tested. The disilicides were produced within a self-aligned process by solid-state reaction of metal film with the silicon substrate. The impurities were implanted into the already formed disilicides. In a wide range of processing regimes (600–1150 °C, 10–120 s) impurity

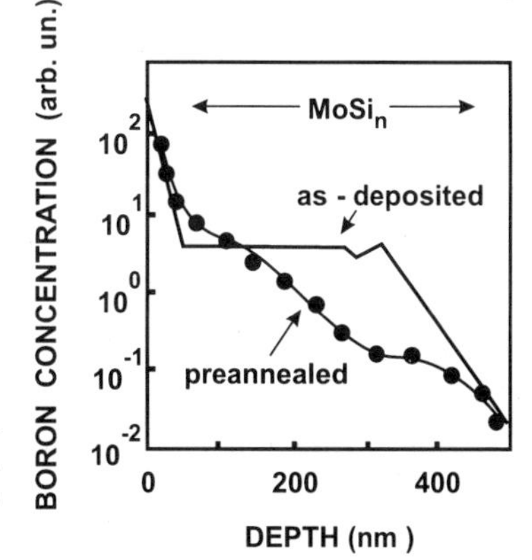

FIGURE 5.15. Boron concentration profiles in as-deposited and preannealed at 1200 °C for 30 s molybdenum silicide films on (100) silicon diffused from spin-on borosilicate glass at 950 °C for 60 min [266].

diffusion was not influenced by the presence of a silicide layer. The impurity profiles in the underlying silicon can be described in terms of diffusion from an infinite source.

In the case of $TiSi_2$ and $TaSi_2$, impurity precipitates in the form of TiB_2, TiAs, TaB_2, and TaAs compounds were observed [269]. They reduce the concentrations of diffusion-active free arsenic and boron in these silicides to levels of 1.6×10^{19} and 4×10^{18} atoms/cm^3, respectively. The same order maximum concentration of the impurity in the silicon was achieved. High concentrations of impurity precipitates were detected in the doped layers [268]. The precipitation is believed to be a thermal-stress-induced process arising from the difference in the expansion coefficients for the $TiSi_2$ and silicon. The impurity local solid solubility decreases with the stress enthalpy and might be considered responsible for the precipitate formation. Precipitates and related defects in the doped layer restrict the use of implanted $TiSi_2$ and $TaSi_2$ as diffusion sources.

In the $CoSi_2$ and WSi_2 impurity sources, no precipitation has been observed. Interface impurity concentrations close to their solid solubility in silicon (about 3×10^{20} As/cm^3 and 8×10^{19} B/cm^3) were obtained with RTP [271]. The use of 100-nm SiO_2 produced from a tetraethyl orthosilicate (TEOS) cap prevented impurity evaporation from the $CoSi_2$ source and resulted in an increase in the maximum boron concentration at the silicon surface, to $(1.5–9) \times 10^{20}$ atoms/cm^3 [270]. Accounting for the low resistivity and good contact properties of $CoSi_2$, it indeed meets requirements for formation of device structures with shallow doped layers in silicon and contacts to them. However, details of the impurity behavior in silicide–silicon structures are still important for further study.

5.6. ELECTROPHYSICAL PROPERTIES OF SILICIDES

Resistivity and Schottky barrier heights for silicon are regarded as a most important property of silicides. In addition, the forbidden band gap, concentration, and mobility of charge carriers are important parameters for semiconducting ones. These parameters are all sensitive to the phase composition and crystalline structure of the silicide layer. The characteristics of silicides synthesized by conventional furnace annealing are well known [1–7], and provide a good basis for comparison with those obtained by RTP.

Variation of the resistivity of a silicide layer as a function of the processing temperature and time is determined by the phase composition of the layer, bulk resistivities of the phases, and structure transformations. Processing temperature appears to be the factor that has the most influence on annealing times of seconds. Figure 5.16 summarizes experimental sheet resistivities of silicides synthesized by RTP at different temperatures. These are typical for titanium, tantalum, tungsten, molybdenum, cobalt, and zirconium silicides. The sheet resistivity decreases as the temperature increases when the richest silicon-containing silicide phase forms, which has a lower bulk resistivity than that produced by diffusion synthesis. For nickel, chromium, and platinum silicides, when the richest silicon-containing phase is produced it has a higher bulk resistivity, and therefore the sheet resistivity increases with processing time.

$TiSi_2$ best illustrates the effects of crystalline structure on resistivity. A phase transition from orthorhombic base-centered modification (C49), with a resistivity on the order of 100 μΩ-cm [36], to a more stable face-centered phase (C54) with lower resistivity, occurs on processing at 600–800 °C.

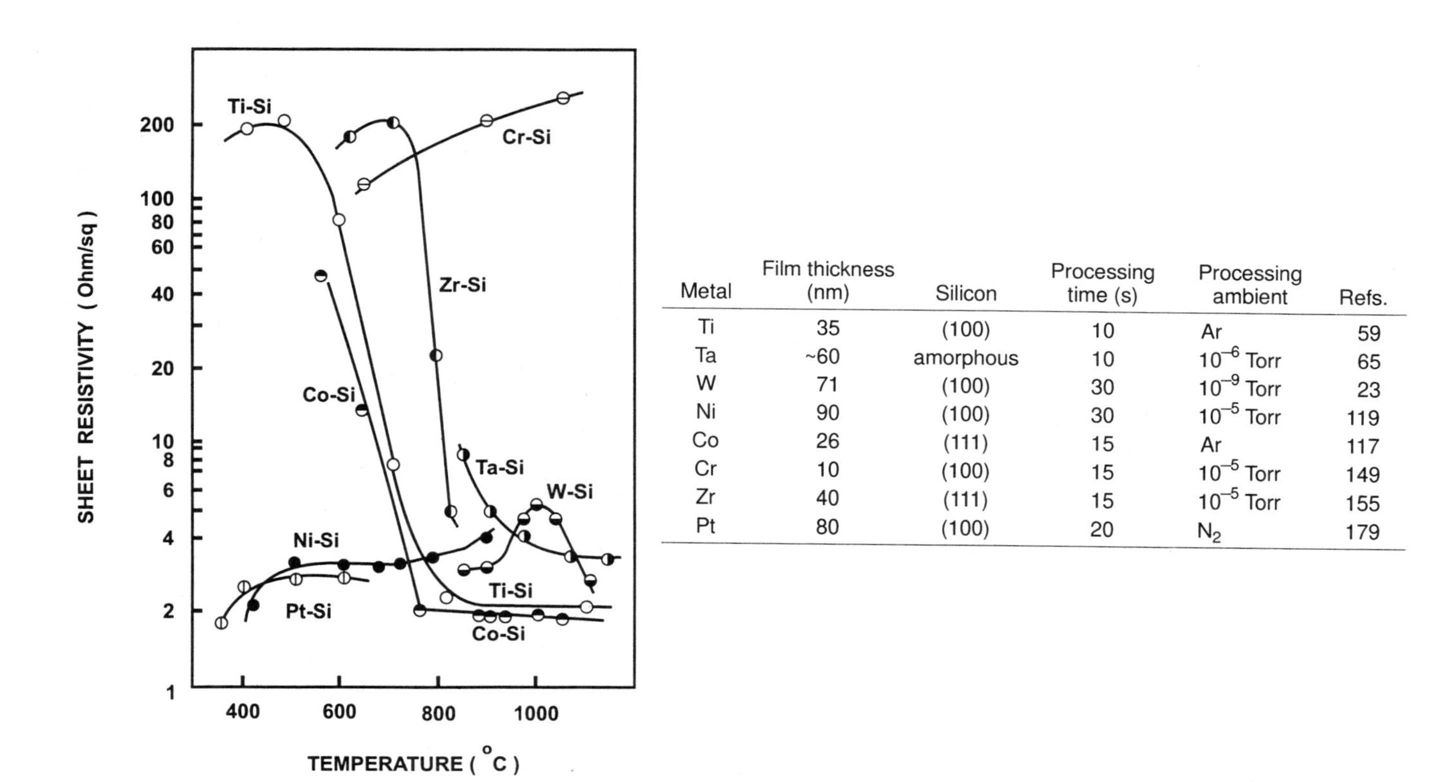

Metal	Film thickness (nm)	Silicon	Processing time (s)	Processing ambient	Refs.
Ti	35	(100)	10	Ar	59
Ta	~60	amorphous	10	10^{-6} Torr	65
W	71	(100)	30	10^{-9} Torr	23
Ni	90	(100)	30	10^{-5} Torr	119
Co	26	(111)	15	Ar	117
Cr	10	(100)	15	10^{-5} Torr	149
Zr	40	(111)	15	10^{-5} Torr	155
Pt	80	(100)	20	N_2	179

FIGURE 5.16. Sheet resistivity of silicide layers formed by rapid thermal processing of metal films on silicon.

Epitaxial silicides always have a lower resistivity than polycrystalline silicides [7]. The experimental data in Fig. 5.16 correspond to the silicon-rich phases with invariable resistivity and stable crystalline structure. The resistivity increases at high temperature as a result of silicide decomposition and mechanical destruction of the layer. Moreover, agglomeration discussed in Section 5.4.2 inevitably restricts a choice of the processing regimes, particularly for silicides thinner than 20 nm.

Silicides produced by RTP have bulk resistivities comparable to those produced by conventional long-time furnace annealing. Their values and recommended processing regimes are summarized in Table 5.13. RTP provides silicides with the lowest resistivity because of reduced contamination in the synthesized layers, by oxygen, carbon, and nitrogen from the annealing ambient. Although there is a challenge of low-resistivity silicides, $TiSi_2$ (C54 phase) and $CoSi_2$ are the best for interconnects in integrated circuits. $TiSi_2$ is frequently employed in industrial processes. However, extremely high chemical activity of titanium to impurities (O, N, B, As), increase of the temperature needed for C49–C54 transition with a decrease of the silicide film thickness, and agglomeration limit the implementation of $TiSi_2$ in submicron technology. $CoSi_2$ was shown [9] to have clear advantages over $TiSi_2$, thus replacing it in promoted applications.

Besides resistivity, other electrophysical parameters of the silicides produced by RTP are usually improved. Schottky barriers of platinum on n-type single-crystal silicon formed in 15–20 s at 500–750 °C have good barrier heights of 0.85–0.86 eV and an ideality factor of 1.04 [176, 275]. The best parameters for PtSi achieved by a two-step conventional furnace anneal are 0.878 eV and 1.03, respectively [276, 277]. These results are in good agreement within experimental uncertainty.

Semiconductor parameters were extensively investigated for chromium silicide layers synthesized by RTP within seconds [150, 151]. That material had previously been determined to be a p-type semiconductor with a band gap of 0.27 eV [278]. Hall effect and optical measurements indicate that the hole concentration and mobility depend significantly on the processing regime, as shown in Fig. 5.17. An increase in the processing temperature or time is followed by a decrease in the carrier concentration,

TABLE 5.13. Bulk Resistivity of the Silicides Formed by Diffusion Synthesis in Thin-Film Metal–Silicon Structures

	Rapid thermal processing			Furnace processing
Silicide	Temperature (°C)	Processing time (s)	Minimum bulk resistivity (10^{-6} Ω-cm)	Minimum bulk resistivity (10^{-6} Ω-cm)
$TiSi_2$	800	10–15	13–15 [12,52,56,245]	13–16 [4]
$TaSi_2$	1100	10	40–50 [65,66,72,74,75]	35–55 [4]
WSi_2	1100	10–30	30–40 [23,87]	~70 [4]
$MoSi_2$	1100	10–15	66–85 [91–94]	90–100 [4]
$NiSi_2$	700	15–30	40–55 [114–119]	50–60 [4]
$CoSi_2$	800	15	18 [117]	18–25 [4]
$ZrSi_2$	820	15	38 [155]	35–40 [4]
Pd_2Si	500	15	28–35 [163,166]	30–35 [4]
PtSi	500 800	20–40 15	31–38 [166,179]	28–35 [4]

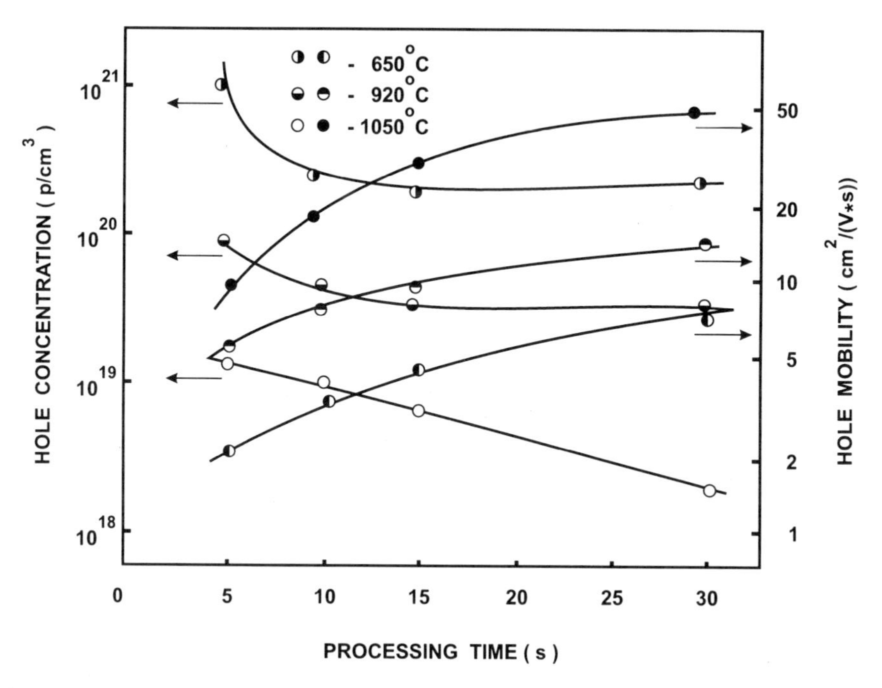

FIGURE 5.17. Hole concentration and mobility in chromium silicide formed by rapid thermal processing of chromium film on (100) silicon [151].

while the mobility monotonically increases. Such a variation appears in the first 15 s of processing when a uniform disilicide is produced. For longer processing the induced temperature becomes the determining factor. In the temperature range of 650–1050 °C the steady-state hole concentration reaches 10^{19}–10^{21} p/cm^3 and has a mobility of 10–50 $\text{cm}^2/$ V·s which correlates well with those observed for conventional furnace annealing [278].

The conductivity of RTP-synthesized chromium silicide layers is typically a function of temperature for doped semiconductors. The "impurity related" conductivity was observed in the temperature range of 20–100 °C and results from the ionization of "impurity" levels at 0.10–0.15 eV. Intrinsic conductivity features are displayed over the temperature range of 320–500 °C. The band gap varies from 0.54 eV for processing at 650 °C to 0.88 eV for processing at 1050 °C. However, optical measurements at room temperature over the wavelength range of 400–1000 cm^{-1} yield a value of 0.35 eV for the 1050 °C-synthesized disilicide [150]. The temperature coefficient of the band gap is minus, at 1.6×10^{-4} eV/K. The band gap obtained is comparable to that previously reported for $CrSi_2$ at 0.24–0.74 eV [278–280] and calculated to be 0.29 eV in Ref. 281. The difference between the results based on electrical and optical measurements is related to the complex zone structure of $CrSi_2$ [151]. In addition, the presence of small amounts of other phases in the disilicide can also influence the results. In conclusion, RTP fabrication of silicides provides a wide range of opportunities for the synthesis of materials with suitable parameters.

6

Rapid Thermal Oxidation and Nitridation

In both ULSI and VLSI the sizes of MOS transistors have been scaled to submicron and micron gate widths such that the requisite dimensions of the gate dielectric films are typically less than 200 Å. Rapid thermal processing (RTP) of thin dielectrics has therefore become a topic of great interest for achieving good electrical properties with a reduced thermal budget for the manufacture of these circuits. This has had particular impact in manufacturing memory devices where very high packing density is required. For example, dynamic random-access memories (DRAMs) [1], electrically programmable read-only memory (EPROM) [2], electrically erasable programmable read-only memory (EEPROM) [3], and greater than 64-Mbit random-access memory (RAM). The presence of oxide charge is clearly demonstrated by the threshold voltage shifts and leakage currents. These effects become more pronounced as the gate oxide thickness decreases and the gate capacitance increases. In EEPROMs Fowler–Nordheim (FN) tunneling current, which is used to charge and discharge the gate of the memory cell, can slowly degrade the gate dielectric properties through trapped charges in the gate dielectric. The resultant shift in the threshold voltage and increase in the leakage current reduces the time a bit of information is held in the memory cell, thus leading to electrical failure of the device. A second advantage of RTP is incorporating nitrogen into the dielectric to reduce boron penetration which occurs when p^+-polysilicon is used in the fabrication of the gate electrode in p-channel MOSFETs.

This chapter reviews RTP of dielectric films, from the rapid thermal oxidation of silicon in oxygen to multistep processes of rapid thermal oxidation and rapid thermal nitridation in NH_3, N_2O, and NO. This also includes a discussion of nitridation of furnace-grown oxides, and nitridation of rapid thermally grown oxides. We begin with a description of the structure of the oxides and the oxidation kinetics, and continue with a review of the effects of various processing ambients and procedures on the electrical properties of the dielectric films. Electrical properties of the dielectric are analyzed with current–voltage (I–V) and capacitance–voltage (C–V) measurements made after fabrication has been completed. Based on these measurements, the interface state density, D_{it}, fixed charge density, Q_f, breakdown field, E_{BD}, and the charge to breakdown, Q_{BD}, are calculated. In some cases measurements are made after dc electrical stress has been performed. Active devices have been built with RTP dielectrics. Measurements of device transconductance and surface channel mobility are discussed. Unless stated otherwise, all dielectrics were grown on lightly doped p-type (100) silicon (Czochralski grown).

6.1. OXIDE GROWTH BY RAPID THERMAL PROCESSING IN OXYGEN

The application of RTP to the oxidation of silicon provides electrical characteristics equal to or even better than furnace oxides and a lower thermal budget than a conventional furnace. In particular, the number of interface states and traps in the bulk are reduced, related in part to the higher growth temperatures used. Rapid thermal oxides have higher reliability and increased radiation hardness than furnace oxides. The equipment for rapid thermal oxidation has essentially the same components as a vacuum RTP system. There are two important factors for oxidation: (1) the ambient should be high-purity oxygen with very low water vapor present, often with evacuation to high vacuum, and (2) the system must be ultraclean.

6.1.1. Interface Structure

The structure of an RTP oxide was examined by high-resolution transmission electron microscopy (HRTEM) [21]. In Fig. 6.1, an amorphous oxide film with abrupt interface can be clearly seen and the thickness is 102 Å. In several studies an ellipsometer was used to measure the dielectric thickness and these measurements were checked against TEM analysis. An assumption needs to be made about the oxide refractive index

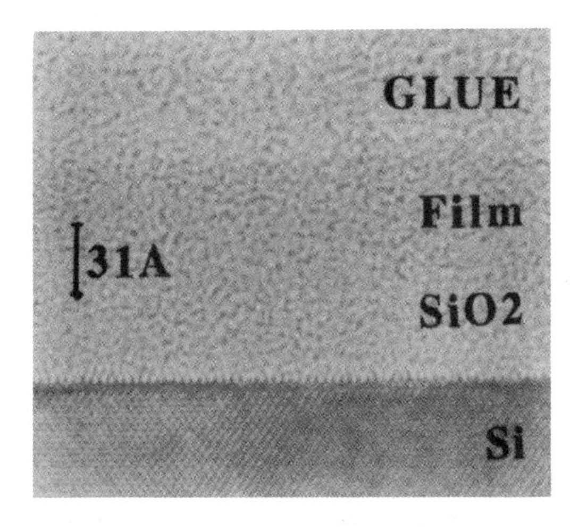

FIGURE 6.1. High-resolution TEM of $Si–SiO_2$ interface thermally oxidized [21].

(i.e., density) when an ellipsometer is used to determine the dielectric film thickness. The equivalent oxide thickness is frequently used as a standard measure for comparison between different dielectric films when the actual thickness is unknown. The definition of equivalent oxide thickness is the thickness of silicon dioxide that would give the same capacitance as the film under test. Ganem *et al.* [5] used a direct method to determine the film thickness through nuclear isotope measurements and converted the data back to equivalent SiO_2 thickness.

The roughness of the silicon/SiO_2 interface has been studied by X-ray diffraction [6] for three types of oxide: a native oxide, a thermally grown oxide of 1200 Å, and a chemically grown oxide. The Czochralski-grown (100) silicon was boron doped with a resistance of 10 Ω-cm. The wafers were cleaned in H_2SO_4/H_2O_2 followed by a 1% HF dip immediately prior to oxidation. The interfacial roughness versus growth temperature is shown in Fig. 6.2. The measurements were taken with a photon beam energy of 7.99 keV focused on a 25 × 20 × 0.5 mm^3 sample. During RTP the oxide interface roughness increases linearly with decreasing growth temperature [7] over the range 800–1200 °C, producing 2.84 Å at 800 °C and 1.76 Å at 1200 °C. Here (100) Si boron doped at 1–2 Ω-cm was used.The thermal oxide interface is approximately 2 Å wide, half that of the native oxide interface width (3.8 Å), which implies that the process of growing the oxide makes the interface smoother. The "RCA-cleaned" sample had the roughest interface (4.6 Å). More work needs to be done on the effect of orientation on interfacial roughness.

6.1.2. Oxidation Kinetics

The oxidation kinetics of silicon have been studied by many researchers over the temperature range 600–1250 °C in pure dry oxygen. Table 6.1 summarizes the work done since Nulman *et al.* [8] first demonstrated rapid thermal oxidation in 1985. Oxidation is carried out in a one-step or two-step process, as shown in a typical time–temperature profile in Fig. 6.3. The first step in the two-step process preheats the chamber, which also provides improved pyrometer temperature readings. The oxides are slightly thicker in this case and Ganem *et al.* [5] found that this thickening was independent of the ambient gas.

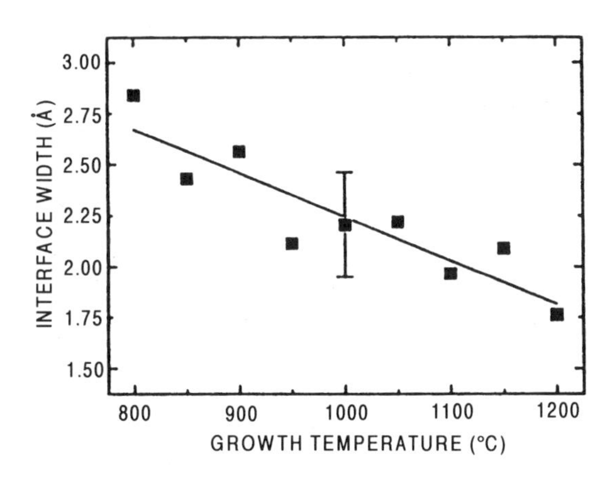

FIGURE 6.2. Growth temperature dependence of interface width. A linear trend with slope equal to –0.002 Å/°C was found. The error bar shown was determined from the discrepancy between two identical measurements of two different pieces from the same wafer at different times [7].

TABLE 6.1. Process Conditions and Electrical Characteristics of Rapid Thermal Oxides Grown in Oxygen

	Process conditions					Electrical properties						
Oxide thickness (Å)	Cleaning	Temperature (°C)	Process time (s)	Uniformity (%)	Gate material	Fixed charge density (cm^{-2})	Interface trap density ($eV^{-1}\ cm^{-2}$)	Leakage current (A/cm^2)	Barrier height (eV)	Charge to breakdown (C/cm^2)	Breakdown field (MV/cm)	Refs.
20–340	RCA	950–1200	5–300	~4 (3" diam.)	—	—	—	—	—	—	—	15
40–130	—	1150	5–30	2 (3" diam.)	Al	1.8×10^{11}	1×10^{10}	—	—	—	13.8	8
—	NH_4:HNO_3:H_2O	—	10–25	—	—	$2\text{–}8 \times 10^{11}$	—	—	—	—	0.8–1.2	22
80–300	—	1050–1150	5–150	5 (4" diam.)	Al	—	$\sim 9 \times 10^{10}$ ($\sim 1 \times 10^{10}$ after RTA)	—	—	—	—	13
101	—	1000–1250	5–60	—	—	$1\text{–}3 \times 10^{11}$	—	—	—	—	13	14
20–300	see text	1000–1200	1–64	—	Al	1.6×10^{10} (90 Å) 9.3×10^{11} (315 Å)	1.8×10^{11} (90 Å) 9.8×10^{11} (315 Å)	—	—	—	—	16,17
80–100	RCA or RCA (HF dip)	950–1150	5–300	—	poly-Si	2.7×10^{11}	3×10^{11}	—	3.29	50 (w/HF dip) 30 (w/o HF dip)	15	29
30–100	—	1000–1200	—	—	poly-Si	—	$3\text{–}9 \times 10^{10}$	—	—	—	11	66
—	piranha/HF	1000–1250	5–65	2 (4" diam.)	Al	—	1.1×10^{10}	—	—	—	10.1 (12.3 after RTA)	21
120	RCA	900–1150	10–600	6	—	—	—	—	—	—	—	18
141–147	—	900–1100	10–500	—	—	5×10^{10}	$< 5 \times 10^{10}$	$< 4 \times 10^{-11}$ (at 5 V)	2.7	—	7.3	26
155	—	1100	60	—	—	—	—	$< 10^{-11}$ (at 5 V)	2.25	—	7(+ve bias) 6.5(–ve bias)	26
100	—	1050	60	—	Al	—	Fig. 6.20	—	1.5 ($+V_g$) 1.92 ($-V_g$)	—	8	77
20–400	RCA HF/ethanol	1060–1240	6–225	—	Al	—	$2\text{–}10 \times 10^{11}$	—	—	3–40	15	25
100	sulfuric/ peroxide, HF dip	1000–1500	13–100	—	Al	—	$2.9\text{–}4.4 \times 10^{10}$	—	—	—	~14.5	71,72

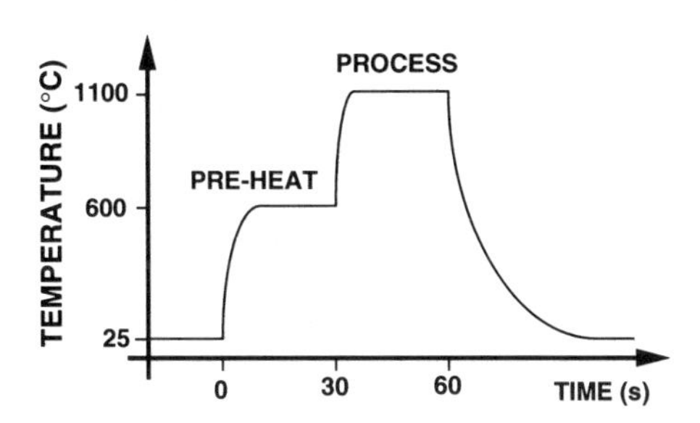

FIGURE 6.3. Typical temperature–time profile in a rapid thermal oxidation process which includes a preheat stage to allow the pyrometer temperature monitoring system to stabilize.

The reaction rate during oxidation increases dramatically with photon exposure [9]. Although radiation is the dominant energy transfer mechanism in both furnace and RTP, the spectral content is different. In RTP the source filament temperature is much higher than that of the wafer, producing significant intensity in the visible and ultraviolet part of the spectrum. As a result, photochemical reactions can occur, such as O_2 dissociation, producing enhanced growth. The use of a low-power-pressure mercury lamp to grow 100 Å of oxide on silicon at 550 °C has been reported by Kazor and Boyd [10]. The reaction rate was approximately five times higher than that of furnace oxidation. The enhancement can be explained in terms of a space charge controlled drift of O^- species to the Si/SiO_2 interface. Low-power-density laser irradiation was shown to increase oxidation rates on (100) and (111) silicon by 3–30% by Young [11, 12]. The enhancement was linearly proportional to the photon flux for energies in the range 2.4–2.7 eV. Enhancement can be considered to result from an increase in the rate of dissociation of O_2 to atomic oxygen and O^- species. An electric field in the oxide will also form via carrier adsorption. These observations support the parallel oxidation models proposed by Han, Helms, and deLarios (see Section 6.2.3). Figure 6.4 shows the percent enhancement as a function of the doping level. At lower doping the enhancement was the greatest, the light causing a large change in the free electron concentration at the interface.

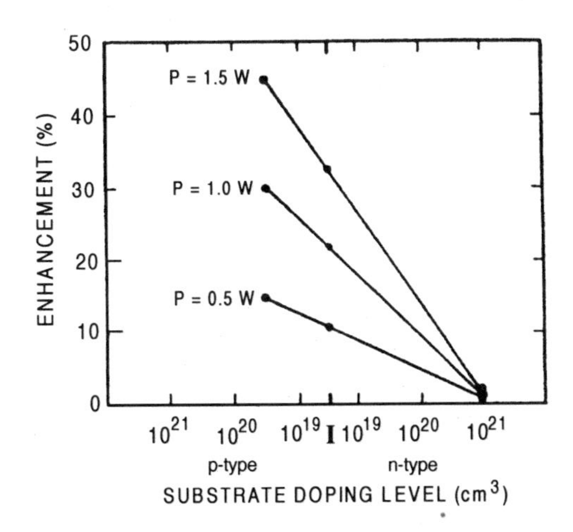

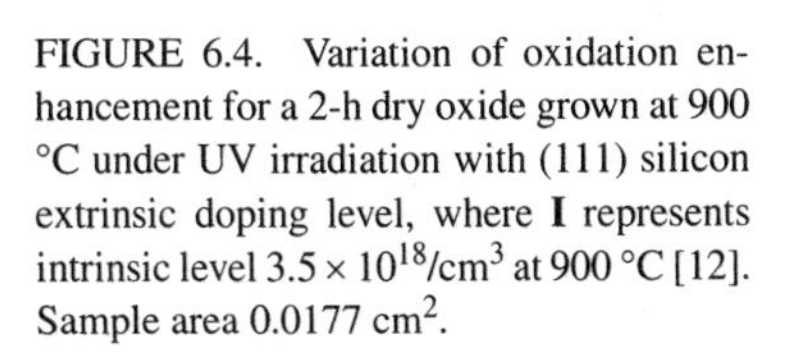

FIGURE 6.4. Variation of oxidation enhancement for a 2-h dry oxide grown at 900 °C under UV irradiation with (111) silicon extrinsic doping level, where **I** represents intrinsic level $3.5 \times 10^{18}/cm^3$ at 900 °C [12]. Sample area $0.0177\ cm^2$.

The oxidation kinetics are more rapid than for a conventional furnace oxidation. Sato and Kiuchi [13] and Tung and Caratini [14] have fit their data to a linear growth model over a limited thickness range. Moslehi *et al.* [15], Ponpon *et al.* [16], Slaoui *et al.* [17], Chiou *et al.* [18], and Ganem *et al.* [5] found a good fit to a linear–parabolic growth with the Deal and Grove model [19]. Figure 6.5 shows the oxide thickness versus the square root of time for (111) and (100) silicon with temperature as a parameter [17]. The oxide thickness was determined with an ellipsometer assuming a constant refractive index of 1.45. Samples were processed with a preheat step to 800 °C for 30 s, which resulted in an initial oxide thickness of about 18 Å before the second step, at the oxidation temperature. The parabolic law was observed for thicknesses over 40 Å, but not for the thinner layers. Based on the data the activation energy was calculated to be 2.26 eV for (100) and 1.99 eV for (111). This agrees fairly well with the value of 2 eV for (100) silicon reported by Sato and Kiuchi [13] and those for furnace oxides in the thin oxide regime [20] at 2.22 eV for (100) and 1.71 eV for (111) silicon. The values for longer times are in good agreement with those reported for furnace oxidation.

In general the growth rates vary between researchers, although most reported a break point between an initial rapid growth and slower continued growth, with film thickness. Tung *et al.* [21] measured on (100) silicon for oxidation times less than 5 s an activation energy of 0.92 eV, but for greater than 5 s it was 1.42 eV, as shown in Fig. 6.6. However, there is considerable variation in the reports from other researchers for (100) silicon: 1.21 eV [15] and 1.24 eV [22]. Possible reasons for these differences are errors in the temperature measurement, differences in initial wafer surface condition, i.e., depending on the cleaning steps, and the chamber design (gas flow, radiation intensity, uniformity, etc.). Finally, accurate measurement of oxide thickness at these small dimensions is difficult. Any surface roughness and inhomogeneity in the film produces light scattering,

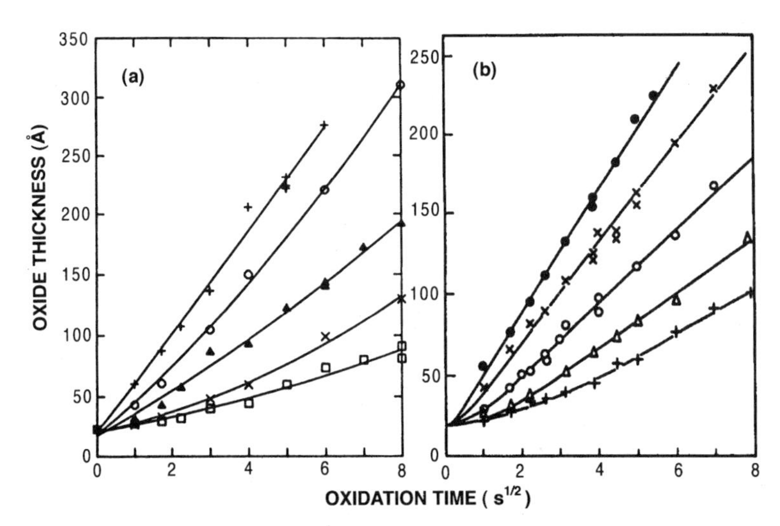

FIGURE 6.5. Oxide thickness versus oxidation time for rapid thermally processed oxides grown at different temperatures. (a) (100) silicon: +, 1200 °C; ○, 1150 °C; ▲, 1100 °C; X, 1050 °C; □, 1000 °C [13]. (b) (110) silicon: ●, 1200 °C; X, 1150 °C; ○, 1100 °C; △, 1060 °C; +, 1000 °C [16].

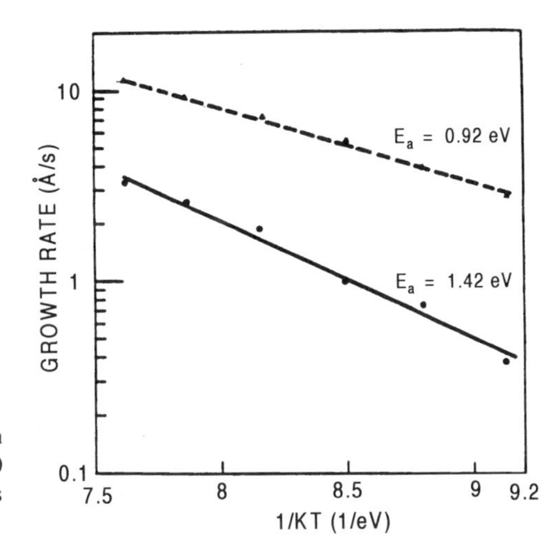

FIGURE 6.6. Activation energy for the oxidation of (100) silicon in oxygen. ---, 0–5 s; —, 5–60 s. Activation energies indicated are a least-squares fit to the experimental data [21].

introducing errors in the thickness measurement. The influence of surface cleaning is perhaps one of the most important parameters in the initial growth of the oxide layer as discussed in the next section.

A two-step oxidation was carried out by Ganem *et al.* [23] to study the oxidation mechanism by first oxidizing in O^{16} and then in O^{18}. The (100) Si wafers were cleaned by rapid thermal cleaning (RTC), and oxidation was conducted under ultradry conditions (< 1 ppm water); other process conditions are given in Table 6.1. Depth profiling analysis of the radiolabeled oxygen, O^{18}, indicated it was concentrated near the Si/SiO_2 interface, demonstrating that the oxidation reactions take place at the interface. The O^{18} therefore was transported as atomic oxygen through the oxide. The oxidation process could be defined by two linear regimes and a parabolic region: an initial fast growth followed by a slower growth and finally transition into parabolic growth for thicker films. In addition, it was found that oxidation at higher temperature produced thicker films. This can be explained by the presence of silicon fragments in the bulk of very thin oxides (< 15 nm) [24]. Summarizing the results the following model is supported: the oxidation takes place forming an inhomogeneous layer containing silicon fragments; growth takes place at the Si/SiO_2 interface with initially fast kinetics. One might propose that at the very beginning of the oxidation process, silicon atoms are injected from the substrate into the oxide layer. If the injection of Si is in excess, some may react to form SiO_2 and some coalesce forming silicon fragments. Cleaning the wafers at high temperatures by RTC apparently causes a higher density of silicon fragments than does chemical cleaning. As the layer thickness increases, a linear region and finally linear–parabolic growth occur corresponding to the Deal and Grove model.

Messaoud *et al.* [25] studied the kinetics over a wide range of temperatures. The thickness was monitored by ellipsometry and the values were found to be in agreement with the Han and Helms model (see Section 6.2.3). Samples cleaned with the RCA process had an initial oxide thickness of 9.4 Å while samples also dipped in 10% HF/ethanol had values of 8.8 Å. Electrical measurements were also made (see next section).

6.1.2.1. Effect of Crystal Orientation. Ohyu *et al.* [26] grew 50-Å oxides on single-crystal silicon of different crystal orientations (111), (110), and (111), and n^+-poly-Si in dry oxygen at 950–1100 °C for 10–500 s. Measured growth rates on single-crystal silicon, n-type, 10 Ω-cm are shown in Fig. 6.7. The rate for oxide growth was (110); (111) > (100) in qualitative agreement with furnace oxides. Irene *et al.* [27] observed a ranking of (110) > (111) > (100) for thin oxides with a crossover in the rates for (110) and (111) for layers thicker than about 180 Å. For thicker films greater than 200 Å, Sun *et al.* [28] observed with oxidation at 1050 °C the ranking (110) > (111) > (100) and no crossover point. Therefore, growth by RTP is in agreement with that observed in furnace oxidation for thickness below 200 Å, but deviates at larger thicknesses.

6.1.2.2. Effect of Wafer Cleaning. The initial condition of the silicon surface greatly influences the kinetics of the oxidation process. The wafer cleaning method is given in Table 6.1. In a study by Ganem *et al.* [5] the effects of different cleaning steps were evaluated on rapid thermal oxidation. Precise measurements of oxygen concentrations were made with nuclear reaction analysis (NRA) and oxygen isotopes. Cleaning in 48% HF solution was most effective in removal of oxide with less than a monolayer forming after 4 h in air (three monolayers after 3 days). However, cleaning in 4% HF in ethanol produced the best stability with a monolayer after 4 h, and only 9 Å of SiO_2 after 3 weeks.

The cleaning step greatly influences the initial oxidation with a 50% increase in oxide growth rate resulting if the native oxide is removed. A study by Moslehi *et al.* [29] showed that RCA cleaning produces a 12-Å film of oxide (as measured by ellipsometry) which was reduced to 8 Å with a HF dip prior to final deionized (DI) water rinse. They compared two RTP systems and found good agreement between systems for processing temperatures below 1100 °C. However, for 1150 and 1200 °C there was a significant error in the predicted values compared to experimental oxide thicknesses (see Section 6.2.3, Figure 6.14).

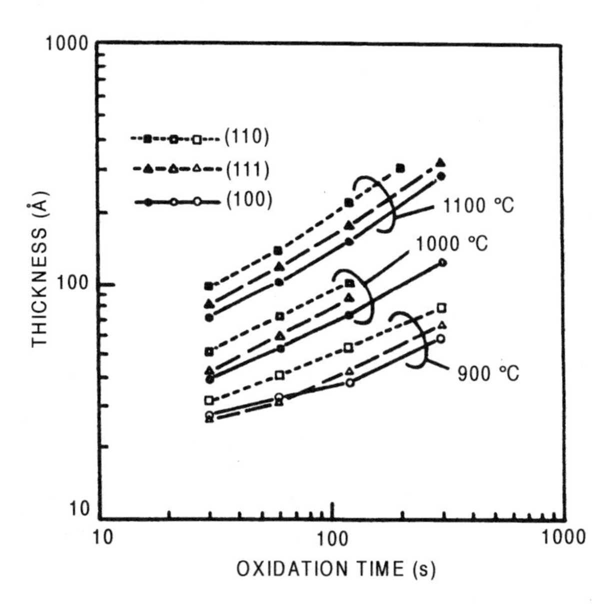

FIGURE 6.7. Crystal orientation dependence of rapid thermal oxidation characteristics of silicon n-type, 10 Ω-cm: ○ and ●, (100); □ and ■, (110); △ and ▲, (111) [26].

The effect of wafer cleaning on oxidation kinetics has been investigated by deLarios *et al.* [30–32]. The effects were studied on (100) silicon for dry furnace oxidation. A retardation of the oxidation rate was observed with a NH_4OH/H_2O_2 preclean with a 25% thinner oxide for oxide thicknesses greater than 200 Å. This is thought to be related to the presence of trace amounts of aluminum metal contaminant on the SiO_2 surface. Removal of 50 Å of oxide in HF/H_2O eliminates the retardation effect. The retardation can be modeled by decreasing the value of the linear rate constant in the Deal and Grove model. Further studies indicated that this effect is decreased above 1000 °C, and disappears for oxidations at 1200 °C. Based on a series of two-step oxidation processes, in which a cleaning step was included to affect the outer surface, it was determined that the effect was caused by the presence of Al on the SiO_2 surface reducing the number of surface sites for the oxidant.

6.1.2.3. Effect of Water Vapor. It is well known that even parts per million water can influence the "dry" oxidation rate [33, 34]. George and Bohling [35] studied the sources of H_2O in an RTP reactor. They measured water with a Meeco analyzer and hydrogen with a Trace Analytical RGD2 residual gas analyzer. They found that the outlet gas had a greater concentration of H_2O than the inlet. When a wafer was processed, the amount increased dramatically in the presence of oxygen, which suggests that hydrogen from the wafer, and other hydrocarbon contamination in the chamber, were reacting with the oxygen during the oxidation process liberating H_2O. This also results in hydrogen being present in the oxide layer as it grows. The effect of hydrogen on the electrical properties of the oxide is also important, as we shall see later in this chapter.

Wet oxidation is more rapid than dry oxidation. Glück *et al.* [36] implemented the wet oxidation with a bubbler. Although a bubbler tends not to be as reproducible as pyrogenic oxidation, no rapid thermal pyrogenic oxide fabrication methods have been reported to date. The wet oxidation process consists of four major steps: purge, preheat, oxidation, and then cooling. The process details are given in Table 6.1. The wet oxides were about two times thicker than the dry oxides. There was a pronounced silicon substrate orientation effect: the rate on (111) silicon was larger than on (100) silicon by a factor of 2.38 compared to 1.47 for the dry oxidation process. The partial pressure of the water is a function of the bubbler temperature. Figure 6.8 shows the oxide thickness versus time at 950–960 °C [37]. The bubbler temperature had a pronounced effect on the oxidation of the (111) surface. In this study the homogeneity was characterized on 4- and 6-inch-diameter silicon wafers. It was found that the homogeneity was a function of the gas flow. In particular, for the side of the wafer next to the input, the oxide was thinner because of convective cooling, but if the silicon guard ring was used this effect was decreased. A homogeneity of 1.4–2% was obtained and reproducibility between wafers of 0.5–0.75% at faster oxidation rates.

6.1.2.4. Effect of Pressure. Parkhutik *et al.* [22] made measurements of the oxygen pressure effect between 10^4 and 3×10^5 Pa. The pressure dependence is shown in Fig. 6.9. This has a parabolic profile indicating an atomic rather than a molecular mechanism of oxygen diffusion into the oxide during the growth.

6.1.2.5. Effect of Stress. Ajuria *et al.* [38] carried out a careful experimental study to examine the effects of annealing on the oxide growth kinetics, particularly in the initial growth regime. Experimental conditions are given in Table 6.2. The reoxidation rates of

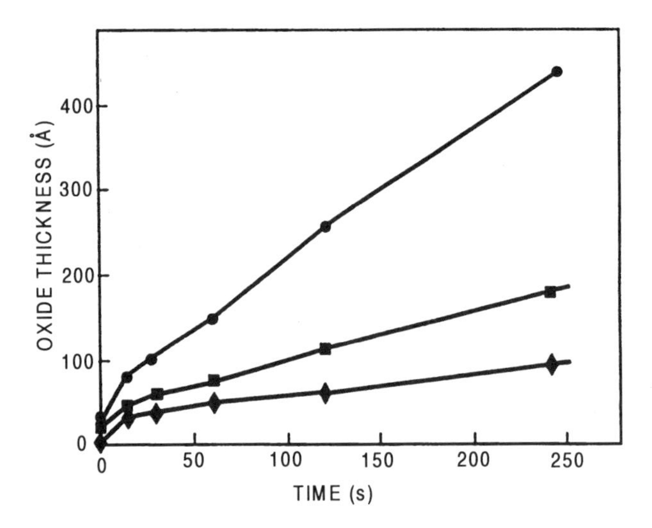

FIGURE 6.8. Time-dependent oxide growth in a wet ambient using a water bubbler at 85 °C for (111) silicon at 950 °C (●), (100) silicon at 950 °C (■) [36], and (100) silicon at 960 °C with hydrogen injection (♦) [37].

thermally grown oxides rapidly decrease with increasing oxide thickness. In contrast, the reoxidation rates of deposited oxides are faster and nearly thickness independent. It was also found that the reoxidation rates could be significantly influenced by inert thermal annealing (see Fig. 6.10A). These results support a stress influence on oxidation rate that may result from enhanced ion transport or bond disruption at the Si/SiO_2 interface. The reoxidation rate at 850 °C as a function of initial oxide thickness is shown for thermal oxides after various annealing processes. Clearly a 30-min argon anneal densified the oxide and decreased the reoxidation rate; however, reoxidation rates for deposited oxides were higher and nearly thickness independent.

The edges of the wafer tend to be cooler than the center because of radiative heat loss from the additional surface area of the wafer perimeter. The temperature gradient across the wafer introduces compressive stress at the center of the wafer and tensile stress toward the edges. The oxidation rate is enhanced by tensile stress so that a thicker oxide results at the edge of the wafer [39]. Figure 6.10B shows the oxide thickness distribution grown

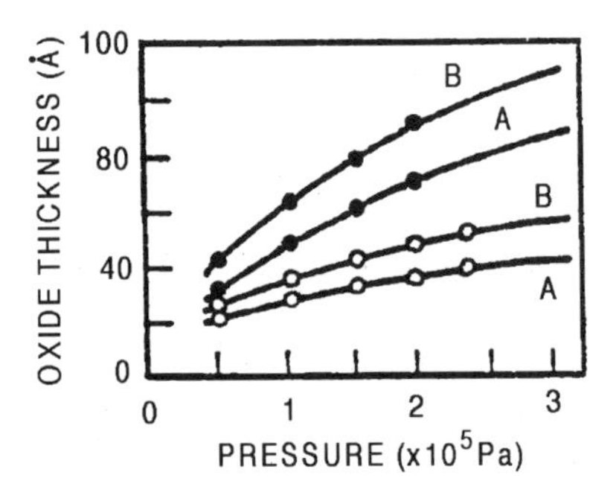

FIGURE 6.9. Oxide thickness versus processing pressure for *p*-type silicon for (○) a dry oxide and (●) a steam oxide, with processing power level as a parameter: A, 25 W/cm^2; B, 35 W/cm^2 [22].

TABLE 6.2. Process Conditions and Electrical Characteristics of Rapid Thermal Oxides Grown in Oxygen and Annealed

	Process conditions			Anneal		Electrical properties					
Oxide thickness (Å)	Cleaning	Temperature (°C)	Process time (s)	Temperature/ time (°C/min)	Gate material	Fixed charge density (cm^{-2})	Interface trap density (eV^{-1} cm^{-2})	Barrier height (eV)	Charge to breakdown (C/cm^2)	Breakdown field (MV/cm)	Ref.
≤100	RCA	900–1050	6/10 (pulsed)	1050/60 (pulses)	—	—	—	—	—	—	74
120	—	1000	25–300	—	—	Fig. 6.19	—	Fig. 6.19	—	6	70
60–100	—	1025	—	800/30	poly-Si	Fig. 6.18	3×10^{10}	3.2	—	—	69
31	RCA	900	43	500/30	Al	1.9×10^{12}	2.5×10^{12}	—	—	—	67
20–50	4%HF dip	900	2–60	400/30	Al	—	$< 10^{12}$	—	—	—	73
200	piranha, dilute HF	1150	60	1150/1 1050/0.6	Al	—	$2–8 \times 10^{10}$	3.1	—	0.1–6	75
	RCA, dilute HF	1150	60	1150 °C	Al	—	$< 1 \times 10^{10}$	3.3		3–25	

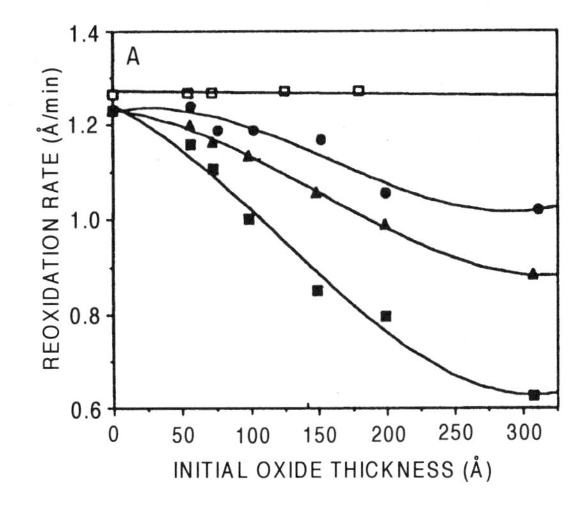

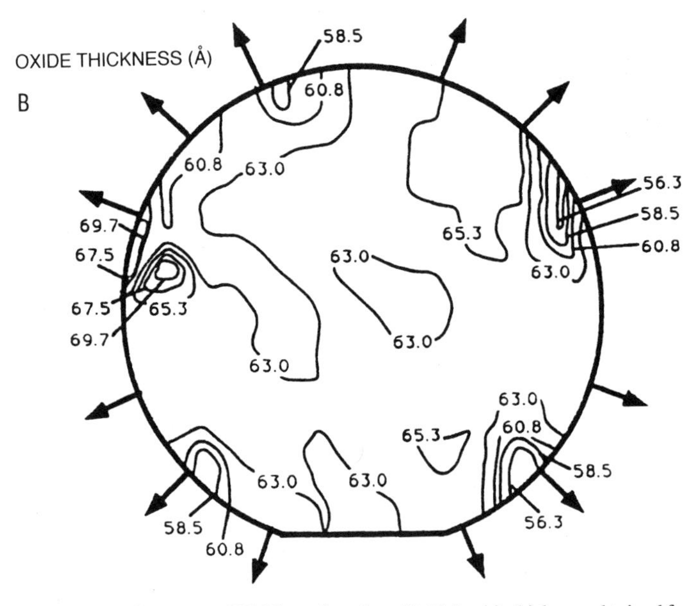

FIGURE 6.10. (A) Reoxidation rates at 850 °C as a function of initial oxide thickness obtained for the following oxides: ■, thermal oxide as-grown at 850 °C; ▲, after annealing at 950 °C for 1 h; ●, after annealing at 1000 °C for 1 h; □, and for an argon densified deposited oxide reoxidized at 1000 °C, 30 min [38]. (B) Oxide thickness distribution for processing at 1100 °C for 30 s in oxygen. Arrows indicate directions of maximum resolved stress, $\theta = (\pi/8)(2i + 1)$ and $\theta = (\pi/4)(2i + 1)$ [39].

at 1100 °C for 30 s in oxygen on (100) silicon. The arrows indicate the directions of maximum stress. The stress increases the oxidation rate on some crystal planes more than others producing the periodic distribution of thicknesses around the perimeter of the wafer. Slip planes occur in the ⟨110⟩ direction along the (111) planes. The slip planes are most noticeable at the edges of the wafer in the radial direction indicating the higher tensile stresses present.

The oxide growth rate is higher because the tensile stress increases the spacing between silicon atoms. These effects are more pronounced in the initial stages of oxide growth and at lower temperatures. At higher temperatures the yield strength of the silicon is reduced and this allows slip to occur pinning the stress at some maximum value and the intrinsic stress is only a factor below the viscous flow point of the oxide, which occurs typically at about 950 °C. Finally, as the oxide becomes thicker an intrinsic stress develops which may influence transport of oxidant through the strained oxide and interface layer.

Tung *et al.* [21] confirmed the observed stress effect on the oxidation rate. They observed that slip lines occurred for temperature above 1075 °C, and developed an oxidation simulation model in which the temperature nonuniformity was included to some extent. When nonuniform growth occurs as a result of temperature uniformity across the wafer, the resultant stresses also influence the oxidation rates. The effect becomes more pronounced at lower oxidation temperatures and larger wafer diameters.

6.1.2.6. Effect of HCl. The addition of HCl to the process gases combined with a multistep process of annealing between rapid thermal oxidation and rapid thermal nitridation showed improvement in the properties of the film, reported by Kim and Jun [40]. Chlorine is widely employed to reduce the apparent interface states at the Si/SiO_2 interface in conventional furnace oxidation. The oxidation growth rate increased with the addition of HCl at a concentration between 0 and 8%, as shown in Fig. 6.11. The effective thickness is plotted which assumes the dielectric constant does not change. The thickness uniformity was excellent at about 2%.

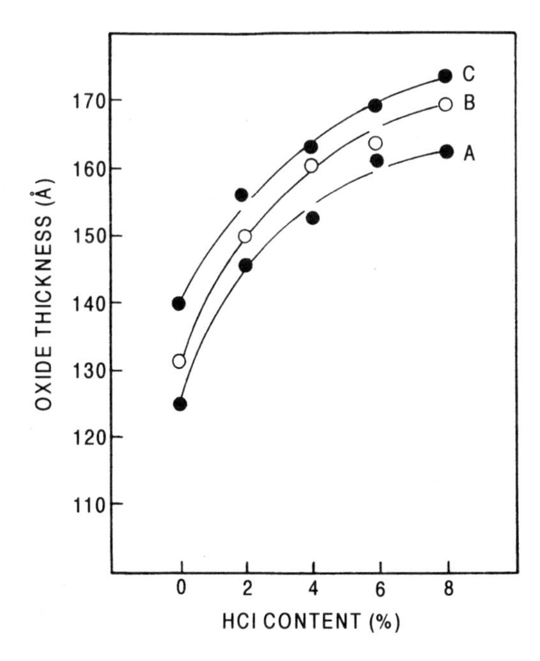

FIGURE 6.11. Effective thickness of the dielectric film as a function of the HCl content in the growth ambient: A, rapid thermal oxidation at 1148 °C for 40 s; B, rapid thermal oxidation and nitridation for 10 s; C, rapid thermal oxidation and nitridation for 120 s [40].

6.1.2.7. Effect of Ozone. Changing the ambient gas to a mixture of ozone and oxygen has a pronounced effect on the initial oxidation kinetics both for ozone produced remotely by a reactor [41] and for ozone produced *in situ* [42].

6.2. MODELS OF THE OXIDATION PROCESS

6.2.1. The Deal and Grove Model of Oxidation

The Deal and Grove model [19] gives an accurate estimate of the oxide thickness in a conventional furnace growth process for oxides thicker than 300 Å. A brief review of the assumptions on which this model is based is given here to illustrate some of the difficulties that arise in modeling of rapid thermal oxidation (see Fig. 6.12A). The model assumes the reactions proceed in the following steps: (1) absorption of oxidizing species at the oxide already formed, (2) diffusion of oxidizing species through the oxide, and (3) reaction of oxygen with silicon at the Si/SiO_2 interface forming SiO_2. Several simplifying assumptions are: (1) there is a single diffusing species in the dielectric film which is responsible for the oxide growth (either molecular oxygen or water), (2) steady-state conditions prevail, and (3) the oxide film is uniform and homogeneous. Two regimes of oxide growth kinetics may be defined: an initial more rapid linear growth region, followed by a slower parabolic growth one. The equation for oxide thickness, x_{ox}, is given by

$$x_{ox}^2 + Ax_{ox} = B(t + t_0) \qquad \text{or equivalently} \qquad \frac{dx_{ox}}{dt} = \frac{B}{2x_{ox} + A}$$

where B is the parabolic rate constant, B/A the linear rate constant, and t_0 the initial time displacement to represent the initial oxide thickness. These parameters are based on the

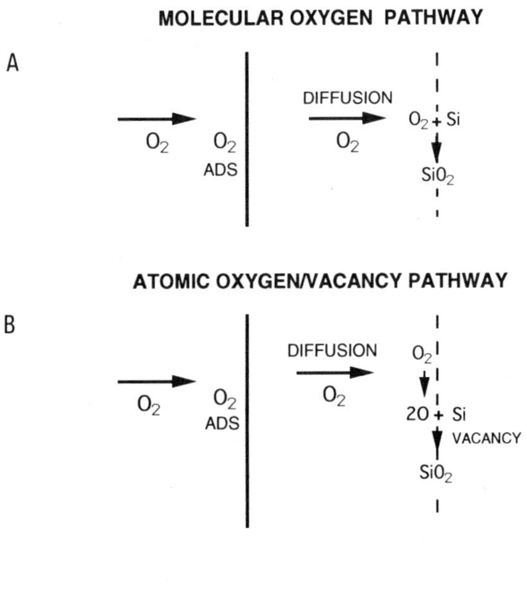

FIGURE 6.12. A schematic representation of the reaction and possible space charge region during rapid thermal oxidation of silicon.

microscopic processes of diffusion of oxidant to the interface and reactions rate at the interface. They are defined as follows:

$$A = 2\,D_{\mathrm{ox}}\left(\frac{1}{K_{\mathrm{s}}}\right), \qquad B = \left(\frac{2D_{\mathrm{ox}}C_{\mathrm{o}}}{N_1}\right) \tag{6.2}$$

where D_{ox} is the diffusivity of oxidant in the growing film, K_{s} the reaction rate, C_{o} the oxidant concentration, and N_1 the number of oxidant molecules incorporated into each unit volume of the oxide formed ($N_1 = 2.25 \times 10^{23}\ /\mathrm{cm}^3$).

This model can be universally applied for a given set of wafer orientation and doping level over a wide range of processing conditions, i.e., temperature, pressure, time with the appropriate choice of A and B. It works well for oxides thicker than 300 Å but for thinner oxides overestimates the thickness. Heavily doped silicon shows an increase in oxidation rate. Boron is preferentially incorporated into the growing film, because of its high segregation coefficient, weakening the bond structure of the SiO_2. Phosphorus has a low segregation coefficient, and piles up at the interface resulting in an increased reaction rate.

6.2.2. Modeling of Thin Oxide Growth

Models for thin furnace-grown oxides are based on including effects that might be in part responsible for the observed enhanced oxidation rate. Other models have been proposed for the thin oxide regime [43]. These include (1) structural effects, (2) space charge, (3) stress, (4) vacancies, and (5) defects. Structural effects such as micropores 10 Å in diameter in the oxide film have been proposed by Samalam [44]. This model works well for temperatures above 800 °C but shows significant deviation at 600 and 700 °C [45]. The oxidation rate is directly related to the oxygen solubility in the surface region. Orlowski and Pless [45] propose both an increase in oxygen solubility and an increase in the reaction rate constant resulting from stress at the interface. This effect exponentially decays as the film thickness increases. Stress will also have an important influence on the oxidation rate through both the transport of oxidant through the interface and the reaction rate at the surface.

Schafer and Lyon [46] model the enhanced oxidation based on space charge effects produced at the Si/SiO_2 interface which promote the formation of charged oxygen species. As SiO_2 forms, a fixed positive charge develops within the oxide layer which produces a space charge layer in the silicon and thus reduces the number of holes, which are necessary for oxidation to take place. Therefore, the oxide rate is decreased. Figure 6.13 shows the reaction velocity calculated from the experimental data, and a least-squares fit to the model versus oxide thickness. The only fitting parameters are maximum reaction velocity, surface charge density, and the decay length. Other electrostatic effects have been suggested including tunneling of electrons through the thin oxide, producing ionization of absorbed oxygen. The oxygen ions are an oxidant with an enhanced diffusion rate to the interface, therefore increasing the oxidation rate. There may be several reasons why these models do not fit well for thin oxides including field assistance, stress effects, and O_2^- to hole-coupled motion effects. Massoud *et al.* [43] propose that an additional oxidation mechanism might be related to self-interstitials in the silicon enhancing the

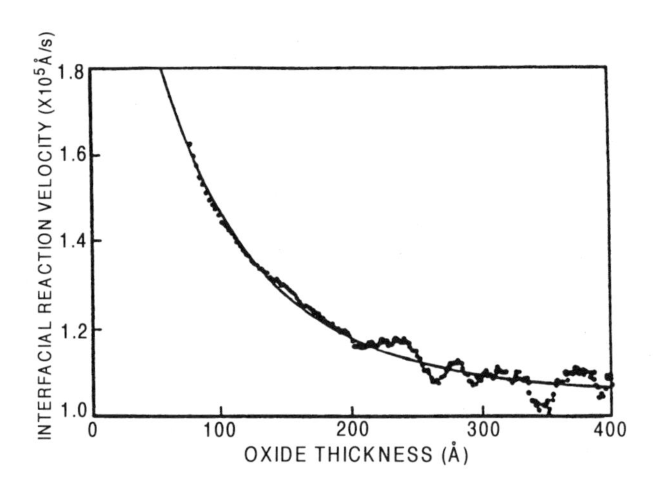

FIGURE 6.13. Interfacial reaction velocity plotted as a function of silicon dioxide thickness. The experimental data are shown as filled circles; the solid line gives the best fit of the model to the data [46].

oxidation rate. Another factor might be atomic oxygen diffusing to the oxidizing interface, and finally, point defects play a role in the initial oxidation rates. An understanding of the increase in oxidation rate in this very thin regime is not complete.

6.2.3. Han–Helms and deLarios Models

A great deal of work has been done on furnace oxidation kinetics for the thin oxide region [47, 48]. In the thin regime there are more complex microscopic processes at work. Massoud and Plummer [49] extend the Deal and Grove model by adding an initial rapid growth region, which decays exponentially with oxide thickness but it is chosen to fit the experimental data. The decay length for the initial growth does not correspond to the initial native oxide thickness. The model was also further modified with an exponential temperature dependence for the initial oxide thickness to provide an excellent fit with experimental data. Therefore, a total of two terms which exponentially decay are added to the Deal and Grove equation to account for the more rapid kinetics in the ultrathin oxide thickness region as follows:

$$x_{ox}(t=0) = x_{initial}$$
$$\frac{dx_{ox}}{dt} = \frac{B}{2x_{ox}+A} + C_1 \exp\left\{-\frac{x_{ox}}{L_1}\right\} + C_2 \exp\left\{-\frac{x_{ox}}{L_2}\right\} \tag{6.3}$$

where the constants were fit to an Arrhenius-type expression as follows:

$$C_1 = C_1^0 \ \exp\left\{\frac{-qE_1}{kT}\right\} C_2 = C_2^0 \ \exp\left\{-\frac{qE_2}{kT}\right\} \tag{6.4}$$

and for growth above 900 °C the parameters are: for (100) silicon $x_{initial}$ = 17 Å, C_1^0 = 3.04 m/min, E_1 = 2.24 eV, L_1 = 12.4 Å, C_2^0 = 6.57 m/min, E_2 = 2.37 eV, and L_2 = 69 Å [20].

This model is used in the SUPREM process simulation tool with $C_2 = 0$, i.e., omitting the term on the far right-hand side of the equation. Massoud and Plummer [49] have also developed a useful expression with parameters to fit to the data for the (111) and (110) planes. A comparison of the Massoud and Plummer modified Deal and Grove model, Eqs. (6.3) and (6.4) with experimental data for rapid thermal oxidation of (111) and (100) silicon at 1150 °C, is given in Fig. 6.14. It can be clearly seen that this model greatly underestimates the oxide thickness. Temperature measurement error is partly responsible for this difference, but the major cause is the more rapid kinetics that occur in this process.

Parallel oxidation mechanisms are proposed for silicon in dry oxygen by Han and Helms [50] and deLarios *et al.* [51]. They consider two oxidation species molecular oxygen and atomic oxygen reacting at vacancy defects, acting independently and assume they are noninteracting. Therefore, the oxidation rate is the sum of the rates for these two species. The following reaction steps are for the atomic oxygen (see Fig. 6.12B): (1) adsorption and incorporation of O_2 below the SiO_2 surface, (2) diffusion of O_2 to the Si/SiO_2 interface, (3) dissociation of O_2 to form O atoms, (4) reaction of O atoms to form SiO_2, presumably at Si interstitials or O vacancy defects, and (5) diffusion of the defect back to the SiO_2 surface. Note the rate constants and activation energies do not depend on the oxygen pressure.

The following is the expression for the oxidation rate assuming steady-state conditions:

$$\frac{dx_{\mathrm{ox}}}{dt} = \left[-G^*\left(1 + \frac{x_{\mathrm{ox}}}{L}\right) + \sqrt{G^{*2}\left(1 + \frac{x_{\mathrm{ox}}}{L}\right)^2 + K * P} \right] \tag{6.5}$$

where the parabolic terms are

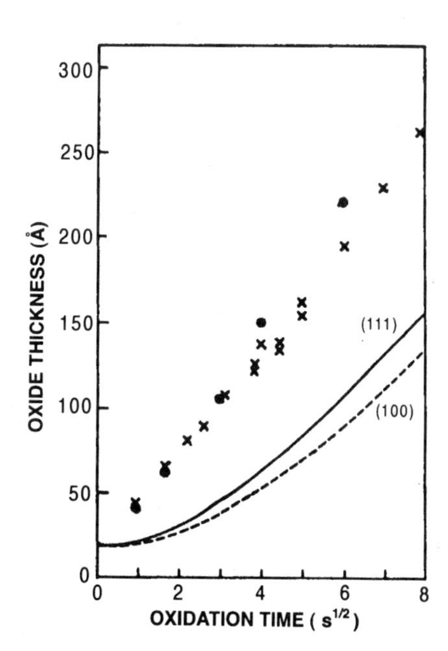

FIGURE 6.14. A comparison between experimental data for (111) (X) and (100) (●) silicon oxidized at 1150 °C, and the Massoud and Plummer model for (111) (——) and (100) (---) [17].

$$C^* = \frac{S_{ox}K_4[O_2]_G}{K_{-4}K_5}, \quad h^* = \frac{K_{-4}K_5}{K_{-4}+K_5}, \quad G^* = \frac{K_2^2(K_3+h^*)}{4N_1^*K_{-3}h^*},$$

$$K^* = \frac{K_2^2K_3}{N_1^{*2}K_{-3}}\frac{C^*}{P}, \quad L = \frac{D_{ox}(K_3+h^*)}{K_3h^*} \tag{6.6}$$

and where D_{ox} is the diffusion constant, S_{ox} the sticking coefficient, C^* the oxygen solubility, N_1^* the volume of oxide produced per reaction between oxidant and vacancy, and the K's the reaction rate constants. The dissociation of O_2 at the Si/SiO_2 interface is also considered and reaction of Si with dissociated O:

$$[O_2] \underset{K_{-2}}{\overset{K_2}{\Leftrightarrow}} 2[O] \qquad Si + 2[O] \underset{K_{-3}}{\overset{K_3}{\Leftrightarrow}} SiO_2 \qquad [O_2]_{AD} \underset{K_{-4}}{\overset{K_4}{\Leftrightarrow}} [O_2] \qquad [O_2]_G \underset{K_5}{\overset{S}{\Leftrightarrow}} [O_2]_{AD} \tag{6.7}$$

where K_4 represents the activation barrier over which an adsorbed oxygen molecule must pass to be incorporated in the film.

For convenience the surface oxygen concentration has been replaced by the ambient pressure, P. Comparing these equations with the Deal and Grove model (6.1) we obtain:

$$B = L\,K^*\,P/G^* \quad \text{and} \quad B/A = K^*\,P/2\,G^* \tag{6.8}$$

The parallel oxidation path relates to molecular oxygen at the Si/SiO_2 interface. The atomic oxygen reaction is rate limited by the availability of vacancies at the interface. If vacancies diffuse to the SiO_2 surface and if the vacancy generation rate is proportional to the oxygen species concentration in the SiO_2 layer, then the two reactions are linked, and we can introduce a parameter β^* to the rate equation as follows:

$$\beta^* = 2\,\alpha'\,D_v/N_1^* \tag{6.9}$$

where α' determines the relationship between the generation of vacancies at the Si/SiO_2 interface and D_v is the vacancy diffusion coefficient. Then the overall equation becomes

$$\frac{dx_{ox}}{dt} = \left[-G^*\left(1+\frac{x_{ox}}{L}\right)+\sqrt{G^{*2}\left(1+\frac{x_{ox}}{L}\right)^2+K^*P}\,\right]\left(1+\frac{\beta^*}{2x_{ox}}\right) \tag{6.10}$$

which can be simplified to the following for thin oxides (for which $x_{ox} >> (KP)^{0.5}\,L/G$)

$$\frac{dx_{ox}}{dt} = \left(-1+\sqrt{1+\frac{K^*P}{G^{*2}}}\right)\left(\frac{\beta^*}{2x_{ox}}\right) \tag{6.11}$$

The values of K^*, G^*, and β^* are chosen to fit experimental data over a wide range of thicknesses, temperatures, and pressures. Arrhenius plots for K^*, G^*, and G^*/L indicate activation energies of 3.3 eV, 1.5 eV, and 1.7 eV, respectively. These are in good agreement with the value of 1.8 eV previously published for B/A in the Deal and Grove model and compare well with the Si–Si bond energy of 1.88 eV. The model also fits data for different methods of surface cleaning such as RCA and RCA with HF dip, and in each case the activation energies are the same, the only difference being in the pre-exponential factor. These models give a good explanation of the kinetics over a range of oxide thicknesses, temperature, pressure, and for different cleaning conditions from several different laboratories. However, a full understanding of the increase in oxidation rate in this very thin regime is not complete at this time.

6.2.4. Modeling of Rapid Thermal Oxidation

To demonstrate the remarkable enhancement in oxidation modeling of kinetic rate observed with RTP we will first make a comparison with the Deal and Grove model [19] which is well known in furnace oxidation. The models that have been developed specifically for RTP are based on given sets of experimental data for a specific reactor chamber. An empirical law is useful to cover experimental data for the unique conditions in a specific reactor. Any model developed cannot be universally applied because every RTP system produces different oxidation rates. One reason for this may be errors in the temperature measurement which have a strong effect on the rate.

Models have been developed that extend the Deal and Grove model to simulate RTP by fitting with new parameters to the experimental data. Rosser *et al.* [52] studied the temperature range 600–1200 °C and a wide range of thicknesses. A good fit for temperatures above 800 °C and for oxide thicknesses greater than 300 Å was found with the appropriately selected values for A, B, and t in Eq. (6.1). The oxide thickness was incorrectly estimated for thicknesses less than 300 Å. Parkhutik *et al.* [22] also found qualitative agreement of experimental data with a Taylor series expansion of (6.1) in the thin oxide regime.

Murali's [53] model is based on the formation of an initial oxygen-rich surface layer in the silicon. The solubility of oxygen is therefore increased directly affecting the reaction rate. Paz de Araujo *et al.* [54] generalize the initial growth enhancement factor derived by Murali for ultrathin oxides over the temperature range 1050–1200 °C with good agreement with experimental data for the initial growth enhancement over the range 10–25 Å.

Ganem *et al.* [5] fit their experimental data to the Deal and Grove model as shown in Fig. 6.15, with B, A, and t_0 parameters listed below for (100) silicon, preheated to 700 °C for 30 s, before the oxidation temperature of 1100 °C. (In calculating the equivalent oxide thickness, a density of 10^{15} atoms/cm^2 was assumed to equal to 0.226 nm of SiO_2.)

$$B = 0.0072\ \mu\text{m}^2/\text{h}, \qquad B/A = 1.9\ \mu\text{m/h},\ t_0 = -0.00236\text{h} \tag{6.12}$$

This value of B was approximately 4 times lower than the values for dry furnace oxidation. B/A is larger by a factor of 4, and since it is proportional to the reaction rate at the interface, we might assume that there is an increase in the reaction rate for RTP. This suggests that the initial oxidation rate is assisted by photons and perhaps charged mobile species at the surface influenced by induced electric fields during the oxidation process.

A space charge assisted oxidant diffusion model has recently been proposed by Kazor [55]. The space charge region within the growing oxide layer is assumed to have a constant planar charge density. Space charge oxidation has been described by Massoud *et al.* [43]. The main source of space charge in conventional oxidation is thermionic emission of electrons from the silicon conduction band into the SiO_2 conduction band. However, during the rapid thermal oxidation high levels of photon irradiation increase the electron concentration in the oxide, resulting in higher ionic oxidant formation. The direction of

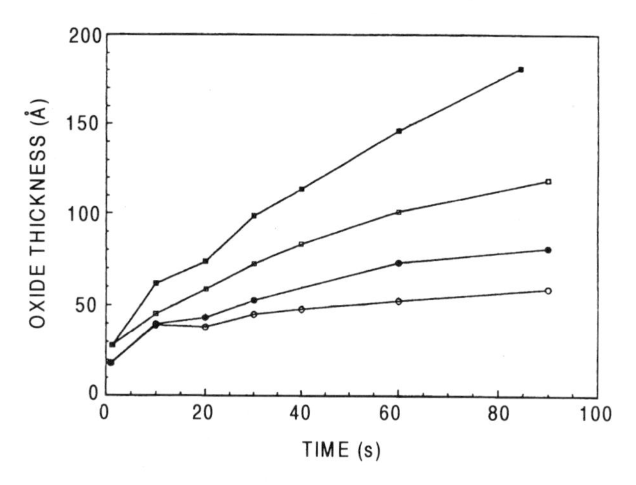

FIGURE 6.15. Kinetics of a two-step rapid thermal oxidation of (100) silicon covered with its natural native oxide with temperature as a parameter: ■, 1150 °C; □, 1100 °C; ●, 1050 °C; ○, 1000 °C [5].

the electric field is toward the SiO_2/Si interface and will therefore enhance the drift of any ionized oxidants toward the interface. Hence [55]

$$\frac{dx_{ox}}{dt} = \frac{B_{eff}}{2x_{ox}} + \frac{D^*\xi}{2\varepsilon} x_{ox} \tag{6.13}$$

where B_{eff} is an effective parabolic rate constant, D^* a constant, ε the dielectric constant, and ξ the effective planar charge density (i.e., the sum of charge sheets present in the oxide). B_{eff} follows the Arrhenius behavior with an activation energy of 2.1–2.4 eV for O^- diffusion based on Moslehi's data [15]. Experimental data indicate that for oxide thicknesses < 300 Å, two regions may be fit to the data: an initial linear growth followed by a slower growth. The transition from linear to parabolic growth will occur at a different oxide thickness, as a function of the growth temperature and conditions. The high level photoinjected electrons may vary from one RTP system to another and this may produce a different magnitude of the space charge region leading to different values for the parameters fit to the model.

6.3. ELECTRICAL PROPERTIES OF SILICON DIOXIDE GROWN IN DRY OXYGEN

The electrical properties of rapid thermally processed oxides are in general superior to those of furnace oxides. A key reason for this is the higher growth temperatures that can be used, which produce fewer defects at the Si/SiO_2 interface. Electrically this results in a lower fixed charge and interface state density.

6.3.1. Electrical Charges at the Si/SiO_2 Interface

To set the stage for the discussion in this section a brief review of the location and type of charges in a Si/SiO_2/metal (or polysilicon) structure is given. A schematic of the interface is given in Fig. 6.16 showing the location and type of charges present both at the interface and in the bulk dielectric. There are four types of charges present: a fixed oxide charge at the interface, Q_f, mobile ionic charge in the bulk, Q_m, interface trap charge, Q_{it}, and bulk trap charge, Q_{ot}. These terms are defined by Deal [56]. There are several factors in the fabrication process that determine the value of these charges.

These charges form two groups, those that are stable and those that are influenced by temperature, time, the application of an external electric field, or tunneling current.

The *fixed oxide charge* is located within the silicon rich transition region up to 35 Å from the Si/SiO_2 interface. It is predominantly positive and has slow trapping times on the order of seconds to months. The concentration of change is a function of the oxidizing ambient, silicon orientation, the maximum growth temperature, the rate of cooling from the growth temperature, and any subsequent annealing of the oxide. The value decreases in order from the (111) to (110) to (100) crystal orientation [57]. The fixed charge decreases with increasing temperature as represented in the Deal triangle. For example, an oxide grown at 850 °C that is subsequently annealed at 1100 °C for sufficient time, will reduce the fixed charge to a level corresponding to growth at 1100 °C. The optimum time for this process is quite critical for higher temperature anneals.

The *mobile ion charge* results from the presence of ionized impurities (Na^+, K^+, Li^+) in the film. These ions move under an applied electric field, particularly at higher temperatures where their mobility is increased.

The *interface trap charge* occurs at the Si/SiO_2 interface and can exchange carriers rapidly with the silicon, through either holes or electrons depending on the type and energy level of the trap. These are fast states with time constants of the order of 1 μs. Therefore, the net charge is a function of the band bending and available carriers at the interface. They arise because of dangling bonds, defects, and impurities at the interface. In general, oxides with a lower fixed charge have a lower interface trap charge. The presence of mobile charge and interface trap charge results in an instability in the threshold voltage during device operation.

Bulk oxide traps are distributed through the film and are immobile. For VLSI devices the bulk trapped charge is not normally a problem unless the oxides are less than 100 Å. However, in ULSI EPROMs in which a Fowler–Nordheim tunneling current through the

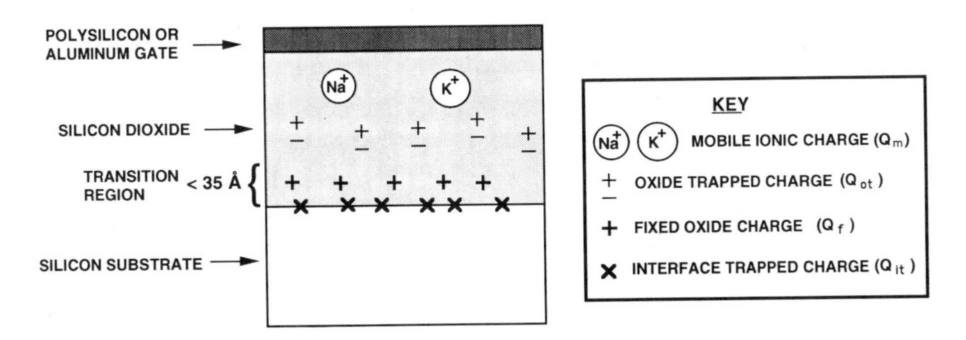

FIGURE 6.16. Schematic diagram indicating the location of electrical defects in a MOS structure.

gate dielectric is used for programming a memory cell, a fraction of the electrons or holes can be trapped within the bulk dielectric. This produces instability in the threshold voltage over time as a result of the trapped charges. Breakdown mechanisms for thin oxides involve the electrons or holes passing through the oxide, and the presence of these traps in the bulk oxide modifies the field resulting in breakdown at lower than expected electric fields. The nonequilibrium conditions give rise to a complex breakdown phenomenon. This is an active area of current research and the reader is referred to pertinent papers in the field for more information [58–61].

6.3.2. Measurement Techniques

Several standard electrical measurements are made to allow a comparison between oxides grown under different conditions. These techniques characterize the charges at the interface or in the bulk, and the dielectric breakdown. An important aspect is to study how these characteristics change with bias stress, i.e., the application of an electric field (~ 10^4 V/cm) and raised temperature (100–250 °C). There is insufficient space here to go into the details of each of these measurement techniques, and the reader is referred to Refs. 62–64.

Low- and high-frequency capacitance–voltage measurements are carried out to determine the oxide fixed charge and the interface trap charge. Shifts in the flat band voltage with bias stress test are related to changes in the oxide trap and interface charge. Large negative shifts in the flat band voltage after a positive bias has been applied to the gate, indicate mobile positive ion contamination in the dielectric, which is not often observed today with modern clean rooms. The interface trap charge can be determined by several techniques including distortion of the C–V curve and a conductance method [64]. Barrier height measurements also give a good indication of the interface trap charge, although accurate measurements are difficult. The interface traps within the band gap are a function of the energy, and in general those in the middle of the gap are most relevant having the higher recombination probability. Therefore, the midgap interface state density, $D_{\text{it-m}}$, is compared among various dielectrics. These measurements are repeated after bias stress tests to determine the shift. No change in the oxide charge and the interface trap density would indicate a stable oxide. The electrical stress tests are an indicator of the performance of the oxide in an active device.

To evaluate the bulk properties of the oxide, I–V measurements are made. The leakage current is not a good measure of the dielectric properties because of a variation in the barrier height between the silicon and gate material. A good measure of the overall quality of the dielectric is given by the charge to breakdown Q_{BD}. This is defined as the integral of the current passed through the oxide until dielectric breakdown occurs. Obviously at higher electric fields the leakage current increases dramatically, and therefore a fixed bias current is normally chosen for the duration of the test, breakdown being indicated by the sudden drop in the applied voltage. These measurements are made with both a positive and a negative bias to indicate any differences in charge injection from the gate and substrate through the dielectric. Two methods are commonly used to determine dielectric breakdown. Direct measurement is carried out with the application of an increasing voltage ramp while monitoring the current which suddenly increases when avalanche breakdown occurs, i.e., time to zero breakdown (TZBD). A second measure of breakdown is time-dependent dielectric breakdown (TDDB) in which a constant current is applied and the time until breakdown measured.

Typical values of the fixed oxide charge for a dry oxide grown at 1100 °C are ~10^{11}/cm^2 for (111) silicon and a factor of three to four lower for (100) silicon [65]. However, the value is a function of the growth temperature, cooling rate, and any postoxidation annealing. The interface trap charge is variable, depending on the given process, contamination, and substrate orientation; however, values of 10^{14}–10^{15} /cm^2 are considered good for furnace oxides. A $D_{it\text{-}m}$ between 10^9 and 10^{11} eV^{-1}cm^{-2} may be measured after oxidation but the important test is the effect of bias stress. With RTP betters control of these values is obtained particularly for thin oxide films which can be grown at a higher temperature than in a furnace. In thin oxides the threshold voltage becomes more sensitive because of the much larger gate capacitance value. Breakdown voltages of 10–12 MV/cm are usual for dielectrics grown in a furnace and can also be achieved by RTP. Care must be exercised, however, in interpreting measurements of breakdown voltage, because if it is a function of the gate area, that indicates the presence of inhomogeneities or even pinholes which break down at a lower electric field. Full discussion of these effects is reviewed by Sze [63] and Nicollian and Brews [64].

6.3.3. Effect of Processing Conditions

Nulman *et al.* [8] studied 40- to 120-Å oxides produced at 1150 °C RTP in pure oxygen. Oxidation cycles are either one- or two-step with a postoxidation anneal step. It was observed that the oxide thickness was larger and the electrical properties were improved with the two-step process. Typical breakdown fields obtained were 13.8 MV/cm and D_{it} of 1×10^{10} eV^{-1}cm^{-2}, before electrical stress. Sato and Kuichi [13] comment that the higher density of interface states compared to furnace oxide and suggest this is related to more interface dangling bonds formed by the more rapid kinetics, typically > 1 Å/s

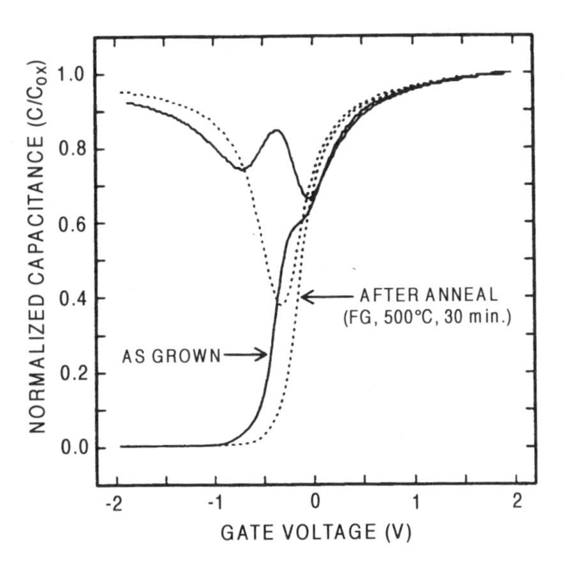

FIGURE 6.17. High-frequency (100 kHz) *C–V* curves for rapid thermal oxidation. The upper curves were taken under intense illumination, which gives results equivalent to a quasistatic *C–V*. The lower curves are conventional high-frequency measurements in the dark [67].

compared to furnace oxides of 0.08 Å/s. They also found that an anneal at 1000 °C reduced the $D_{it\text{-}m}$ from 5×10^{10} to $1\text{–}1.6 \times 10^{10}$ $eV^{-1}cm^{-2}$. Fukuda *et al.* [66] observed an improvement in the oxide quality with higher growth temperatures.

Stathis *et al.* [67] used *C–V* and electron paramagnetic resonance measurements to characterize interface defects in 30-Å oxide films. A very narrow interface state peak was observed, which was removed by a forming gas anneal. Process conditions are given in Table 6.2. Figure 6.17 shows the *C–V* measurement and the peak in the low frequency and width in the ledge in the high-frequency *C–V* (100 kHz) can be clearly seen to be removed by an annealing step at 500 °C for 30 min. EPR measurements provide an independent measure of interface states and also distinguish the energy. P_{b0} and P_{b1} centers were observed with a total spin density of 1.8×10^{12} /cm^2 and after annealing the P_{b1} has disappeared. This is an amphoteric state with two levels, one above and one below the midgap; however, other work demonstrates a broader peak than reported here [68].

Lo *et al.* [69] studied the effect of thickness on the charge trapping properties of oxides 60–100 Å thick which had undergone a postdeposition anneal for 30 min at 800 °C (see Table 6.2). Plots of J/E^2 indicate a barrier height of 3.2 eV. The oxide trapping is demonstrated in Fig. 6.18 for carriers injected from the poly-Si gate. Initially positive charge trapping occurs at hole traps, which is strongly dependent on the oxide thickness. As bias stress proceeds at 10 mA/cm^2, electron trapping occurs as the dominant mode, but is again dependent on oxide thickness. Fresh samples have similar *C–V* curves independent of oxide thickness; however, after bias stress, $D_{it\text{-}m}$ increases, with the best resistance to stress shown by the thinner oxides (60 Å). A direct interfacial bond-breaking model is proposed here, in which electrons are accelerated in the gate electric field and in a thicker oxide there is a larger defect generation volume.

Similar improvements were also observed in a study of 120-Å oxides grown at 1000 °C by Eftekhari [70]. Figure 6.19 shows that annealing of the oxides *in situ* after growth in an ambient resulted in an increase in the breakdown strength, with a maximum barrier height for a 1000 °C anneal, and a decrease in the fixed charge density. Eftekhari also showed that the barrier height was decreased with increasing annealing temperature as a result of a change of the charge states at the Si/SiO_2 interface. He proposed that the

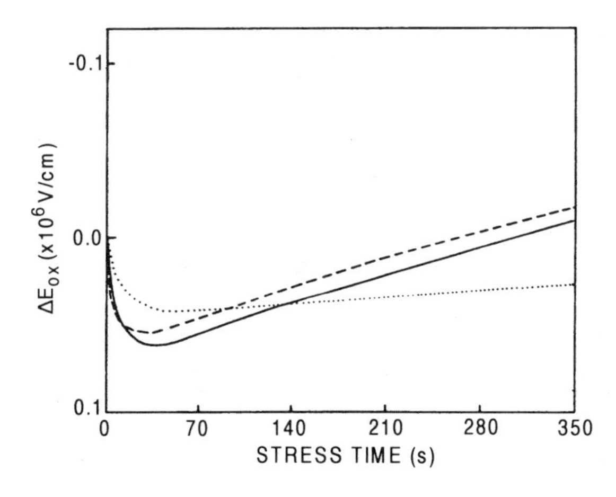

FIGURE 6.18. Change of applied oxide field ($\Delta E_{OX} = \Delta V_g / x_{OX}$) as a function of stress time under F-N electron injection at ~10 mA/cm^2 with electrons injected from the gate electrodes with oxide thickness as a parameter: ——, 100 Å; – – –, 80 Å; ·······, 60 Å [69].

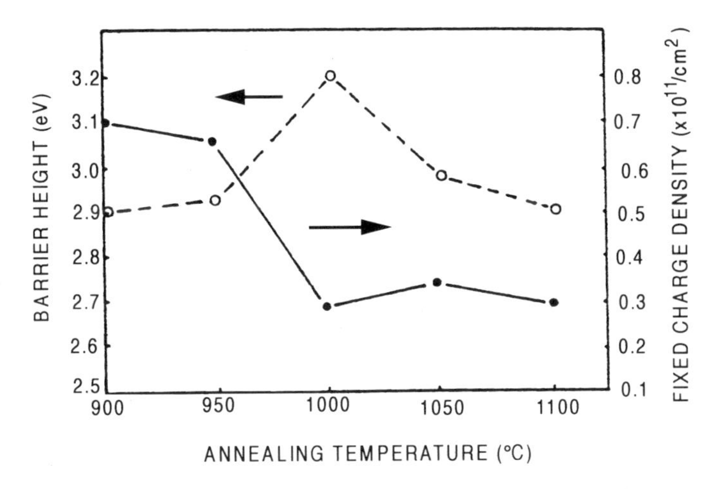

FIGURE 6.19. Variation of fixed charge density (●) and barrier height (○), for a 120-Å rapid thermal oxide grown at 1000 °C, as a function of the annealing temperature for a fixed annealing time of 30 s [70].

decrease in breakdown strength at higher growth temperatures was related to the introduction of additional defects in the oxide.

Fonseca and Campabadal [71, 72] measured the electrical characteristics of RTP oxides and found that the higher growth temperatures produced a lower fixed charge, interface trap density, and TZBD. They found contradictory results, however, for TDDB. In particular, for oxidation at 1000, 1100, and 1150 °C, TDDB at 95% failure was t_{bd} = 18, 23, and 28 s, respectively. These results indicate that dielectric strength (i.e., TZBD) does not uniquely determine the breakdown quality of the dielectric, and the higher-temperature oxides had a greater TDDB. They postulate that the reason for the higher resistance to electrons in TDDB tests is the lower mechanical stress in the oxides grown at a higher temperature. However, with a voltage ramp test like TZBD electrons are injected into the dielectric grown at lower temperature which block additional incoming electrons, thereby delaying the damage. Thus, TDDB is a better measure of the wear of the dielectric rather than TZBD which is more of an instantaneous measure of dielectric breakdown. A major influence on the results was the perimeter area of the electrode and therefore it is important to compare samples with the same perimeter/area ratio. In addition, carrier lifetimes were measured. An anneal at 1050 °C following the 110-s oxidation produced longer carrier lifetimes. This was in contrast to a furnace anneal which degraded the carrier lifetime.

Rapid thermally grown oxides 20–50 Å thick were characterized by fabricating tunnel diodes by Depas *et al.* [73]. A distinct feature was observed in the *C–V* curves indicating the presence of interface states. The surface potential was found from the Berglund integral:

$$\Psi_S = \int_{V_{fb}}^{V} \left(1 - \frac{C_{LF}}{C_{ox}}\right) dV \qquad (6.14)$$

where C_{LF} is the low-frequency capacitance and C_{ox} the oxide capacitance. The surface potential is necessary to calculate the tunneling current. Assuming that the electrons tunnel from the accumulation layer through a trapezoidal barrier into the metal appears to fit the data for the thicker oxides. For thinner oxides the current is too large to measure the gate capacitance. The density of interface traps was determined using the conductance method. A very small increase in trap density with decreasing oxide thickness was observed. The number of interface traps was reduced from 10^{12} cm^{-2}eV^{-1} to approximately 10^{10} cm^{-2}eV^{-1} by a postmetallization anneal at 400 °C for 30 min in 10% H_2/N_2. The electron mass was determined at 0.3 ± 0.02 based on fitting to the I–V curve. The stability of tunneling current was also increased after annealing as a result of reduction of water-related electron traps in the oxide layer. Uniform pinhole-free oxide layers were grown.

6.3.3.1. Pulsed Processing. Singh [74] studied oxides < 100 Å thick and found that an anneal in argon increased the oxide thickness and reduced fixed charge and interface traps. The processing involved oxidation for 60 s at 900 °C followed by four cycles of oxidation pulses 5 s long at 900 °C and anneal pulses of 60 s at 1050 °C. It is possible that the role of interstitial Si atoms in oxidation could explain the enhanced oxidation rate of an additional 10 Å. The alternating oxygenation followed by annealing could allow relaxation of stress in the film resulting in increased oxidant diffusivity or the reduction of positive charge at the interface. Also, interstitials at the interface could account for enhanced oxidation rates.

6.3.3.2. Effects of Wafer Cleaning. The effect of wafer cleaning and processing conditions was studied by Moslehi *et al.* [29]. Results of the electrical measurements are summarized in Table 6.1. The $D_{it\text{-}m}$ and V_{fb} were a function of the growth temperature and cleaning process used. When a HF dip was included before the final DI water rinse in the RCA cleaning process the V_{fb} became a function of the growth temperature, having its lowest value at the highest growth temperature (1150 °C); however, the D_{it} was higher with the HF dip. It must be noted that a large D_{it} can introduce errors in the calculation of V_{fb}. For example, a wafer cleaned with a HF dip and oxidized to a 77-Å-thick film at 1150 °C had a D_{it} of 8.9×10^{11} eV^{-1} cm^{-2} which was reduced to 1.5×10^{11} eV^{-1} cm^{-2} after a further RTA of 45 s at 1038 °C in Ar. No attempt was made to evaluate the barrier height, F_B, for the devices with different oxide thicknesses.

6.3.3.3. Interface Roughness. A recent study by dos Santos Filho *et al.* [75] examined the importance of interface roughness on the oxide quality. Two different annealing cycles were examined, a pulsed anneal and a slow ramp, each of them with the same maximum temperature (see Table 6.2). The films were characterized electrically and the oxide was removed by a dilute HF etch, which is known not to influence the silicon roughness [76] and the interfacial roughness measured with an atomic force microscope (AFM) and laser light scattering. Oxides prepared by a slow cool ramp had improved electrical characteristics, such as charge to breakdown. At the interface nanometric protrusions up to 2.5 nm high and 100 nm wide were observed; however, oxides prepared by pulsed thermal annealing had interface protrusions up to 2 nm high and 5 nm wide. These measurements indicate a correlation of electric breakdown field and charge to breakdown with the microroughness at the interface.

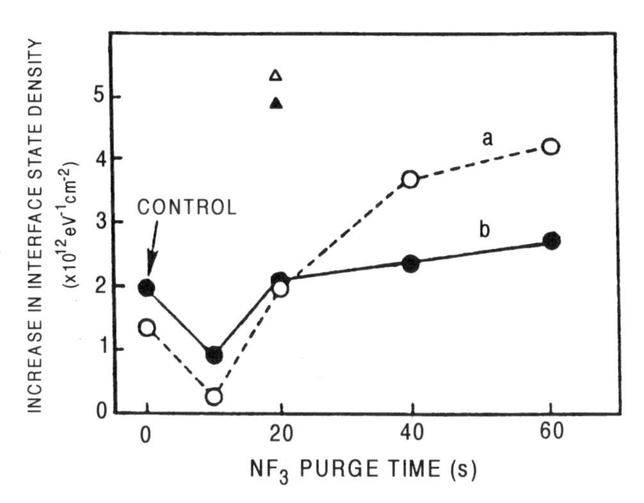

FIGURE 6.20. Interface state generation at a surface potential of 0.2 eV (i.e., the characteristic peak density) in fluorinated oxides after Fowler–Nordheim injection at a current density of 17 μA/cm^2 for 300 s, as a function of the NF_3 purge time: (a) positive gate bias and (b) negative gate bias. Both control and fluorinated oxide samples were processed in NF_3 at 900 °C for 20 s prior to rapid thermal oxidation (Δ, ▲) and are shown for comparison [77].

6.3.3.4. Effect of Halogens. Previous work on furnace oxidation indicated that the dielectric breakdown was increased by the addition to the oxide layer of fluorine in small quantities in the parts per billion range. Lo *et al.* [77] carried out rapid thermal oxidation in oxygen with dilute NF_3 added. Process conditions are given in Table 6.1. SIMS depth profiles indicate that little fluorine is present at the interface. Initially the fixed charge is increased over control oxides grown without fluorine. However, the fluorinated oxides did show an enhanced resistance to ΔD_{it} for both bias polarities under high-field stress tests. Figure 6.20 shows the results after a stress test of 300 s at a current density of 17 μA/cm^2. The improvements are thought to reflect the reduction of stress at the interface by the formation of Si–F bonds, although excessive F incorporated in the dielectric results in substitution of the –O– bridging bonds and leads to surface states and reduced breakdown voltages. Further work on liquid-phase oxidation by Lu and Hwu [78] indicates that fluorinated gate oxides have greater radiation hardness.

6.3.3.5. Effect of Nitrogen. Hames *et al.* [79] studied the effect of the addition of small amounts of nitrogen (between 10 and 1000 ppm) to the oxygen during oxidation. This had little effect on the electrical properties, except for a trend of increasing the frequency of low-field breakdowns, and an increase in the D_{it}. Even 10 ppm N_2 produced a degraded D_{it}.

6.3.4. Oxides Grown on Polycrystalline Silicon

The formation of thin dielectrics with excellent electrical properties is important for DRAMs. The electrical properties of rapid thermal oxide grown by Ohyu *et al.* [26] on poly-Si were compared with furnace oxide. The barrier height was higher in the case of

TABLE 6.3. Process Conditions and Electrical Characteristics of Furnace Oxides after Rapid Thermal Annealing and/or Nitridation[a]

	Process conditions				Electrical properties				
Process	Temperature (°C)	Time (s)	Thickness (Å)	Gate material	Fixed charge density (cm^{-2})	Interface trap density ($eV^{-1}cm^{-2}$)	Charge to breakdown (C/cm^2)	Breakdown field (MV/cm)	Refs.
RTA and RTN	800 900–1200	— 3–300	80–120	Al	$\sim 2 \times 10^{11}$	$\sim 5 \times 10^{10}$	—	~14	82
RTN and RTO or RTA	900–1150 1150 1050–1150	15–60 60–200	50–120	Al	Fig. 6.22	$\sim 10^{10}$	—	—	83,84,89,90
RTN	900–1200	10–60	100	poly-Si	$1–5 \times 10^{11}$	—	20.4	~11	85
RTN and RTO	1100 1150 1100	60 10–50 0–60	~86	poly-Si	—	$\sim 5 \times 10^{10}$	—	—	91
Anneal RTN (25–100% NH_3) RTO and RTA (N_2, Ar, Ar/O_2)	950 950–1150 1150	4 6–60 60	~80	poly-Si	—	$\sim 10^{11}$	24	~13	87
RTN	1150	45, 90	150	poly-Si	—	3.6×10^{11}	—	>12	86
RTN and RTO	1050, 1150 1050, 1150	45 45	125	poly-Si	—	—	—	—	96
RTN and RTO	900 1000	40 30	120	poly-Si	—	—	—	—	98
RTN	950	30 multiple cycles	125	Al	—	shift = 1.2×10^{11}	2.7	11	88
RTN and RTO	900 1050 (N_2O or O_2)	20 45	75	poly-Si	11×10^{11}	7×10^{10}	14–20 (negative) 40–70 (positive)	~14	93

[a] All RTAs carried out in Ar and all RTNs in NH_3 unless otherwise indicated.

the rapid thermal oxide. The TDDB (50% failure) was greater for the rapid thermal oxide than the furnace oxide under positive bias. Other characteristics are listed in Table 6.3.

6.4. RAPID THERMAL PROCESSING OF FURNACE OXIDES AND NITRIDATION OF SILICON IN AMMONIA

Furnace oxides are normally grown at a lower temperature to obtain better control of the thickness; however, the final maximum temperature the oxide is exposed to largely determines the fixed charge at the interface, as given by the Deal triangle. Therefore, the dielectric properties can be improved by fast radiative annealing at a higher temperature. A second process that has been used to improve the properties of thin oxides is a nitridation step. For a conventional furnace anneal, the increase in thermal budget makes it attractive to examine RTP as an alternative nitridation process. Annealing in ammonia or nitrous oxide at a high temperature should achieve this.

6.4.1. Rapid Thermal Nitridation

An excellent review of investigations on rapid thermal nitridation of furnace oxides is given by Massoud [80]. The advantages are: (1) improvement in effective barrier to dopant diffusion, (2) reduction of interface trap generation with electrical stress, (3) improved resistance to plasma and radiation damage, (4) increased charge to breakdown demonstrating greater endurance to hot-electron injection, and (5) a higher dielectric constant resulting in larger gate capacitance. However, the nitridation process also introduces defects which increase the fixed charge, act as electron traps, and also increase the gate leakage current.

Nitridation of silicon was studied by Moslehi and Saraswat [81] in ammonia. Nitridations were performed on (111) and (100) silicon at 950–1230 °C for periods of 30 s to 2 h. For processing times of less than 5 min the RTP was used, while for longer times an rf-heated furnace. The nitride thickness was measured with an ellipsometer assuming a dielectric constant of 2.0 for the film, and by AES, RBS, and TEM. The growth kinetics of the nitride indicated a weak dependence on the pressure, crystal orientation, and substrate doping level. The independence of the process on pressure and crystal orientation is indicative of a diffusion-limited growth process. The nitride thickness increased to a limiting value as the silicon nitride film acts as a diffusion barrier to further film growth.

Moslehi and Saraswat [81] also carried out nitridation of furnace oxides in ammonia. They also observed an initial increase of film thickness and weak orientation dependence. Figure 6.21 shows an AES depth profile of the 405-Å furnace oxide after nitridation in ammonia for various times between 30 s and 2 h. The AES measurements indicate significant buildup of nitrogen at the interface after short processing intervals, with a saturation after 15 min. Further nitridation produces a change in the stoichiometry of the film but no detectable increase in film thickness. There is also a small shift in the position of the nitrogen peak which can be attributed to the growth of an interfacial oxide layer. Also after 10 min of processing the surface nitrogen concentration saturates very rapidly. These results demonstrate the fast diffusion of the ammonia through the oxide and its fast

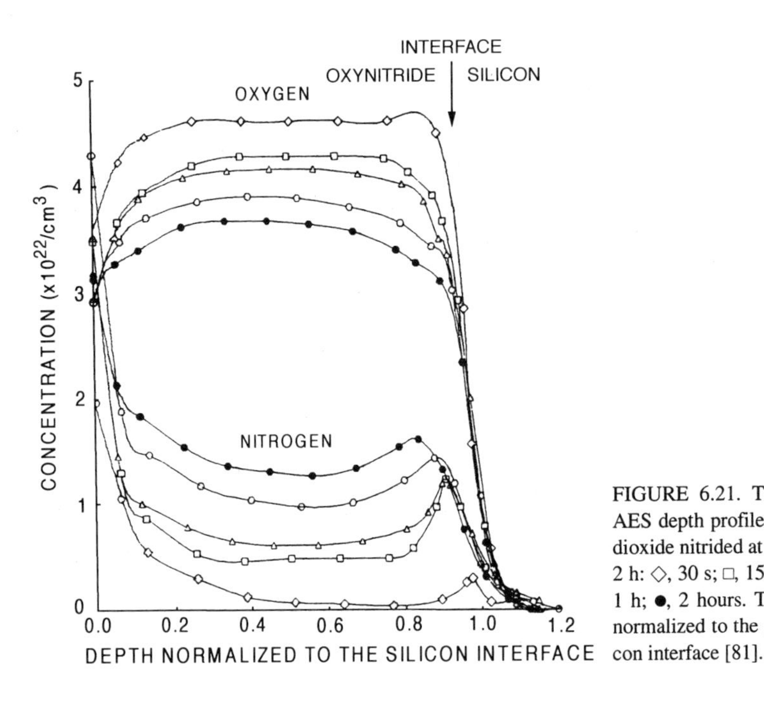

FIGURE 6.21. The concentration–AES depth profiles for 405-Å silicon dioxide nitrided at 1100 °C for 30 s to 2 h: ◇, 30 s; □, 15 min; △, 30 min; ○, 1 h; ●, 2 hours. The depth profile is normalized to the position of the silicon interface [81].

reaction at the interface and at the surface of the oxide. The nitridation of the bulk oxide takes place more slowly.

Nulman and Krusius [82] first carried out nitridation of furnace-grown oxides in ammonia by RTP and also found a nitrogen-rich surface layer at short nitridation time, and formation of silicon oxynitride layer at longer times. The reverse bias breakdown voltage increased with nitridation time. *C–V* measurements indicated that there were positive charges introduced at the interface and the $D_{\text{it-m}}$ was also reduced by up to 70% (from 10^{11} to 5×10^{10} eV^{-1} cm^{-2}). However, annealing might have the same effect (Table 6.3).

Hori *et al.* [83, 84] found the capacitance increased with nitridation time in ammonia indicating an increase in the dielectric constant of the film. Auger measurements show nitrogen is present in the film and therefore an oxynitride is beginning to form. *C–V* measurements indicated a fixed positive charge at the interface for nitridation times up to 15 s, which almost disappeared for longer nitridation times. The D_{it} and Q_{f} were found to reach a peak with process time (see Fig. 6.22). Here the oxide was processed at 950–1150 °C for 10 to 300 s. This turnaround effect in electrical parameters with processing time was accounted for by a two-factor model, which is discussed in the following section. Overall electrical properties were comparable or better than for furnace oxides at shorter processing times and for higher temperatures. Electrical measurements carried out by Shih *et al.* [85] on rapid thermal nitridation of furnace oxides under conditions indicated in Table 6.3 indicate a turnaround effect which they attribute to an interfacial strain relief model. Above 950 °C the oxide becomes a viscous layer and the stress is reduced.

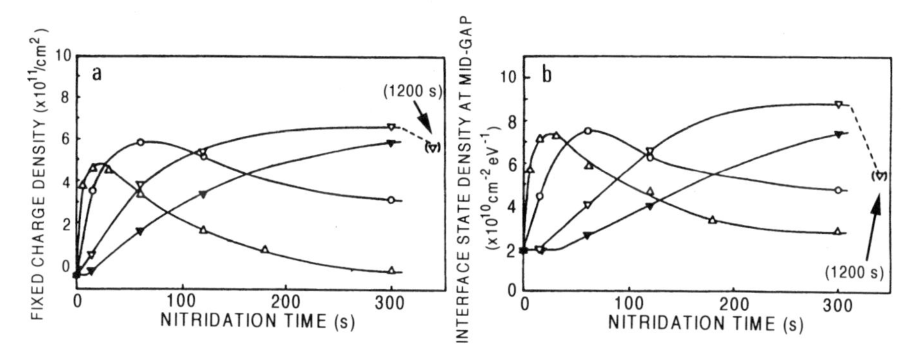

FIGURE 6.22. Characteristics of 80-Å-thick nitrided oxides. (a) Fixed charge density versus nitridation time and (b) interface state density at midgap versus nitridation time, for several nitridation temperatures: △, 1150 °C; ○, 1050 °C; ▽, 950 °C; ▼, 900 °C. Both vary in a similar way for a given nitridation temperature: at first both increase, reach respective maxima, and then decrease gradually. Value after processing for 1200 s at 950 °C is also indicated (▽) [84].

Liu *et al.* [86] made a comparison between rapid thermal nitridation and furnace nitridation, with thermal oxide as a control. The furnace oxide was grown at 1000 °C in an O_2/Ar ambient, ratio 1:50, and the furnace nitridation was 1 h at 1000 °C. Rapid thermal nitridation was carried out at 1150 °C for 45 and 90 s (see Table 6.3). The dielectric films were all 150 Å thick and a self-aligned polysilicon gate completed the MOS-capacitor fabrication. There was a shift in V_{fb} observed in accumulation with oxides, but not the nitrides, indicating a fixed positive charge. Also, an increase in the charge to breakdown, and suppression of high-field injection-induced interface state generation were observed. The lightly nitrided films gave the best results with a decrease in the trap generation rate, charge to breakdown, and D_{it}.

6.4.1.1. Effect of Pre- and Postnitridation Anneals. Wright *et al.* [87] added a pre- and postnitridation anneal to improve the electrical properties of the film. The process details in Table 6.3 indicate that the prenitridation anneal had very minimal effect but the postnitridation anneal improved properties for anneals in N_2, Ar, and Ar/O_2 mixtures. The Q_{BD} was improved with annealing and greatest for the higher processing temperatures. The improvements are believed to be related to the removal of hydrogen from the film. An oxygen anneal followed by an inert gas anneal is slightly better than a single annealing step.

6.4.1.2. Multiple Rapid Thermal Nitridation. Furnace oxides were studied by Wu and Hwu [88] in N_2O at the process parameters given in Table 6.3. Both the Q_{BD} and E_{BD} increased with the number of rapid thermal nitridation cycles. The MOS capacitors also had improved radiation resistance with a radiation-induced $\Delta D_{it\text{-}m}$ of 2×10^{-10} eV^{-1} cm^{-2} and ΔV_{fb} of 0.01 V on exposure to 1 Mrad of Co-60. The improvements are believed to be caused by nitrogen incorporation at the Si/SiO_2 interface suppressing interface trap generation during voltage stress (at 1 mA/cm^2). SIMS data indicate that the SiN/Si ratio increases with the number of cycles. Wu and Hwu concluded that three cycles at 950 °C for 30 s gave the best results.

6.4.2. Reoxidation of Nitrided Oxides

The values of $D_{\text{it-m}}$ and Q_f were observed first to increase, then decrease with processing time. Hori and Iwasaki [89, 90], carried out further studies indicating that this turnaround effect was not a function of the nitrogen incorporation in the film alone and that the effects of reoxidation were also important. Auger and SIMS were used to determine the nitrogen, oxygen, and hydrogen concentration in the dielectric films.

Hydrogen incorporation during the nitridation process introduces a large number of electron traps. Subsequent reoxidation or annealing has been partially successful at reducing hydrogen incorporation but with an increase in the thermal budget. Shih and Kwong [91] nitrided furnace oxides, grown at 850 °C in an oxygen/HCl ambient, and measured the dielectric constant of the film with an ellipsometer. After reoxidation of up to 30 s the film thickness did not change although the dielectric constant did. This indicates the removal of Si–N bonds at the interface. However, the film thickness increased with rapid thermal oxidation times longer than 30 s, indicating new oxide growth at the interface. TEM measurements confirmed the oxide thicknesses. Relating these results to the electrical measurements, the $D_{\text{it-m}}$ and ΔV_{fb} movement on electrical stressing were reduced by rapid thermal reoxidation.

In summary, the rapid thermal nitridation induced physical defects that act as interface states in Fowler–Nordheim tunneling and are probably caused by distorted Si–O and Si–N bonds. Interfacial oxidation effectively annihilates these defects. The existence of nitrogen at the interface allows the regrowth of an interfacial oxide with fewer strained bonds. The two-factor model of Hori *et al.* [92] correlates the charge trapping with the physical properties. This model for interface generation can explain the turnaround behavior in D_{it}. During nitridation, –H, –OH species are produced, which are well known to produce electron traps, and they may be included in the film. The atomic hydrogen is also important because the interface state generation mechanism involves interaction between hot electrons and weak Si–SiO_2 bonds at the interface in the form of (1) hydrogen-containing centers and (2) strained Si–O bonds. The ΔV_{fb} is proportional to the concentration of hydrogen [90], with constant of proportionality $K_H = 9.3 \times 10^{-22}$ V/cm^3:

$$\Delta V_{\text{fb}} = K_H\,[\text{H}] \tag{6.15}$$

and ΔV_{fb} is not dependent on the initial nitrogen concentration, $[\text{N}]_{\text{init}}$. This suggests that nitrogen incorporation is not responsible for the generation of electron traps and this picture is consistent with the formation of an oxygen-rich layer at the Si/SiO_2 interface which is not involved in the reduction of electron trapping. After reoxidation the nitrogen concentration near the surface is decreased and the nitrogen-rich layer near the interface moves from its initial location with movement of the Si/SiO_2 interface location. However, the nitrogen is essential for immunity against interface-state generation, possibly through a reduction of strained Si–O bonds. The result is a complicated relation between the starting nitride conditions and reoxidation on ΔD_{it} [92]:

$$\Delta D_{\text{it-m}} = \frac{F_{\text{ox}}}{(1 + K_N[\text{N}]_{\text{int}}^{m_1})}\,[\text{H}]^{m_2} \tag{6.16}$$

were F_{ox} and K_N are constants and m_1 and m_2 are fit parameters, estimated at 2 and 2.5, respectively, for the dielectrics they studied. This empirical equation predicts that a nitrided oxide with a small amount of hydrogen reduces $\Delta D_{it\text{-}m}$ to a lower value than a thermal oxide! The hydrogen decreases rapidly on reoxidation and approaches a level comparable to that observed in a furnace oxide.

SIMS measurements indicate that hydrogen present in the process is reduced on reoxidation. However, annealing in nitrogen produced the same effect on hydrogen concentration as reoxidation. Therefore, the loss of hydrogen is controlled mainly by an out-diffusion mechanism. Some caution is necessary when examining SIMS data for hydrogen because it is a very small and mobile species, but qualitative results are obtained. Also, AES results indicate that annealing in nitrogen reduces the nitrogen content of the film much more slowly than annealing in oxygen. Electrical properties including Q_f increase with rapid thermal nitridation time. Higher reoxidation temperature produces best results. The D_{it} was low for all samples except the 900 °C, 30 min anneal. When the starting nitridation was heavy, the reduction of ΔV_{fb} by reoxidation proceeds more slowly.

Han *et al.* [93] reoxidized furnace oxides nitrided in ammonia with the process conditions shown in Table 6.3. The furnace oxide was grown at 950 °C. The interface endurance was studied by monitoring the $D_{it\text{-}m}$ with injected charge. The samples reoxidized in N_2O gave the best results showing smaller electron trap generation rates under both polarities. They speculate that the effective removal of hydrogen from the interface is responsible for the improved electrical properties. Subsequent reoxidation of an N_2O nitrided oxide has no beneficial effect [94].

Okada *et al.* [95] studied the effect of first growing 40 Å of furnace oxide and then carrying out RTP in N_2O/O_2. They found that the charge to breakdown was higher than for processing in only N_2O/O_2 and the distribution in oxide thickness for the 100% N_2O process was greatly reduced. The reader is referred to Section 6.7 which discusses growth of oxides in N_2O/O_2 mixed ambients.

6.4.3. *Model of Defect Structures in Nitrided Furnace Oxides*

Yount *et al.* [94] used electron spin resonance spectroscopy (ESR) to identify point defects in N_2O nitrided and NH_3 nitrided silicon dioxide films. The furnace oxides 181 Å thick were grown at 900 °C and subsequently nitrided for 20 min either in a N_2O at 900, 950, or 1000 °C or in an NH_3 ambient at 1100 °C. ESR measurements are able to identify the defect centers in the dielectric. A P_b center is an unpaired electron on a silicon atom bonded to three other silicon atoms at the interface. Spin resonance studies have demonstrated that these centers are responsible for the majority of interface states. Figure 6.23 shows the number of P_b defects produced under different processing conditions, after exposing to UV radiation for 24 h which provides a simulation of the effect of hot-carrier or radiation damage in the dielectric. The oxide film shows the greatest shift compared to the nitrided films indicating its insensitivity to defect generation. The nitrided film has the highest number of initial defects. This has been postulated to reflect the greater concentration of nitrogen and hydrogen at the interface. Observations were also made of E' centers which are the dominant hole trapping mechanism in these dielectrics. An E'

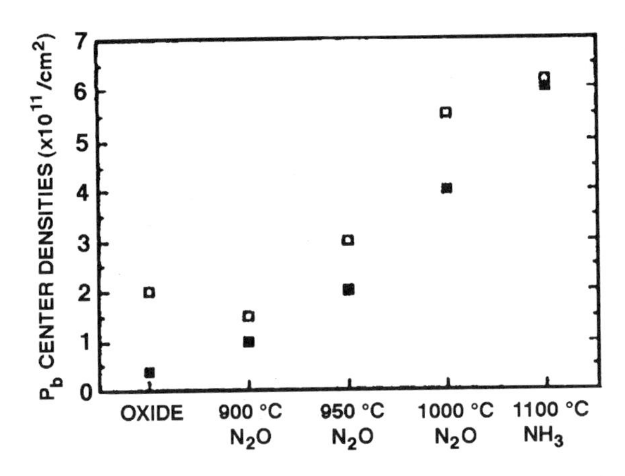

FIGURE 6.23. ESR measurements showing the density of unpaired electron P_b centers in as-processed (virgin) (■), and UV-illuminated (□), dielectrics. P_b centers are the primary source of interface states in the Si/SiO_2 system [94].

center consists of an unpaired electron in an sp^3 hybridization orbital of a silicon atom backbonded to three oxygen atoms and is usually associated with a positively charged oxygen vacancy. None were detected in the as-prepared samples for all processing conditions studied. However, after UV irradiation E' centers appeared in all but the samples nitrided in ammonia. The levels were between 0.5 and $1.5 \times 10^{17}/cm^3$. There was a striking difference between the two spectra for the N_2O and NH_3 processed films. These features indicate that there are N–H bonds present in the NH_3 processed samples but absent from the N_2O processed samples and explain the greatly reduced electron trapping in the N_2O processed dielectrics.

6.4.4. *Effectiveness of Dielectric as a Diffusion Barrier*

Measurements were made to evaluate the effectiveness of rapid thermal nitridation of furnace oxides [96] as a diffusion barrier to boron. After the first set of *C–V* measurements the samples were annealed at 900 °C for 10 min in N_2. Three subsequent anneals under the same conditions were made and the *C–V* measurement repeated after each one. It was found that the furnace oxide showed significant shift in the V_{fb} of 3 V; however, the reoxidized nitrided furnace oxide film had less than 200 mV change. The lack of boron penetration into the channel could reflect either the nitrogen-rich region acting as a diffusion barrier or a large change in the boron segregation coefficient between the silicon and dielectric film. The former is more likely.

6.4.5. *Electronic Devices Using Reoxidized Nitrided Oxides*

Hori *et al.* [97] show the practical use of these films in *n*- and *p*-channel 0.5-μm MOSFET devices. Reoxidized nitrided furnace oxide films were processed at 910 °C for 15–150 s and measurements were made of g_m after bias stressing. It was found that the small signal transconductance, g_m, was little affected, although there was a significant reduction in shift in the threshold voltage for the reoxidized nitrided oxides compared to the furnace oxides. CMOS devices were also studied by Hori *et al.* [97] N-MOSFETs were fabricated by Wright *et al.* [87] with 80- and 160-Å gate oxides, and the surface mobility was measured. The hole mobility decreased by 10% after rapid thermal nitridation and reoxidation. A postprocess anneal in argon led to a further decrease, although excessive rapid thermal oxidation time degraded the properties.

Reoxidation of nitrided oxides in MOSFET structures was studied by Joshi and Kwong [98]. Gate induced drain leakage (GIDL) measurements showed the leakage currents caused by band to defect tunneling promoted by interface states under low-voltage operation. Electrical stress was carried for 2 h at a given V_G with constant V_D = 7 V. Measurements of leakage current were made at negative V_G with the channel turned off. They concluded that degradation of the interface occurs via electron traps after rapid thermal nitridation but is remarkably improved with reoxidation. They also reported measurements of Δg_m which were a maximum for a given V_G with the rapid thermal oxidation/rapid thermal nitridation films.

Cheng *et al.* [99] built layered structures on poly-Si by first nitriding the poly-Si in ammonia at 850 °C for 60 s. After that a 30-Å-thick layer of silicon nitride was deposited and a low-pressure oxidation was performed at 850 °C, 0.5 Torr for 30 min, followed by a poly-Si top electrode. The effective dielectric oxide thickness was 37.2 Å using a dielectric constant of 3.7. The cumulative failure rates under voltage ramp tests were greatly improved.

6.5. NITRIDATION OF RAPID THERMALLY GROWN OXIDES IN AMMONIA

In the previous section we considered the rapid thermal nitridation of furnace oxides and it makes logical sense to further reduce the thermal budget in growing the dielectric film by rapid thermal oxidation. In fact, RTP has the advantage of better control for oxidation at short time intervals. The second advantage is that a higher temperature can be used in the growth process, therefore reducing the dielectric fixed charge as given by the Deal triangle. In this section we will consider nitridation in ammonia and in the next section processing in nitrous oxide, including the direct growth of oxynitride films.

6.5.1. Structure of Dielectric Film

Naito *et al.* [100] grew very thin rapid thermal oxides and then annealed them in an ammonia ambient. Process details are given in Table 6.4. AES depth profiles indicate the presence of nitrogen at the interface and throughout the thickness of the SiO_xN_y film. This is confirmed by XPS measurements which also show that the binding energy of the Si–O and Si–N bonds do not correspond to stoichiometric SiO_2 and Si_3N_4.

Tung *et al.* [21] studied rapid thermal nitridation of rapid thermally grown oxides. Auger measurements indicated that a SiO_xN_y structure results with a peak in the nitrogen concentration near the Si/SiO_2 interface and at the surface of the film (see Fig. 6.24). The concentration of nitrogen in the film increases with the nitridation time. This observation is in agreement with the observed increase in dielectric constant of the film of up to 20%, assuming a constant film thickness. Process details are given in Table 6.4.

Iberl *et al.* [101] also carried out reoxidation of rapid thermally nitrided rapid thermal oxides and made SIMS, RBS, and NRA measurements to investigate the film

TABLE 6.4. Process Conditions and Electrical Characteristics of Rapid Thermal Oxides Nitrided in Ammonia

	Process conditions				Electrical properties				
Process	Temperature (°C)	Time (s)	Thickness (Å)	Gate material	Fixed charge density (cm^{-2})	Interface trap density ($eV^{-1}cm^{-2}$)	Charge to breakdown (C/cm^2)	Breakdown field (MV/cm)	Refs.
				Single-crystal silicon substrate					
RTO	1100	—	80	Al	$1–5 \times 10^{11}$	—	—	—	100
RTN	800–1150	5–300							
RTO	1000–1250	5–60	96	Al	6.4×10^{11}	$\sim 1.1 \times 10^{10}$	—	14.6	21
RTA	1100	60							
RTN	1150–1250	5–150							
RTC	700–1000	0–200	82–98	poly-Si	—	1.1×10^{11}	30	11	118
RTO	1000–1200	7–30							
RTN	1100	30–60							
RTO	1100	0–30							
RTC	1276	—	250	poly-Si	—	$> 3 \times 10^{10}$	—	—	101
RTO	1100	40–80							
RTN	1050	30–120							
RTO	1150	60							
RTO	900	—	105	poly-Si	—	—	8 (negative) 30 (positive)	9–9.5	102
RTN	1000	10							
RTO	1150	60							

RTC	—	—	—	poly-Si	$<10^9$	—	—	19	111
RTO (O_2/H_2)	1150	45							
RTN	1050	60							
RTO	1150	60							
RTCVD poly-Si	700	150							
RTA	850	30	80	—	—	1×10^{11}	—	13	40
RTO	1148	40							
RTA	800	30							
RTN	1150	10–120							
RTO	1050–1150	—	100	poly-Si	—	1×10^{11}	17–22 Fig. 6.25 [$J = 1$ A/cm^2]	7–12	103
RTN	950–1200	1–60							
RTA	1050–1150 (in O_2 or N_2)	60							
RTO	—	—	230	Al	—	$<10^{11a}$	—	—	104
RTN	1000–1100	90							
RTO	1000–1100	30							
				Polycrystalline silicon substrate					
RTO	1100	60	~175	poly-Si	—	—	—	3–4	105,106
RTN	1000	20							
RTO	1100	60							

[a]Function of processing gas pressure.

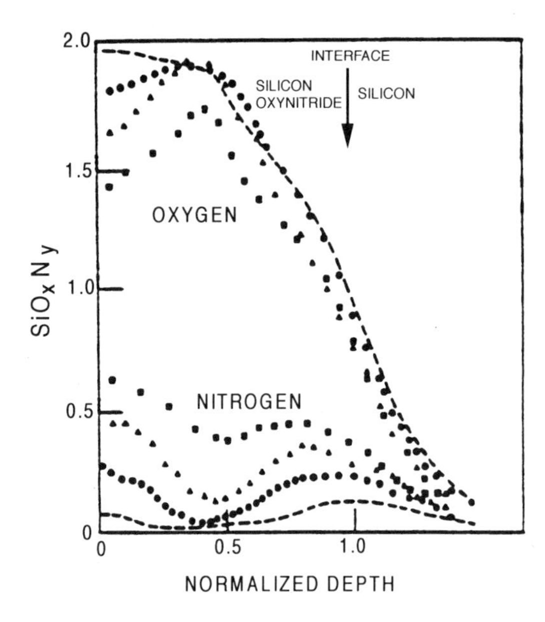

FIGURE 6.24. Auger depth profiles for nitrogen and oxygen of silicon after rapid thermal oxidation in oxygen at 1120 °C, 20 s, rapid thermal anneal in argon at 1100 °C, 60 s, and rapid thermal nitridation in ammonia for various times: –, 5 s; ●, 30 s; ▲, 120 s; ■, 150 s [21].

stoichiometry. The cleaning step was carried out *in situ* by heating the wafer to 1276 °C in the RTP chamber (see Table 6.4 for other processing parameters). The SIMS and NRA data indicated that nitrogen diffused toward the bulk Si interface producing a peak concentration at the interface.

6.5.2. Electrical Properties

An increase in breakdown voltage with nitridation time and temperature was observed. Even a relatively lightly nitrided wafer (less than 5%) results in a dramatic improvement in the immunity against interface-state generation. However, longer nitridation times showed less improvement in electrical properties. Measurements by Iberl *et al.* [101] indicated that it was important to have the top layer as nitrogen free as possible to reduce leakage currents.

The reoxidation effect in nitrided oxide films was investigated by Dutoit *et al.* [102]. They measured the charge to breakdown and found that there was a turnaround in the charge to breakdown with nitridation time, in a similar manner to that for nitrided furnace oxides in the two-factor model of Hori and Iwasaki [90]. This model is related to the release of hydrogen from the interface and suggests a similar mechanism may be important here. The goal of this study was to optimize the RTP processing temperatures and times for reduction in the electron and hole trap generation rate, and maximize Q_{BD}, while minimizing threshold voltage shifts. It was found that the nitridation time was a key factor in maximizing Q_{BD}, and that the higher temperatures (1150 °C) were preferred for the reoxidation step. They also measured the low-field breakdown by the TZBD and found the best results were for lightly nitrided wafers which produced the highest

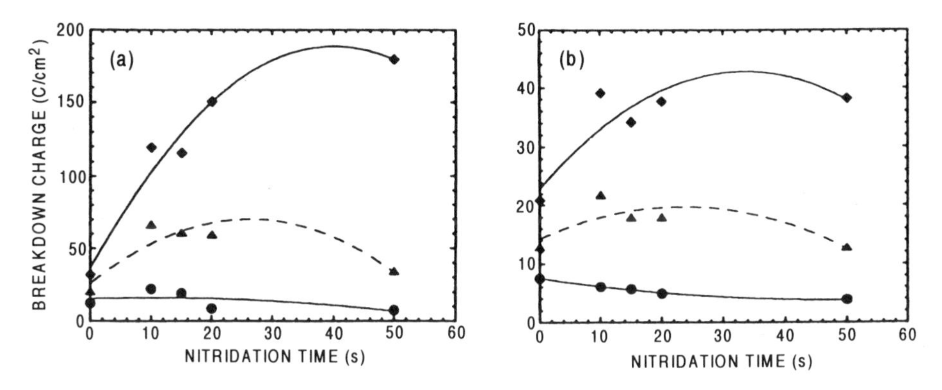

FIGURE 6.25. Breakdown charge versus nitridation time for (a) positive and (b) negative gate bias, with injection current density as a parameter: ♦, 0.1 A/cm^2; ▲, 1 A/cm^2; ●, 10 A/cm^2. (Nitridation: 1000 °C in NH_3; reoxidation: 1150 °C, 60 s; gate area: 2×10^{-5} cm^2) [103].

endurance of the dielectric film under high-field stress. This confirms that the loss of hydrogen is controlled mainly by an out-diffusion mechanism.

Dutoit *et al.* [103] went on to study nitridation in either ammonia or nitrous oxide and found that the two nitridation processes showed similar results. Processing in N_2O is discussed in the next section. For the ammonia-based process an important parameter is the nitridation time. The charge to breakdown was a function of the nitridation time, and different for positive and negative bias stress, as shown in Fig. 6.25. There is a maximum for short nitridation times which decreases slightly with increasing processing temperature. The reason for the asymmetry in the Q_{BD} is not well understood. Finally, reoxidation of the film at 1150 °C for 60 s further increased the Q_{BD}.

Lu *et al.* [104] investigated the effect of changing the time, pressure, and temperature in rapid thermal processing on the radiation hardness. A fairly thick oxide was used for this study (230 Å) and an aluminum gate was annealed in nitrogen for 15 min at 400 °C. A turnaround behavior was observed similar to the Hori model for $D_{\text{it-m}}$ and ΔV_{fb}. The processing conditions that produced the lowest initial $D_{\text{it-m}}$ did not correspond to the best radiation hardness. Excellent radiation hardness was achieved for a low nitrogen pressure (250 mTorr) during the reoxidation process, a high reoxidation temperature (1050 °C) and a long reoxidation time (100 s). However, the sample with the best radiation hardness was processed at 550 mTorr, 1050 °C, and $\Delta D_{\text{it-m}}$ of 5.1×10^{10} eV^{-1}cm^{-2}.

6.5.3. *Polysilicon Substrates*

Rapid thermal oxide layers grown on polysilicon were investigated with nitridation, reoxidation, and postoxidation anneals by Lo *et al.* [105, 106]. The electrical properties improved with nitridation. The electron trapping was greater than for an oxide grown directly on bare silicon wafers. This suggests that the surface roughness of the polysilicon plays an important role in defining the electrical characteristics. The lower electrode injection was insensitive to nitridation whereas the upper poly-Si electrode injection was sensitive to nitridation. This may partly reflect the change in surface roughness. Reoxidation was also found to reduce electron trapping. However, prolonged reoxidation degraded the properties. The effect of nitridation was to change the slope of the $I-V$ curve indicating that electron trapping is taking place during the leakage current application, i.e., the trapped electrons introduce a negative charge which opposed the applied field thus reducing the current flow. With a negative V_G a higher leakage current was observed. The trap generation rate was measured at a leakage current of 100 $\mu A/cm^2$. The trap centroid moved closer to the lower poly-Si electrode.

On polysilicon Itoh *et al.* [107] built stacked layers of n^+-poly-Si/Si_3N_4/n^+-poly-Si using rapid thermal nitridation on the bottom poly-Si electrode to reduce SiO_2 and after LPCVD of Si_3N_4 RTP at 1100 °C was carried out in pure O_2 for 40 s. The dielectric leakage currents were measured and compared to wafers in which a conventional furnace oxidation was done instead of the RTP. Stacked dielectrics produced by RTP of Si_3N_4 films have been demonstrated to produce low leakage currents [108]. The use of plasma-assisted nitrides has also been examined by Ma *et al.* [109].

6.5.4. *Use of Reoxidized Nitrided RTP Oxides in Electronic Devices*

A study of nitridation of rapid thermal oxides was reported by Yoneda *et al.* [110]. Ultrathin oxides are grown in trench capacitor structures for high-density DRAMs. The trench capacitors were 1×1 μm^2 and 3 μm deep. At the corners and base of the trenches there can be oxide thinning and electric field enhancement as a result of the sharp corners present. Both nitrided rapid thermal oxidation films 80 Å thick and reoxidized nitrided rapid thermal oxidation films were examined and compared to conventional thermally grown oxides 50 Å thick. There was little advantage of reoxidation indicated by the electrical measurements.

Active devices have been reported by Dutoit *et al.* [102] who demonstrating the feasibility of rapid thermally grown oxides in a 1-kbit memory array in 3-μm CMOS with

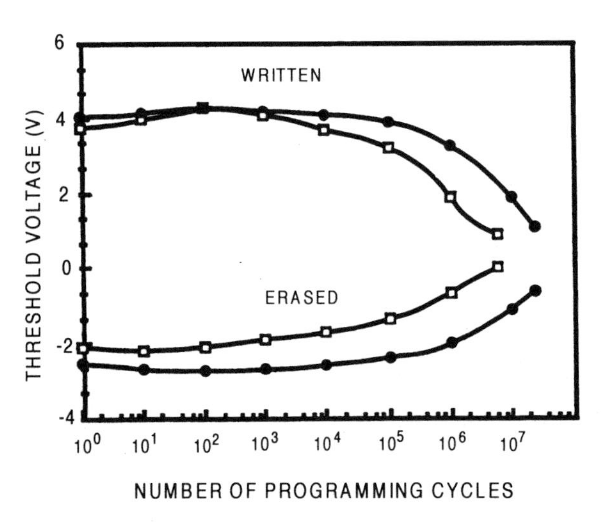

FIGURE 6.26. Endurance characteristics of two EEPROM cells: one with a furnace-grown oxide (□), the other with an optimized RTP oxynitride (●) (programming pulses: 18 V, 2 ms, rise time: 100 μs) [102].

self-aligned contacts. The endurance characteristics of two EEPROM cells, one with a furnace-grown oxide and the other with an optimized RTP oxynitride, are shown in Fig. 6.26. The endurance of the nitrided gate dielectrics was over 10^6 cycles.

RTP has been used for the entire process by Fröschle *et al.* [111] including rapid thermal cleaning in a H_2/Ar ambient before rapid thermal oxidation in O_2/H_2/Ar. Both RTA and rapid thermal nitridation were used and finally a polysilicon gate was deposited by RTCVD in SiH_4/H_2 at 700 °C with flow rates of 255 and 1000 sccm (standard cubic centimeters per minute), respectively. The uniformity was 3%. Electrical properties are given in Table 6.4.

6.6. RAPID THERMAL PROCESSING OF RAPID THERMAL OXIDES IN NITROUS OXIDE

Nitridation of rapid thermally grown oxides improves the electrical properties of the dielectric according to the processing parameters in Table 6.5 [112, 113]. Nitridation of rapid thermal oxides in N_2O followed by rapid thermal oxidation in oxygen was compared with rapid thermal oxidation in N_2O and rapid thermal oxidation in oxygen. The advantage of using N_2O instead of ammonia is that hydrogen, which will degrade the electronic properties of the interface, is not introduced into the process. A subsequent reoxidation in O_2 has also been found to further improve the electrical performance. All of the processes involving N_2O produced improved Q_{BD} and $D_{it\text{-}m}$ over the rapid thermal oxidation in oxygen. This results from the nitrogen inclusion into the oxide layer at the interface reducing interface traps. In addition, furnace oxides were prepared and compared with rapid thermal oxidation/rapid thermal nitridation (in NH_3) samples. The effects of charge injection on the $D_{it\text{-}m}$ and ΔV_{fb} were evaluated. The nitrided oxides had a higher bias stress endurance for D_{it} than the furnace oxides and the higher process temperatures gave the best results (1000–1200 °C).

Yoon *et al.* [114, 115] found that processing in N_2O gave improved results based on a reduction in the stress-induced leakage current after a nitridation step. Details of dc and ac stress measurements are summarized in Table 6.5. Thin films were nitrided in N_2O by Bellafiore *et al.* [116]. Both the furnace-nitrided and RTP-processed samples showed nitrogen accumulation at the Si/SiO_2 interface as measured by AES with a nitrogen peak concentration of between 2 and 4 at. %. The measured D_{it} indicates the level is higher in RTP films than in furnace oxides. The effect of nitridation decreases the D_{it} in both cases and greatly improves the endurance to bias stress tests. Measurements of the ΔV_{fb} and Q_{BD} change with bias stressing indicating that the higher fraction of N_2O dielectric has the improved properties over the preoxidation layer (see Fig. 6.27). A negligible effect is observed on the breakdown electric field. Fukuda and Nomura [117] have examined the properties of nitridation of RTP-grown silicon dioxide with process details as shown in Table 6.5. The gate voltage shift under bias stress testing at 100 mA/cm^2 was less for the nitrided films than the RTP oxidized films.

Effect of Wafer Cleaning. Fukuda *et al.* [118] examined the effect of rapid thermal cleaning before rapid thermal oxidation in oxygen. Cleaning was carried out in either (1) H_2 at 700–1100 °C for 20–200 s or (2) 1% HCl in Ar at 700–1000 °C for 20–200 s. The

TABLE 6.5. Process Conditions and Electrical Characteristics of Rapid Thermal Oxides Nitrided in N_2O

	Process conditions				Electrical properties				
Process	Temperature (°C)	Time (s)	Thickness (Å)	Gate material	Interface trap density ($eV^{-1}cm^{-2}$)	Interface trap density after bias stress ($eV^{-1}cm^{-2}$)	Charge to breakdown (C/cm^2)	Breakdown field (MV/cm)	Refs.
RTO	1100	5–28	80–100	poly-Si	2×10^{11}	—	50	—	117,119
RTN	1100	0–25							
RTO	950–1100	20–120	30–100	poly-Si	5×10^{10}	4×10^{11}	20 (negative) 80 (positive) ($J = 0.2$ A/cm^2)	15.5 (oxidation only 14.9)	114,115
RTN									
RTO	1050	50	70–90	poly-Si	—	Fig. 6.27	14–20 (negative) 26–40 (positive) ($J = 0.1$ A/cm^2)	15	116
RTN	1100	30							
RTO	1100	30	50–100	poly-Si	—	—	—	—	117
RTN	1100	30							

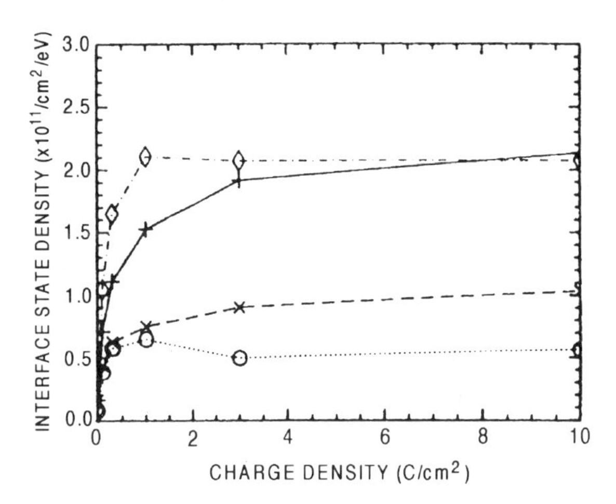

FIGURE 6.27. Variation of midgap interface state density as a function of charge injection from substrate. Interface state density calculated from high-frequency and quasistatic C–V data, substrate injection at constant current density of 10 mA/cm^2. Oven samples are p-type substrate and RTP samples are n-type. +, RTO; X, RTO and RTN; ◇, furnace oxide; ○, furnace oxynitride [116].

thickness uniformity was excellent at ±1%. The effect of cleaning was important in increasing the breakdown strength with little difference between the two techniques. Fukuda *et al.* found that excessive cleaning or cleaning at temperatures above 1100 °C produced degraded properties. The effect of electrical stress under Fowler–Nordheim injection at a current density of 10 mA/cm^2 on D_{it} was studied with quasistatic high-frequency C–V measurements. All samples showed an increase in D_{it} with stress. However, the rapid thermal nitridation in N_2O produced the least change. Similar results were obtained for the ΔV_{fb}. It is proposed that the N_2O nitridation produces a smaller change in the electron and hole trapped charge because of the lower density of broken bonds at the Si/SiO_2 interface.

Effect of Process Cycling. Fukuda et al. [119] fabricated composite dielectric films comprised of multiple layers of oxide and oxynitride. First a rapid thermal oxide was grown in oxygen, and then a rapid thermal oxide was grown in a N_2O ambient. The fraction of N_2O was varied from 50% to 100% in argon. The interface state density was reduced and the total charge to breakdown was greater than 30 C/cm^2, with less than 5% nitrogen at the interface.

Electronic Devices Fabricated with Dielectrics Nitrided in N_2O

The effect on device performance of devices with nitrided dielectric was evaluated by fabricating n- and p-MOSFETs by Okada *et al.* [95]. The devices comprise a 40-Å gate oxide and single drain with 0.4-μm drawn gate length, and 15-μm drawn gate width. Figure 6.28 shows the degradation in g_{max} after 1000 s of stress at 30 μA/μm for the n-MOSFET. The maximum transconductance was evaluated in the linear region of device operation with a drain source voltage of 0.1 V and a gate voltage of 3.3 V. The dielectric thickness was determined from C–V measurements assuming a dielectric constant of 3.9. No degradation in the current drive was observed in the saturation region for either n- or

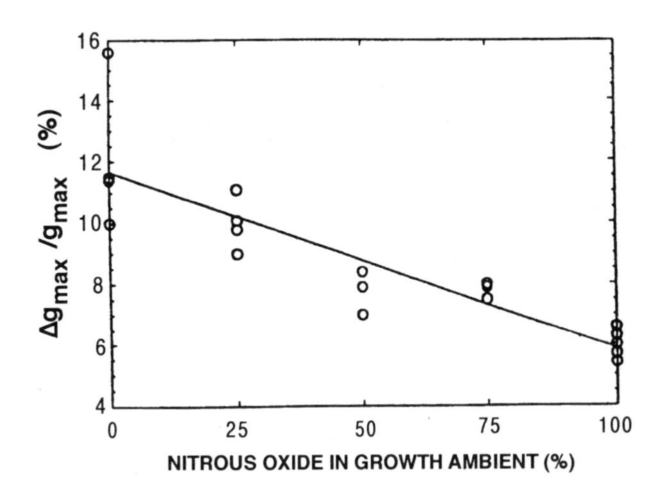

FIGURE 6.28. The degradation of g_{max} for an *n*-MOSFET after a bias stress of 30 μA/μm gate length for 1000 s. Gate length 0.4 μm and gate width 15 μm [95].

p-MOSFETs. The larger scatter in the data for the 100% N_2O-processed films may be related to the poor thickness uniformity.

6.7. GROWTH OF OXYNITRIDES IN NITROUS OXIDE AND NITRIC OXIDE

The advantage of direct oxynitridation in nitrous oxide or nitric oxide, rather than nitridation in ammonia and reoxidation of a rapid thermally grown oxide, is that hydrogen, which is known to be a source of carrier traps and interface states, is not present. Also, a lower thermal budget is achieved and better control of nitrogen concentration can be achieved through varying the nitrous oxide/oxygen ratio in the chamber.

6.7.1. *Film Composition*

There have been several studies of the film composition by an array of surface analysis techniques, as listed in Table 6.6. Hwang *et al.* [120] showed that the oxide film contains oxygen and nitrogen by a SIMS depth profile. Auger measurements indicate that the total nitrogen concentration is less than 4% in the film and concentrated at the Si/SiO_2 interface with no detectable nitrogen in the bulk dielectric [121]. The SIMS depth profile measured by Okada *et al.* [122] is shown in Fig. 6.29. They also demonstrated that the nitrogen interface concentration increased with growth time and temperature, as expected. These concentrations are confirmed by the work of Kuiper *et al.* [123] in which nitrogen concentrations at the interface were extracted from elastic recoil detection (ERD) and AES depth profiles. In both cases nitrogen is concentrated near the interface at $1.4 \times 10^{15}/cm^2$ which is equivalent to approximately 2 at. % and there is less than 0.2 at. % N in the bulk dielectric.

TABLE 6.6. Process Conditions and Structure of Dielectrics Grown in N_2O and NO

Process conditions					Analysis				
Process	Gas	Temperature (°C)	Time (s)	Thickness (Å)	Technique	Nitrogen concentration at the interface (at. %)[a]	Nitrogen concentration in the bulk (at. %)[a]	Comments	Refs.
				Rapid thermal processing					
RTN	N_2O	1100	40	60	SIMS	8	—	—	120
RTC RTN	— N_2O	700–1100 950–1200	20–200 —	65	AES	< 4	—	—	121
RTN	N_2O	1150	100–300	200–300	ERD	~2 ($1.4 \times 10^{15}/cm^2$)	< 0.2	interface thickness 20Å	123
RTN RTA	$N_2O + O_2$ Ar	1100	80–140 60	~100	SIMS	0.89	0.1	Fig. 6.29	95,122,146
RTN	N_2O	1150	30, 100, 300	—	SIMS & AES	~0.5	low	—	126
RTN	N_2O	800–1200	—	~100	NRA	6.6	—	—	125
RTN	N_2O	900–1100	—	20–450	NRA	1 ($8 \times 10^{14}/cm^2$)	low	Fig. 6.30	124
RTN	NO[b] O_2/NO[b]	1150 950/1150	300 90/300	38	XPS	4–5 4	4–5 1.5	Fig. 6.32	127,128,135
RTN RTA	$N_2O + O_2$ N_2	1000 1000	— 60	50–150	XPS	$\frac{[N]}{[N]+[O]} = 0.06$	$\frac{[N]}{[N]+[O]} = < 0.01$	—	134
				Furnace processing					
—	$N_2O + O_2$	1050–1150	1–5 min	~100	AES	3–5	—	—	145
—	$N_2O + O_2$	950–1100	15–60 min	30–450	AES	4	~1	interface thickness ~14–20Å	131
—	N_2O	1000	—	~100	SIMS	1.1	—	—	95,139
	$N_2O + O_2$	1000	—	50–150	XPS	$\frac{[N]}{[N]+[O]} = 0.025$	$\frac{[N]}{[N]+[O]} = 0.02$	—	134
—	N_2O	—	35–95	80–160	SIMS	0.5	—	—	130

[a] Function of oxidation temperature.

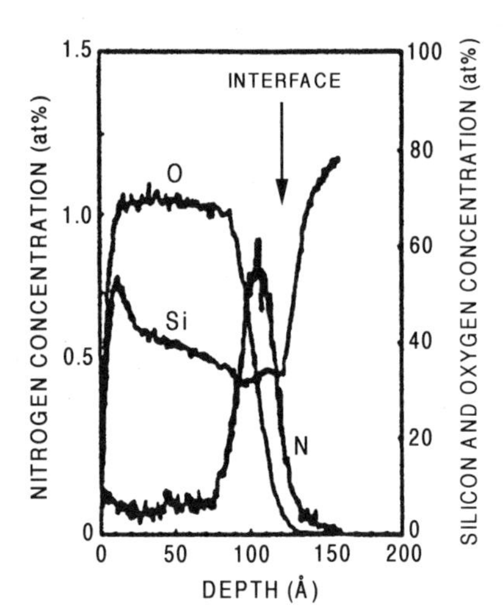

FIGURE 6.29. SIMS depth profile of the RTP oxynitride film which was grown on silicon at 1100 °C in N_2O [122].

An absolute technique for the measurement of concentration is based on nuclear isotopes. Tang *et al.* [124] measured absolute values of the nitrogen concentration by nuclear reaction analysis. The $^{14}N(d,\alpha_{0,1})^{12}C$ nuclear reaction induced by 1.1-MeV incident D^+ ions was used, which has a sensitivity of ~6×10^{13} atoms/cm^2 and is quantitative. Standards of Ta_2O_5 were used for oxygen and Si_3N_4 for nitrogen. To obtain depth profile information, samples were etched in a 1% HF solution. Figure 6.30 shows the ^{14}N areal density as a function of film thickness with growth temperature as a parameter. The concentration increases with processing temperature. These results were confirmed by Green *et al.* [125] using nuclear reaction analysis of Si oxidized in N_2O. The N content of the N_2O oxides increased linearly with oxidation temperature, but in

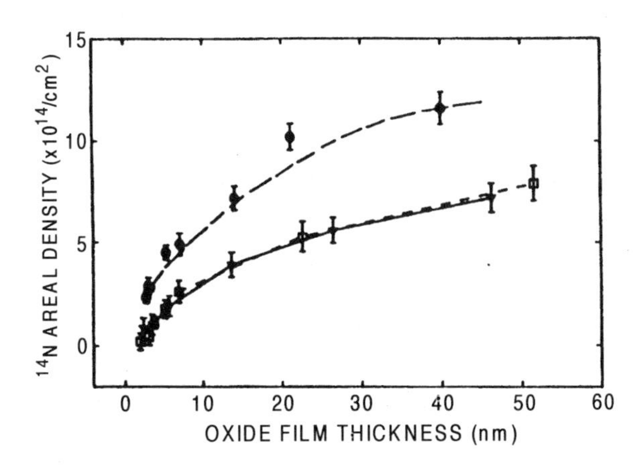

FIGURE 6.30. Total ^{14}N areal density as a function of oxynitride film thickness for temperatures of 900 °C (▼), 1000 °C (□), and 1100 °C (●). The smooth curves are drawn as a guide only [124].

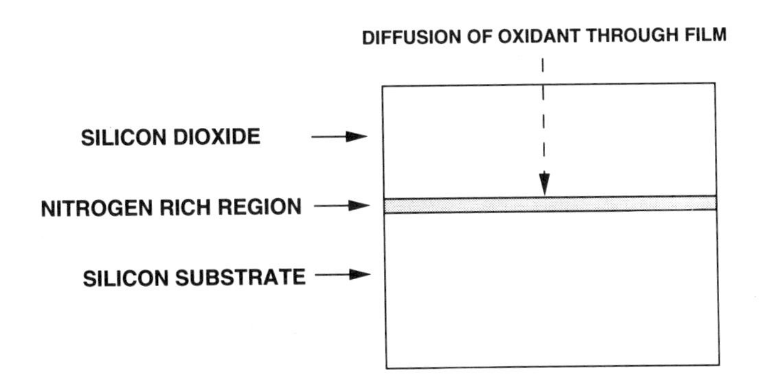

FIGURE 6.31. Schematic cross section of the dielectric two-layer model for oxynitridation in N_2O.

general was less than 1 at. %; for example, an oxidation at 1000 °C in N_2O contained 7×10^{14} N/cm^2 (1 at. %).

In contrast to the above measurements, Weidner and Krüger [126] found that concentrations of nitrogen increased at the interface with processing time, reaching a maximum of about 0.5 at. % based on SIMS and AES measurements. In general there is a wider variation in the data from SIMS and AES than from NRA and XPS techniques. In any case it is clear that the concentration of nitrogen is high enough to form at least a monolayer of SiO_xN_y at the Si/SiO_2 interface. Figure 6.31 is a simplified model of the dielectric structure.

The distribution of nitrogen in the dielectric film was strongly affected by processing in NO as opposed to N_2O. XPS measurements by Yao *et al.* [127, 128] indicate that nitrogen was distributed throughout the dielectric at a level between 4 and 5 at. %. Figure 6.32 shows the concentration for two films both grown at 1150 °C. The nitrogen concentration for the N_2O-grown ambient is clearly observed near the Si/SiO_2 interface, while the NO-grown sample has nitrogen throughout the film and at a much higher concentration. Measurements were also made for samples oxidized in pure oxygen and annealed in NO. Nitrogen was piled up at the interface at a peak concentration of 4–5 at. %, depending on processing temperature.

Kuiper *et al.* [123] measured the hydrogen content and found $1–1.5 \times 10^{15}$/cm^2, similar to values observed for dry oxidation and attributable to surface contamination. When 10–10,000 ppm hydrogen was added to the N_2O, the N concentration generally remained unchanged [126]. X-ray measurements by Weidner and Krüger [126] indicate that from the proportions of nitrogen and oxygen we obtain possibly a β-crystoblite with a FWHM of 5.5 nm.

6.7.1.1. Interfacial Roughness. The roughness of the interface was found to be approximately 20 Å. The interfacial roughness is an important factor and influences the electrical properties at the interface. The smoothness of the SiO_2 interfaces is greater the higher the N content. The oxidation temperature is an important factor. Figure 6.33 shows the interfacial width extracted from X-ray scattering intensity for 100-Å-thick oxides grown in pure N_2O [125]. The interfaces produced are smoother than those formed by growth in pure oxygen.

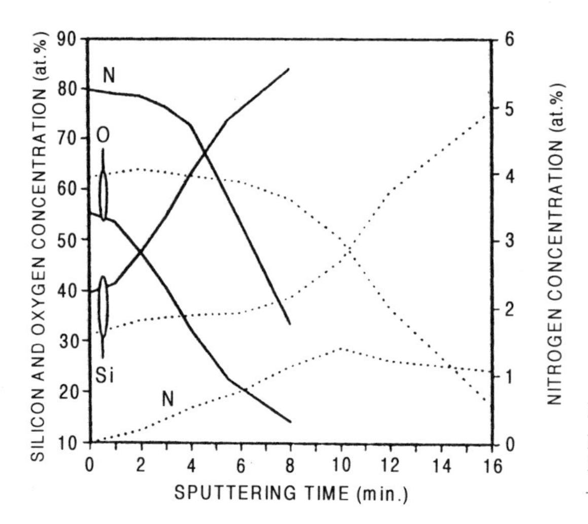

FIGURE 6.32. XPS depth distribution profiles for samples grown at 1150 °C for 5 min with gas ambient as a parameter: ——, NO; ---, N_2O [127].

6.7.1.2. Furnace Growth. Several studies of oxide growth in nitrous oxide have been undertaken, as listed in Table 6.6. Similar results are observed with a buildup of nitrogen at the interface for both RTP and furnace processing [122]. However, Carr and Buhrman [129] also compared the results to a furnace-grown oxide in N_2O and found that the nitrogen concentration was distributed throughout the furnace-grown sample, in contrast to the above study. It is not clear why these differing results were obtained. To shed light on these effects, an experiment by Saks *et al.* [130] showed that if an oxide is processed in N_2O and then in O_2 the nitrogen peak is no longer at the interface but within the dielectric film. If the ambient is switched back to N_2O, the peak moves from the bulk to the interface again [130]. Similar results were obtained by Chao *et al.* [131] for furnace

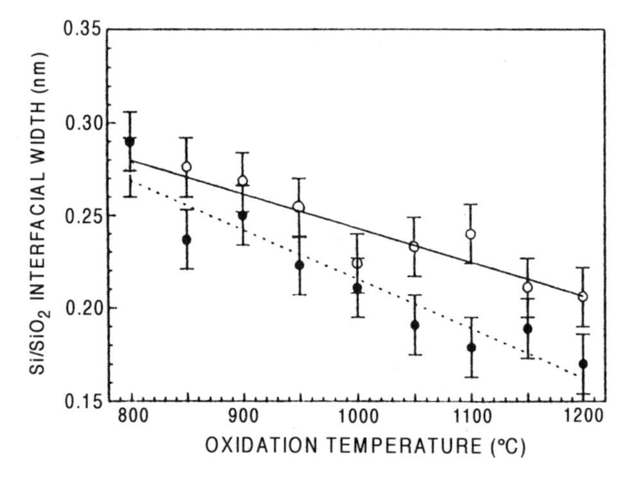

FIGURE 6.33. Si/SiO_2 interfacial width as a function of oxidation temperature for ~10-nm oxides grown in O_2 (○, solid line) and N_2O (●, dashed line) [125].

oxides. A two-layer model is proposed to account for the reduction in the kinetics and through Auger depth profiling measurements N is observed at the interface. The nitrogen-rich layer is approximately 14–20 Å thick and has a refractive index of 1.77. FTIR measurements indicate the presence of Si–O and Si–N bonds. Chu *et al.* [132] show there is a limit in the growth in pure N_2O. A 64-Å oxide was produced after 2 h at 950 °C in pure N_2O. There was little difference in the growth rate on (100) and (111) silicon. The growth rate is much less dependent on the silicon crystal plane than for growth in pure oxygen.

6.7.2. Reaction Model

Vasquez and Madukar [133] proposed that interfacial strain contributes to the nitrogen pileup at the interface. The bonding energy of Si–N is 4.56 eV, Si–O 8.43 eV, Si–Si 3.39 eV, and Si–H 3.2 eV. Therefore, Si–O bond formation should be favored and the relative reaction rates will be a function of the concentration of oxygen and NO_x at the silicon surface. It is well known that N_2O thermally decomposes by the following reaction:

$$N_2O = N_2 + O \tag{6.17}$$

and the following secondary reactions can occur:

$$N_2O + O = N_2 + O_2$$

$$N_2O + O = 2NO$$

$$O + O + M = O_2 + M \tag{6.18}$$

where M is a molecule that catalyzes the formation of molecular oxygen. One of the three species O, O_2, or NO must be predominant in the oxidation process. The O is very reactive and therefore is most likely at low concentrations in a furnace oxidation when the N_2O breaks down before reaching the silicon surface. This suggests that NO is the dominant oxidation mechanism. Therefore, some nitrogen is trapped by Si dangling bonds during the process of oxidation by reaction of the NO with silicon. For RTP, however, the decomposition of the N_2O takes place much closer to the wafer surface and there may be a significant concentration of O present. Oxidation continues through the diffusion of the oxidant to the reaction interface.

The observation that nitrogen is mainly concentrated at the interface may be interpreted in two ways. First, consider Si atoms that diffuse through the nitride layer and become oxidized. The larger concentration of oxygen over nitrogen at the interface results from the higher concentration or reactivity of the oxidizing species at the interface. An alternative explanation assumes that there is an equilibrium throughout the oxide layer. At the outer surface some nitrogen is trapped by Si dangling bonds and when the process of oxidation continues these nitrogen atoms are replaced by O atoms, so that some nitrogen is incorporated in the bulk of the oxide. The low concentration of nitrogen in the bulk of the oxide formed by RTP suggests there is a chemical mechanism that removes nitrogen previously incorporated in the film and replaces it with oxygen as the film grows.

To gain some understanding of why the nitrogen remains concentrated at the interface a series of XPS measurements were undertaken by Carr and Buhrman [129]. They compared furnace- and RTP-grown films and found remarkable differences in the kinetics for growth in pure N_2O. The composition and chemical bonding of the dielectric films were derived from XPS data. They show an increased nitrogen concentration at the interface of the RTP film, whereas the furnace-processed films show a more uniform distribution and the percentage of nitrogen was also different. This is in contrast to the studies discussed above which show a buildup of nitrogen at the interface for both processes. A chemical model for the reaction has been proposed by Saks *et al.* [130] based on reaction energies calculated from semiempirical quantum-mechanical calculations. The following reaction can take place which has a lower activation energy of approximately 1 eV:

$$\text{Si–N–Si} + \text{NO} \Rightarrow \text{Si–O–Si} + \text{N}_2 \tag{6.19}$$

and a second reaction that might take place is less probable because of the higher activation energy (~ 3 eV):

$$\text{Si–N–Si} + \text{O}_2 \Rightarrow \text{Si–O–Si} + \text{NO} \tag{6.20}$$

The activation energies for the back reactions were estimated at ~7.2 and ~7.4 eV, respectively. These energies will be modified by the particular bonding structure of the film which makes it difficult to estimate activation energies accurately. The assumptions made in this case are that both the oxygen and nitrogen bond to the silicon divalently. This model is compared with data from furnace oxidations in N_2O. Further measurements indicate that the absence of UV radiation in the furnace further strengthens the conclusion that UV radiation is not involved in nitrogen removal [134]. Carr *et al.* conclude that exposure of the wafer to atomic oxygen at typical processing temperatures can readily reduce nitrogen that is incorporated in the dielectric film. To summarize there are significant differences observed in the nitrogen concentration in the film for the furnace and the RTP processes. This leads us to propose a parallel reaction path present under UV irradiation.

Based on the limited effect that exposure to hydrogen has on the interface nitrogen concentration, a model is proposed in which the nitrogen is incorporated from NO_x species dissociated at the interface and favors the Si–O binding sites and defect sites even in the presence of hydrogen [126]. Both Si≡N and Si–N=H_2 bonds were observed in the grown and annealed films by Yao *et al.* [135].

6.7.3. *Kinetics*

Ishikawa *et al.* [136] examined the growth of silicon dioxide in nitrous oxide at fairly low temperatures (200–500 °C) for 0.5 h. They found oxidation rates much lower than for furnace oxidation; typically a film 40 Å thick was produced after 30 min at 300 °C. It is believed that the irradiation causes the N_2O to break down into nitrogen and atomic oxygen which reacts with the silicon. The growth of oxide in N_2O is slower than in pure oxygen as demonstrated by Ting *et al.* [137] and some researchers report self-limiting growth even at high temperatures [138] (Fig. 6.34). Also, they observed a parabolic

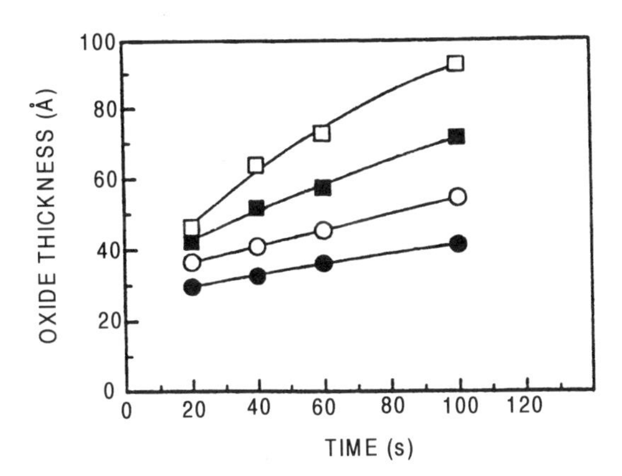

FIGURE 6.34. Thickness of N_2O oxides using ellipsometry, as a function of oxidation time for different growth temperatures: □, 1100 °C; ■, 1050 °C; ○, 1000 °C; ●, 950 °C [138].

growth rate approximately one-quarter that observed in dry oxygen, although they did not observe the thickness limitation.

An advantage of mixing an ambient of N_2O and O_2 is that the film thickness is not limited to about 100 Å, because of self-limited growth. It was found that oxygen addition increased the growth rate and improved the thickness uniformity. The growth kinetics observed fit a simplified Deal and Grove model if the gas-phase transport is neglected. Okada *et al.* [95] assumed diffusion-limited growth and a parabolic law after 5 min of growth, which resulted in an activation energy of 1.42 eV. Yoon *et al.* [138] examined rapid thermal oxidation in N_2O with rapid thermal cleaning. Okada *et al.* [139] made SIMS depth profiles as a function of the nitrous oxide/oxygen gas composition and temperature. They found that higher concentrations of nitrogen were present at the interface for higher processing temperatures and higher fraction of N_2O in the ambient. In addition, they carried out reoxidation of the film in a conventional furnace. The films with a greater concentration of nitrogen at the interface were slower to increase thickness, i.e., most resistant to reoxidation. The growth rate and uniformity increased with addition of oxygen to the ambient gas.

More recent measurements by Grant and Hsieh [140] in pure nitrous oxide for longer times and over a wider temperature range indicate growth that is in agreement with the Deal and Grove model with modified parameters. Figure 6.35 shows the growth rate under different oxygen/nitrous oxide ratios at a temperature of 950 °C. It is observed that the oxide thickness is greater than the higher proportion of oxygen that is present.

These observations lead us to postulate a two-layer model, shown in Fig. 6.31. The nitrogen and nitrogen-related species must diffuse through the oxynitride to react at the substrate/dielectric interface. Thus, nitrogen is incorporated into strained Si–O bonds and dangling Si bonds to form Si–N bonds.

6.7.3.1. Effect of Pressure. The effect of pressure was also studied. Figure 6.36 shows the oxidation rate in pure oxygen and in pure nitrous oxide. As the pressure is reduced, the oxidation rate decreases. Films grown in N_2O at reduced pressure also had improved thickness uniformity compared to films grown at 1 atm [141].

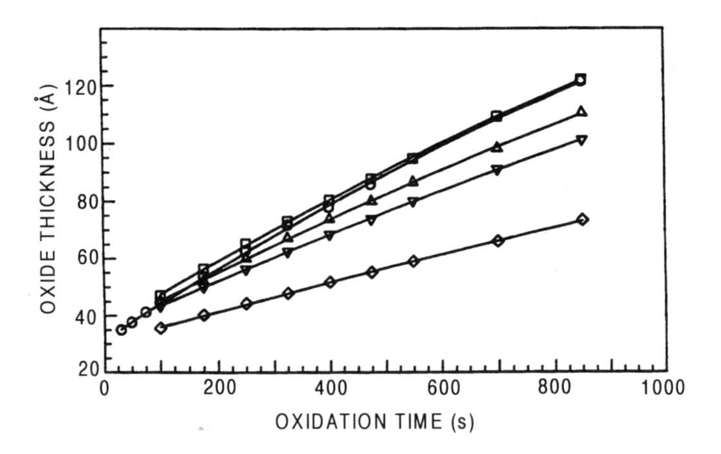

FIGURE 6.35. Oxide thickness versus oxidation times for different O_2 / N_2O ratios for a wafer temperature of 950 °C: ○, 100%; □, 75%; △, 50%; ▽, 35%; ◇, 0% [140].

6.7.3.2. Properties as a Diffusion Barrier. Rapid thermally grown oxynitride films show excellent properties as a diffusion barrier to boron compared to rapid thermally grown oxides in oxygen [142]. The segregation coefficient was also estimated to be lower for the oxynitrides than for the oxides. Green *et al.* [125] showed that a small amount of nitrogen, about one monolayer, can retard boron penetration through the oxide by two orders of magnitude. Figure 6.37 shows the measured boron penetration as a function of the nitrogen content.

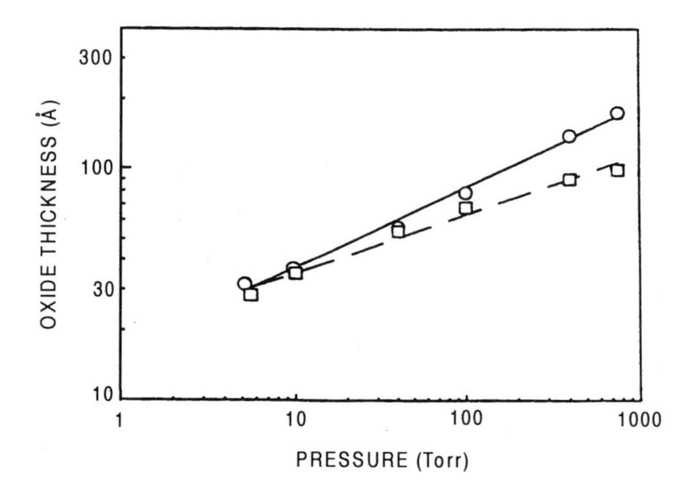

FIGURE 6.36. Oxide thickness as a function of oxygen pressure for oxides grown on (100) silicon at 1050 °C for 180 s in (□) N_2O and (○) O_2 [28].

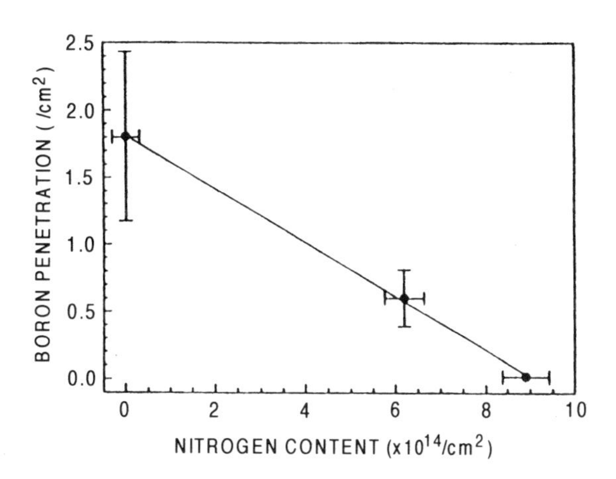

FIGURE 6.37. Boron penetration into silicon as a function of nitrogen content of N_2O oxides. Boron penetration is the area under carrier concentration versus depth profile curves. Boron implant was 5×10^{15} B^+/cm^2 at 10 keV followed by an anneal at 1000 °C for 1 h (B/cm^3, determined by spreading resistance) [125].

6.7.4. Electrical Characteristics

Interface state density is difficult to compare between samples because it is a function of the bias stress. A more effective measure of oxide quality for comparing different processes is the total charge to breakdown, Q_{BD}. There are significant improvements in the electrical properties of the films processed in N_2O. In particular, there was a reduction in interface state generation under X-ray radiation and under constant current stress, less electron trapping, and the charge to breakdown increased [142]. Process details are given in Table 6.7. Similar results were obtained by Ahn *et al.* [143] and Ting *et al.* [144] who observed improvements in the charge trapping and interface state generation with electrical bias stress. In pure N_2O the higher growth temperatures led to reduced flat-band voltage shift and lower levels of charge trapping on bias stress tests [145]. The resistance to interface state generation is shown in Table 6.8 after bias stressing. There is a distinct improvement over the furnace oxide and RTP oxide grown in pure oxygen. The D_{it} for the nitrided oxide is a function of the nitridation time. The nitrogen concentration in the film was compared for RTP wafers and those processed in a conventional furnace. The charge to breakdown was measured. The RTP samples also had the better Q_{BD} and significant nitrogen at the interface only, while the furnace-processed samples has a more uniform distribution of nitrogen throughout the layer and poor Q_{BD}.

6.7.4.1. Effect of Varying N_2O/O_2 Ratio. Several studies examined the effect of adding O_2 to the N_2O ambient during RTP. Chu *et al.* [146] studied mixed nitrous oxide/oxygen ambients, and compared the results to an oxygen ambient. The N_2O-grown films had a greater time to breakdown which was attributed to the incorporation of nitrogen at the Si/SiO_2 interface reducing the interface state density. Okada *et al.* [147] found good control of the nitrogen concentration at the Si/SiO_2 interface: between 0 and 1 at. % by varying the N_2O/O_2 ratio. The incorporation of nitrogen at the interface improves the Q_{BD} and suppresses D_{it} generation under Fowler–Nordheim electrical stress. Measurements of the D_{it} with *C–V* indicated ΔV_{fb} shifts negative with increasing initial nitridation, but the levels here are lower than those reported for reoxidized nitrided films because of the lower initial nitrogen concentration. Therefore, they concluded that a light

TABLE 6.7. Process Conditions and Electrical Properties of Dielectrics Grown in N_2O and O_2

	Process conditions					Electrical properties				
Process	Gas	Temperature (°C)	Time (s)	Thickness (Å)	Gate material	Interface trap density ($eV^{-1}cm^{-2}$)	$\Delta D_{it\text{-}m}$ after bias stress test ($eV^{-1}cm^{-2}$)	Charge to breakdown (C/cm^2)	Breakdown field (MV/cm)	Refs.
RTO	N_2O	200–500	30 min	20–45	—	2×10^{11}	—	—	—	136
RTN	N_2O	1100	40	60	poly-Si	—	—	850 ($J = 50$ mA/cm^2)	8	142
RTO	$N_2O + O_2$	1050–1150	40–320	~100	poly-Si	—	—	—	—	146
RTC		700–1100	20–200	65	poly-Si	2×10^{10}	2×10^{11} ($Q = 1$ C/cm^2)	—	—	137,144
RTN	N_2O	950–1200	—							
RTO	$N_2O + O_2$		80–140	100	Al	$< 10^{10}$	—	1	—	147
RTA	Ar		60							
RTO	N_2O	1100	—	100	poly-Si	—	—	—	—	155
RTO	N_2O or O_2	1100–1200	—	83–96	poly-Si		1×10^{11} (2×10^{11} after anneal) ($Q = 2$ C/cm^2)	—	—	145
RTA	N_2	1100	10							
		950	900							
RTO	N_2O or O_2	950–1100	20–120	30–100	poly-Si	4×10^{11}	—	—	15.5 (Ox. only 14.9)	138
RTN	N_2O									
RTO	$N_2O + O_2$	850–1100	50–850	30–180	—	~2×10^{11}	—	—	~12	140
RTO	$N_2O + O_2$	1100	—	~100	poly-Si	—	—	1–3 (positive) 3–8 (negative) ($J = 10$ mA/cm^2)	> 5	95
RTN	N_2O	950	30	112	Al	—	3.2×10^{11} ($Q = 0.01$ C/cm^2)	2.7 ($J = 1$ mA/cm^2)	11	152,153
RTO	$N_2O + O_2$	1120	—	80	poly-Si	—	—	Fig. 6.38	—	148
RTO	$N_2O + O_2$	800–900	4–12	~20	Ag	2.5×10^{10} (1.2×10^{10} after anneal)	10×10^{10}	—	—	154
RTO	$N_2O + O_2$	1150	—	80	poly–Si	—	—	40–70 (positive)	—	151
RTO/RTN	$O_{2\,+}\,O_2/N_2O$	1100, 1150	—	80	poly–Si	—	—	Fig. 6.39		

TABLE 6.8. Electrical Properties of Oxynitrides Grown in Nitrous Oxide

Oxide thickness (Å)	Growth conditions	Shift in flat band voltage after bias stress (mV)		Shift in midgap density of states after bias stress ($\times 10^{10}$ eV^{-1} cm^{-2})		Ref.
		$Q_{inj} = 0.1$ C/cm^2	$Q_{inj} = 1$ C/cm^2	$Q_{inj} = 0.1$ C/cm^2	$Q_{inj} = 1$ C/cm^2	
—	Control oxide	−27	−65	50	130	143
—	950 °C N_2O	−8	−35	20	80	
—	1000 °C N_2O	−2	−10	7	30	
83	1150 °C N_2O	−350	−340	8	13	145
80	1150 °C N_2O and 1100 °C N_2 anneal	−320	−320	9	27	

nitridation does not degrade the interface. Okada *et al.* [95] studied the effect of the gas ambient on the interfacial nitrogen concentration. They found similar results for the charge to breakdown. The Q_{BD} is a function of N_2O concentration in the process (measured by the current step method). The structural and electrical properties were also examined by Bauer and Burte [148]. By adjusting the N_2O/O_2 ratio it is possible adjust the N_2 content in the dielectric film, as shown in Fig. 6.38. Higher Q_{BD} values were obtained for electron injection from the gate than for electron injection from the Si substrate. Excellent results were obtained for an O_2/N_2O ratio of 3:1. The nitrogen-rich layer is responsible for the higher Q_{BD}. These values are higher than those reported by Okada *et al.* [95].

6.7.4.2. Effect of Reduced Pressure. Films grown in pure nitrous oxide at reduced pressure have a lower value of D_{it}. At 40 mTorr the D_{it} was reduced from 1.4×10^{10}/ $eV^{-1}cm^{-2}$ before electrical stress to 1.9×10^{10} $eV^{-1}cm^{-2}$ and after electrical stress at a current density of 100 mA/cm^2 for 100 s from 1.1×10^{12} $eV^{-1}cm^{-2}$ for growth at atmospheric pressure to 1.9×10^{11} $eV^{-1}cm^{-2}$ at 40 mTorr growth [28]. Values of 1.4×10^{11} $eV^{-1}cm^{-2}$ at 50 mTorr were less than those (3.5×10^{11}/eV/cm^2) at 1 atm.

6.7.4.3. Effect of Temperature. Eftekhari [149] measured the fixed charge density and leakage current for films grown in N_2O. These two properties both decrease for higher growth temperature. The effect of bias stress is more noticeable for the oxides processed at the lower temperatures of 1050 and 1100 °C.

6.7.4.4. Thickness Uniformity. Thickness uniformity has been studied by Lange *et al.* [150] who grew RTP oxides in N_2O with uniformities better than 2%. Overall the charge to breakdown of the N_2O oxides was ten times that for the O_2 oxides. Wrixon *et al.* [151] carried out an extensive study under the process conditions shown in Table 6.7. Growth in pure N_2O has the advantage of a higher Q_{BD} but poor thickness uniformity. A mixed O_2/N_2O ambient has improved uniformity and a higher growth rate. Thickness measurements were performed on a 10 × 10 point scan of 4-inch-diameter wafers. The uniformity is defined here as the maximum value minus the minimum value. Typical thickness uniformities achieved for a pure N_2O process were between 10 and 15 Å, which was reduced to 7–10 Å for a mixed N_2O/O_2 ambient at 1150 °C. For comparison, oxides

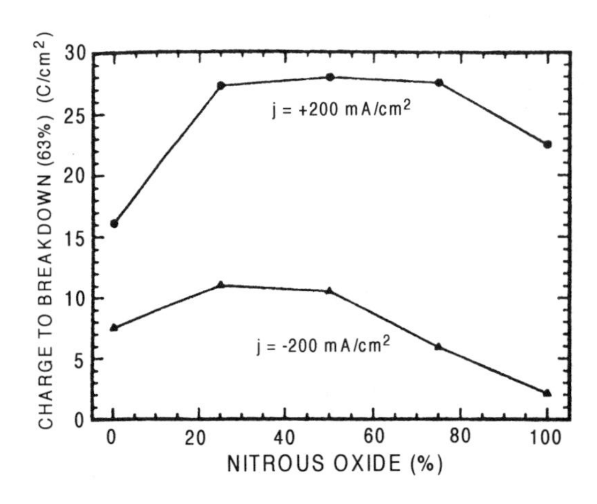

FIGURE 6.38. Dependence of the Q_{BD} (63%) values on the N_2O content of the oxidizing atmosphere after a constant injection with a current density of +200 mA/cm^2 (●) and −200 mA/cm^2 (▲) [148].

grown in pure O_2 at 900 °C in a furnace had a uniformity of 2 Å and by RTP a value of 6 Å.

6.7.4.5. Annealing and Multiple Process Cycling. The effect of multistep processing was to improve the electrical properties, specifically Q_{BD} and E_{BD}. Both two-step [139] and multiple cycling [152, 153] to the process temperature were studied. SIMS data indicated that SiN/N ratios increased with the number of cycles studied. Bienek and Nényei [145] found that the value and uniformity of the D_{it} degraded with a postprocessing anneal (see Table 6.8). The uniformity of $D_{it\text{-}m}$ after a bias stress test with charge injection of 2 C/cm^2 was between 12×10^{10} and 18×10^{10} eV^{-1}cm^{-2}. The variation in $D_{it\text{-}m}$ over the wafer of 3-inch diameter was less than 16%. These values are significantly less than previously reported. This is in contrast to the work of Eftekhari [154] showing a fourfold drop in $D_{it\text{-}m}$ after annealing.

6.7.4.6. Multiple Step Process. The thickness variance was greatly reduced for a two-step process in which the first step is in oxygen, and the second step is in N_2O. For a one-step process the minimum variation was obtained at the smaller N_2O concentration of 25%, the balance being oxygen. The charge to breakdown was greatly enhanced for positive bias with the addition of N_2O to the growth chamber. Charge to breakdown values are shown for the two-step process in Fig. 6.39 [151].

6.7.4.7. Active Devices. Tunnel diodes were used to characterize the dielectric film by Eftekhari [154] and the barrier height was determined. The interface state density was reduced by a factor of four and even further after annealing. The reverse bias current of the diode was also reduced by an order of magnitude. The barrier height was affected by the method of oxidation and the annealing step. The mean barrier height measured was 0.43 eV.

6.7.4.8. Radiation Resistance. Advanced lithography techniques based on synchrotron radiation are being successfully employed for fine geometry and high-aspect-ratio fabrication. The durability of these RTP dielectric films was studied by Arakawa *et al.* [155] because exposure to X rays may produce electrical defects in the oxide. It was found that the $D_{it\text{-}m}$ for the film exposed to radiation density of 540–2700 mJ/cm^2 was approximately ten times higher than for unexposed samples, i.e., 13×10^{10} eV^{-1}cm^{-2}. However, it was independent of energy exposure, and the defects were eliminated by a

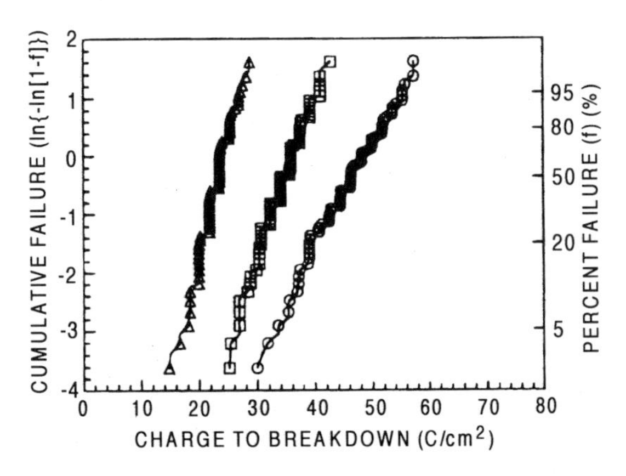

FIGURE 6.39. Weibull distributions of measured Q_{BD} values under positive bias for 80-Å oxynitrides grown by the two-step process of initial growth in O_2 followed by growth in N_2O (J = 2 A/cm^2). Samples are: ○, 40 Å (O_2) 40 Å (N_2O); □, 50 Å (O_2) 30 Å (N_2O); △, 65 Å (O_2) 15 Å (N_2O) [151].

hydrogen anneal at 400 °C for 30 min. Bias stress was not studied by Arakawa *et al.* Radiation resistance was also reported by Wu and Hwu [153]. The process is shown in Table 6.7. They found that by repeating the oxidation process several times the radiation hardness was increased over that for the one-time process. The radiation-induced $D_{it\text{-}m}$ shift was 7×10^{10} $eV^{-1}cm^{-2}$ after exposure to 1 Mrad from a Co-60 source for a single process. After two cycles the $D_{it\text{-}m}$ shift was only 4×10^{10} $eV^{-1}cm^{-2}$ after exposure.

In summary, a higher growth temperature is preferred for better thickness uniformity and a higher Q_{BD}. In the mixed N_2O and O_2 ambient a higher concentration of N_2O results in enhanced electrical properties, but poorer thickness uniformity. There is little improvement in Q_{BD} for positive bias stress testing, i.e., with electrons injected from the substrate. However, the oxynitride film shows improved resistance to charge trapping.

6.7.5. *Models for the Shift in* D_{it}

A model for the changes in traps in terms of a broken bond model has been proposed. Silicon has strain defects at the interface, Si–O–Si, Si–H, Si–OH and Si– (dangling bonds) and dimers. Hydrogen passivates the dangling bonds. The energies of hot electrons (~2–3 eV) can break weak bonds generating negative defects at the interface. The following reactions produce hole traps:

$$O_3^{2-}\ Si\bullet + h^+ \rightarrow O_3^{2-}\ Si^+$$

or

$$Si_3^{2-}\ Si\bullet + h^+ \rightarrow Si_3^{2-}\ Si^+ \qquad (6.21)$$

Injection of an electron into the interface is captured by these stressed sites:

$$O_3^{2-}\ Si^+ + e^- \rightarrow O_3^{2-}\ Si\bullet \qquad (6.22)$$

Therefore, the D_{it} is strongly dependent on the high-field injected defects and the interface stress defects. Both of these are minimized by the Si–N formation. Hole and electron trapping and injection are also minimized.

The improvements in electrical properties are thought to be caused by nitrogen incorporation into the film which releases interfacial strain and suppresses interfacial trap generation during voltage stress. Rapid thermal nitridation in NH_3 introduces $-NH_x$, –H, and –OH groups into the interface which produce electron traps. Therefore, improvements in electrical properties are gained with a nitrous oxide ambient because of the absence of hydrogen. Ting *et al.* [121] proposed that the trap states mainly come from Si–O weak bonds at the Si/SiO_2 interface. There are less such bonds if these bonds are formed during higher-temperature processing. The electrical properties showed small electron trapping with N_2O films related to the formation of interface Si–N bonds. The D_{it} was initially comparable in both films. However, under electrical stress, at 1 mA/cm^2, the N_2O film was superior. A reasonable model for the improved endurance was proposed as follows: interfacial strain relaxation occurs through the formation of SiO_xN_y by nitrogen incorporation into the Si/SiO_2 interface.

Degradation under negative bias is higher than under positive bias because of the greater number of traps created [148]. Figure 6.40 shows a sketch of the gate/insulator/

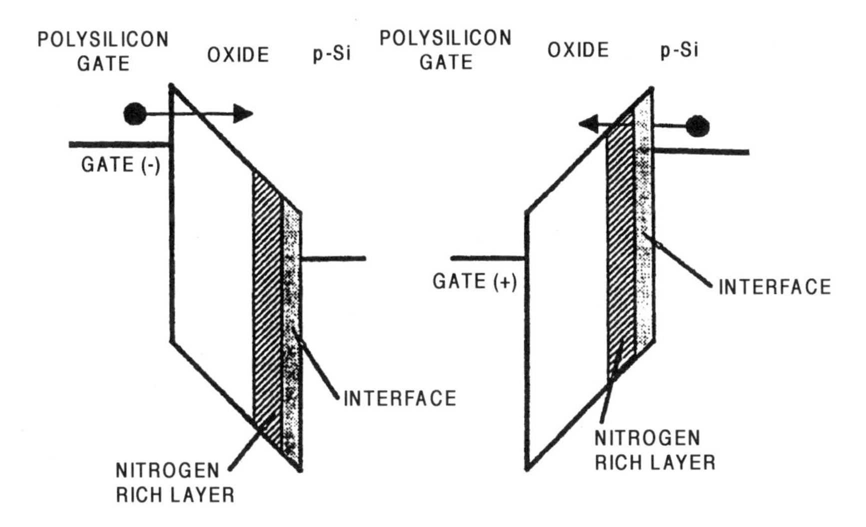

FIGURE 6.40. Schematic energy-band diagram with a nitrogen-rich layer at the SiO_2 / Si interface showing electron injection from polysilicon gate and Si substrate [148].

silicon interface under positive and negative bias. When the electron is accelerated from the gate, it gains kinetic energy and may gain enough energy to generate traps in the nitrogen-rich layer. However, for electrons tunneling from the Si substrate they first pass through the nitrogen-rich layer and have not gained sufficient kinetic energy to produce traps in the nitrogen-rich layer. Note that the Q_{BD} for negative bias stress is highest for the lower values of nitrogen content at the interface (Fig. 6.38). Interface smoothing during nitridation may also play a role in increasing the breakdown field. This is an area of active research.

6.8. CONCLUDING REMARKS

Oxidation kinetics are more rapid for RTP in oxygen than for a conventional furnace process. The increased growth rate is in part caused by photoassisted oxidation, particularly in the thickness range less than 200 Å. Temperature uniformity also affects oxide thickness through induced stress which can have a significant impact on the oxidation kinetics. For furnace oxides the Han–Helms and deLarios models consider two parallel reaction paths involving molecular oxygen and atomic oxygen. These models fit a greater range of oxidation conditions and wider thickness range than the Deal and Grove model. However, for RTP the model parameters must be chosen empirically. A detailed understanding of the reaction mechanisms responsible for the accelerated growth has not been developed.

Much of the initial work on RTP oxides focused on nitridation of furnace oxides. Although this was effective in increasing the breakdown field for these oxides, there was an increase in interface charge (D_{it}) under electrical bias stress which translated into a lower charge to breakdown. This may be related to hydrogen incorporation at the interface which can be effectively controlled by reoxidation. Although this chapter has not been an exhaustive review of publications on this topic, most researchers do agree that a light nitridation followed by reoxidation leads to improved properties, reaching a Q_{BD} of greater than 30 C and $E_{BD} > 13$ MV/cm.

More recent work has focused on the use of nitrous oxide ambients for either nitridation of rapid thermal oxides or growth directly on the silicon. Growth in nitrous oxide is self-limiting; however, addition of oxygen to the ambient was found to improve the uniformity and properties. The improvements in charge to breakdown and interface state density appear to be strongly related to the amount of nitrogen at the silicon/dielectric interface. It appears that the higher concentration of nitrogen is beneficial. The broken-bond model in which the defects are reduced by nitrogen incorporation at the interface lends support to this conclusion. However, there are other factors that strongly influence the electrical properties, such as surface roughness, which are also modified by the RTP conditions. These topics need further study to fully understand the interfacial properties of RTP dielectrics. The best charge to breakdown observed has been $Q_{BD} \sim 450$ C for oxides processed in N_2O/O_2 mixtures.

Early work on electrical measurements reports D_{it} and E_{BD}; however, the critical test of dielectric behavior is to examine the charge to breakdown and the shift in interface state density with a bias stress test. An initially low D_{it} can be degraded with a bias stress. The bias stress test is more representative of the dielectric reliability in device operation because the electric field is applied for a much longer duration. Rapid thermal oxides have high reliability and radiation hardness than furnace oxides.

CMOS and EEPROM devices have been built with materials processed by RTP, indicating this is a very useful technology. Excellent-quality ultrathin dielectric films have been made and integrated into practical devices with an improved effective small signal transconductance. The deployment of this technology into manufacturing will increase in the future as the thermal budget and device geometries decrease further. Our understanding of the physical/chemical processes that take place during RTP is still not complete and although much of the data presented in this chapter have been empirical, further work is needed to gain a better understanding of the detailed microscopic mechanisms involved in RTP.

7

Rapid Thermal Chemical Vapor Deposition

Rapid thermal chemical vapor deposition (RTCVD) is a relatively new development in rapid thermal processing. Although it is being applied to a range of materials, the potential for further development is great. The earliest work was on single-crystal epitaxial silicon by Gibbons and colleagues [1] who referred to the process as limited reaction processing (LRP). One of the principal advantages of RTCVD is that sharp transitions are obtained between layers of differing composition or doping, while exposing the substrate to a much lower thermal budget than a furnace low-pressure chemical vapor deposition (LPCVD) process. The maximum temperature reached is typically higher, however, providing good crystal quality and dopant activation. Applications of lightly doped epitaxially grown silicon on heavily doped silicon substrates include the fabrication of CMOS devices with reduced latch-up, and high-frequency bipolar junction transistors (BJTs) with a lightly doped base region on a heavily doped buried layer.

A second important area for the application of RTCVD is the growth of Ge_xSi_{1-x} heterostructures on silicon. RTCVD demonstrates the ability to form alternating layers of semiconductor materials with sharp transitions in the stoichiometry. This is of paramount importance for strained-layer Si/Ge_xSi_{1-x}/Si heterostructures providing the ability to tailor the band gap due to a reduction in the indirect band gap relative to the unstrained layers. Such materials offer the promise for new devices such as field effect transistors (FETs), avalanche photodiodes, heterojunction bipolar transistors, infrared detectors, and resonant tunneling diodes. If the thickness of the strained layer is too large, relaxation takes place, and for that reason it is necessary to precisely define the layer thickness and stoichiometry.

One of the key advantages of RTCVD is that a cold wall system is used which generates fewer particulates. A second advantage is that higher deposition rates are observed and because the wafer is heated to a higher temperature improvements in film quality are achieved. RTCVD also reduces the problems of autodoping and interdiffusion through a lower thermal budget than conventional LPCVD.

Apart from the reduced thermal budget of RTCVD, advantages of lower-cost equipment and less-demanding vacuum requirements compared to molecular beam epitaxy (MBE) are often realized. A typical RTCVD system is shown in Fig. 7.1. The important difference between this system and those discussed in the previous chapter is the additional gas handling equipment necessary for the introduction of the source gases. However, the maximum base pressure will be limited by the elastomeric seal, if used between the quartz window and vacuum system chamber.

Important properties in CVD films include the defect density, contamination levels, surface appearance, and minority carrier lifetimes, among others. An excellent review of RTCVD has been published by Hoyt [2]. There is insufficient space here to discuss the nucleation, chemical reaction kinetics, and gas dynamics of CVD and therefore the reader is referred to excellent books by Sherman [3] and Sivaron [4] for a discussion of chemical vapor deposition. In this chapter we will review the deposition processes for several materials and discuss their properties briefly.

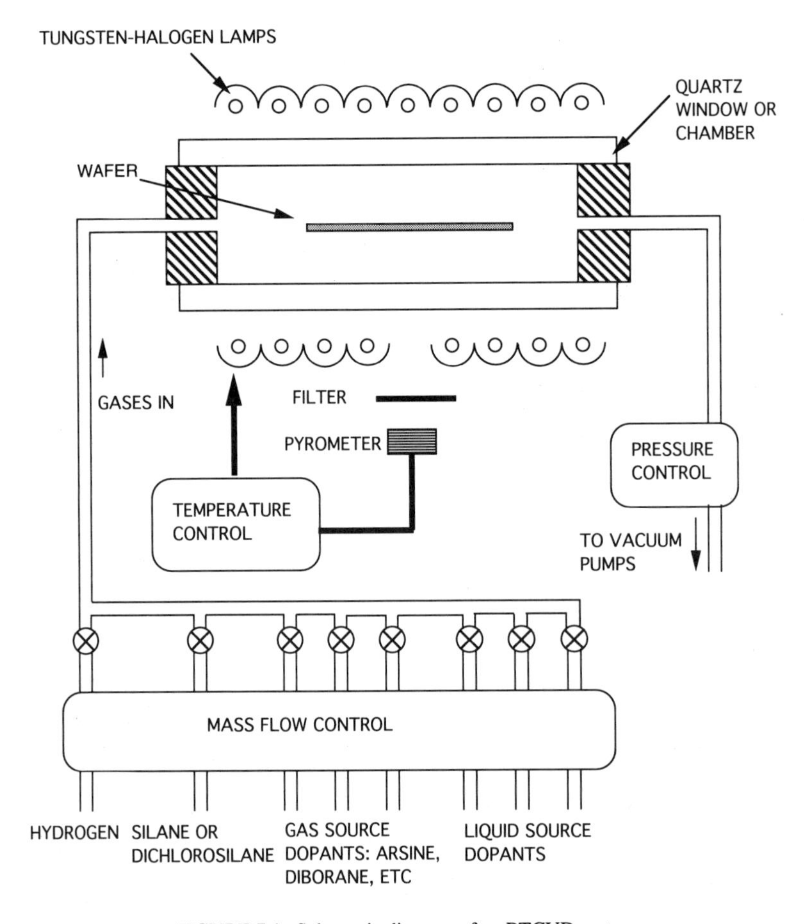

FIGURE 7.1. Schematic diagram of an RTCVD system.

To obtain good-quality films, cleaning is a very important step. Normally an RCA-based cleaning followed by an *in situ* cleaning step is required to obtain reproducible surfaces.

7.1. DEPOSITION OF SILICON

The interest in RTCVD is motivated by the drive toward low-thermal-budget processing. There is a great deal of interest in the growth of lightly doped silicon on heavily doped substrates for CMOS circuits, bipolar devices, and for heterostructures for advanced heterojunction bipolar transistors (HJBTs) in which the ability to control doping profiles precisely is required. Conventional epitaxial growth of high-quality material is by MBE, at a higher temperature, which is a slower and more expensive fabrication option. The nucleation and growth of epitaxial silicon films has been studied in great detail in conventional reactors [5].

7.1.1. Single-Crystal Silicon

The first report of RTCVD was that of Gibbons *et al.* [1] who proposed it as an approach to produce steplike change in doping concentration in ultrathin silicon epitaxial film growth. The transitions between layers of different doping concentrations were obtained by cycling the wafer to the growth temperature in a different gas ambient. A second advantage of this process is that the substrate was exposed to a much lower thermal budget than in a furnace process, hence reducing dopant diffusion, even though the maximum temperature reached was higher providing good dopant activation. Details of the processing parameters are given in Table 7.1. To minimize autodoping from the substrates an oxide layer was deposited on the reverse side of the wafer. An *in situ* preclean was performed in hydrogen and then HCl/H_2 mixture to remove carbon and oxygen contamination from the silicon surface. The chamber was purged with the process gas, dichlorosilane, before the process was carried out. The heating rate was > 250 °C/s so that the wafer reached the growth temperature typically in under 3 s. Undoped films between 0.1 and 3.0 μm thick were grown on highly antimony-doped substrates with a concentration change between 10^{16} and 10^{18} atoms/cm^3 in a thickness of 30 nm, as indicated by secondary ion mass spectrometry (SIMS) measurements. This transition is comparable to that achievable with MBE. Gibbons *et al.* report that the impurity concentration step for arsenic-doped substrates was not as abrupt as for the antimony-doped substrates.

Gronet *et al.* [6] grew epitaxial silicon films in silane. They consisted of alternating layers of heavily and lightly doped silicon with excellent layer delineation, as shown in the SIMS profile in Fig. 7.2. Multiple layers were deposited sequentially and *in situ* by changing the gas composition between each high-temperature growth step. The boron concentration changes by four orders of magnitude over 40 nm demonstrating excellent control of both layer thickness and doping profile. However, care must be taken in interpreting the SIMS profile because there is spreading in the distribution produced by ion/atom collisions in the substrate. Electrical measurements were also carried out with Van der Pauw structures. In some cases the SIMS data were significantly higher than the

TABLE 7.1. Processing Conditions for RTCVD of Epitaxial Silicon

	Cleaning		Process		
Substrate	Temperature/time/ pressure (°C/s/Torr)	Gas	Temperature/time/ pressure (°C/s/Torr)	Gas	Refs.
(100) and (111) Si, <0.02 Ω-cm	1120/10–30	1% HCl in H_2	920–980/15–180/—	SiH_2Cl_2	1
(100) and (111) Si, <0.02 Ω-cm	1150/60/500	H_2	850–950/60/4.2 850–950/5–10/4.2	SiH_4 5.2 ppm B_2H_6 in H_2	6,9,24 (Fig. 7.2)
(100) Si doped 2×10^{18}/ cm^3 Sb	1200/30/250	H_2	900–1000/—/6	0.8% SiH_2Cl_2 in H_2 with AsH_3 (or B_2H_6) in H_2	25
(100) Si, 10–20 Ω-cm, *n*-type	1000/25/—	1% HCl in H_2	550–900°C	1% SiH_4 in H_2	8
(100) and (111) *p*-type	950/1/—	H_2	650–1070/—/1.8–2	0.1–10% SiH_2Cl_2 in H_2 or SiH_4 in H_2	26,27,33,34 (Fig. 7.9)
(100) Si *n*-type	—	—	450–700/300/0.3–3 Note: UHV base pressure ~5×10^{-9}	SiH_4 in H_2	15 (Fig. 7.5)
Silicon	>1000/<60/<10	H_2	850–1050/60/4.2	0.5–10% SiH_2Cl_2 in H_2 and HCl	14,35 (Fig. 7.4)
(100) Si 14–22 Ω-cm	800–1150/—	H_2	800/—/5	5% SiH_2Cl_2 in H_2 with 100 ppm AsH_3 (or B_2H_6) in H_2	11,17–20,22 (Fig. 7.10)
(100) Si 14–22 Ω-cm, *p*-type	[a]760/15/240m Torr —/60/80 mTorr	10% Si_2H_6 in H_2	650–850/—/10–100 Note: UHV base pressure ~3×10^{-9}	10% Si_2H_6 in H_2 with 100 ppm AsH_3 (or B_2H_6) in H_2	23,28,38 (Fig. 7.6, 7.11)
(100) Si *n*-type	—	H_2	450–800/—/5	SiH_4 in Ar_2	32 (Fig. 7.8)

[a]Wafers cleaned in standard RCA process and immediately prior to loading individually dipped in 5% HF solution for 20s and rinsed for 10s in DI water.

electrical data, indicating inactive boron to be present in the structures. Crystals showed < 10 defects/cm^2 for lightly doped layers and ~3000 defects/cm^2 for p^+ epitaxial layers, as measured by inspection after a defect-selective Schimmel etch [7] . Green *et al.* [8] have also deposited epitaxial heterostructures with abrupt interfaces. Process details are given in Table 7.1.

Meyerson [9] also grew Si homoepitaxially at temperatures greater than 750 °C in silane and by conventional CVD under ultrahigh vacuum (UHV) conditions. He achieved low defect density (< 10^3/cm^2) and excellent film uniformity of ±2% both within a wafer and from wafer to wafer. This is a result of the dominance of surface reaction kinetics at low pressures and where the homogeneous component of silane pyrolysis is negligible.

The abruptness of the interface produced was examined by Turner *et al.* [10] in great detail by taking SIMS profiles at different incident ion energies (see Fig. 7.3). The decay

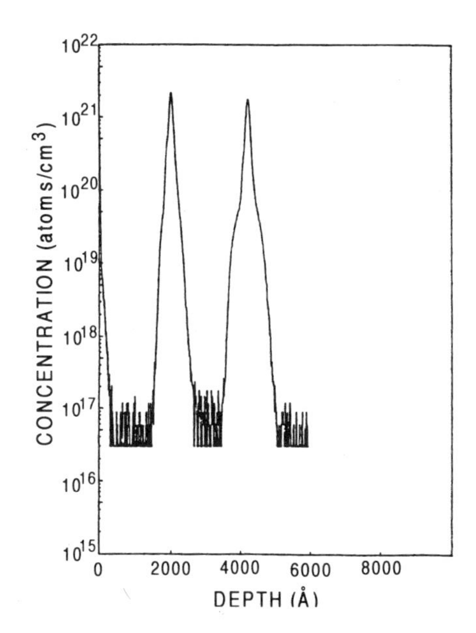

FIGURE 7.2. SIMS profile of boron concentration in a multilayer sample consisting of layers of doped and undoped silicon. The incident beam was oxygen at 10 kV with a substrate bias of –5 kV (Charles Evans and Associates). The first p^+ pulse deposited shows a slight amount of diffusion caused by the thermal cycles of subsequent layers [6].

length is different for the leading and the trailing edge of the profile because of broadening resulting from different physical processes at each interface. At the trailing edge the recoil against an incident oxygen atom is the predominant mechanism producing the effect. However, at the leading edge it is caused by a two-atom process in which an incident oxygen atom collides with a silicon atom, which then recoils against a boron atom in the substrate. If one assumes a random walk for the boron atom, then good agreement between

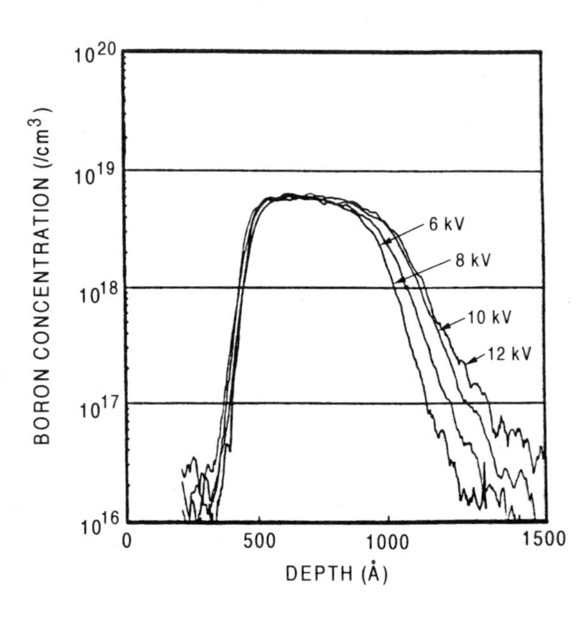

FIGURE 7.3. SIMS depth profiles of boron doping in RTCVD epitaxial silicon film grown at 900 °C. The spectra are taken at various primary beam energies [10].

the calculated diffusion lengths and the interface decay was found. This strongly suggests that the abruptness of the interface is limited by the diffusion of the dopant during the epitaxial growth process. Therefore, SIMS data must be carefully interpreted. Using energy-dependent measurements an improved estimate of the hyperabrupt interface was obtained by extrapolating the decay length against energy to the intercept at zero energy.

7.1.2. *In Situ Wafer Cleaning*

The state of the surface onto which the film nucleates and grows has an important influence on the quality of the epitaxial layer grown. Green *et al.* [8] used a system with a base pressure of 10^{-7} Torr, and the wafers were preheated to 600 °C (to stabilize the pyrometer temperature monitoring) followed by an *in situ* cleaning step in 1% HCl in hydrogen at 1000 °C. This provides a very clean surface for epitaxy without the need for UHV conditions employed for MBE film growth. The maximum base pressure is, however, limited by the elastomeric seals and chamber outgassing, although much of the work has used UHV systems. Hsieh *et al.* [11] review wafer cleaning techniques and distinguish between *ex situ* wafer cleaning to remove contaminants, and *in situ* cleaning to expose an atomically clean surface for epitaxy. Hydrogen annealing is the most commonly used *in situ* cleaning step, often at 1000 °C (see Table 7.1). Hsieh *et al.* studied a range of temperatures and chamber base pressures, and found that the defect density in the epitaxial layer grown was greater for the higher-temperature hydrogen anneals and higher base pressures. At higher processing temperatures the surface was rougher as a result of etching of the silicon by the hydrogen. With a lower base pressure and 800 °C, a featureless surface indicating fewer defects was obtained. Studies have confirmed that the cleaning is still effective at temperatures as low as 850 °C. However, for a base pressure in the millitorr range an anneal of > 1 min is required at 1000 °C. Sanganeria *et al.* [12] examined the effect of *in situ* cleaning involving HF treatment followed by a bake at 750 °C for 15 s in hydrogen at 240 mTorr in a UHV environment. The surface can be passivated against further contamination by exposure to a low-pressure Si_2H_6 environment. Following this the oxygen level remains below the detection limit of SIMS even after exposure to air at 10^{-6} Torr in the load lock for 10 min. If the Si_2H_6 exposure was not carried out, there was substantial oxygen and carbon contamination on the wafer while in the load lock.

7.1.2.1. Effect of Oxygen and Carbon Contamination. This was investigated by Green *et al.* [13] in a dichlorosilane process between 550 and 900 °C using SIMS analysis. They found that oxygen incorporation is mainly dictated by temperature and the impurities in the source gas, and secondarily by impurities from the process chamber. Values as high as $10^{20}/cm^3$ were observed at low deposition temperature. However, at high temperature this was reduced to $5 \times 10^{17}/cm^3$ and the carbon contamination was about $10^{18}/cm^3$ from either the chamber or source gas. This suggests that point-of-use purifiers for the gases may be employed to advantage. Any oxygen was mainly present as precipitates based on measurements of IR spectra. Higher deposition temperature had a lower oxygen content probably reflecting the volatility of SiO on the silicon surface. The presence of these impurities was also detected via the degraded characteristics in a *p–n* junction diode (as a result of reduction in carrier lifetime). For example, the ideality factor increased from 1.0 for a deposition at 800 °C, to 1.6 for a deposition at 650 °C.

7.1.2.2. Kinetics. Green *et al.* [8] deposited films from silane of up to 1.5 μm at growth rates from 2 Å/s at 550 °C to 20 Å/s at 900 °C. Growth at temperatures greater than 850 °C had diffusion-limited kinetics. In general, interfacial cleanliness and film purity increased with increasing deposition temperature. SIMS profiles indicate there was less oxygen but more carbon in the epitaxial layers than in the Czochralski-grown silicon substrate (carbon ~ $10^{17}/cm^3$ and oxygen ~ $10^{18}/cm^3$). Films deposited at < 750 °C exhibited a defect density of < 200 $/cm^2$.

Growth in dichlorosilane was reported by Lee *et al.* [14]. The *in situ* preclean was very critical for obtaining good-quality films as indicated by too short a cleaning time producing poor films. The processing temperature is also important because at less than 900 °C polysilicon forms. The concentration of dichlorosilane in the reaction chamber causes increased growth up to a critical concentration, above which stacking faults develop. However, this value increased with temperature and good films formed at 1000 °C if the dichlorosilane concentration was below 2%. The growth rate temperature dependence shown in Fig. 7.4 demonstrates a mass transport reaction rate-limited regime at high temperatures and a surface reaction-limited regime at lower temperatures. The transition occurs at about 950 °C. At temperatures above 975 °C the growth rate increases linearly with the volume concentration of dichlorosilane as indicated by the linear dependence of epilayer thickness with flow rate.

Liehr *et al.* [15] examined the reaction kinetics for the growth in silane at temperatures between 450 and 700 °C, and at pressures of 0.3–3 mTorr using a novel "freeze-out" method. Silicon wafers, *n*-type (100) (Sb doped), were cleaned by the RCA method and followed by a HF dip and deionized water rinse. After initial outgassing in the process chamber the wafers were heated to 300 °C and then to the process temperature. The

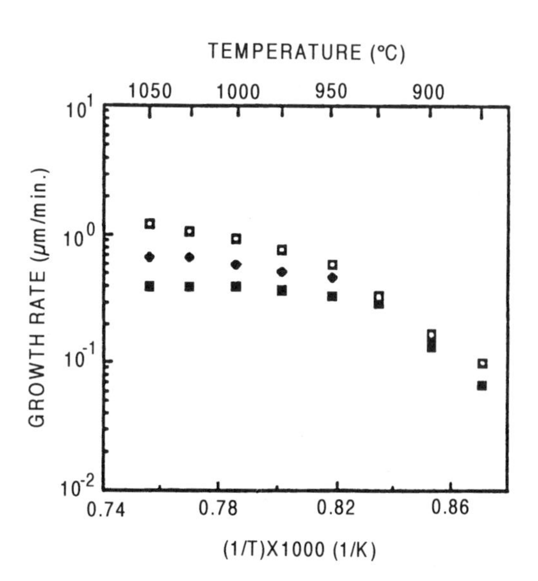

FIGURE 7.4. Temperature dependence of epitaxial growth of silicon at a pressure of 6 Torr, flow of 1200 sccm, as a function of the SiH_2Cl_2 concentration: □, 10%; ♦, 5%; ■, 2% [14].

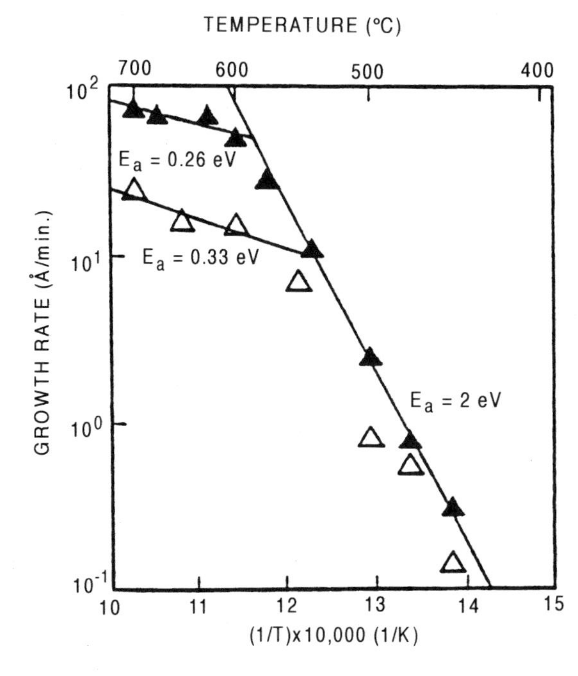

FIGURE 7.5. Growth rates of epitaxial silicon on n^+-Si (100) from silane. Two growth conditions are shown: ▲, 3 mTorr; Δ, ~0.3 mTorr. Film thickness was determined by SIMS. Activation energies were determined from a least-squares fit to the data [15].

"freezing out" method involves rapidly cooling the wafer, and then measuring a desorption spectrum. The process was repeated at different temperatures. The results confirm that the hydrogen on the silicon surface is predominantly in the form of Si–H groups. Monolayer coverage was found for temperatures up to ~ 500 °C, with partial coverage at higher temperatures. The silane adsorption at the growth temperature is therefore influenced by the number of available sites, i.e., the hydrogen desorption can be the rate-limiting step at lower growth temperatures. The growth rates were calculated based on SIMS depth profiles referenced to the Sb doping in the substrate (see Fig. 7.5). An activation energy of ~ 0.33 eV is observed at higher temperatures, in good agreement with the LPCVD process in the reaction rate-limited regime. However, at lower temperatures the rate is limited by the desorption of hydrogen from the surface producing an effective activation energy of 2.0 eV. The growth rate is much lower than that for dichlorosilane (Fig. 7.4). The reader is referred to Section 7.1.5 on selective epitaxy for more discussion of growth kinetics.

A model based on detailed reaction mechanism has been proposed by Hierlemann *et al.* [16] which includes two reaction pathways. First, dichlorosilane is adsorbed onto the silicon surface and decomposes, producing H_2, HCl, and $SiCl_2$. Second, decomposition of dichlorosilane takes place in the gas phase at temperatures above 850 °C, generating $SiCl_2$ which directly reacts with the surface. Both reactions were necessary for good agreement between theory and experiment.

7.1.2.3. Effect of Doping. Hsieh and colleagues [17–20] have shown that good-quality epitaxial films can be grown at lower temperatures if dopant is introduced. Films heavily doped with arsenic or boron epitaxial layers were grown at 800 °C. Undoped ones had a rough-grained surface, indicating the presence of polycrystalline silicon. A signifi-

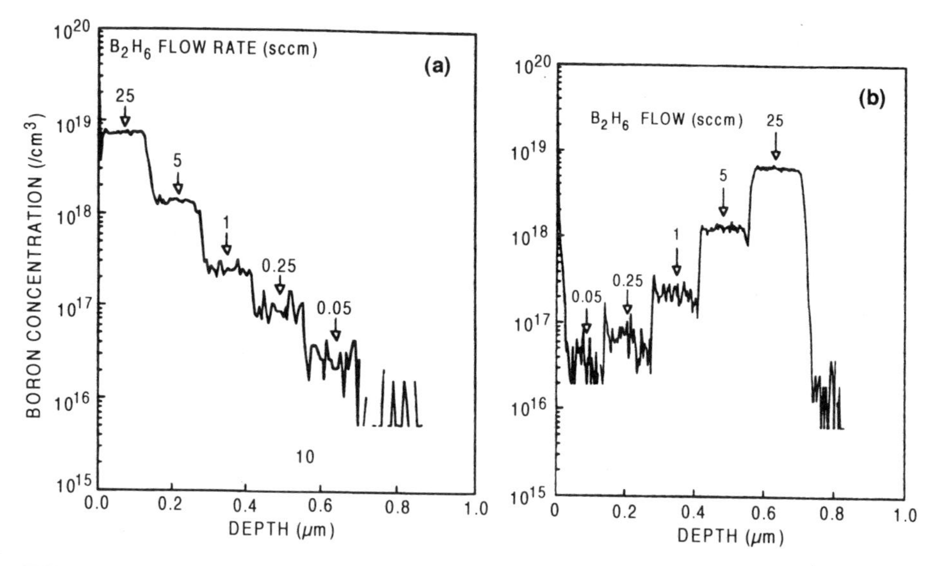

FIGURE 7.6. Stepped boron profiles through multilayer epitaxial film deposited at 800 °C in (a) normally graded profile and (b) retrograde profile. Growth rate 0.35 μm/min, disilane flow 20 sccm, hydrogen flow 180 sccm, pressure 80 mTorr, temperature 800 °C [23].

cant reduction in the surface defects was observed as the concentration of AsH_3 increased in the gas flow from 1 to 7.5 vol %, i.e., the dopant improves the surface morphology. Uniform doping in the layers as deduced from SIMS profiles is achieved. Hsieh *et al.* discuss solid-phase epitaxy as being a factor in the improved epilayer quality. Previous studies by Jeon *et al.* [21] have shown that at doping levels higher than 3×10^{18} atoms/cm^3, epitaxial growth is possible at lower temperatures. They commented that at very high doping levels, stress may also play a role in inhibiting the epitaxial process. Similar effects were reported by Hsieh *et al.* [22] for boron doping. The best-quality films were deposited with 20% diborane producing films with boron concentrations of 5×10^{19} atoms/cm^3. Sanganeria *et al.* [23] studied the boron incorporation in epitaxial silicon grown in Si_2H_6 with 500 ppm B_2H_6 in UHV conditions. They were able to demonstrate growth at low temperatures over the range of doping levels shown in Fig. 7.6 for both retrograded and graded doping profiles. The growth rate was found to be independent of the B_2H_6 flow rate from 0.25 to 25 sccm in a 20 sccm disilane and 180 sccm hydrogen flow. In addition, the boron incorporation in the film was found to be insensitive to the deposition temperature and a weak function of the growth rate.

7.1.3. Electrical Characteristics of Epitaxial Silicon

The electrical properties of 2 to 3-μm-thick epitaxial films were evaluated by Sturm *et al.* [24]. In particular, the generation lifetimes were measured by the deep depletion recovery of MOS capacitors. Crystal defects and impurity generation recombination centers can lower the lifetime by many orders of magnitude. Metal gates were defined on a steam furnace oxide and the gate voltage was held constant to maintain the depletion volume constant. The generation lifetimes were 1.4–4.3 μs for the *n*-type and 14–94 μs for the *p*-type layers. This compares favorably with values measured for bulk silicon 49–87 μs for *n*-type and 144–203 μs for *p*-type. In addition, *p–n* junction diode properties were measured including the ideality factor and breakdown voltage. Silicon devices with low lifetimes should give ideality factors equal to 1, and in this work the p^+–n devices were 1.05 and the n^+–p were 1.10. Reverse leakage currents were 7 pA at 5 V reverse bias. The relatively good results obtained for these films indicate the excellent crystalline quality obtained suitable for devices. King *et al.* [25] also examined the electrical properties of epitaxially grown *p–n* junctions. Process details are given in Table 7.1 for the formation of consecutive layers of *n*-type (3–4 × 10^{16}/cm^3 As) and *p*-type (7 × 10^{17}/cm^3 B). The diodes had ideality factors of 1.01, low reverse leakage (3.5 nA/cm^3 at –5 V), and a sharp transition to avalanche breakdown at –22 V.

Mathiot *et al.* [26] characterized the interfacial imperfections by *C–V* measurements. Process details were given by Regolini *et al.* [27]. The layers were grown for 2-min intervals in different gas ambients to produce modulated doping profiles and gold deposited to form Schottky diodes. Based on *C–V* measurements the best results were obtained for dichlorosilane with an epitaxial layer doping of 10^{13}/cm^3, an interface charge density of about 10^9/cm^2, compared to 10^{10}/cm^2 for silane, and a minority carrier lifetime of 300–400 μs. This method can also be applied for the examination of doping profiles, provided that a full solution of Poisson's equation is made.

Doped films were also grown by Sanganeria *et al.* [28] who used disilane as the source gas in a UHV system and with processing conditions given in Table 7.1. MOS capacitors were fabricated to evaluate the quality of the epi-Si through generation lifetime determined by measurements of time-dependent relaxation of the MOS-cap from deep depletion into inversion. The carrier generation lifetime was not a strong function of the growth temperature in the range 200–400 μs which is comparable to those achieved with low-temperature epi-Si.

7.1.4. Deposition of Polycrystalline Silicon

The early work on polysilicon focused on selective deposition, which is discussed in the next section. Ren *et al.* [29] demonstrated deposition of polysilicon on (100) silicon with 10% silane–argon mixtures at 575–850 °C and at pressures of 1–5 Torr. The deposition rate exceeded 1000 Å/min, much greater than for LPCVD process. At temperatures below ~600 °C the films were an amorphous structure. The growth kinetics were reaction rate limited indicated by an activation energy of 1.65 eV, although above 780 °C a decrease in activation energy was observed. The surface roughness of the film was evaluated by the UV reflectance spectrum and grain structure by cross-sectional transmission electron microscopy (XTEM). Columnar grain structure was observed. The surface roughness was a function of the deposition temperature. One might suspect temperature to be a determining factor; however, SIMS and Auger electron spectroscopy (AES) spectra indicated that there was a 3–4% oxygen concentration in the film. This may be responsible for improved surface roughness over a conventional furnace process.

Oxygen has a pronounced effect on the crystalline structure of the polysilicon film. Xu *et al.* [30, 31] demonstrated that the addition of N_2O to the growth ambient improved

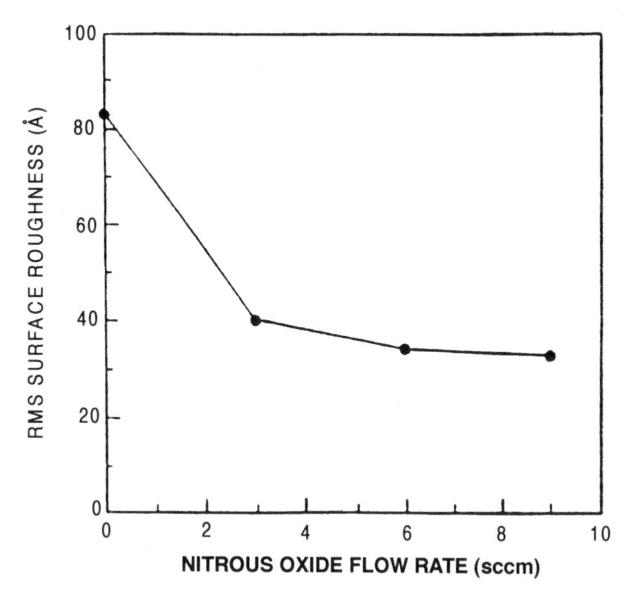

FIGURE 7.7. Root-mean-square surface roughness of the epitaxial film 3500 Å thick as a function of the nitrous oxide flow rate. Pressure 4 Torr, temperature 700 °C, and silane flow rate 30 sccm [30].

the surface morphology by the incorporation of oxygen into the film particularly at grain boundaries. It is believed that the oxygen slows the growth of the silicon grains, thereby allowing a finer grain structure to form. The films were characterized by atomic force microscopy (AFM) and TEM. Typical roughness values observed were under 100 Å. The oxygen concentration was measured by SIMS and AES and found to be < 10% distributed through the film thickness. Figure 7.7 shows the root-mean-square surface roughness for a 3500-Å-thick film as a function of the nitrous oxide flow rate. The key advantage of films grown with oxygen is that the grain structure is stable even after subsequent annealing (≥ 900 °C). For a conventionally grown LPCVD polysilicon film, even if a low deposition temperature is used to produce an amorphous film, doping by implantation and subsequent annealing will enlarge the grain size.

The use of quadrupole mass spectrometry for process monitoring controlled by differential pumping to an RTCVD system was reported by Tedder *et al.* [32]. This provides an *in situ* monitor of the reactant flow, reactant depletion, and gas reactant by-products. The growth of poly-Si over the range 450–800 °C in silane was studied. Figure 7.8 shows the relative partial pressures of SiH_2^+ representative of reactant concentration during poly-Si growth at 750 °C. In general, the dynamic performance of RTCVD can be monitored including the stoichiometry and chamber cleanliness. Figure 7.8 shows the change in process ambient as growth occurs.

7.1.5. Selective Layer Growth

To produce selective epitaxial growth, chlorine must be added to the growth ambient. This may be in the form of HCl or dichlorosilane. A window is opened in the silicon

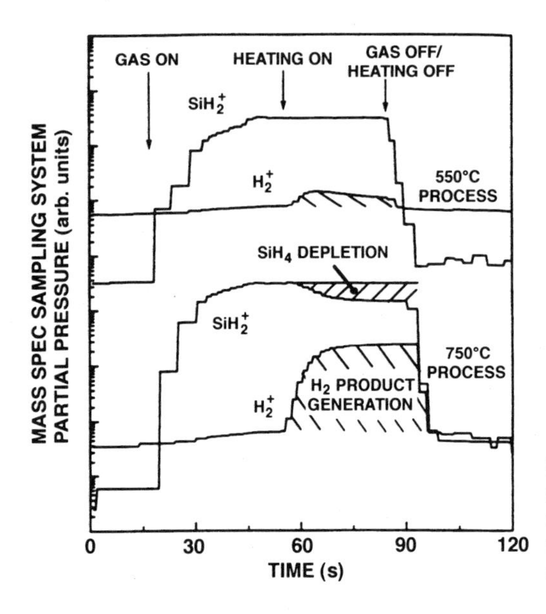

FIGURE 7.8. Fragmentation pattern of silane caused by ionization, using a thoriated Ir filament operating at 70 eV ionization energy at the entrance of a quadrupole mass spectrometer [32].

dioxide exposing a region of single-crystal silicon. The basis of selective growth is the control of deposition conditions such that single-crystal silicon occurs in the seed window but the etching activation of HCl removes poly-Si nuclei from the dielectric surface.

Selective epitaxial growth was reported by Regolini *et al.* [27, 33, 34] in both silane and dichlorosilane on silicon with respect to silicon dioxide. Process details are given in Table 7.1. Figure 7.9 shows the growth rate plotted versus inverse temperature for both silane and dichlorosilane ambient. Below 650 °C the growth rate was low, up to 800 °C it was reaction rate limited, and above 800 °C it was diffusion rate limited. The apparent activation energy was 2.56 eV. This suggests a different dissociation reaction than that normally proposed (formation of $SiCl_2$ and H_2 which has an activation energy of 1.43 eV), i.e., decomposition of the dichlorosilane as follows:

$$SiH_2Cl_2 \rightarrow SiHCl + HCl \tag{7.1}$$

which has a dissociation energy of 2.65 eV.

Full selectivity is obtained throughout the temperature range for the dichlorosilane process, but only above 950 °C for the silane process. The activation energy obtained with the silane process was 1.65 eV and with the addition of HCl was 3.22 eV. Selective growth was only obtained again at lower temperatures. The growth rate was a linear function of the HCl partial pressure, decreasing with the addition of HCl.

Lee *et al.* [35] examined growth without the addition of HCl in a dichlorosilane process. They obtained selective growth at 1050, 1000, 975, and 950 °C with no Si nucleation on the oxide. However, there was roughness along the Si/SiO_2 interface, at 950 and 975 °C, and noticeable faceting of (111) and (311) surfaces. Smooth surfaces resulted at 1000 and 1050 °C. Ishitani *et al.* [36] describe the mechanism of (311) facet formation adjacent to the SiO_2/Si interface, provided an acute angle was present between the epi-Si

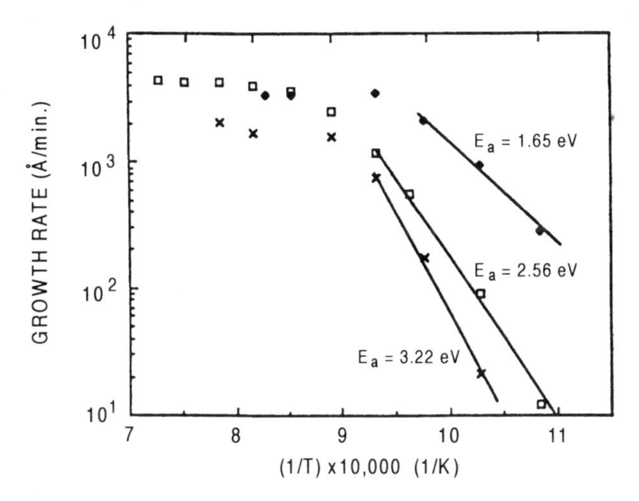

FIGURE 7.9. Growth rate as a function of inverse absolute temperature for epitaxial growth of silicon: ♦, 40 sccm of silane; □, 80 sccm of dichlorosilane; X, 20 sccm of silane and 2 sccm of HCl. In all cases total flow is 2 slm in hydrogen. Activation energies are indicated calculated from a best fit [33].

and oxide wall. The facet width observed was typically approximately equal to the epi-Si film thickness. This results from Si atoms near the oxide sidewalls that tend to nucleate at lower-energy planes, rather than the (100) substrate orientation. Also, they found from the growth rate (100) > (111) > (311). At higher temperatures the lateral diffusion of Si nuclei enhances the lateral growth rate which fills in any undercut in the Si/SiO_2 interface which can be produced during the H_2 cleaning step carried out at 1150 °C. However, a lower-temperature cleaning step at 1000 °C was found to minimize this problem.

Hsieh *et al.* [11] also report on selective growth of epitaxial Si in a process in which the addition of HCl to the dichlorosilane was examined. It was found that epi-Si could be selectively grown into openings in thick oxide layers without HCl when the dichlorosilane concentration was less than 2%. Addition of HCl decreases the growth rate (see Fig. 7.10). The surface is smooth, with the exception of distinct faceting at the Si/SiO_2 interface, particularly at higher deposition temperatures. Normally, (311) planes are observed when the oxide mask is aligned along the [110] direction. The faceting occurs because the arriving silicon atoms tend to grow on the planes with a lower surface energy. Also, (111) faceting planes are observed. At higher temperatures (> 1000 °C) a higher density of epitaxial defects, predominantly twins, was observed.

A study of selective growth by Ye *et al.* [37] showed that selective layer growth can only occur when the reaction system is near thermodynamic equilibrium where supersaturation is low. They consider a three-component system of Si–H–Cl with equilibrium between these three components in the gas phase and reduced pressure in the millibar regime. The reaction rate constants are used to calculate the supersaturation against pressure for the Cl/H ratio, over the range 0.01 to 0.1. When the supersaturation is high, etching effects can be neglected and blanket deposition is more likely to occur. When it is low, we must consider the HCl concentration which promotes the formation of $SiCl_2$ over the concentration of $SiCl_4$. The ratio of partial pressures $P(HCl)/P(Si)$ explains why

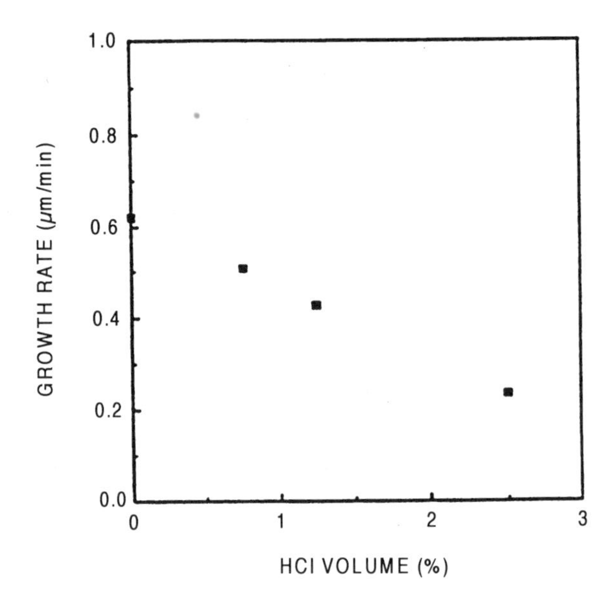

FIGURE 7.10. Dependence of growth rate on HCl volume concentration for undoped selective epitaxial growth of silicon with deposition parameters of 1050 °C and 1.5% dichlorosilane volume concentration [11].

the selectivity is increased for a process with reduced dichlorosilane flow. It also explains why higher growth rate occurs at higher dichlorosilane pressures. This is, however, not a particular advantage for RTP because of the typically slow growth rates with dichlorosilane.

Recently, Violette *et al.* [38] have demonstrated selective deposition at lower temperatures and examined the effect of removing HCl from the process. Experiments were carried out in a UHV chamber with cryopumping to obtain a very clean system. Point-of-use gas purifiers were also used to remove contaminants such as oxygen and water vapor. Features were defined in 100-nm-thick thermal SiO_2 with the windows aligned to the ⟨110⟩ directions in the substrate. Process details are given in Table 7.1. The source gas Si_2H_6 was chosen to increase the deposition rate over silane, without the presence of chlorine. The growth kinetics show reaction rate-limited behavior at lower temperatures, with an activation energy of 2.04 eV, greater than that observed before, for silane because of partial coverage of the surface by hydrogen. To demonstrate the effect of pressure on nucleation on the SiO_2, films were grown at 760 °C and pressures of 100, 40, and 10 mTorr. The results are presented in Fig. 7.11. Nucleation takes place at the higher pressures; however, at 10 mTorr there is good selectivity between the Si and SiO_2. This is in agreement with the polysilicon growth with dichlorosilane in which improved selectivity is observed at lower pressures, discussed in the next section. Violette *et al.* speculate about the critical adatom density that gives rise to nucleation in comparison with the surface mobility and formation of a stable nucleus. An incubation period can be defined that refers to the time before nucleation occurs on the SiO_2. It was found that the incubation does not depend on temperature, indicating that even though the dimension of critical nucleus is an inverse function of temperature (based on nucleation theory), this suggests that the selectivity is degraded at higher temperatures. Also at the higher

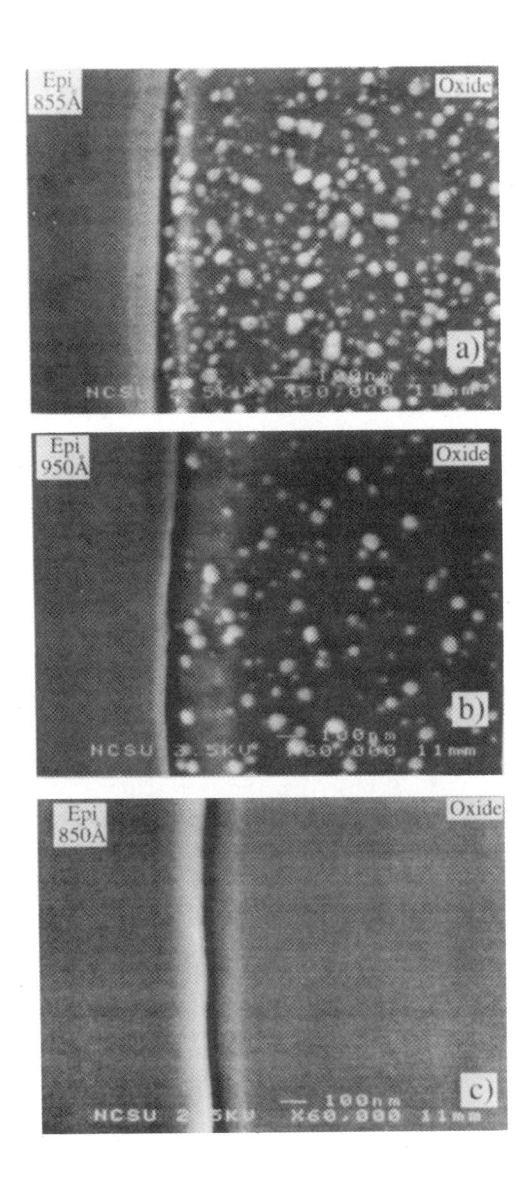

FIGURE 7.11. SEM micrographs showing the reduction in nucleation density with deposition pressure at 760 °C: (a) 100, (b) 40, (c) 10 mTorr [38].

TABLE 7.2. Processing Conditions for RTCVD of Polycrystalline Silicon

	Cleaning			Process	
Substrate	Temperature/time/ pressure (°C/s/Torr)	Gas	Film thickness (nm)	Temperature/time/ pressure (°C/s/Torr)	Gas
(100) Si *n*-type[a]	—	—	300	580/—/1.5	SiH_4 and B_2H_6 in H_2
(100) Si *p*-type[b]	1025/40/6	H_2	200–450	775–1000/—/6	$SiCl_2H_2$ in H_2 with AsH_3
1000 Å or 400 Å SiO_2 on (100) Si[c]	—	—	250	575–850/—/1–5	10% SiH_4 in Ar
SiO_2 on (100) Si[d]	—	—	300–350	700–750/—/4–5	10% SiH_4 in Ar with N_2O

[a]From Ref. 40. [b]From Ref. 39. [c]From Ref. 29. [d]From Ref. 30.

temperatures SiO is more volatile. At this point further measurements need to be made for a comparison of the incubation period during deposition in dichlorosilane.

7.1.5.1. Selective Growth of Polysilicon. Selective deposition of polysilicon was demonstrated for the first time by Hsieh *et al.* [39] on silicon with respect to silicon dioxide. The advantage of selective growth is that deposition is self-aligned to an oxide mask, and submicron linewidths have been defined. The silicon layer was grown with a dichlorosilane process, details of which are given in Table 7.2. The effect of temperature on film quality was pronounced; at temperatures above 950 °C single-crystal material was grown, whereas at 800 °C polysilicon formed. An Arrhenius plot indicates reaction rate-limited growth up to 900 °C, and at higher temperatures the reaction rate is a weak function of temperature. The selectivity is excellent for a 2% dichlorosilane in hydrogen at 6 Torr (flow 1000 sccm); however, when the dichlorosilane was increased to 5%, some nucleation on the oxide region was observed. The average polycrystalline silicon grain size was 0.03 μm for a deposition at 775 °C and increased to 0.2 μm at 850 °C. Figure 7.12 shows contact openings filled with As-doped polysilicon deposited at 850 °C. The introduction of arsine, above 5 ppm, decreases the growth rate. The adsorption of arsine on the surface has been shown to decrease growth in LPCVD processes. The growth rate increases with dichlorosilane flow rate between 25 and 150 sccm as expected.

7.1.6. Fabrication of Active Devices

Limited reaction processing has been applied to the growth of polysilicon films by Sturm *et al.* [40] for the gates of *n*- and *p*-channel MOSFETs. The devices were produced entirely by RTP. First, an RTCVD epitaxial silicon layer was deposited, which was rapid thermally oxidized, and then a polysilicon gate deposited. The growth conditions for the epitaxial layer are given in Table 7.2. The oxidation process was carried out at 1150 °C and 500 mTorr in oxygen for 60 s. The polysilicon was deposited with silane and diborane at 580 °C and 1.5 Torr to a thickness of 0.3 μm. The active devices yielded a surface mobility of 490 $cm^2/V{\cdot}s$. All of the steps have been carried out by RTP so the final threshold voltage adjustment would have to be carried out with an implant through the polysilicon gate. This work demonstrates integration of many of the processes for IC fabrication by RTP.

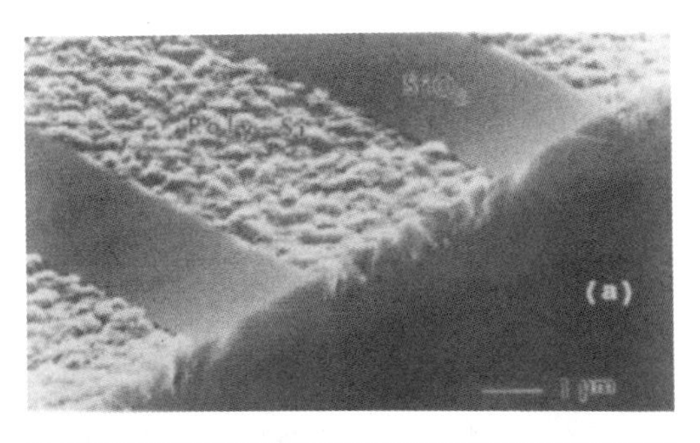

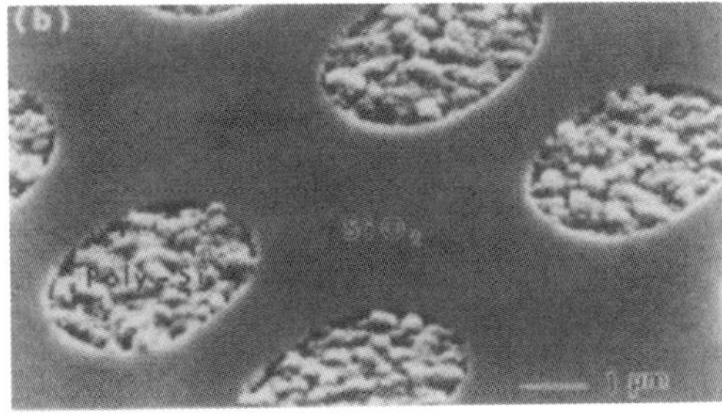

FIGURE 7.12. SEM micrograph showing the oblique view of polycrystalline silicon after deposition at 850 °C (a) interconnect lines and (b) contact filling [39].

RTP has been used for the entire process by Fröschle *et al.* [41] including rapid thermal cleaning in a H_2/Ar ambient before rapid thermal oxidation in O_2/H_2/Ar. Both RTA and rapid thermal nitridation were used and finally a polysilicon gate was deposited by RTCVD in SiH_4/H_2 at 700 °C with flow rates of 255 and 1000 sccm, respectively. The uniformity was 3%. Electrical properties are given in Table 6.4. This demonstrates the possible use of RTP for integration of several fabrication steps, and is amenable to clustering into a single vacuum system.

7.2. DEPOSITION OF GERMANIUM, Ge_xSi_{1-x}, AND Si/Ge_xSi_{1-x}/Si HETEROLAYERS

There has been considerable interest in growing high-quality epilayers of Ge_xSi_{1-x} [42]. The application of RTCVD to SiGe layers was first demonstrated by Gronet *et al.* [43]. In their work 15-nm-thick layers of SiGe alloy were fabricated with abrupt interfaces which were not possible with LPCVD because of diffusive mixing at the interface. Strained-layer superlattices of Si/Ge_xSi_{1-x}/Si were grown *in situ* by changing the gases in the chamber while the wafer was cool, with a typical interruption in growth cycle of 60 s. The Ge_xSi_{1-x} layers are strained because of the 4.2% mismatch between the silicon and germanium lattices. Process details are given in Table 7.3. The first layer was grown with 27 sccm silane and 0.7 sccm germane and 3 lpm hydrogen. The samples were analyzed by Auger depth profiling and Rutherford backscattering spectroscopy. The film thickness and germanium fraction were 3.7 μm and 9%, respectively. Figure 7.13 shows

TABLE 7.3. Processing Conditions for RTCVD of Epitaxial Ge_xSi_{1-x} Heterostructures

	Cleaning		Process		
Substrate	Temperature/time/ pressure (°C/s/Torr)	Gas	Temperature/time/ pressure (°C/s/Torr)	Gas	Refs.
(100) Si	1150–1200/30–120/—	H_2	900/60/2–4	SiH_4 and GeH_4 in H_2	43
(100) Si 14–22 Ω-cm	1000/60/—	H_2	1000/60/5	$SiCl_2H_2$ and GeH_4 in H_2	50
(100) Si	—	—	500–800/—/2.5	$SiCl_2H_2$ and GeH_4 in H_2	61
(100) Si 38–44 Ω-cm, *p*-type	100–0/60/—	H_2	900/—/— Note: growth in UHV chamber	$SiCl_2H_2$ and GeH_4 in H_2	44
(100) Si *p*-type	—	—	600–700/—/6	$SiCl_2H_2$,GeH_4, and B_2H_6 or PH_3 in H_2	47
(100) Si 14–22 Ω-cm	1000/45/—	H_2	900/15–20/5	$SiCl_2H_2$ and GeH_4 in H_2	51
(100) Si	1020/30/—	H_2	550–1000/—/200Pa	SiH_4/GeH_4 in H_2	48
(100) Si 5–21 Ω-cm	—	H_2	800–900/—/5	$SiCl_2H_2$/GeH_4 and AsH_3 in H_2	53
(100) Si 20 Ω-cm	—	—	600–700/—/4	$SiCl_2H_2$/GeH_4 and B_2H_6 in H_2	68
(111) Si & "V"-grooved (100) Si	850/—/—	1% HCl in H_2	Si: 850/—/— GeSi: 700/—/—	— $SiCl_2H_2$ / GeH_4 in H_2	49
Si	900/60/—	H_2	600–700/—/—	SiH_4 / GeH_4 in H_2	55
(100) Si	—	—	Si: 750/—/— SiGe: 620/—/0.01 Si: 750/—/—	SiH_4 / GeH_4 in H_2	58–60, 71
(110) and (100) Si	—	—	700/—/6 625/—/6	$SiCl_2H_2$ $SiCl_2H_2$ / GeH_4 in H_2	71
(100) Si		—	500–550/—/6 Note: growth in UHV chamber	SiH_4 and GeH_4 in H_2	70
(100) Si	800/—/50 1000/—/50	H_2	800/300/8 and 50	$SiCl_2H_2$ / GeH_4 in H_2	52

a typical Auger measurement of the alloy layers contain 17% germanium alternating with silicon. Based on high-resolution TEM, the interface region is abrupt with a transition width of 20 Å, approximately five monolayers.

Because of the lattice mismatch between Ge and Si there is strain introduced with growth of GeSi alloys. The upper limit to the mechanical stability is given by the Matthews–Blakeslee expression. Above this thickness the layer is in a metastable state and relaxation can occur during growth or postdeposition annealing.

Mechanically stable Ge_xSi_{1-x} films have been deposited by Green *et al.* [44] under the processing conditions presented in Table 7.3. These are on or below the Matthews–Blakeslee mechanical equilibrium curve. The oxygen concentration was a strong function of the deposition temperature as indicated by SIMS measurements for two layers; the one

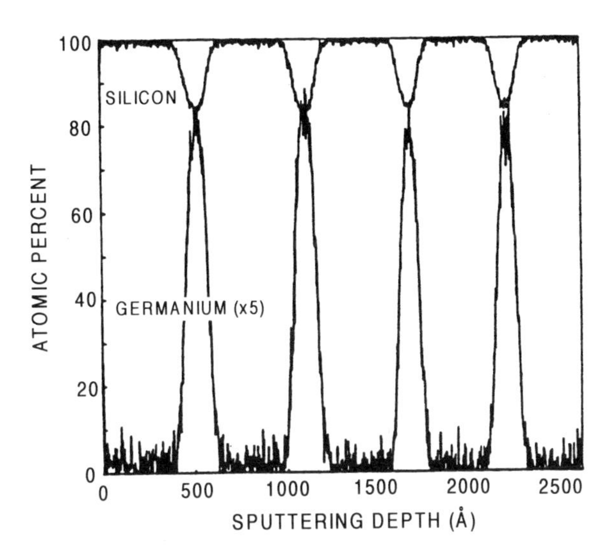

FIGURE 7.13. Sputtering Auger spectra for Si/SiGe heterostructures showing nominal alloy layer germanium content of 17%. Note the germanium data were increased by a factor of 5. The Auger yield was calibrated using RBS data from a number of thick, single alloy layers deposited on silicon (Charles Evans and Associates) [43].

grown at 650 °C had 6×10^{20} atoms/cm^3 whereas the film grown at 900 °C had 4×10^{17} atoms/cm^3. The Si–Ge interdiffusion coefficient was measured as a function of temperature, as shown in Fig. 7.14. The experimental data are compared with published values for the diffusion coefficient [45, 46]. All films were capped with about 1000 Å of silicon. The activation energy calculated is 4.89 ± 0.52 eV for 15% Ge and 4.64 ± 0.43 for 20% Ge. The diffusion coefficient calculated from the figure is 3.0×10^{-17}cm^2/s at 950 °C and hence the interdiffusion is 33 Å for a 1-h anneal. Therefore, the postdeposition anneal at 950 °C does not introduce significant diffusion.

Sturm *et al.* [47] applied RTP to the growth of 100-Å layers in GeSi heterostructures. They changed the stoichiometry of each layer by switching the temperature at a fixed gas ambient, and found that the germanium fraction was lower at higher growth temperatures. Process details are given in Table 7.3. From an Arrhenius equation fit to the growth versus

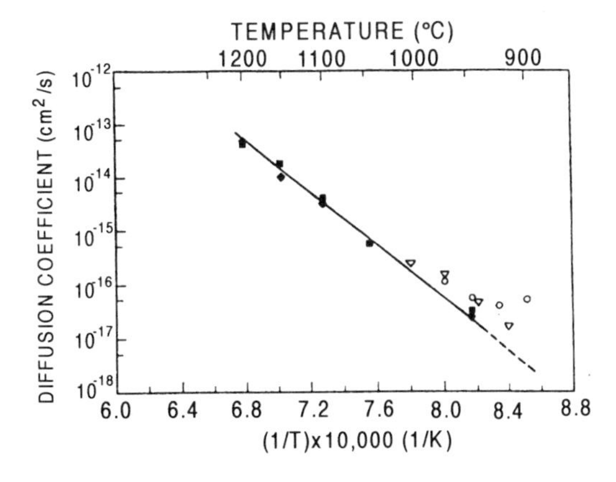

FIGURE 7.14. Diffusion coefficients determined by ion scattering as a function of reciprocal temperature for Si–Ge interdiffusion in 15% (■) and 20% (♦) Ge alloy films. For comparison the results of Van de Walle *et al.* [45] (▽) and Höllander *et al.* [46] (○) are plotted.

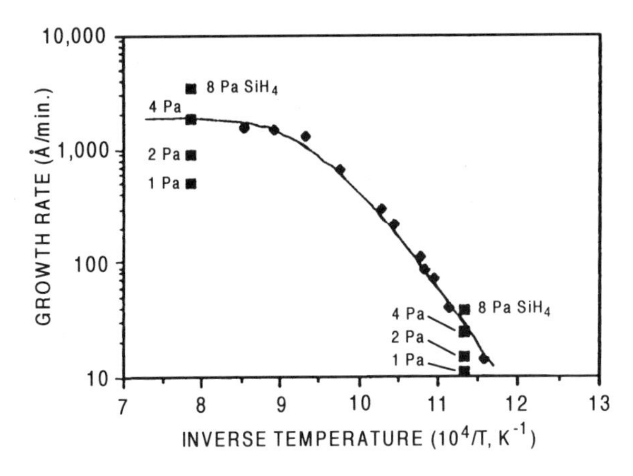

FIGURE 7.15. Silicon growth rate from 2% silane in hydrogen at 200 Pa versus inverse temperature. The model prediction is compared with the experimental data of Si (100) epitaxial growth. —— model results; ♦, experiment; ■, various silane pressures [48].

temperature, the activation energy was found to be 2 eV. Layer abruptness was 10 Å. The infrared transmission through the substrates was also measured and found to give an accurate estimate of the wafer temperature even when the alloy film was grown.

Dutartre *et al.* [48] studied growth in silane/germane mixtures over the temperature range 850–1000 °C and obtained films with no dislocations or stacking faults. They found a low apparent activation energy and a growth rate proportional to silane partial pressure (Fig. 7.15) and an activation energy of 1.95 eV. These values agree well with the activation energy for hydrogen desorption from a Si (100) surface and suggest this is the rate-limiting step: the Langmuir–Hinshelwood adsorption mechanism. The dramatic change of deposition rate with addition of germanium suggests a catalytic effect, which allows reasonable growth rates at lower temperatures and good layer-to-layer control in GeSi heterostructures.

Howard *et al.* [49] studied deposition on (111) surfaces in dichlorosilane/germane mixtures. They observed a strain relief mechanism from TEM of open-ended stacking faults in $Si_{0.85}Ge_{0.15}$. Substrates were etched with a set of V-shaped grooves 2.5 μm apart as shown in Fig. 7.16. On (001) portions of the wafer, defect-free strained layers formed.

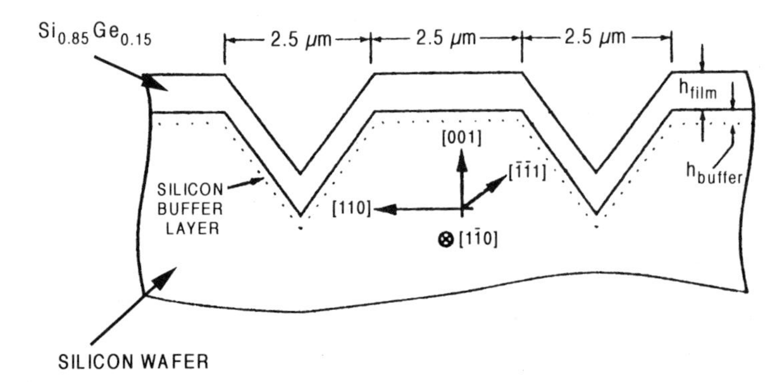

FIGURE 7.16. A [110] cross-sectional schematic view of the anisotropically etched nonplanar silicon substrate and film showing dimensions and crystal directions [49].

On a (111) surface defects are observed ~ 27 nm from the $Si_{0.85}Ge_{0.15}/Si$ interface, which were identically oriented, and uniformly distributed in the strained layer, with a density of 4–5 × 10^{10} /cm^2. This is consistent with the critical strain energy concept for planar defects.

Multilayers of GeSi have been deposited by Jung *et al.* [50] and characterized by XTEM and TEM. There was no evidence for nonuniform stoichiometry. The dislocations were concentrated mainly at the interface resulting in a density of 10^6/cm^2 at the deposition temperature of 900 °C. TEM indicated that the dislocations are not necessarily in the same plane and are capable of strong interactions (see Fig. 7.17). Later work by Jung *et al.* [51] on strained Ge_xSi_{1-x} layers indicated that if the thickness of the strained layer was too large, relaxation occurred by formation of misfit dislocations at the interface. Their work used a germane/dichlorosilane gas mixture at 5 Torr to grow 1000-Å-thick layers at 900 °C, with an *in situ* hydrogen prebake (process conditions are given in Table 7.3). The dichlorosilane/germane ratios were varied from 19:1 to 95:1 at a total flow rate of 1 lpm. Structures of GeSi and pure silicon of five and three layers respectively were formed in which all layers were planar and had grown epitaxially, although twinning planes and a high density of dislocations were present at the interfaces and through the top layer. However, there were no threading dislocations in the other four layers. The short misfit dislocations at the interface were aligned with the ⟨011⟩ directions, indicating partial relaxation. Dislocation loops confined to the interfaces varied from 2000 to 100 Å in diameter. The layer thicknesses were 1060, 1100, 920, 1040, and 750 Å. Auger depth profiling was used to measure stoichiometry of the film. Some pileup of germanium occurred at the first interface, which may be an artifact caused by the higher sputtering

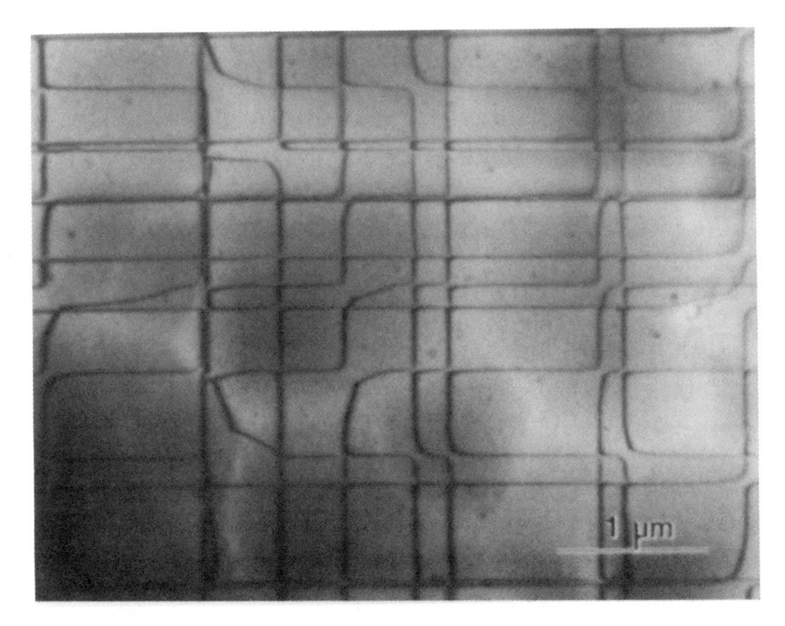

FIGURE 7.17. Plan-view TEM micrograph showing misfit dislocation network at $Ge_{0.07}Si_{0.93}/Si$ interface [50].

yield of germanium over silicon, although no pileup was observed at other interfaces in the multilayer structure. A more plausible explanation is that the defects in the top layer allowed an enhanced diffusion rate for germanium to the first interface. In the three-layer structure multiple threading dislocations were observed indicating significant three-dimensional growth.

Multilayers were also grown by Watson *et al.* [52] as relaxed layers. Growth rates were calibrated by Rutherford backscattering measurements. Several step and graded structures were studied including grading rates of 10, 20, and 40% per μm for 5- and 10-step increments. Growth rate was approximately 4 nm/s at 50 Torr. In each sample, a capping layer of $Si_{0.7}Ge_{0.3}$ 1 μm thick completed the procedure. X-ray diffraction measurement indicated the layers were fully relaxed, and TEM showed an estimated threading dislocating density of $\sim 1 \times 10^7/\mathrm{cm}^2$.

In situ doping of Ge_xSi_{1-x} with As to lower the growth temperature and improved the layer quality has been reported by Jung *et al.* [53] Considerable surface roughness was observed with the undoped samples; however, *in situ* dopant improved the film epitaxy quality although the growth rate was decreased. Epitaxial films were grown with defect densities on the order of $10^3/\mathrm{cm}^2$. The germanium film fraction and strain were determined by high-resolution X-ray diffraction. The spectra measured with a $Cu_{K\alpha 1}$ source are shown in Fig. 7.18. Film strains were 94, 99, and 99%, normalized with respect to the fully strained film for the same Ge fractions of 8, 6, and 13%, respectively. The two satellite peaks in the spectrum correspond to the GeSi layer thickness. The growth conditions of each of the samples are listed in the caption.

Effects of capping layers were found to increase the thermal stability of the films [54]. A strained Ge_xSi_{1-x} layer capped with an unstrained silicon epitaxial layer was fabricated. The Ge_xSi_{1-x} layer was grown at 628 °C with a Ge mole fraction of 0.224 ± 0.006 to a thickness of 490 ± 50 Å. The silicon capping layer was grown at 700 °C with

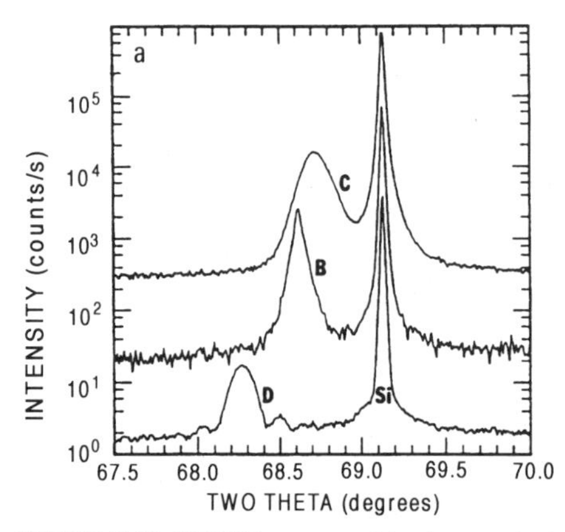

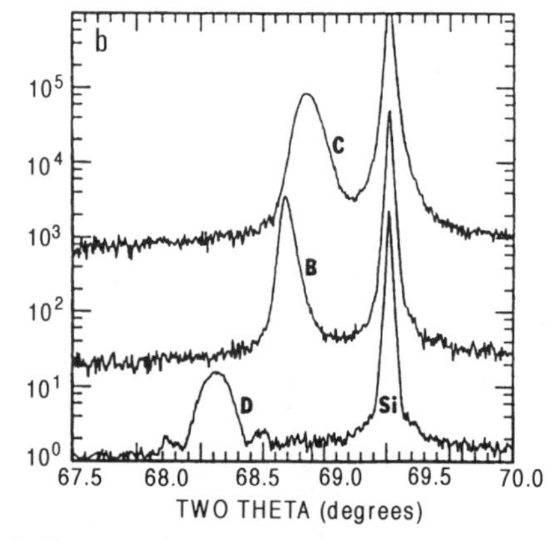

FIGURE 7.18. HRXRD spectra of *in situ* arsenic-doped Ge_xSi_{1-x} epitaxial layers for (a) (004) symmetric reflection and (b) (224) asymmetric reflection. Samples were processed as follows: B = 900 °C/180 s, 8% Ge, junction depth 4250 Å; C = 900 °C/180 s, 6% Ge, junction depth 5250 Å; D = 800 °C/180 s, 13% Ge, junction depth 900 Å. In all cases the ambient arsine was 5 ppm except for sample C which was 1 ppm [53].

a 1.6% dichlorosilane in hydrogen, producing a growth rate of 17 Å/min. Very few dislocations were observed in the alloy layer. However, when the uncapped layer was annealed at 850 °C, relaxation defects appeared. Defects were created by nucleation, followed by propagation and interactions. The capping layer enhances the stability of the strained Ge_xSi_{1-x} layer by inhibiting one of these processes. The minimum silicon cap thickness required to prevent defects forming in a 500-Å-thick Ge_xSi_{1-x} layer was estimated to be approximately 500 Å.

Glowacki and Campidelli [55] reported epitaxy of silicon and SiGe in a UHV system with *in situ* spectroscopic ellipsometry to monitor film growth. Depositions were carried out at a pressure of 2 Torr: at temperatures above 600 °C diffusion-limited growth occurs and at lower reaction rate limited growth temperatures with an activation energy of 1.736 eV. This is attributed to hydrogen desorption. Figure 7.19 compares kinetics for Si and $Si_{0.85}Ge_{0.15}$ growth at a pressure of 0.74 mTorr for the same silane flow for UHV-CVD [56]. A model proposed by Dutartre *et al.* [57] is based on reactive adsorption of silane with a sticking coefficient independent of temperature and hydrogen desorption following a first-order reaction. A sticking coefficient that fits the experimental data is 3×10^{-3}, which is lower than that observed in hot wall UHV-CVD. The increase in growth rate at low temperature is attributed to the catalytic desorption of hydrogen by germanium and the lower growth rate at high temperature to a reduction in the silane sticking coefficient. Selective growth with respect to oxide was observed with the addition of chlorinated species (HCl, SiH_2Cl_2) at the expense of growth rate.

Gu *et al.* [58–60] have deposited SiGe alloy films and used the process conditions given in Table 7.3. Raman spectroscopy revealed oxygen and carbon present at concentrations of 1.0×10^{18} and $1.5 \times 10^{18}/cm^3$, respectively. Figure 7.20 shows the Raman shift of the Si–Si and Ge–Ge peak against Ge concentration as a function of growth temperature. These show the three principal optical phonon modes: Ge–Ge 300/cm, Ge–Si 400/cm, and Si–Si 500/cm. In this alloy the number of Si–Si and Ge–Ge bonds are proportional to $(1-x_{Ge})^2$ and x_{Ge}^2, respectively, where x_{Ge} is the fractional Ge content. Increased Ge content increases the peak to a higher frequency and the Si–Si peak to a

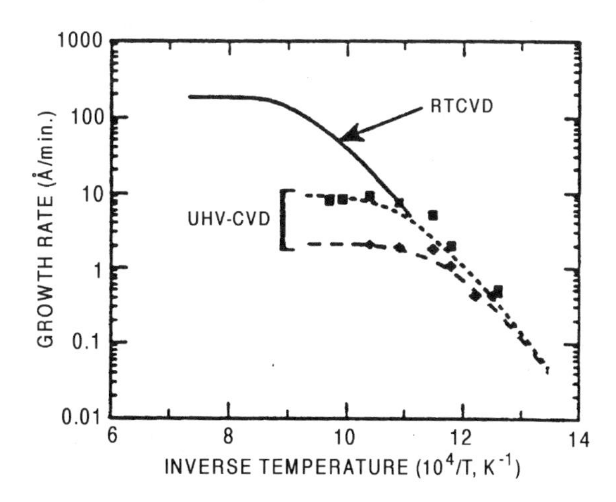

FIGURE 7.19. Silicon growth rate from silane in hydrogen for RTCVD at a total pressure of 2 Torr versus inverse temperature (solid line). Partial pressure of silane 30 mTorr, flow rate 40 sccm [56]. Experimental data of Si ⟨100⟩ epitaxial growth in UHV-CVD grown in pure silane at a pressure of 0.74 mTorr and silane flow rate of 20 sccm (■), and pressure of 0.19 mTorr and silane flow rate of 6 sccm (♦). The model predictions are compared with the dashed curves [55].

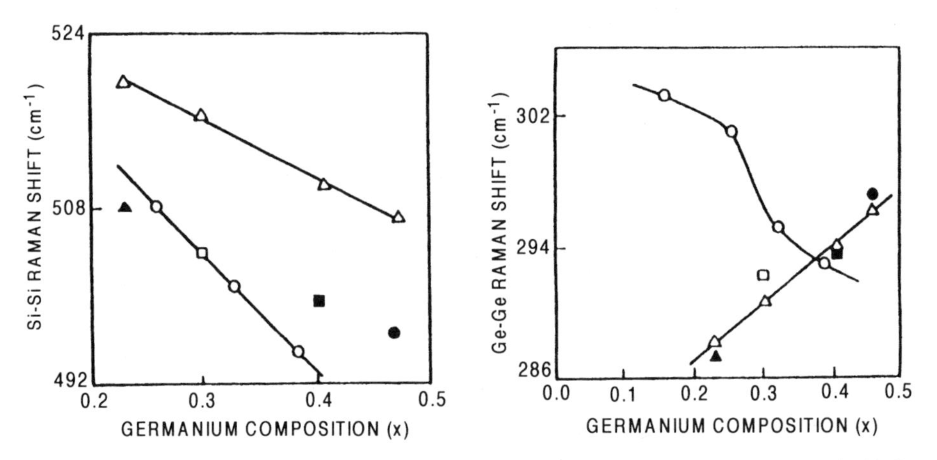

FIGURE 7.20. The Raman shift of the Si–Si and Ge–Ge Raman peak against Ge concentration in the Si–Ge layers as a function of growth temperature: ○, 620 °C; ●, 650 °C; ■, 675 °C; □, 710 °C; ▲, 750 °C; △, calculation. The calculated results shown are for fully strained Si–Ge alloy [58].

lower frequency. The homogeneity of the GeSi films may also change with temperature: At higher temperatures the mobility of adatoms is increased and a more uniform distribution obtained. To summarize the Ge is more homogeneous in the samples with higher Ge contents.

Selective deposition was demonstrated by Zhong *et al.* [61] with selectivity to silicon with respect to silicon dioxide. The selectivity appeared to be a function of the germane concentration. The deposition rate is enhanced by the addition of germane. Figure 7.21 shows the germane content of the epitaxial layer versus the germane flow rate. The dotted line indicates the boundary between selective and nonselective growth. However, contaminants such as oxygen and water vapor may play a role in defining the location of this

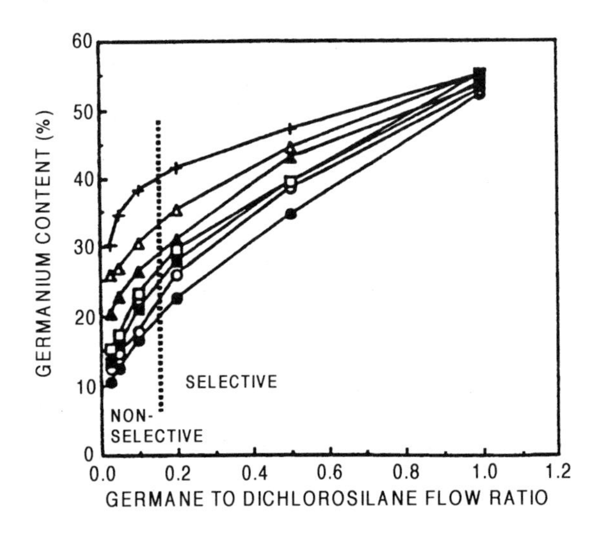

FIGURE 7.21. Effect of the germane/dichlorosilane ratio on the percent germanium in the film with deposition temperature as a parameter: ●, 800 °C; ○, 750 °C; ■, 700 °C; □, 650 °C; ▲, 600 °C; △, 550 °C; +, 500 °C. The boundary between selective and nonselective growth is indicated by the dotted line [61].

boundary. This demonstrates that selectivity can be achieved without the addition of HCl. When the growth temperature is below 650 °C, the percentage of germanium in the alloy is less sensitive to the percentage of germane in the growth ambient. These measurements suggest that to decrease the germane content of the film, while still maintaining selective deposition, higher temperatures are preferred.

7.2.1. Growth of $Si_{1-x-y}Ge_xC_y$ Epitaxial Layers

In the previous section the growth of SiGe heterostructures was discussed; however, the addition of carbon-containing layer will allow the growth of SiGe layer with well-controlled misfit strain. Regolini *et al.* [62] have demonstrated epitaxial growth of Ge/Si layers with carbon added for strain reduction because of the smaller atomic volume of the carbon (see Table 7.4). In the work of Fischer and Zaumseil [63] on pseudomorphic $Si_{1-x}Ge_x$ ($0.15 < x < 0.35$) and $Si_{1-y}C_y$(0.008–0.016) thin layers (~ 100 nm) on (100) silicon substrates, the lattice constant was measured *in situ* by X-ray diffraction. An anneal above 1000 °C resulted in interdiffusion of Ge and Si, strain relief, and a reduction of the Ge concentration in the layer. For the strained $Si_{1-y}C_y$/Si system, nucleation and diffusion-controlled growth of SiC nanocrystals was found to be the dominant mechanism, above 800 °C, decreasing the amount of C atoms in the substitutional sites and thus relieving the strain in the layer. It was shown that these layers behave similarly to GeSiC layers. Namely, a 1% C content caused a similar absolute deformation as did a 10% Ge content. Further, relaxation occurs through the introduction of misfit dislocations, interdiffusion of Si/Ge atoms, and diffusion of C atoms toward growing SiC precipitates. Improvements in two-dimensional growth are produced by the addition of a small amount (1% C_3H_8 in hydrogen) to growth at a temperature of 900 °C. However, the problem of low solubility of C in Si and possibility of SiC precipitates occurs. Mi *et al.* [64] speculated

TABLE 7.4. Processing Conditions for RTCVD of Epitaxial $Si_{1-x-y}Ge_xC_y$ Heterostructures

	Cleaning		Process			
Substrate	Temperature/time/ pressure (°C/s/Torr)	Gas	Film thickness (nm)	Temperature/time/ pressure (°C/s/Torr)	Gas	Ref.
(100) Si	—	—	100	Si: 900/—/—	SiH_4	62
			110–390	SiGeC: 650/—/—	$Si(CH_4\ CH_2)_4/Si(CH_3)_4$	
(100) Si	—	—	80	550/—/—	$Si(CH_4\ CH_2)_4/Si(CH_3)_4$	72
(100) Si	1100	H_2	130–170	900/—/1	SiH_4 1% GeH_4 in H_2 1% C_3H_6 in H_2	64
(100) Si	1000/40/—	H_2	—	Si: 900/—/—	SiH_4	67
			150	SiGeC: 525–650/—/<10	$SiH_4GeH_4/SiCH_6$ in H_2	
(100) Si	1050/30/—	H_2	—	Si: 900/—/—	—	65
			80–150	SiGeC: 550—600/—/1.5	$SiH_4/GeH_4/SiCH_6$ in H_2	
			70	Si: 700/—/—	—	

that the growth at 900 °C was controlled by the adsorption of Ge and C atoms, which tend to compete with Si for available sites. They concluded that C_3H_8 scavenges the oxygen and acts as a surfactant to inhibit three-dimensional growth. X-ray rocking curves and PL spectra indicate excellent interface quantity which is the result of the removal of residual oxygen from the growth chamber.

Mi *et al.* [65] have produced $Si_{1-x-y}Ge_xC_y/Si$ heterostructures with methyl silane ($SiCH_6$) as the source gas. This produces a wider range of process conditions than C_3H_8 added to silane. Defect-free alloy layers with compositions of 10, 11, 20 at. % Ge and up to 2.2 at. % C were produced. The lattice parameters were tailored so that the strain was minimized. A Si buffer layer was grown followed by growth of the alloy layers, and some samples had a Si capping layer (grown at 700 °C). Abrupt interfaces were observed and there was no detectable variation of the concentration within the alloy layer and no SiC precipitation. Figure 7.22 shows the variation of C concentration and strain as a function of the ratio

$$P_m = \frac{P(SiCH_6)}{P(SiH_4) + P(GeH_4) + P(SiCH_6)}$$

for three sets of samples. The growth rate does not follow the expected behavior as reported for $Si_{1-x}Ge_x$ growth at low Ge concentration where the growth is proportional to $P(GeH_4)/P(SiH_4)$. Both tensile and compressive strained films were produced. The maximum tensile value was –0.7% for binary and –0.35% for ternary alloys. Efficient C

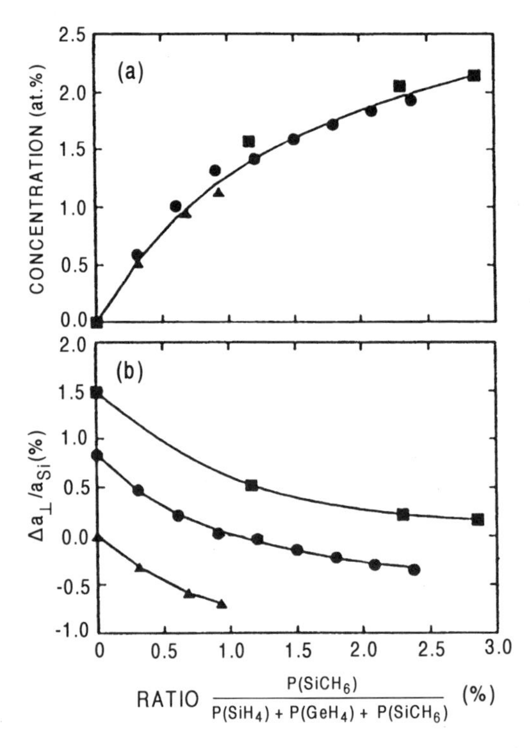

FIGURE 7.22. Carbon concentration in Si/SiGeC/Si multilayer structures (a) and strain (b) as a function of $SiCH_6$ content (P_m) in the precursor gases [65]. ■, 20 at % Ge; ●, 11 at % Ge; and ▲, 0 at % Ge.

incorporation has been achieved with a high temperature (600 °C) and lower growth rate (2.7 nm/min). Methyl silane begins to decompose at 600 °C to stable hydrocarbons; however, it is likely that a surface decomposition takes place at lower temperature with C incorporation into the film.

Atzmon *et al.* [66] reported growth of films with excellent crystal quality using C_2H_4 in an atmospheric system with a Ge:C ratio of 8.5:1 and obtaining up to 2 at. % C in the film. Amorphous films were produced when the Ge:C ratio was less than 3:1.

Bodnar and Regolini [67] also used methyl silane to grow epitaxial SiGeC layer on a Si buffer layer (see Table 7.4). Two methods were used to characterize the C content Fourier transform infrared (FTIR) and X-ray diffraction. The substitutional C content was calculated from the absorption peak at 605 cm^{-1} associated with ^{12}C localized vibration mode: There was good agreement between these methods within 5%.

$$y = 4.606 \times 10^{-4} \frac{(\text{Abs}_{\text{max}} - \text{Abs}_{\text{min}})}{\ell}$$

where ℓ is the layer thickness and Abs_{max} and Abs_{min} are, respectively, the maximum and minimum infrared absorbance. Based on X-ray diffraction the d_{004} atomic distance was used to determine the C content. At high mass flows the C substitution fraction saturates and the layer is made up of a mixture of substitutional C and amorphous SiC. The effect of annealing at 800 °C was examined. SiGeC strain relaxation occurs by O and C diffusion, which leads to a small decrease in the substitutional C content, and formation of oxide and disordered Si–C bonds. At 900 °C after a 15-μm anneal all of the substitutional C has diffused and some precipitation of amorphous SiC occurs.

7.2.2. *Polycrystalline Si_xGe_{1-x} Films*

Doped polycrystalline Si_xGe_{1-x} films were deposited by Sanganeria *et al.* [68] onto 1000 Å of thermal silicon dioxide, coated with a thin layer of polysilicon (< 20 Å) to provide nucleation sites. A low growth temperature of 600–700 °C was used. Films were doped with boron which had no measurable effect on the deposition rate, although it is well known that boron enhances LPCVD of polysilicon. The boron concentration was found to exceed the solid solubility limit in the films because of the nonequilibrium nature of the growth. Films were characterized by conductivity and Hall effect measurements. Postdeposition RTA at 1000 °C produced a small increase in the carrier mobility and film conductivity reaching a minimum value of 1 Ω-cm.

7.2.3. *Pure Germanium Films*

Pure germanium films have been deposited selectively onto silicon with respect to silicon dioxide [69] from germane in hydrogen at 425 °C with a deposition rate of 800 Å/min. The motivation for this study was to find an alternative interconnect technology to silicides such as TiSi which consume a fraction of the substrate when formed by reacting the thin-film metal with the silicon substrate. Deposition at a pressure of 3 Torr produced an apparent activation energy of 1.81 eV. It was also demonstrated that the Ge could be deposited onto the SiO_2 layer, if first a thin seed layer of polysilicon was grown.

Recent work on epitaxial germanium growth onto silicon has been reported by St. Amour and Sturm [70]. Thin 0.3–14 monolayer films of germanium were deposited in germane on silicon (100) and characterized by Auger and RBS measurements. Samples were grown with a thin layer of germanium sandwiched between layers of Si. Figure 7.23a shows the aerial density of germanium per layer as a function of the deposition time. The amount of germanium in each layer was determined by integration of the RBS peaks. The Auger analysis was carried out at 1 keV in a neighboring vacuum chamber connected via a sample transfer stage to the growth chamber. Figure 7.23b shows that the decay of the Si(LMM) intensity peak corroborates the RBS measurements. A decay length of 2.2 monolayers indicates layer growth (two-dimensional Stranski–Krastonov growth) and slower decay indicating intermixing, alloying, or island growth after three monolayers. The slower growth of the first monolayer at 500°C takes place with island growth (Figure 7.23a).

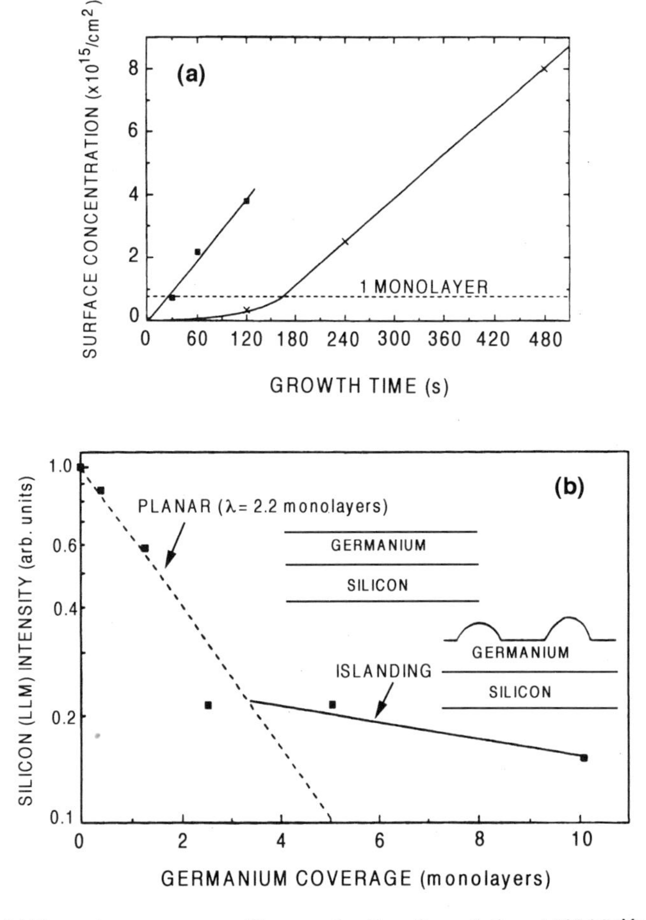

FIGURE 7.23. (a) Germanium coverage on silicon as a function of growth time at 500 °C (X) and 550 °C (■). (b) Silicon substrate Auger intensity as a function of germanium coverage for growth at 550 °C [70].

7.2.4. Electrical Properties, Fabrication of Active Devices, and Photoluminescence Measurements

Active devices were fabricated by Green *et al.* [44]. A unity gain frequency, f_T, of 11 GHz was obtained for an HJBT with an emitter area of $1.5 \times 5\ \mu m^2$ and subcollector of $30 \times 30\ \mu m^2$. After growth the device was not exposed to a temperature greater than 950 °C. A schematic of the device is shown in Fig. 7.24a and the characteristics in Fig. 7.24b indicate a voltage gain of 15 was obtained.

Heterojunction transistors were fabricated by Sturm *et al.* [47] with up to 50 periods of the superlattice. HJBTs with narrow-band-gap base regions and emitter dimensions of

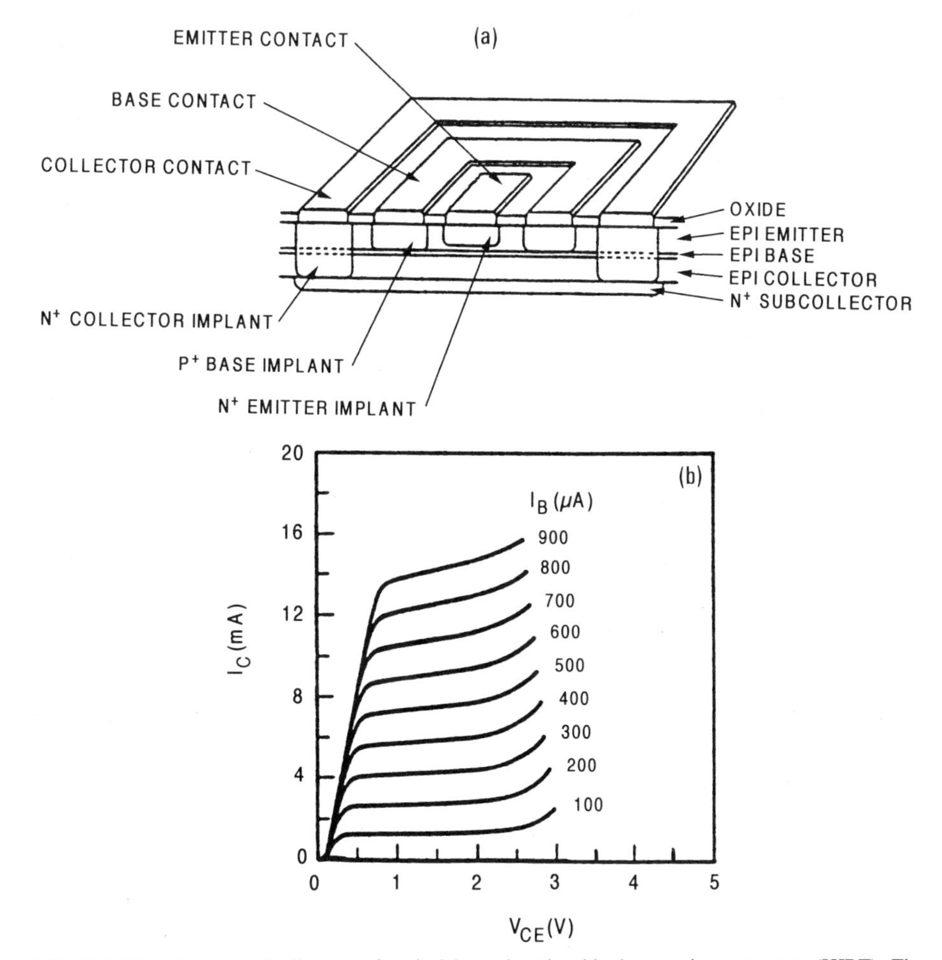

FIGURE 7.24. (a) Schematic diagram of typical heterojunction bipolar transistor structure (HJBT). The epitaxial layer structures are as follows: 1000 Å emitter boron doped 10^{16} atoms/cm^3; 350 Å base 15% germanium in silicon boron doped 5×10^{18} atoms/cm^3; 3000 Å collector arsenic doped 10^{17} atoms/cm^3. (b) Common emitter characteristics of an HJBT grown *in situ* at 900 °C [44].

$60 \times 60\ \mu m^2$ were fabricated. Gummel plots were made at room temperature (collector and base current versus base–emitter voltage). The lifetimes observed were low, probably because of oxygen incorporation in the epitaxial film.

Quantum wells have been fabricated by Liu *et al.* [71] on strained SiGe layers on (110) silicon. The thickness of the quantum wells were measured by XTEM and abrupt interfaces within 3–4 Å were present. If the substrates were annealed at 900 °C after growth, relaxation was observed. Photoluminescence spectra were taken at temperatures of 4 and 77 K with an argon-ion laser. The band gap was extracted from the spectra taken at 77K using an electron–hole plasma model. Stronger effects were observed on the (110) than the (100) silicon substrates. A quantum confinement energy of up to 110 meV was observed by varying the well width from 133 Å to 17 Å.

A series of photoluminescence measurements on SiGeC films on (100) silicon by Boucard *et al.* [72] show two distinct features at low pumping intensities: a deep-level broad peak around 800 meV and a well-resolved band-edge luminescence. The 800 meV peak observed in the as-grown samples was strongly enhanced by an anneal at 700 °C. Shown in Fig. 7.25b at high pumping intensity are several peaks including an intense peak at 1.038 eV ascribed to the no-phonon (NP) transition, and a transverse-optical (TO) Si–Si phonon replica at 0.98 eV. At 77 K the band-edge luminescence at 1.1 eV is attributed to an electron-hole. A blueshift in the band gap for the SiGeC alloys with respect to the GeSi is observed, which increases with the carbon content of the film. The mechanism of strain relaxation can be excluded. Table 7.4 gives the process conditions for samples 80 nm thick, with a 60-nm Si capping layer. Low temperatures around 650 °C gave rise to minimal SiC precipitate formation.

Bermond *et al.* [73] have studied photoluminescence of fully and partially relaxed SiGe layers. Fabrication at 800 °C with 0–20% Ge has been repeated by Regolini *et al.* [62]. A lifetime of 9 μs for a photoluminescence excitonic transition indicates that the

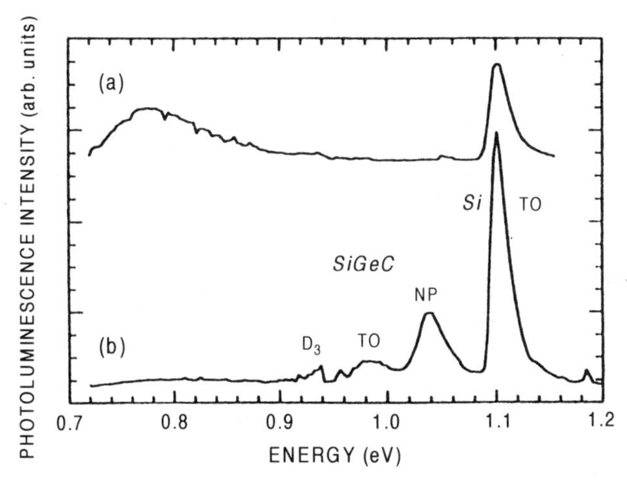

FIGURE 7.25. (a) 77 K deep-level photoluminescence spectrum of a $Si_{0.84}Ge_{0.155}C_{0.005}$ alloy. Ar^{2+} excitation: 50 mW; laser spot: 3-mm diameter. (b) Band-edge luminescence of the sample. Ar^{2+} excitation: 50 mW; laser spot ~ 0.5-mm diameter [72].

thin strained SiGe layer has excellent electrical properties, compared to 4 μs for the substrate. A near-band-gap excitonic photoluminescence (PL) was observed with a 514.5-nm green line cw argon-ion laser at powers between 0.1 and 50 W/cm^2 and samples held in a cryostat at temperatures of 4–300 K. A detailed study of the effect of layer thickness and temperature was made. Aluminum Schottky barriers were made and the ideality factor measured as a function of strain relief. A value of 1.3 was found for small thicknesses (≤ 100 nm) increasing to 2.0 at a thickness of 200 nm. This is related to the introduction of generation-recombination traps caused by the strain relaxation in thicker films. For thicknesses greater than a critical value, the PL spectrum changes with four addition D lines at 0.805, 0.874, 0.934, and 0.995 eV due to dislocations. The location of these D lines was independent of the Ge content, suggesting their origin is the Si or SiGe/Si interface. The presence of a Si capping layer was found to annihilate the high surface recombination velocity in the GeSi layer, resulting in an absence of photoluminescence in very thin uncapped layers. Measurement of the NP lines makes determination of the indirect band gap possible by adding the dissociation energy of the free exciton. The equation for photoluminescence intensity

$$I(\hbar\omega) = \sqrt{\hbar\omega - E_o}\, \exp\left[-\frac{\hbar\omega - E_o}{kT}\right]$$

fit the data, where E_o is the sum of the free exciton dissociation energy, 14.7 and 4.15 meV and the band gap energy. The variation in band gap with germanium content, at 90 K, in fully strained layers was found to be fit by the following equation:

$$E_g = 1.171 - 1.01\, x_{Ge} + 0.835\, x_{Ge}^2$$

Warren *et al.* [75] made Hall measurements from 300 to 20 K and magnetotransport measurements. Magnetotransport measurements at 1.6 K show well-defined Shubnikov de Haas oscillations confirming the presence of a two-dimensional hole gas at each Si/SiGe interface. Charge transfer Hall measurements were made as a function of Ge content and spacer layer thickness. Similar measurements have also been made by Jiang *et al.* [76] and Ruolian *et al.* [77].

7.3. DEPOSITION OF OXIDES AND NITRIDES

In the previous chapter the RTP of dielectric films through rapid thermal oxidation, rapid thermal nitridation, and reoxidation was described. The films produced had excellent electrical properties; however, because of the need for a reoxidation and/or annealing steps, it is appropriate to examine RTCVD as a technique with a lower overall thermal budget. Furthermore, an additional motivation is that a uniform N and O concentration profile is produced throughout the film, as opposed to a concentration built up at the interface.

7.3.1. Growth and Structure

Ultrathin stacked dielectric films of Si_3N_4/SiO_2 were prepared by Ting *et al.* [78]. The first layer was a rapid thermally grown oxide, grown as described by Ting *et al.* in

TABLE 7.5. Processing Conditions and Electrical Characteristics of RTCVD Dielectrics

		Process			Electrical properties				
Dielectric film	Thickness (Å)	Temperature/time/pressure (°C/s/Torr)	Gas	Gate material	Interface trap density (eV^{-1} cm^{-2})	Interface trap density after bias stress (eV^{-1} cm^{-2})	Leakage current (A/cm^2)	Breakdown field (MV/cm)	Refs.
SiO_2 and	40	RTO: 1050/—/—	O_2	poly-Si	$\sim 2 \times 10^{10}$	—	10^{-5} at 10 MV/cm	—	78
Si_3N_4	30	RTCVD: 850/—/1	SiH_4/NH_3 1:40 (diluted in N_2)						
SiN_xO_y	80–200	800/—/3	SiH_4,N_2O, in Ar	poly-Si	2×10^{10}	—	—	~12	80
SiO_2 and	100	RTO: 1000/20–120/760	O_2	poly-Si	$\sim 2 \times 10^{10}$	—	10^{-5} at 10 MV/cm	—	79
Si_3N_4	40	RTCVD: 700–900/ 360–900/760	SiH_4/NH_3						
SiN_xO_y	55–85	RTCVD: 800–950/15/ 2.5–3	SiH_4 (10% in Ar), N_2O, NH_3	—	6×10^{10}	—	—	—	84
SiN_xO_y	70–200	RTA: 800/—/3	SiH_4, N_2O, NH_3	poly-Si	$0.3–2 \times 10^{10}$	$\sim 10^{11}$ (see Fig. 7.29)	2×10^{-11}	—	82,83
SiN_xO_y	70–100	PECVD: (30W RF power) 300/15/— RTCVD: 700/115/3	SiH_4 (10% in Ar), O_2, N_2O	Al	$3–5 \times 10^{10}$	—	—	>10	85
SiN_xO_y	60[a]	RTCVD: 800/1000/20 RTO: 1000/30/—	SiH_4, NH_3, N_2O, O_2	poly-Si	—	—	—	12.8–15.3	88
SiN_xO_y		RTCVD: 800/—/3	SiH_4 (10% in Ar), NH_3, N_2O		7×10^{10} (3 at.% N)	—	—	—	87

[a]Equivalent oxide thickness.

the previous chapter (see Ref. 121 of Chapter 6). The second layer 30 Å of nitride was deposited by RTCVD at 850 °C, 1 Torr, in a silane/ammonia gas mixture; other process details are presented in Table 7.5. The electrical properties are discussed in Section 7.3.2.

Lee and So [79] studied a stacked layer of RTO and RTCVD Si_3N_4 grown *in situ*. The dielectric films were typically 100 Å thick. Wafers were cleaned by a HF dip prior to deposition. Figure 7.26 shows the refractive index and deposition rate as a function of the gas flow ratio and the growth temperature. The deposition rate increased with temperature but decreased with NH_3/SiH_4 ratio and also the refractive index saturated to a constant value. FTIR and AES measurements indicated N and O species throughout the film.

RTCVD of oxynitride films has also been studied by Xu *et al.* [80] in the thickness range 80–200 Å in a silane/nitrous oxide gas mixture (see Table 7.5). An *ex situ* RCA and HF dip cleaning step were used but there was no *in situ* high-temperature cleaning step. Figure 7.27 shows the silicon/oxygen atomic ratio dependence as a function of the silane to nitrous oxide flow rate ratio. For ratios greater than 4%, silicon-rich oxides are formed, and for less than 4%, the atomic ratio of silicon to oxygen is approximately 50%. The reaction between silane and nitrous oxide proposed is

$$SiH_4 + 4N_2O \rightarrow SiO_2 + 4N_2 + 2H_2O \tag{7.2}$$

In addition, Auger depth profiles suggest a uniform composition throughout the film thickness (see inset of Fig. 7.27). The temperature dependence (not shown) indicates an activation energy of 1.76 eV at 2% silane/nitrous oxide and 1.54 eV at 4% silane/nitrous oxide. The deposition rate is quite high at 55 Å/min at 800 °C with a ratio of 2%; this is two orders of magnitude higher than for dry thermal oxidation processes. Electrical properties are discussed in the next section.

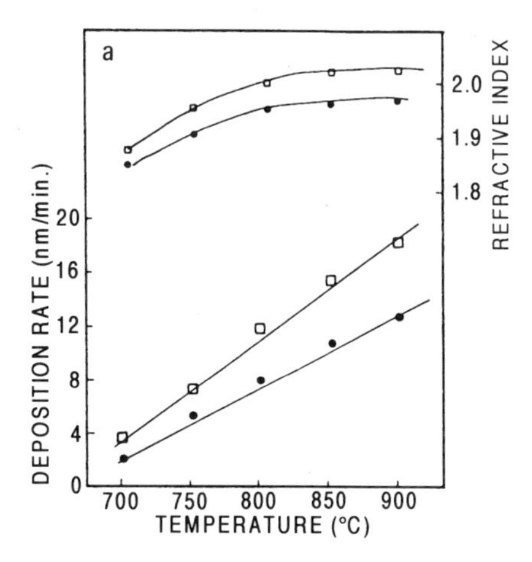

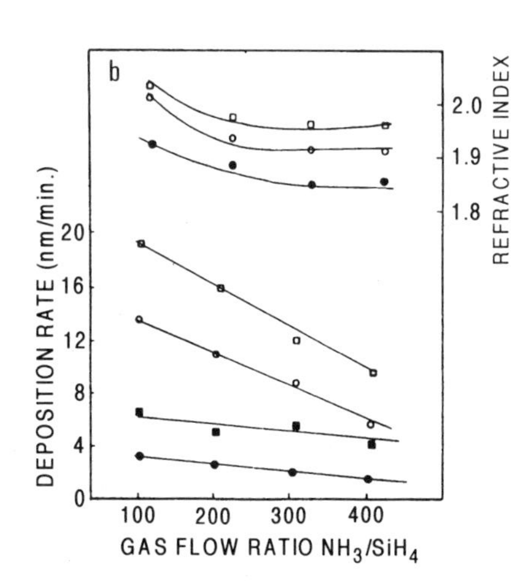

FIGURE 7.26. (a) Effects of the deposition temperature on the deposition rate and the refractive index (total pressure 1.01×10^5 Pa, $SiH_4 \approx 4$ sccm), for the following NH_3/SiH_4 ratios: (●) 300; (□) 100. (b) Effect of the NH_3/SiH_4 input gas ratio on the deposition rate and the refractive index for the following temperatures: (●) 700 °C, SiH_4 = 3 sccm; (■) 700 °C, SiH_4 = 7 sccm; (○) 800 °C; (□) 900 °C [79].

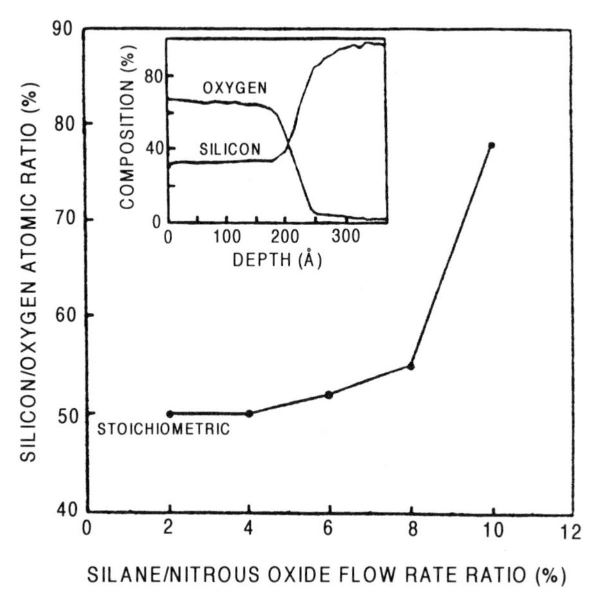

FIGURE 7.27. Silicon/oxygen atomic ratio of RTCVD oxide film as a function of silane/nitrous oxide flow rate ratio. The deposition temperature was 800 °C. The inset is the Auger electron spectroscopy profile of an RTCVD silicon dioxide film with a silane/nitrous oxide flow rate ratio of 2% [80].

Growth of SiO_2 in a silane/oxygen gas mixture at 450°C was studied by Cobianu *et al.* [81]. Wafers were cleaned prior to deposition in 100% HNO_3, boiling in HNO_3, respectively, and dipped in 1% HF solution. A critical partial pressure was observed for a sharp transition from no film to high deposition rates (see Fig. 7.28). Similar, critical silane partial pressures are observed at all pressures producing a quite narrow useful process window. The explosion limit for silane at a total pressure of 0.50 mTorr (0.67 mbar) is 0.167 mTorr (0.223 mbar) at room temperature, and 0.140 mTorr (0.187 mbar) at the deposition temperature of 450 °C. Therefore, significant gas-phase reactions can take place. However, uniform layers were produced at a total pressure of 3.375 mTorr (4.5 mbar), silane pressure of 0.059 mTorr (0.070 mbar) for an O_2/SiH_4 ratio of 11 producing a deposition rate of 130 nm/min (uniformity 5%). The chamber walls may participate in the reaction process; however, without *in situ* characterization one cannot rule out current models of silane oxygen reactions in which intermediates are produced in the gas phase, and the final SiO_2 product is formed on the wafer surface.

Later work was done by Xu *et al.* [82] and McLarty *et al.* [83] with different process gases silane/ammonia/nitrous oxide. Auger analysis indicates that the stoichiometry of the SiN_xO_y changes with ammonia/nitrous oxide gas flow ratio, with increasing nitrogen at higher values of ammonia flow. Since the relative strength of the N–O bond is much higher than the Si–H and N–H bonds, the decomposition of nitrous oxide is slower than for ammonia and silane at a given temperature. This is why an increase in the ratio increases the nitrogen content but reduces that of oxygen; however, the film growth rate decreases. Films deposited at 800 °C and 3 Torr were characterized by ellipsometry. Variations in refractive index were observed and a higher nitrogen content than LPCVD Si–N–O films, which show higher oxygen content at the dielectric/silicon interface. The

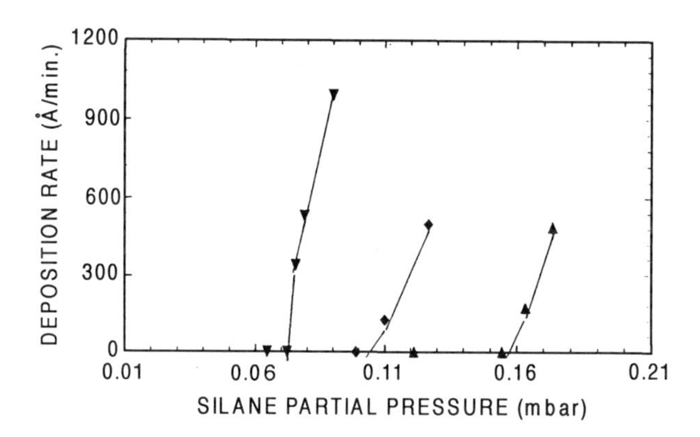

FIGURE 7.28. Deposition rate as a function of partial pressure of silane with total pressure as a parameter (▲, 0.6 mbar; ♦, 1 mbar; ▼, 4.5 mbar) (1 mbar = 0.75 mTorr). Deposition temperature 450 °C, oxygen/silane mole ratio 2 [79].

authors presented TEM and Auger data demonstrating that the film is a random mixed matrix of Si, N, and O rather than a two-phase mixture of SiO_2 and Si_3N_4.

McLarty *et al.* [84] deposited dielectrics with 10% silane in $Ar/N_2O/NH_3$ at flow rates of 4/100/0–100 sccm, respectively, at 3 Torr. Changes in the flow rate ratio of NH_3/N_2O led to an increase in the atomic ratios and decrease in Si–O–N deposition rate. However, in this work the deposition rates are greater than for conventional LPCVD. Auger depth profiles indicated a homogeneous composition for all flow rates and consistently no oxygen concentration built up at the Si–O–N/Si interface compared to conventional LPCVD films.

Stacked gate dielectrics were formed by plasma-assisted low-temperature processing combined with RTCVD by Misra *et al.* [85]. A two-step process of plasma-enhanced chemical vapor deposition (PECVD) followed by RTCVD produces a thin (0.5–0.6 nm) SiO_2 layer of high quality followed by deposition of the SiO_2 film, performed *in situ* in a UHV system. Substrates were cleaned by the RCA process and a 1:30 $HF{:}H_2O$ dip for 15 s with other process details given in Table 7.5. *In situ* AES measurements after the PECVD process demonstrate that carbon was effectively removed from the wafer surface.

The optical and surface properties of RTCVD oxides grown in mixtures of SiH_4/NH_3 at 750 °C on InP substrates were studied by Lebland *et al.* [86]. Deposition rate was a function of the SiH_4/NH_3 ratio and found to be 100 Å/s at a ratio of 10. Structural properties were measured by nuclear reaction analysis (NRA) using $^{14}N(d,a)^{12}C$ at 830 keV and RBS indicating dense films of uniform thickness. Stoichiometry was N/Si = 1.31 and was independent of thickness. AFM measurements indicate an isotropic surface roughness of 2.2 nm RMS value for a film thickness of 100 nm. Before imaging the samples were cleaned by boiling in trichloroethylene and rinsing in methanol to obtain reproducible results. FTIR measurements were used to obtain a quantitative determination of the Si–H and N–H adsorption bonds and indicated little detectable oxygen content. These dense pure films were confirmed with a low etch rate in buffered HF solutions of 8 Å/s.

7.3.2. Electrical Properties

Ultrathin stacked dielectric films of Si_3N_4/SiO_2 were prepared by Ting *et al.* [78]. No annealing or reoxidation of the film was performed after the deposition process. The equivalent thickness at the oxide/nitride derived from high-frequency *C–V* measurements was 58 Å. A polysilicon gate was deposited by LPCVD, and immediately doped by $POCl_3$, and lithography, RIE etching, and a forming gas anneal at 450 °C completed the structure. Improvements in the low-field breakdown were significant compared to thermal oxides. The flat band instabilities, usually associated with charge trapping at the Si_3N_4/SiO_2 interface, were also reduced. However, the amount of charge trapping and the leakage currents are comparable to those observed with pure SiO_2. According to Ting *et al.,* the reduction in charge trapping is caused by the nitride layer being so thin that the traps at the oxide/nitride interface can no longer hold their charge. The advantage of this structure is that the defect density has been reduced while low charge trapping is obtained. A small negative shift in the flat band voltage (V_{fb}) indicates the presence of a positive fixed charge, which has previously been reported as associated with nitrogen incorporation at the Si/SiO_2 interface (see Chapter 6).

The electrical properties of the films produced by Xu *et al.* [80] provide a promising dielectric technology with a reduced thermal budget. In particular, the interface trap density of the polysilicon/SiO_xN_y/Si films is lower than or comparable to the furnace-grown oxide. The D_{it} is independent of the oxide thickness. An anneal at 1080 °C in argon had no noticeable effect on D_{it}. It is believed that the hydrogen content is lower than for RTP nitrided and oxidized films discussed in the previous chapter. However, the deposition with an ammonia/nitrous oxide ratio of 4% was silicon rich and inferior in its D_{it} at $>10^{11}$ eV^{-1} cm^{-2} and high-resolution XTEM indicated a rougher Si/SiO_2 interface for these samples. The breakdown follows Fowler–Nordheim tunneling and catastrophic breakdown field histograms were grouped around 12 MV/cm. No breakdown occurred below 9 MV/cm, indicating an intrinsic breakdown mechanism rather than pinholes.

Lee and So [79] fabricated silicon dioxide from SiH_4/NH_3 gas mixtures and observed a shift in ΔV_{fb} to the negative with increasing NH_3/SiH_4 ratio. It is believed that more N–H bonds produce negative ΔV_{fb} shifts as a result of positive fixed charge in the interface. Residual hydrogen also plays an important role in determining charge defects. The fixed charge was remarkably reduced after annealing at 1000 °C in an Ar ambient for 20 min with a reduction in the N–H bonds at the interface.

Silicon oxynitride films grown by Xu *et al.* [82] show excellent electrical properties. Leakage currents had a weak $1/T$ dependence at all applied gate voltages 8–12 V indicating Fowler–Nordheim tunneling. Typical values are given in Table 7.5. The interface trap density was determined by simultaneous measurement of quasistatic and 100-kHz *C–V* curves. D_{it} is a function of the ammonia/nitrogen ratio, as shown in Fig. 7.29. A minimum value of $< 3 \times 10^9$ $eV^{-1}cm^{-2}$ was observed for a ratio less than 40%. The higher-nitrogen-content film is more stable but has a higher initial D_{it}. A considerable amount of hydrogen is introduced into the film during growth from the thermal decomposition of silane. The higher nitrogen content of the film may block hydrogen incorporation during the *in situ* deposition of the polysilicon gate electrode. Relaxation of stress at the dielectric/silicon interface may also play a role in improving the reliability of the

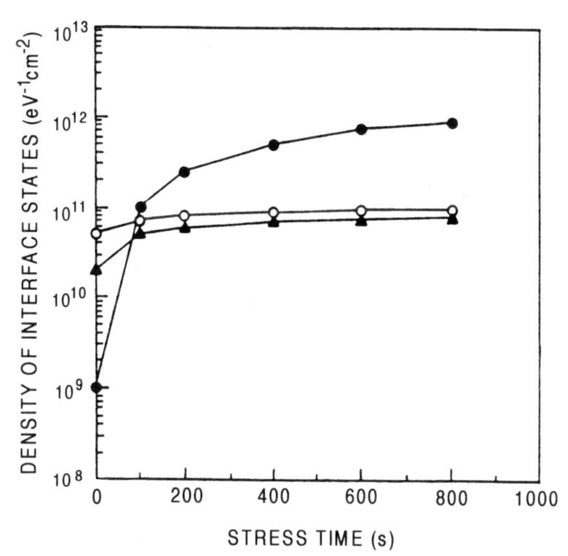

FIGURE 7.29. Interface trap density as a function of Fowler–Nordheim stress time for Si–N–O and control oxide films. The polysilicon gate is biased at a negative voltage and the injection current density is 10^{-5} A/cm^2. Ammonia/nitrous oxide flow rate ratios: ●, 20%; ▲, 100%; ○, a thermal oxide [82].

dielectric with increased nitrogen concentration (as discussed in the previous chapter). Hill *et al.* [87] also studied oxynitride dielectric in active devices. Charge pumping measurements were made and the peak current was found to increase with increasing nitrogen content in the film, agreeing with the observed increase in D_{it} with nitrogen. The shift to more negative voltages also indicates the presence of a positive charge at the interface. The threshold voltage change for *n*-channel devices was more negative with increasing atomic percent of nitrogen, but small shifts were observed for *p*-channel devices.

Plasma-assisted low-temperature growth of SiO_2 was found to lower the D_{it} and produce a high E_{BD}. All wafers subjected to the plasma process showed an improvement in D_{it}. However, no significant effect was seen with the use of an 800 °C vacuum anneal prior to RTCVD of the SiO_2. Wafers not subject to the PECVD process did show differences particularly if a HF dip was included, which resulted in a high D_{it} and ΔV_t. To summarize, a low D_{it} was observed consistently and found to be independent of the processing temperature, over the range 300 or 400 °C. Preliminary results on device performance of FETs have shown comparable performance to thermal oxide gates.

A stacked structure with RTCVD Si_3N_4 was investigated by Yoon *et al.* [88] for the fabrication of DRAMs. It has been reported that reoxidation of silicon nitride reduces defects such as pinholes in the film as demonstrated by a reduction of the leakage current. After the silicon nitride is RTCVD deposited with a silane/ammonia gas mixture, a reoxidation process is carried out in N_2O or O_2. Process details are listed in Table 7.5. The leakage currents for the nitride films were less than those for the oxide films. The breakdown voltage after bias stress current injection of 1 mA/cm^2 was 15.3 MV/cm under positive gate bias and 12.8 MV/cm for negative gate bias for the nitride film versus 8.3–8.9 MV/cm for the oxide film. A Weibull plot indicated an effective breakdown of 16 MV/cm and the t_{bd} at 10 MV/cm was ~10 years. Charge trapping, measured by

$\Delta I_g/I_{g,0}$ as a function of time under constant bias stress of 12 MV/cm, shows less drift for the nitride film than for rapid thermal oxidation. The reduction in charge trapping, through a reduction of Si–H bonds in the nitride, therefore extends the t_{bd} of the dielectric via a reduction in the rate at which the field builds up.

7.3.3. *Active Devices*

Hill *et al.* [87] studied the formation of oxynitride in active devices with gate lengths of 1–6 μm. The wafers were treated with an anneal in oxygen at 900 °C for 15 s before an *in situ* polysilicon deposition. The SiH_4/N_2O ratio was 1.28% and the NH_3/N_2O ratios employed were 0, 2, 5, 20, 50, and 100%. Electrical performance was compared against a thermal oxide grown at 900 °C in pure oxygen. A correlation between the nitrogen content and electrical properties was found. Nitrogen was shown to be uniformly distributed through the film, although hydrogen concentration was minimal at the interface. Electrical performance indicated that for less than 3 at. % nitrogen an RTCVD dielectric is viable for VLSI (see Table 7.5). Charge pumping experiments showed a shift to negative gate potential characteristic of positive oxide charge. Both *n*- and *p*-channel devices were fabricated and the g_m values decreased with increasing nitrogen contents. Hot carrier stress measurements demonstrated no further reduction in the interface state generation rate for atomic percentages of nitrogen greater than 3 at. %.

McLarty *et al.* [83] made MOSFETs with dielectrics grown in mixtures of $SiH_4/N_2O/NH_3$. Transconductance (g_m) values decreased but an increase in high-field breakdown was observed. Devices with 50–200 Å SiN_xO_y gate dielectric were fabricated on 0.1 Ω-cm (100) silicon. Wafers were RCA cleaned prior to RTCVD. An *in situ* polysilicon gate was formed and a postdeposition anneal at 900 °C in O_2 for 15 s was carried out. AES measurements for various NH_3/N_2O ratios between 20 and 100% show that Si content was independent. Deposition uniformity on 4-inch-diameter wafers was 10%. Electrical properties such as D_{it} were independent of NH_3/N_2O ratio and did not change with annealing. Hot carrier stress tests showed an important effect dependent on the NH_3/N_2O ratio. At V_{DS} = 5 V and V_G = 2 V which is below the Fowler–Nordheim tunneling limit, there was no difference between the annealed and nonannealed samples. However, for NH_3/N_2O = 1, the ΔV_t variation was higher before annealing and even after annealing higher than for NH_3/N_2O = 0.2. This indicates again that H may be playing an important role in trap formation via water-based traps. Therefore, reducing the NH_3/N_2O total flow should produce films with a lower atomic percentage of nitrided oxide.

7.4. DEPOSITION OF SILICON CARBIDE

Silicon carbide has a wide band gap making it an attractive material for high-temperature semiconductor device applications. The application of SiC to advanced HJBTs will require very short processing times at high temperature and the *in situ* growth of other layers will facilitate multilayer device fabrication. Steckl and Li [89] have deposited epitaxial β-SiC on (100) silicon at 1200 °C. The growth is a two-step process; first a high-temperature carbonization of the silicon substrate is carried out at 1300 °C, followed by reaction of the silane and propane at 1200 °C (process details are given in

Table 7.6). After several minutes, film thicknesses up to 130 nm were produced. The films appear smooth with a roughness of less than 12 nm. The structural properties are a cubic crystal as indicated by diffraction patterns with a lattice spacing of 4.4 Å. Twins which originate at the interface were observed to extend into the film. At high magnification, pitting of the surface was observed consisting of inverted rectangular pyramidal voids in the silicon subsurface region. The growth effects of propane flow, ramp rate, and temperature were examined. Void-free films were produced under optimal conditions of temperature and Si/C ratio in the gas phase. For low propane flow rates (2 sccm) isolated whitish particles were observed with no evidence of voids; however, at higher flow rates (7–10 sccm) films with pyramidal voids are formed, and at 20 sccm few voids are observed, and isolated whitish deposits. These effects can be explained in terms of nucleation site density theory. Under high propane flow rates and with a large number of nucleation sites a continuous film grows rapidly. Once the silicon surface is sealed the reaction must take place by diffusion of either Si or C through the surface layer. The growth rate essentially stops because the diffusivities are small. On the other hand, at a low density of nucleation sites the growth is dendritic and therefore three-dimensional growth takes place. The optimum growth temperature was found to be 1200 °C based on X-ray diffraction data of the film structure. The ramp rate was found to have little effect on growth rate but a pronounced effect on the crystallinity. The optimum ramp rate appeared to be 25 °C/s with decreasing crystallinity at higher ramp rates (typically 50 °C/s). Later work by Li and Steckl [90] examined the nucleation and growth mechanism by AFM, scanning electron microscope (SEM), and TEM characterization. In most cases, rectangular voids were formed with a continuous SiC film over the void. The number of voids increased, and their size decreased with increasing hydrocarbon content. This occurs because the SiC nucleation density is primarily determined by the hydrocarbon partial pressure. The experimental results suggest the following growth process: (1) SiC nuclei form, (2) SiC nuclei grow as material is consumed and Si trenches form, (3) emergence of Si voids and lateral extension, and (4) sealing of a void and formation of a continuous film over the void. Highly oversaturated C_3H_8 gives rise to a continuous film growth of limited thickness.

TABLE 7.6. Processing Conditions for RTCVD of β-SiC

	Cleaning			Process	
Substrate	Temperature/time/ pressure (°C/s/Torr)	Gas	Film thickness (nm)	Temperature/time/ pressure (°C/s/Torr)	Gas
(100) Si *n*-type[a]	25/120/760 1150–1200/120/760	H_2 1.5% HCl in H_2	130	Carbonize: 1300/120–150/760 25/120/760 SiC: 1200/240/760	0.73% C_3H_8 in H_2 H_2 0.86% SiH_4/0.73% C_3H_8 in H_2
—[b]	1200/120/760	HCl in H_2	10	1300/90/—	C_3H_8 (5% in H_2)
(100) Si[c]	—	—	—	1000–1040/—/—	SiH_4/C_3H_8 in H_2
(100) Si *p*-type[d]	900/600/10^{-6}	—	500	Carbonize: 1200/70/2.5 SiC: 1150/—/2.5	C_3H_8 in H_2 SiH_4 (5% in H_2)/ C_3H_8 (5% in H_2)

[a]From Refs. 89, 90, 94. [b]From Ref. 91. [c]From Ref. 93. [d]From Ref. 95.

Ultrathin SiC films were studied by Li and Steckl [91]. Films were grown to a thickness of 10 nm by carbonization at 1300 °C (see Table 7.6). Wafers were cleaned in dilute HF, rinsed in water, and blown dry before an *in situ* cleaning step at 1200 °C in an HCl/H_2 ambient. The crystallinity of the SiC was found to be strongly dependent on the carbon content in the growth ambient. There was a transition flow rate of propane which separates single-crystal from polycrystalline growth. The region for void-free growth is at significantly higher flow rates. At these flow rates, uniform, smooth, polycrystalline SiC is produced. AES measurements (uncorrected) indicated a Si/C ratio of 55:45. HRTEM images show that the SiC{111} planes are generally well aligned to the Si lattice, with every fourth Si plane coinciding with every fifth SiC. The corresponding ratio of bulk lattice constants is 5.431 Å/4.359 Å. These results support the theory that the initial SiC "buffer layer" grown during carbonization accommodates the large lattice mismatch by gradually changing composition through the layer. Mogab and Leamy [92] have developed a model for SiC growth on Si. They propose that the growth is limited by out-diffusion of Si through pores in the SiC film. When the hydrocarbon partial pressure is increased beyond a critical value, the pores in the SiC are sealed, and further growth is greatly reduced. A high surface density of nucleation sites, randomly oriented, will produce a polycrystalline film; however, the same degree of correlation or orientation preference occurs for single-crystal growth.

Chiew *et al.* [93] have grown β-SiC at lower temperatures (1000–1040 °C). SIMS profiles indicate stoichiometric layers at a gas flow ratio $C_3H_8/SiH_4 = 100$. However, the carbon content can be adjusted in two ways: by increasing the propane flow rate, or by increasing the pressure while maintaining the same flow ratio (see Fig. 7.30). No significant interfacial oxide was detected by SIMS. The critical process parameter is partial pressure of water vapor, above which silicon monoxide forms. Even at high pumping rates the oxygen content is $\sim 2 \times 10^{18}/cm^3$. However, oxygen content is also a function of the propane partial pressure and at high C content the O content is increased as a result of gettering of water by the hydrocarbon species.

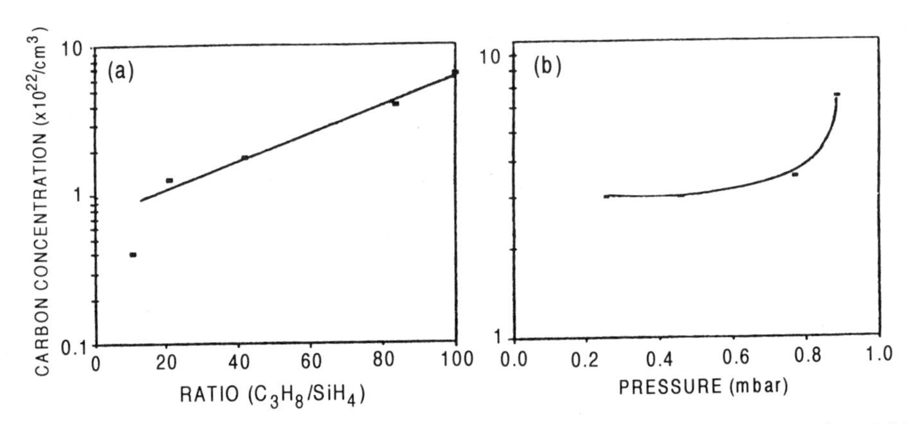

FIGURE 7.30. Carbon content in layers deposited at 1040 °C as a function of (a) gas flow rate ratio and (b) total process pressure at a gas flow ratio $C_3H_8/SiH_4 = 100:1$ [93].

Active Devices

Heterojunction diodes of SiC/Si have been demonstrated by Yih *et al.* [94]. Planar diodes were grown in propane defined selectively with SiO_2 as the growth mask and Ni contacts. The planar diodes had an ideality factor of 1.36 and a reverse breakdown voltage of 50 V. Mesa diodes were also fabricated at a lower temperature (900 °C) with a methyl silane precursor (CH_3SiH_3) and reactive ion etching in NF_3 and CHF_3/O_2. They had an ideality factor of 1.05 and a reverse breakdown voltage of 150 V. Hwang *et al.* [95] fabricated a β-SiC/Si heterojunction backward diode with epitaxial growth on (100) Si. Process conditions are listed in Table 7.6. Prior to growth, substrates were heated to 900 °C for 10 min in vacuum 10^{-6} Torr to remove the native oxide. The substrate was first carbonized and then β-SiC was grown, nickel and aluminum were evaporated to form ohmic contacts, and mesa diodes were defined by plasma etching in O_2 (30%) CF_4 (70%) at room temperature. The electrical characteristics of the backward diode were excellent. The turn-on voltage for forward bias was high, up to 4 V, in part because of the wide band gap of SiC (2.2 eV). Under reverse bias the depletion region was larger than the 1-μm film thickness. This will retard the tunneling of carriers from the SiC layer to Si substrate so that only a diffusion current remains. A figure of merit, ξ, defined as

$$\xi = \frac{d^2I/dV^2}{dI/dV}$$

was found to be insensitive to temperature from room temperature to 180 °C. This makes the diode suitable for microwave detection applications at high temperatures.

7.5. DEPOSITION OF OTHER MATERIALS

RTCVD has been applied to the deposition of several other materials including silicides and other semiconductors. CVD permits good control of the material composition in addition to the film thickness and the uniformity. However, precursors are needed for the components which are stable at room temperature and can undergo pyrolysis on the substrate.

7.5.1. Gallium Arsenide Heterostructures

The first work reported on RTCVD was the deposition of GaAs by Reynolds [96]. Using (100) GaAs substrates (Czochralski grown), alternating layers of GaAs and AlGaAs layers were fabricated by the thermal control technique in which the wafer was heated to the growth temperature for short periods and the gas ambient was changed at low temperature. Metal organic precursors were used including: trimethylarsenic (TMA), tetraethylgallium (TMG), trimethylaluminum, trimethylindium and diethylzinc. Process conditions are given in Table 7.7. Typical growth rates were 2–3 μm/h. The $Al_xGa_{1-x}As$ ($0 \le x \le 1$) were smooth and featureless whereas the $In_yGa_{1-y}As$ ($0 \le y \le 0.22$) were characterized by a cross-hatched pattern indicating dislocations caused by lattice mismatch with the substrate. Alternating layers of $GaAs/Al_xGa_{1-x}As$ up to 11 layers were fabricated. Surfaces were characterized by SIMS, AES, and RBS measurements. An abrupt interface within 50 Å and few crystal defects were detected. However, some crystal defects were detected at the interface believed to be related to oxygen incorporation. SIMS analysis indicated the presence of carbon in the film. Substrate electrical properties were characterized by Hall effect and *C–V* measurements of Schottky diodes. The typical background doping was 0.2–1 × $10^{17}/cm^3$ and mobility was 2500–3000 cm^2/V in the GaAs. Thermal precracking of the trimethylarsenic was attempted to reduce the contamination and resulted in background doping of 4 × $10^{16}/cm^3$ and electron mobilities of 3000–3500 cm^2/V. The optimum electron mobility was ~3000 cm^2/V with no precracking at a given TMA/TMG ratio of 6.5 and growth temperature of 670 °C. With precracking the minimum mobility was ~4000 cm^2/V at a TMA/TMG ratio of 2 and 650 °C growth temperature.

Katz *et al.* [97] grew $In_{0.53}Ga_{0.47}As$ films between 150 and 500 nm thick with metal organic precursors, lattice matched on InP substrate (see Table 7.7). The wafers were cleaned in chloroform, acetone, methanol, water, sulfuric acid, and nitrogen blown dry before loading in the chamber. Excellent-quality films with smooth morphology and uniform stoichiometry were produced. The InGaAs/InP interface was abrupt, with some slight interdiffusion of the Ga, as shown in Fig. 7.31. AES and TEM measurements support these findings. RBS channeling spectra estimated the fractional amount of disorder by an approximate minimum yield (χ_{min}) of 3.6% which is essentially equal to a bare InP wafer. The films produced were always *p*-type at ≤ 5 × $10^{15}/cm^3$ and carrier mobilities of ~75 $cm^2/V{\cdot}s$ at 300 K.

Self-aligned locally diffused W(Zn) *p*-type contacts on InGaAs/InP structures were made by Katz *et al.* [98] for photonic devices. A layer of InGaAs on InAs was previously grown and windows in a SiO_2 masking layer were defined. Substrates were cleaned and W(Zn) films deposited for a dopant source under the process conditions shown in Table 7.7. A further *in situ* anneal step at 500 °C for 10 s was carried out to form the ohmic contact. As-deposited films showed rectifying behavior with a Schottky barrier height of 0.31 eV.

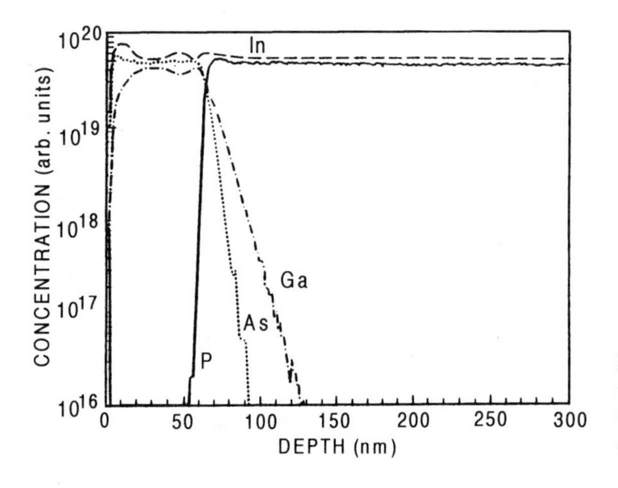

FIGURE 7.31. SIMS depth profile of undoped RT-LPMOCVD $In_{0.53}$-$Ga_{0.47}As$ layer grown on Si–InP (100) substrate at 535 °C for 10 min [97].

TABLE 7.7. Processing Conditions for RTCVD of Various Materials

		Cleaning		Process			
Material	Substrate	Temperature/time/ pressure (°C/s/Torr)	Gas	Film thickness (nm)	Temperature/time/ pressure (°C/s/Torr)	Gas	Refs.
GaAs $Al_xGa_{1-x}As$ $In_yGa_{1-y}As$	GaAs	700/300/—	tetramethylarsine	—	670/—/0.5 720/—0.5 —/—/—	tetramethylarsine, tetramethylgallium, tetraethylaluminum, tetramethylindium, diethylzinc, and SiH_4 in H_2	96
InGaAs	(100) InP	—	—	150–500	450–550/ 30–600/1–5	tertiarybutylarsine, trimethylindium, trimethylgallium and in H_2	97
InP	InP	—	—	~50	475–550/ 30–600/1–5	tertiarybutylphosphine, tetramethylindium, diethylzinc in H_2	99,100
W/Zn	InGaAs/InP	—	—	40–50	450–550/ 20–80/2–2.5	WF_6/diethylzinc	98
$TiSi_2$	(100) Si	—	H_2	—	800/60/1–10	SiH_4 and $TiCl_4$ in H_2	101,105
β-$FeSi_2$	(111) Si *p*-type	1000/30/1	H_2	—	750–850/ 60–600/—	SiH_4, He in H_2 and $FeCl_3$ in Ar	106,107
TiN_x	(100) InP	—	—	10–100	300–450/ —/3–15	NH_3/tetrakis dimethyl amido titanium in H_2	109
Y-Ba-Cu-O	MgO or $SrTiO_3$	—	—	—	600/60—then 850/30 min/—	see text	110,111

7.5.2. Indium Phosphide

Metal organic chemical vapor deposition (MOCVD) of Zn-doped InP on InP substrates has been reported by Katz *et al.* [99, 100]. This material will be important for photonic and electronic devices. A key process is the controlled incorporation of Zn for well-defined p–n junctions. Process details are given in Table 7.7 with tertiarybutylphosphine, trimethylindium, and diethylzinc as the source gases for P, In, and Zn, respectively, and hydrogen as the carrier gas. InP (Fe-doped semi-insulating) substrates were cleaned in chloroform, acetone, methanol followed by H_2SO_4, H_2O_2, and DI water. RBS, TEM, and X-ray rocking curve measurements indicated that defect-free epitaxial films were produced. Oxygen and carbon content were determined by SIMS and found to be 1.5×10^{17} and $9 \times 10^{17}/cm^3$, respectively. The carbon behaves as a donor in this material; however, this may not all be electrically active. Hall effect measurements at room temperature with Hg–In alloyed contacts indicated p-type material and acceptor doping level was a function of the growth temperature and gas composition. The deposition kinetics were evaluated from RBS and TEM thickness measurement. The maximum growth rate of 130 nm/min was achieved at 550 °C and 3 Torr, although layers could be grown at temperatures as low as 475 °C.

7.5.3. Titanium Disilicide

Titanium disilicide has been deposited by Regolini *et al.* [101] onto silicon. In Chapter 5 we reviewed in detail the properties of the silicides produced by solid-state diffusion in thin-film metal–silicon structures. However, in this case the CVD process directly forms the silicide. The reactants are $TiCl_4$ and SiH_4, diluted in hydrogen. Additional process details are given in Table 7.7. The polycrystalline titanium disilicide formed exhibits a stable C54 structure. Three different situations arise as a function of the partial pressure of silane, either (1) selective growth, (2) rough surface, or (3) consumption of the silicon substrate in the reaction. Clearly formation of the disilicide without consumption of the Si substrate is preferred. The $TiSi_2$ grains appear to nucleate and grow between islands of silicon grown epitaxially on the substrate. Later work by Mercier *et al.* [102, 103] on the selective deposition indicated that films of 15–20 $\mu\Omega$-cm could be made with a roughness of 200–300 Å. They found that the composition of the reactants had to be carefully adjusted for good selectivity and to minimize the substrate consumption. The $TiCl_4$ has a tremendous reactivity with the silicon substrate and a competing reaction with the silane. Multicycle processing was examined, but after four cycles it was found that selectivity was lost. This might be understood in terms of the following reaction:

$$TiCl_4 + H_2 \rightarrow TiCl_2 + 2\,HCl \tag{7.3}$$

SiH_2 is in general a breakdown product of the silane [104] which subsequently reacts with the HCl as follows:

$$2HCl + SiH_2 \rightarrow SiCl_2 + 2H_2 \tag{7.4}$$

The $SiCl_2$ migrates to the available silicon and this is the basis of the selective growth. However, at some point the direct dissociation of $SiCl_2$ can take place on the silicon dioxide (with H_2), and addition of HCl opposes this reaction again increasing selectivity. Therefore, the way to maintain selectivity while minimizing silicon consumption is to carry out a two-step process, first a thin layer of TiSi with silicon consumption, then a process in which the nucleation barrier is lower for the deposition of $TiSi_2$ on $TiSi_2$.

Regolini *et al.* [105] show the redistribution of the dopant in the substrate during the growth of titanium disilicide. Three cases were examined: boron, phosphorus, and arsenic, ion implanted and then annealed at 950 °C for 10 min. The redistribution of the dopants was examined by SIMS. Boron does not segregate at the interface although some redistribution of the dopant was observed. After etching the SiO_2 and $TiSi_2$ in HF silicon islands were observed along with a few particles of TiB_2. Phosphorus-doped samples showed some interface segregation and a decrease in the concentration level by evaporation. Arsenic exhibits different behavior: It diffuses from the substrate into the silicide layer. Grain size was also studied in this work. Elongated platelets parallel to the silicon surface were observed with lateral dimensions below 0.5 μm. To increase the number of nuclei formed on the surface a pretreatment by an argon plasma was used. The final layer thickness is about 0.1 μm and the surface is smoother than untreated surfaces.

7.5.4. *Iron Disilicide*

Regolini *et al.* [106, 107] also studied the growth of β-$FeSi_2$ selectively on silicon with respect to SiO_2. β-$FeSi_2$ is a semiconductor with a band gap of about 0.87 eV. The source for iron is a chloride formed in an external reactor in which high-purity iron is exposed to chlorine. Other process details are presented in Table 7.7. After the film growth no postdeposition annealing was carried out. The RBS spectra indicated epitaxial growth and a stoichiometric composition, while high-resolution TEM shows an abrupt interface with the (202) planes spaced at 3.07 Å aligned to the (111) silicon planes spaced 3.13 Å, i.e., a 1.9% mismatch. The film morphology was studied extensively by Berbezier *et al.* [108]. The selective deposition with respect to the SiO_2 was a function of the gas composition. Figure 7.32a shows an SEM of a 190-nm-thick film with 0.5-μm minimum feature size. It was found that the silicon substrate atoms are also consumed by the growing silicide, although silane concentration can be increased to minimize this effect. At some point the selectivity is lost. The electrical properties of these films are of interest for infrared detector devices. The material is *p*-type showing an acceptor activation energy of 50 meV, as illustrated in Fig. 7.32b where the conductivity is also shown.

7.5.5. *Titanium Nitride*

RTCVD of TiN has been investigated by Katz *et al.* [109] as an alternative to reactive sputtering or reactive evaporation onto InP substrates. Using tetrakis dimethyl amido titanium (DMATi) as a precursor provides a low deposition temperature of 300–450 °C and typical growth rates of 1–3 nm/s. Wafers were cleaned in 10:1 H_2O:HF and degreased in warm chloroform, acetone, and methanol before heating to 500 °C in the chamber. Process details are given in Table 7.7.

The deposition rate increased with temperature ($E_a \sim 0.15$ eV), gas flow rate, and a stable process window was identified. Films were characterized by AES, SEM, TEM, and SIMS. Excellent defect-free morphologies were obtained. The TiN phase was inde-

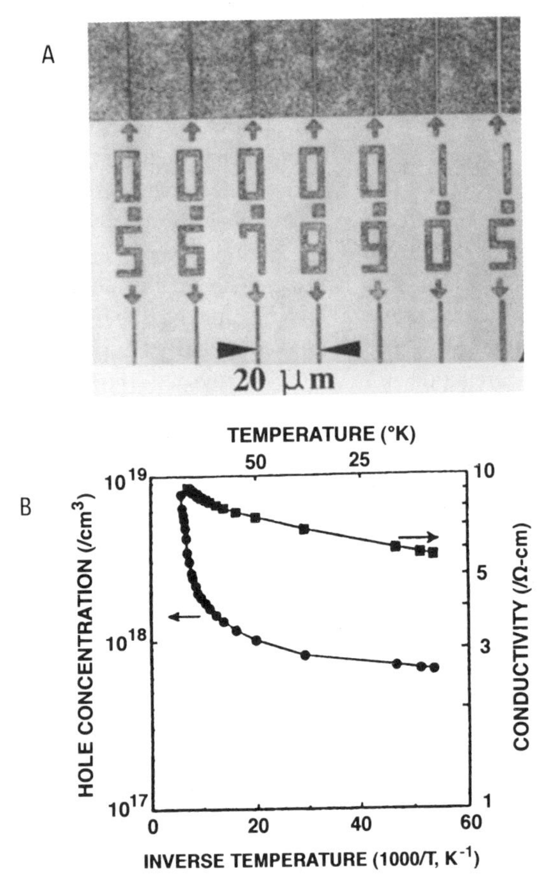

FIGURE 7.32. (a) Optical micrograph of selective deposition of β-$FeSi_2$ on ⟨111⟩ p-type silicon. Positive and negative lines are etched in silicon dioxide (bright) with a minimum linewidth of 0.5 μm. (b) Hole concentration and conductivity as a function of reciprocal temperature for a 1000-Å layer of β-$FeSi_2$ grown at 850 °C by RTCVD in 60 s [107].

pendent of the addition of NH_3 and amount of H_2 dilution. An abrupt TiN/InP interface was observed within two to three atomic layers. However, the addition of small amounts of NH_3 led to a significant improvement in film purity. Figure 7.33 shows the variation in film thickness and resistivity with NH_3 flow rate. Films were compressive with a stress of 0.05–2 GPa. Features were etched using SF_6 reactive ion etching and a photoresist mask. The selectivity with respect to the InP substrate was effectively infinite. On some wafers a p^+ $In_{0.53}Ga_{0.47}As$ layer was grown by MOCVD to facilitate electrical characterization of the film. The resistivity was in the range 400–800 μΩ-cm. Selective growth with respect to SiO_2 was also demonstrated.

7.5.6. Y–Ba–Cu–O Films

High-temperature superconducting films have been grown by a variety of techniques. In particular, MOCVD at 850 °C has been found to produce good films. However, to make

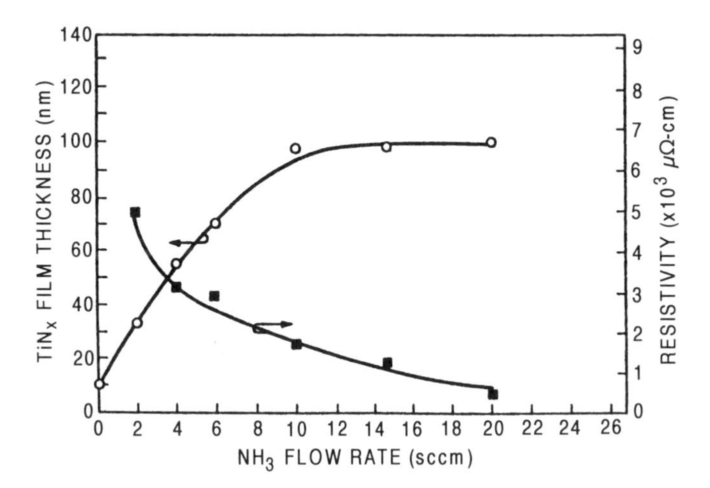

FIGURE 7.33. TiN_x film thickness and resistivity as a function of the NH_3 flow rate (sccm) into the reactive gas mixture. Pressure 1.45 Torr, time of 15 s with Ti source at 40 °C, flow rate of DMATi 19 sccm and hydrogen 750 sccm [109].

practical use of these films in electronic devices and circuits one would like to lower the deposition temperature for process compatibility with integrated circuits. Singh *et al.* [110, 111] deposited films of Y–Ba–Cu–O onto MgO and $SrTiO_3$ using the following β-diketonate metal chelate chemical precursors: 2,2,6,6-tetramethyl-3,5-heptanediono-yttrium [$Y(TMHD)_3$], -barium [$Ba(TMHD)_2$], -copper [$Cu(TMHD)_2$] and nitrous oxide or oxygen as a source for oxygen. Growth takes place in a two-step process, as listed in Table 7.7, followed by an anneal in oxygen at 400 °C and atmospheric pressure for 4 h. After growth, films were characterized by conductivity and X-ray diffraction. The film deposited on MgO showed a transition to superconductivity at 84 K, and predominantly *c*-axis with trace amounts of CuO and Y_2BaCuO_3.

7.6. CONCLUDING REMARKS

Although RTCVD is a relatively new process for the growth of thin films, it has had great success over the past 10 years. Many materials have now been deposited although most of the work has focused on epitaxial growth of silicon, SiGe heterostructures, SiO_xN_y, and β-SiC. The key advantages are control of material stoichiometry, higher deposition rates, lower thermal budget, and lower particulates in a cold wall system.

Excellent-quality lightly doped epitaxial silicon has been fabricated with defect densities < 200 /cm^3, abrupt control of dopant profiles within 30 nm, and minority carrier lifetimes of ~300 μs. Films have been grown in silane, disilane, and dichlorosilane. Selective deposition against silicon dioxide has been effective over a range of process conditions, particularly with dichlorosilane. The silicon structure is a function of temperature; at low temperatures (< 600 °C) amorphous silicon forms, at higher temperatures up to ~ 900 °C polysilicon forms, and at > 950 °C single-crystal silicon forms. The surface condition of the wafer is important in defining the crystal quality; in addition, the presence of any impurities in the gas or chamber can introduce defects. An *in situ* hydrogen anneal is often performed at 1000 °C for up to 1 min. Doping with As or B increases the growth rate. For polysilicon growth, the grain size and surface roughness are strongly influenced by oxygen present in the film. Selective deposition can be obtained under a specific range of growth conditions, for example in the case of silane with the addition of HCl or in the case of dichlorosilane at < 2% concentration in hydrogen. Selective growth has been demonstrated for both single-crystal silicon and polysilicon.

Strained-layer $Si/Ge_xSi_{1-x}/Si$ heterostructures have been fabricated with hyper-abrupt interfaces on the order of 10 Å. Up to 50-layer structures have been reported in addition to relaxed layers, polycrystalline films, pure germanium films, and selective growth. It was found that the addition of As dopant decreased the growth rate and improved the crystal quality. Material has been good enough for fabrication of electronic devices. With the addition of carbon, $Si/Si_{1-x-y}Ge_xC_y/Si$ films have been grown with lower strain. However, the problem of SiC precipitates has to be avoided with careful process control. Active devices have been made in epitaxially grown silicon and SiGe heterostructures, including *n*- and *p*-MOSFETs.

The CVD of dielectric in the form of silicon oxynitride, silicon nitride, and stacked oxide/nitride structures has been studied by several groups. Annealing at high temperature in vacuum was found to improve the dielectric properties. The presence of hydrogen in the Si/SiO_2 interface arises from the source gases and has an important influence on the interface states. Overall results indicate that the electrical properties are comparable to furnace-grown oxides, but grown with a reduced thermal budget.

SiC has a wide band gap making it an attractive choice for advanced heterostructures and high-temperature devices. Void-free films have been produced under optimum conditions when an initial buffer layer is grown. Active devices have been fabricated. Other materials deposited by RTCVD include single-crystal GaAs, GaAlAs, titanium nitride, polycrystalline titanium disilicide, β-$FeSi_2$, and superconducting YBCO films.

The RTCVD processes discussed in this chapter have been developed for materials with electronic applications. RTCVD deposition processes are compatible with the integrated circuit fabrication process. In the future, by combining sequences of several RTP processes in a clustered tool, improved device performance and yield will be realized through process integration.

References

CHAPTER 1

1. M. Von Allmen, Coupling of beam energy to solids, in *Laser and Electron Beam Processing of Materials*, edited by C. W. White and P. S. Peercy (Academic Press, New York, 1980), pp. 6–19.
2. I. B. Khaibullin and L. S. Smirnov, Pulsed annealing of semiconductors. Status report and unsolved problems (review), *Sov. Phys. Semicond.* **19**(4), 353–367 (1985).
3. G. H. Vineyard, General introduction, *Discuss. Faraday Soc.* **31**, 7–23 (1961).
4. J. E. E. Baglin, R. T. Hodgson, J. M. Neri, and R. Fastov, Test for nonthermal transient annealing in silicon, *J. Appl. Phys. Lett.* **43**(3), 299–301 (1983).
5. *Handbook of Thin Film Technology*, edited by L. I. Maissel and R. Glang (McGraw–Hill, New York, 1970).
6. P. Sigmund, Theory of sputtering yield of amorphous and polycrystalline targets, *Phys. Rev.* **184**(2), 386–416 (1969).
7. G. Carter and D. G. Armour, The interaction of low energy ion beams with surfaces, *Thin Solid Films* **80**(1), 13–29 (1981).
8. C. Steinbruchel, A simple formula for low-energy sputtering yields, *Appl. Phys. A* **36**(1), 37–42 (1985).
9. S. Kakac and Y. Yener, *Heat Conduction* (Hemisphere, Washington, DC, 1985).
10. *Thin Films—Interdiffusion and Reactions*, edited by J. M. Poate, K. N. Tu, and J. W. Mayer (Wiley, New York, 1978).
11. V. E. Borisenko, V. V. Gribkovskii, V. A. Labunov, and S. G. Yudin, Pulsed heating of semiconductors, *Phys. Status Solidi A* **86**(2), 573–583 (1984).
12. C. Hill, The contribution of beam processing to present and future integrated circuit technologies, in *Laser–Solid Interactions and Transient Thermal Processing of Materials*, edited by J. Narayan, W. L. Brown, and R. A. Lemons (North-Holland, Amsterdam, 1983), pp. 381–392.
13. T. O. Sedgwick, Short time annealing, *J. Electrochem. Soc.* **130**(2), 484–493 (1983).
14. V. E. Borisenko, S. N. Kornilov, V. A. Labunov, and Y. V. Kucherenko, Oborudovanie dlya impul' snoii termoobrabotki materialov poluprovodnikovoii elektroniki intensivnim nekogerentnim svetom, *Zarubejnaya Elektronnaya Tekhnika* **6**, 45–65 (1985).
15. V. A. Labunov, V. E. Borisenko, and V. V. Gribkovskii, Impulrsnaya termoobrabotka materialov poluprovodnikovoii elektroniki nekogerentnim svetom, *Zarubejnaya Elektronnaya Tekhnika* **1**, 3–57 (1983).
16. R. A. Powell, T. O. Yep, and R. T. Fulks, Activation of arsenic implanted silicon using an incoherent light source, *Appl. Phys. Lett.* **39**(2), 150–152 (1981).
17. D. B. Zvorikin and Y. I. Prokhorov, *Primenenie Luchistogo IK-Nagreva v Elektronnoii Promishlennosti* (Energiya, Moscow, 1980).
18. A. F. Burenkov, F. F. Komarov, M. A. Khumahov, and M. M. Temkin, *Tables of Ion Implantation Spatial Distribution* (Gordon & Breach, New York, 1986).

19. E. Rimini, Energy deposition and heat flow for pulsed laser, electron and ion beam irradiation, in *Spring College on Crystalline Semiconducting Materials and Devices*, edited by E. Rimini (International Center for Theoretical Physics, Trieste, 1984), pp. 15–49.
20. A. S. Okhotin and A. S. Pushkarskii, *Teplofisicheskie Svoistva Poluprovodnikov* (Atomisdat, Moscow, 1972).
21. P. G. Merli, Computer simulation of electron beam annealing of silicon, *Optic* **56**(3), 205–222 (1980).
22. S. S. Strelchenko and V. V. Lebedev, *Soedineniya* A^3B^5, *Sprayochnik* (Metallurgiya, Moscow, 1984).
23. J. R. Meyer, M. R. Kruer, and F. T. Bartoli, Optical heating in semiconductors: Laser damage in Ga, Si, InSb and GaAs, *J. Appl. Phys.* **51**(10), 5513–5522 (1980).
24. P. Y. Lin and J. C. Maan, Optical properties of InSb between 300 and 700 K. I. Temperature dependence of the energy gap, *Phys. Rev. B* **47**(24), 16274–16278 (1993).
25. J. M. Cole, P. Humphreys, and L. G. Earwaker, A melting model for pulsed laser heating of silicon, *Vacuum* **34**(10/11), 871–874 (1984).
26. E. P. Donovan, F. Spaeren, D. Turnbull, J. M. Poate, and D. C. Jacobson, Heat of crystallization and melting point of amorphous silicon, *Appl. Phys. Lett.* **42**(8), 698–700 (1983).
27. Y. I. Nissim, CW laser annealing and CW laser assisted diffusion in gallium arsenide, Ph.D. thesis, Stanford University (1981).
28. G. G. Bentini and L. Correra, Analysis of thermal stresses induced in silicon during xenon arc lamp flash annealing, *J. Appl. Phys.* **54**(4), 2057–2062 (1983).
29. A. Lietoila, A. B. Gold, and J. F. Gibbons, Temperature rise induced in Si by continuous xenon arc lamp radiation, *J. Appl. Phys.* **53**(2), 1169–1172 (1982).
30. M. O. Lampert, J. M. Koebel, and P. Siffert, Temperature dependence of the reflectance of solid and liquid silicon, *J. Appl. Phys.* **52**(8), 4975–4976 (1981).
31. G. E. Jellison, D. H. Lowndes, D. N. Mashburn, and R. F. Wood, Time-resolved reflectivity measurements on silicon and germanium using a pulsed excimer KrF laser heating beam, *Phys. Rev. B* **34**(4), 2407–2415 (1986).
32. G. E. Jellison and D. H. Lowndes, Time-resolved ellipsometry measurements of the optical properties of silicon during pulsed excimer laser irradiation, *Appl. Phys. Lett.* **47**(7), 718–721 (1985).
33. Z. Y. Gotra, S. A. Oseredko, and Y. V. Bobitskii, Impulsnii lazernii otjig ionno-implantirovannih poluprovodnikovih materialov, *Zarubejnaya Elektronnaya Tekhnika* **6**, 3–77 (1983).
34. G. E. Jellison and F. A. Modine, Optical absorption of silicon between 1.6 and 4.7 eV at elevated temperatures, *Appl. Phys. Lett.* **41**(2), 180–182 (1982).
35. I. W. Boyd, T. D. Binnie, J. I. B. Wilson, and M. J. Colles, Absorption of infrared radiation in silicon, *J. Appl. Phys.* **55**(8), 3061–3063 (1984).
36. J. C. Sturm, P. V. Schwartz, and P. M. Garone, Silicon temperature measurement by infrared transmission for rapid thermal processing applications, *Appl. Phys. Lett.* **56**(10), 961–963 (1990).
37. H. S. Carslaw and J. C. Jaeger, *Conduction of Heat in Solids* (Oxford University Press, London, 1975).
38. W. E. Boyce and R. C. DiPrima, *Elementary Differential Equations* (Wiley, New York, 1965).
39. A. R. Mitchell, *Computational Methods in Partial Differential Equations* (Wiley, New York, 1969).
40. F. Ferrieu and G. Auvert, Temperature evolution in silicon induced by a scanned CW laser, pulsed laser or an electron beam, *J. Appl. Phys.* **54**(5), 2646–2649 (1983).
41. D. L. Kwong and D. M. Kim, Pulsed laser heating of silicon: The coupling of optical absorption and thermal conduction during irradiation, *J. Appl. Phys.* **54**(1), 366–373 (1983).
42. G. Edelin, An analytical solution to Fick's equations in the case of the dissociative diffusion mechanism, *Phys. Status Solidi B* **98**(2), 699–708 (1980).
43. V. A. Labunov, N. I. Sterjanov, V. E. Borisenko, V. A. Pilipenko, L. D. Buiko, and V. N. Topchiii, Prostranstvennoe raspredelenie temperaturi v plastine kremniya pri impulsnom otjige izlucheniem ksenonovih lamp, *Elektronnaya Tekhnika, Ser. 6, Materiali* **8**, 17–21 (1982).
44. J.-T. Lue and C.-C. Chao, The regrowth and impurity diffusion processes in the arc annealing of ion-implanted silicon, *J. Appl. Phys.* **53**(2), 984–987 (1982).
45. G. G. Bentini, Solid state annealing of ion implanted silicon by incoherent light pulses and multiscan electron beam, *Radiat. Eff.* **63**(2), 125–131 (1982).
46. G. K. Celler, M. Robinson, and L. E. Trimble, Spatial melt instabilities in radiatively melted crystalline silicon, *Appl. Phys. Lett.* **57**(9), 868–870 (1983).

47. T. E. Seidel, D. J. Lischner, C. S. Pai, and S. S. Lau, Temperature transient in heavily doped and undoped silicon using rapid thermal annealing, *J. Appl. Phys.* **57**(4), 1317–1321 (1985).
48. G. F. Cembali, P. G. Merli, and F. Zignani, Self-annealing of ion-implanted silicon: First experimental results, *Appl. Phys. Lett.* **38**(10), 808–810 (1981).
49. R. A. McMahon, H. Ahmed, R. M. Dobson, and J. D. Spreight, Characterisation of multiple-scan electron beam annealing method, *Electron. Lett.* **16**(8), 295–297 (1980).
50. M. Lax, Temperature rise induced by a laser beam, *J. Appl. Phys.* **48**(9), 3919–3924 (1977).
51. G. Bentini, L. Correra, and C. Donolato, Defects introduced in silicon wafers during rapid isothermal annealing: Thermoelastic and thermoplastic effects, *J. Appl. Phys.* **56**(10), 2922–2929 (1984).
52. R. Henda, E. Scheid, and D. Bielle-Daspet, A 3-dimensional temperature uniformity model for a thermal processing furnace, in *Rapid Thermal and Integrated Processing III*, edited by J. J. Wortman, J. C. Gelpey, M. L. Green, S. R. J. Brueck, and F. Roozeboom (MRS, Pittsburgh, 1994), pp. 419–424.
53. J. Blake, J. C. Gelpey, J. F. Moquin, J. Schlueter, and R. Capodilupo, Slip free rapid thermal processing, in *Rapid Thermal Processing of Electronic Materials*, edited by S. R. Wilson, R. Powel, and D. E. Davies (MRS, Pittsburgh, 1987), pp. 265–272.
54. K. Hehl and W. Wesch, Calculation of optical reflection and transmission coefficients of a multi-layer system, *Phys. Status Solidi A* **58**(1), 181–188 (1980).
55. J. P. Colinge and F. Van de Wiele, Laser light absorption in multilayers, *J. Appl. Phys.* **52**(7), 4769–4771 (1981).
56. P. K. Bhattacharya, S. K. Gupta, and A. G. Wagh, Temperature profiles in a two-component heterogeneous system heated with a CW laser, *Radiat. Eff.* **63**(5), 197–204 (1982).
57. M. A. Korolev, S. G. Tatevosyan, and V. A. Khaustov, Analiz temperaturnogo raspredeleniya v mnogosloiinih strukturah s lokalnim istochnikom tepla, *Elektronnaya Tekhnika, Ser. 3, Mikroelektronika* **2**, 58–60 (1987).
58. E. M. Sparrow and R. D. Cess, *Radiation Heat Transfer* (Cole Publishing, Belmont, CA, 1970).
59. P. Baeri, G. Foti, and J. M. Poate, Phase transitions in amorphous Si produced by rapid heating, *J. Appl. Phys.* **50**(5), 3783–3792 (1979).
60. K. Kubota, C. E. Hunt, and J. Frey, Thermal profiles during recrystallization of silicon on insulator with scanning incoherent light line sources, *Appl. Phys. Lett.* **46**(12), 1153–1155 (1985).
61. V. E. Borisenko and S. G. Yudin, Temperaturnie polya v plenochnih mikroelektronnih strukturah pri luchistom nagreve, *Ing. Fiz. Jurn.* **60**(5), 794–800 (1991).
62. *Tablitsi Fizicheskih-Velichin. Spravochnik*, pod red. I. K. Kikoina (Atomizdat, Moscow, 1976).
63. J. Levoska, T. T. Rantala, and J. Lenkkeri, Numerical simulation of temperature distributions in layered structures during laser processing, *Appl. Surf. Sci.* **36**(1), 12–22 (1989).
64. A. Lietoila, Laser processing of high dose ion implanted Si: The solid phase regime, Ph.D. thesis, Stanford University (1981).
65. J. A. Thornton and D. W. Hoffman, Internal stress in titanium, nickel, molybdenum, and tantalum films deposited by cylindrical magnetron sputtering, *J. Vac. Sci. Technol.* **14**(1), 164–168 (1977).
66. K. N. Tu, K. Y. Ahn, and S. R. Herd, Silicide films for archival optical storage, *Appl. Phys. Lett.* **39**(11), 927–929 (1981).
67. D. V. Hoffman and J. A. Thornton, Effects of substrate orientation and rotation on internal stresses in sputtered metal films, *J. Vac. Sci. Technol.* **16**(2), 134–137 (1979).
68. J. R. Jimenez, Z.-C. Wu, L. J. Schowalter, B. D. Hunt, R. W. Fathaner, P. J. Grunthaner, and T. L. Lin, Optical properties of epitaxial $CoSi_2$ and $NiSi_2$ films on silicon, *J. Appl. Phys.* **66**(6), 2738–2741 (1989).
69. A. Borghesi, L. Nosenzo, A. Piaggi, G. Guizzetti, C. Nobili, and G. Ottaviani, Optical properties of tantalum disilicide thin films, *Phys. Rev. B* **38**(15), 10937–10939 (1988).
70. R. G. Long, M. C. Bost, and J. E. Mahan, Optical and electrical properties of semiconducting rhenium disilicide thin films, *Thin Solid Films* **162**(1), 29–40 (1988).
71. A. Borghesi, G. Guizzetti, L. Nosenzo, and A. Stella, Optical properties of PtSi and Pt_2Si, *Thin Solid Films* **140**(1), 95–98 (1986).
72. T.-J. Shieh and R. L. Carter, RAPS—A rapid thermal processor simulation program, *IEEE Trans. Electron Devices* **36**(1), 19–24 (1989).

73. N. Akiyama, Y. Inoue, and T. Suzuki, Critical radial temperature gradient inducing slip dislocations in silicon epitaxy using dual heating of the two surfaces of a wafer, *Jpn. J. Appl. Phys.* **25**(11), 1619–1622 (1986).
74. N. C. Schoen, Thermoelastic stress analysis of pulsed electron beam recrystallization of ion-implanted silicon, *J. Appl. Phys.* **51**(9), 4747–4751 (1980).
75. L. Correra and G. G. Bentini, Thermal profiles and thermal stresses introduced on silicon during scanning line shaped beam annealing, *J. Appl. Phys.* **54**(8), 4330–4337 (1983).
76. B. A. Boley and J. H. Weiner, *Theory of Thermal Stresses* (Wiley, New York, 1960).
77. A. Kelly and G. W. Groves, *Crystallography and Crystal Defects* (Longman, London, 1970).
78. R. Von Labusch, Berechnung des peirlspotentials im diamantgitter, *J. Physica Status Solidi* **10**(2), 645–657 (1965).
79. A. R. Chaudhuri, J. R. Patel, and L. G. Rubin, Velocities and densities of dislocations in germanium and other semiconductor crystals, *J. Appl. Phys.* **33**(9), 2736–2746 (1962).
80. H. Suzuki, A note on the Peierls force, *J. Phys. Soc. Jpn.* **18**(Suppl. 1), 182–186 (1963).
81. T. Suzuki and H. Kojima, Dislocation motion in silicon crystals as measured by the Laue x-ray technique, *Acta Metall.* **14**(8), 913–924 (1966).
82. K. Sumino, I. Yonenaga, and A. Yusa, Mechanical strength of oxygen doped float zone silicon crystals, *Jpn. J. Appl. Phys.* **19**(12), L763–L766 (1980).
83. W. Schroter, Yield point and dislocation mobility in silicon and germanium, *J. Appl. Phys.* **54**(4), 1816–1820 (1983).
84. B. Reppich, P. Haasen, and B. Ilschner, Kriechen von siliziumeinkristallen, *Acta Metall.* **12**(11), 1283–1288 (1964).
85. H. Siethoff and P. Haasen, Yield point and dislocation mobility in silicon, in *Lattice Defects in Semiconductors*, edited by R. R. Hasiguti (University of Tokyo Press, Tokyo, 1968).
86. J. R. Patel and A. R. Chaudhuri, Macroscopic plastic properties of dislocation-free germanium and other semiconductor crystals: I. Yield behavior, *J. Appl. Phys.* **34**(9), 2788–2799 (1963).
87. I. Yonenaga and K. Sumino, Dislocation dynamics in the plastic deformation of silicon crystals, *Phys. Status Solidi A* **50**(2), 685–693 (1978).
88. J. Doerschel, F.-G. Kirscht, and R. Baehr, Plastische verformung von siliziumeinkristallen unterschiedlicher ausgangsversetzungsdichte im streckgrenzenbereich, *Krist. Tech.* **12**(11), 1191–2000 (1977).
89. V. E. Borisenko, A. M. Dorofeev, and K. D. Yashin, Osobennosti obrazovaniya linii skolJeniya v plastinah monokristallicheskogo kremniya pri impulsnoii termoobrabotke nekogerentnim svetom, *Elektronnaya Tekhnika, Ser. 6, Materiali* **7**, 3–5 (1985).
90. B.-J. Cho and C.-K. Kim, Elimination of slips on silicon wafer edge in rapid thermal process by using a ring oxide, *J. Appl. Phys.* **67**(12), 7583–7586 (1990).
91. G. L. Young and K. A. McDonald, Effect of radiation shield angle on temperature and stress profiles during rapid thermal annealing, *IEEE Trans. Semicond. Manuf.* **3**(4), 176–182 (1990).
92. R. Kakoshke, E. Bubmann, and H. Foll, Modeling of wafer heating during rapid thermal processing, *Appl. Phys. A* **50**(2), 141–150 (1990).
93. J. F. Gongste, T. G. M. Oosterlaken, G. C. J. Bart, G. C. A. M. Janssen, and S. Radelaar, Deformation of Si (100) wafers during rapid thermal annealing, *J. Appl. Phys.* **75**(6), 2830–2836 (1994).
94. Y. M. Cho, A. Paulraj, T. Kailath, and G. H. Xu, A contribution to optimal lamp design in rapid thermal processing, *IEEE Trans. Semicond. Manuf.* **7**(1), 34–41 (1994).
95. N. I. Sterjanov, V. A. Pilipenko, V. A. Gorushko, A. F. Volkodatov, and E. F. Lobanovich, Issledovanie strukturnogo sovershenstva ionno-legirovannih sloev kremniya posle impulsnvgo opticheskogo otjiga, *Elektronnaya Tekhnika, Ser. 6, Materiali* **4**, 49–53 (1983).
96. Y. P. Sinkov, Modelirovanie i optimizatsiya na EVM ustroistva impulsnoi termoobrabotki poluprovodnikovih plastin izlucheniem lineinih galogennih lamp, *Elektronnaya Tekhnika, Ser. 1, Elektronika SV* **3**, 36–41 (1984).
97. R. P. S. Thakur and R. Singh, Thermal budget consideration in rapid isothermal processing, *Appl. Phys. Lett.* **64**(3), 327–329 (1994).

CHAPTER 2

1. L. C. Kimerling and J. L. Benton, Defects in laser-processed semiconductors, in *Laser and Electron Beam Processing of Materials,* edited by C. W. White and P. S. Peercy (Academic Press, New York, 1980), pp. 385–396.
2. A. V. Dvurechenskii, G. A. Kachurin, E. V. Nidaev, and L. S. Smirnov, *Impulsnii Otjig Poluprovodnicovih Materialov* (Nauka, Moscow, 1982).
3. A. Mesli, J. C. Muller, and P. Siffert, Deep levels subsisting in ion implanted silicon after various transient thermal annealing procedures, *Appl. Phys. A* **31**(3), 147–152 (1983).
4. B. Hartiti, A. Slaoui, J. C. Muller, and P. Siffert, Thermal annealing of eximer-laser-induced defects in virgin silicon, *Mater. Sci. Eng. B* **4**(1–4), 257–260 (1989).
5. W. O. Adekoya, J. C. Muller, and P. Siffert, Annealing kinetics during rapid and classical thermal processing of a laser induced defect in n-type silicon, *Appl. Phys. Lett.* **49**(21), 1429–1431 (1986).
6. W. O. Adekoya, J. C. Muller, and P. Siffert, Rapid thermal annealing of electrically-active defects in virgin and implanted silicon, *Appl. Phys. A* **42**(3), 227–232 (1987).
7. G. Pensl, M. Schulz, P. Stolz, N. M. Johnson, J. F. Gibbons, and J. Hoyt, Electronic defects in silicon after transient isothermal annealing, in *Energy Beam–Solid Interactions and Transient Thermal Processing,* edited by J. C. C. Fan and N. M. Johnson (North-Holland, Amsterdam, 1984), pp. 347–358.
8. N.-E. Chabane-Sari, L. Thibaud, S. Kaddour, M. Berenguer, and D. Barbier, Hole trap level generation in silicon during rapid thermal annealing: Influence of substrate type and process conditions, *J. Appl. Phys.* **71**(7), 3320–3324 (1992).
9. R. Langfeld, H. Baumann, and K. Bethge, Time-resolved defect production and annealing during electron-beam processing of silicon, *Appl. Phys. A* **33**(4), 251–254 (1984).
10. J. T. Borenstein, J. T. Jones, J. W. Corbett, G. S. Oehrlein, and R. L. Kleinhenz, Quenched-in defects in flashlamp-annealed silicon, *Appl. Phys. Lett.* **49**(4), 199–200 (1986).
11. D. Barbier, M. Remram, J. F. Joly, and A. Laugier, Defect-state generation in Czochralski-grown (100) silicon rapidly annealed with incoherent light, *J. Appl. Phys.* **61**(1), 156–160 (1987).
12. E. Susi, A. Poggi, and M. Madrigali, Electrical properties of rapid thermal annealing induced defect in silicon, *J. Electrochem. Soc.* **142**(6), 2081–2085 (1995).
13. W. O. Adekoya, J. C. Muller, and P. Siffert, Electrical effects of surface and deep states induced in n-type silicon by rapid thermal processing, *Appl. Phys. Lett.* **50**(18), 1240–1242 (1987).
14. A. Mesli, E. Courcelle, T. Zundel, and P. Siffert, Process-induced and gold acceptor defects in silicon, *Phys. Rev. B* **36**(15), 8049–8062 (1987).
15. Y. A. Kapustin, B. M. Kolokolnikov, A. A. Sveshnikov, and V. P. Zlobin, Electrical properties of defects formed as a result of pulsed photon annealing of silicon, *Sov. Phys. Semicond.* **22**(9), 1078–1079 (1988).
16. D. Mathiot, Cobalt related levels in P and (P+B) doped n-type silicon: Possible observation of the (CoB) pair, *J. Appl. Phys.* **65**(4), 1554–1558 (1989).
17. B. Hartiti, V. T. Quat, W. Eichhammer, J. C. Muller, and P. Siffert, Gettering of gold by rapid thermal processing, *Appl. Phys. Lett.* **55**(9), 873–875 (1989).
18. B. Hartiti, W. Eichhammer, J. C. Muller, and P. Siffert, Activation and gettering of intrinsic metallic impurities during rapid thermal processing, *Mater. Sci. Eng. B* **4**(1–4), 129–132 (1989).
19. B. Hartiti, J. C. Muller, and P. Siffert, Defect generation and gettering during rapid thermal processing, *IEEE Trans. Electron Devices* **39**(1), 96–104 (1992).
20. W. Eichhammer, V. T. Quat, and P. Siffert, On the origin of rapid thermal process induced recombination centers in silicon, *J. Appl. Phys.* **66**(8), 3857–3865 (1989).
21. W. Eichhammer, M. Hage-Ali, R. Stuck, and P. Siffert, Rapid thermal process-induced recombination centers in ion implanted silicon, *Appl. Phys. A* **50**(4), 405–410 (1990).
22. Y. Tokuda, N. Kobayashi, A. Usami, Y. Inoue, and M. Imura, Spatial distributions of induced traps in silicon by rapid thermal processing, *J. Cryst. Growth* **103**(1–4), 297–302 (1990).
23. K. G. Barraclough, P. J. Asby, J. G. Wilkes, and L. T. Canham, *Oxygen, Carbon and Nitrogen in Silicon, in Properties of Silicon* (INSPEC, London, 1988), pp. 279–324.
24. *Handbook of Semiconductor Wafer Cleaning Technology, Science, Technology, and Applications,* edited by W. Kern (Noyes Publications, Park Ridge, NJ, 1993).

25. L. Mouche, F. Tardif, and J. Derrien, Mechanisms of metallic impurity deposition on silicon substrates dipped in cleaning solution, *J. Electrochem. Soc.* **142**(7), 2395–2401 (1995).
26. A. Chantre, M. Kechouane, and D. Bois, Quenched-in defects in cw laser irradiated virgin silicon, in *Defects in Semiconductors II* (North-Holland, Amsterdam, 1983), pp. 547–551.
27. E. R. Weber, Transition metals in silicon, *Appl. Phys. A* **30**(1), 1–22 (1983).
28. P. M. Mooney, L. J. Cheng, M. Suli, J. D. Gerson, and J. W. Corbett, Defect energy levels in boron-doped silicon irradiated with 1-MeV electrons, *Phys. Rev. B* **15**(8), 3836–3843 (1977).
29. Y. H. Lee, K. L. Wang, A. Jaworowski, P. M. Mooney, L. J. Cheng, and J. W. Corbett, A transient capacitance study of radiation-induced defects in aluminum-doped silicon, *Phys. Status Solidi A* **57**(2), 697–704 (1980).
30. E. R. Weber and N. Wiehl, Transition metal impurities in silicon, in *Defects in Semiconductors II* (North-Holland, Amsterdam, 1983), pp. 19–32.
31. *Semiconductor Silicon*, edited by H. R. Huff, R. J. Kriegler, and Y. Takeishi (The Electrochemical Society, Pennington, 1981), p. 331.
32. G. D. Watkins, The interaction of irradiation-produced defects with impurities and other defects in semiconductors. EPR studies in silicon, in *Radiation Effects in Semiconductor Components* (Journess D'Electronique, Toulouse, 1967), pp. A1–A9.
33. L. C. Kimerling, Defect states in electron bombarded silicon: Capacitance transient analysis, in *Radiation Effects in Semiconductors*, Conf. Ser. No. 31 (Institute of Physics, London, 1977), pp. 221–230.
34. H. Kitagawa, H. Nakashima, and K. Hashimoto, Energy levels and solubility of electrically active cobalt in silicon studied by combined Hall and DLTS measurements, *Jpn. J. Appl. Phys.* **24**(3), 373–374 (1985).
35. H. Nakashima, Y. Tsumori, T. Miyagawa, and K. Hashimoto, Deep impurity level of cobalt in silicon, *Jpn. J. Appl. Phys.* **29**(8), 1395–1398 (1990).
36. K. Kakishita, K. Kawakami, S. Suzuki, E. Ohta, and M. Sakata, Iron-related deep levels in n-type silicon, *J. Appl. Phys.* **65**(10), 3923–3927 (1989).
37. C. T. Sah and C. T. Wang, Experiments on the origin process-induced recombination centers in silicon, *J. Appl. Phys.* **46**(4), 1767–1776 (1975).
38. V. T. Quat, W. Eichhammer, and P. Siffert, Electron diffusion length in rapid thermal processed p-type silicon, *Appl. Phys. Lett.* **53**(20), 1928–1930 (1988).
39. P. Omling, E. R. Weber, L. Montelious, H. M. Alexander, and J. Michel, Electrical properties of dislocations and point defects in plastically deformed silicon, *Phys. Rev. B* **32**(10), 6571–6581 (1985).
40. A. R. Peaker and E. C. Sidebotham, Dislocation-related deep levels in silicon, in *Properties of Silicon* (INSPEC, London, 1988), pp. 221–224.
41. V. T. Quat, W. Eichhammer, and P. Siffert, Inhomogeneous defect activation by rapid thermal processes in silicon, *Appl. Phys. Lett.* **54**(13), 1235–1237 (1989).
42. C. J. Polard and J. D. Speight, Scanning electron beam annealing of oxygen donors in Czochralski silicon, in *Laser–Solid Interactions and Transient Thermal Processing of Materials*, edited by J. Narayan, W. L. Brown, and R. A. Lemons (North-Holland, Amsterdam, 1983), pp. 413–418.
43. Y. Tokuda, N. Kobayashi, A. Usami, Y. Inoue, and M. Imura, Thermal donor annihilation and defect production in n-type silicon by rapid thermal annealing, *J. Appl. Phys.* **66**(8), 3651–3655 (1989).
44. A. G. Ital' yantsev, V. N. Mordkovich, and E. M. Temper, Role of athermal processes in pulsed annealing of ion-implantation-doped Si films, *Sov. Phys. Semicond.* **18**(5), 577–578 (1984).
45. F. P. Korshunov, N. A. Sobolev, V. A. Solodukha, and V. A. Varhapovich, Ispolzovanie impulsnogo otjiga dlya uluchsheniya parametrov radiatsionno obrabotannogo kremniya i struktur na ego osnove, *Izv. Akad. Nauk BSSR Ser. Fiz. Mater. Nauk* **4**, 66–71 (1987).
46. F. P. Korshunov, N. A. Sobolev, V. A. Sheraukhov, T. N. Gaponyuk, V. K. Shesholko, and V. V. Gribkovskii, Pulsed annealing of nuclear-transmutation-doped silicon, *Sov. Phys. Semicond.* **21**(11), 1192–1195 (1987).
47. L. Bischoff and R. Kogler, Rapid thermal annealing of neutron transmutation doped silicon, *Phys. Status Solidi A* **113**(2), K185–K188 (1989).
48. J. Bourgoin and M. Laannoo, *Point Defects in Semiconductors II. Experimental Aspects* (Springer-Verlag, Berlin, 1983).

49. R. Singh, J. Mavoori, R. P. S. Thakur, and S. Narayanan, Importance of the photoeffects in rapid isothermal processing, in *Rapid Thermal and Integrated Processing III*, edited by J. J. Wortman, J. C. Gelpey, M. L. Green, S. R. J. Brueck, and F. Roozeboom (MRS, Pittsburgh, 1994), pp. 437–443.
50. R. L. Cohen, J. S. Williams, L. C. Feldman, and K. W. West, Thermally assisted flash annealing of silicon and germanium, *Appl. Phys. Lett.* **33**(8), 751–755 (1978).
51. R. Klabes, J. Matthai, M. Voelskow, G. A. Kachurin, E. V. Nidaev, and H. Bartsch, Flash lamp annealing of arsenic implanted silicon, *Phys. Status Solidi A* **66**(1), 261–266 (1981).
52. J. Narayan and R. T. Young, Flame annealing of arsenic and boron implanted silicon, *Appl. Phys. Lett.* **42**(5), 466–468 (1983).
53. J. S. Williams, W. L. Brown, H. J. Leamy, J. M. Poate, J. W. Rodgers, D. Rousseau, G. A. Rozgonyi, J. A. Shelnutt, and T. T. Sheng, Solid-phase epitaxy of implanted silicon by cw Ar ion laser irradiation, *Appl. Phys. Lett.* **33**(6), 542–544 (1978).
54. G. Gotz, H.-D. Geller, and M. Wagner, Pulse laser induced high-temperature solid-phase annealing of arsenic implanted silicon, *Phys. Status Solidi A* **73**(1), 145–151 (1982).
55. D. Kirilov, R. A. Powell, and D. T. Hodul, Raman scattering study of rapid thermal annealing of As implanted Si, *J. Appl. Phys.* **58**(6), 2174–2179 (1985).
56. M. Miyao, A. Polman, W. Sinke, and F. W. Saris, Electron irradiation-activated low-temperature annealing of phosphorus-implanted silicon, *Appl. Phys. Lett.* **48**(17), 1132–1134 (1986).
57. L. Csepregi, J. W. Mayer, and T. W. Sigmon, Regrowth behavior of ion-implanted amorphous layers on ⟨100⟩ silicon, *Appl. Phys. Lett.* **29**(2), 92–93 (1976).
58. P. Germain, K. Zellama, S. Squelard, J. C. Bourgoin, and A. Gheorghin, Crystallization in amorphous germanium, *J. Appl. Phys.* **50**(11), 6986–6994 (1979).
59. K. Zellama, P. Germain, S. Squelard, J. C. Bourgoin, and P. A. Thomas, Crystallization in amorphous silicon, *J. Appl. Phys.* **50**(11), 6995–7000 (1979).
60. J. Narayan, Interface structures during solid-phase-epitaxial growth in ion-implanted semiconductors and crystallization model, *J. Appl. Phys.* **53**(12), 8607–8614 (1982).
61. J. C. Bourgoin and R. Asomoza, Solid phase growth of silicon and germanium, *J. Cryst. Growth* **69**(2/3), 489–498 (1984).
62. H. S. Chen and D. Turnbull, Specific heat and heat of crystallization of amorphous germanium, *J. Appl. Phys.* **40**(10), 4914–4915 (1969).
63. F. Spaepen, On the configurational entropy of amorphous Si and Ge, *Philos. Mag.* **30**(2), 417–422 (1974).
64. J. C. C. Fan and C. H. Anderson, Transition temperatures and heats of crystallization of amorphous Ge, Si and $Ge_{1-x}Si_x$ alloys determined by scanning calorimetry, *J. Appl. Phys.* **52**(6), 4003–4006 (1981).
65. J. M. Poate, Some thermodynamic properties of amorphous Si, *Nucl. Instrum. Methods Phys. Res. B* **209/210**(1), 211–217 (1983).
66. E. P. Donovan, F. Spaepen, D. Turnbull, J. M. Poate, and D. C. Jacobson, Heat of crystallization and melting point of amorphous silicon, *Appl. Phys. Lett.* **42**(8), 698–700 (1983).
67. L. Csepregi, E. F. Kennedy, J. W. Mayer, and T. W. Sigmon, Substrate orientation of the epitaxial regrowth rate from Si-implanted amorphous Si, *J. Appl. Phys.* **49**(7), 3906–3911 (1978).
68. T. W. Sigmon, Recrystallization of ion implanted amorphous and heavily damaged semiconductors, *IEEE Trans. Nucl. Sci.* **28**(2), 1767–1770 (1981).
69. A. Lietoila, A. Wakita, T. W. Sigmon, and J. F. Gibbons, Epitaxial regrowth of intrinsic, P-doped and compensated (P+B-doped) amorphous Si, *J. Appl. Phys.* **56**(3), 4399–4405 (1982).
70. G. L. Olson, S. A. Kokorowski, J. A. Roth, and L. D. Hess, Laser-induced solid phase crystallization in amorphous silicon films, in *Laser–Solid Interactions and Transient Thermal Processing of Materials*, edited by J. Narayan, W. L. Brown, and R. A. Lemons (North-Holland, Amsterdam, 1983), pp. 141–153.
71. C. Licope and Y. I. Nissim, Impurity induced enhancement of the growth rate of amorphized silicon during solid-phase epitaxy: A free-carrier effect, *J. Appl. Phys.* **59**(2), 432–438 (1986).
72. L. Csepregi, E. F. Kennedy, T. J. Gallagher, J. W. Mayer, and T. W. Sigmon, Reordering of amorphous layers of Si implanted with ^{33}P, ^{75}As, and ^{11}B ions, *J. Appl. Phys.* **48**(10), 4234–4240 (1977).
73. W. O. Adekoya, M. Hage-Ali, J. C. Muller, and P. Siffert, Dose effects during solid phase epitaxial regrowth of boron-implanted, germanium-amorphized silicon induced by rapid thermal annealing, *Appl. Phys. Lett.* **53**(6), 511–513 (1988).

74. I. Suni, G. Goltz, M. G. Grimaldi, and M. A. Nicolet, Compensating impurity effect on epitaxial regrowth rate of amorphized Si, *Appl. Phys. Lett.* **40**(3), 269–271 (1982).
75. W. O. Adekoya, M. Hage-Ali, J. C. Muller, and P. Siffert, Direct evidence of recrystallization rate enhancement during rapid thermal annealing of phosphorus amorphized silicon layers, *Appl. Phys. Lett.* **50**(24), 1736–1738 (1987).
76. S. A. Kokorowski, G. L. Olson, and L. D. Hess, Kinetics of laser induced solid phase epitaxy in amorphous silicon films, *J. Appl. Phys.* **53**(2), 921–926 (1982).
77. E. F. Kennedy, L. Csepregi, J. W. Mayer, and T. W. Sigmon, Influence of ^{16}O, ^{12}C, ^{14}N, and noble gases on the crystallization of amorphous Si layers, *J. Appl. Phys.* **48**(10), 4241–4246 (1977).
78. S. U. Campisano, Impurity and concentration dependence of growth rate during solid epitaxy of implanted Si, *Appl. Phys. A* **29**(3), 147–149 (1982).
79. J. S. Williams and R. G. Elliman, Role of electronic processes in epitaxial recrystallization of amorphous semiconductors, *Phys. Rev. Lett.* **51**(12), 1069–1072 (1983).
80. F. F. Komarov, V. S. Solovyev, V. S. Tishkov, and S. Y. Shiryaev, Thermal recrystallization of silicon amorphous layers after argon, oxygen and nitrogen ion implantation, *Radiat. Eff.* **69**(3/4), 179–189 (1983).
81. J. W. Mayer, L. Eriksson, and J. A. Davis, *Ion Implantation of Semiconductors* (Academic Press, New York, 1970).
82. G. Carter and W. A. Grant, *Ion Implantation of Semiconductors* (Arnold, London, 1976).
83. J. F. Gibbons, Ion implantation, in *Handbook on Semiconductors*, edited by T. S. Moss, Vol. 3, edited by S. P. Keller (North-Holland, Amsterdam, 1980), pp. 599–640.
84. *Ion Handbook for Materials Analysis*, edited by J. W. Mayer and E. Rimini (Academic Press, New York, 1978).
85. A. Nylandsted Larsen, G. Sorensen, F. Nielsen, L. D. Nielsen, and V. E. Borisenko, An experimental investigation of ion-implantation combined with laser and incoherent light annealing and laser-induced diffusion for the production of solar cells, in *Photovoltaic Power Generation*, Vol. 3, edited by R. Van Overstraeten and W. Palz (Reidel, Dordrecht, 1983), pp. 52–61.
86. L. D. Nielsen, A. Nylandsted Larsen, and V. E. Borisenko, Incoherent light annealing of phosphorus implanted silicon with solar cell production in view, *J. Phys. (Paris)* **44**(10), c5/381–c5/385 (1983).
87. A. Nylandsted Larsen and V. E. Borisenko, Behaviour of implanted arsenic in silicon crystals subjected to transient heating with incoherent light, *Appl. Phys. A* **33**(1), 51–58 (1984).
88. V. E. Borisenko and D. I. Zarovskii, Ten-second annealing of implanted silicon by low-energy electrons, *Sov. Phys. Semicond.* **18**(10), 1192–1194 (1984).
89. V. E. Borisenko, V. V. Gribkovskii, A. G. Dutov, V. A. Kolosov, and K. E. Lobanova, Povedenie implantirovannoii surmi v kremnii pri sekundnoii termoobrabotke necogerentnim svetom, *Fiz. Tekh. Poluprov.* **20**(4), 778 (1986).
90. A. Nylandsted Larsen and L. Correra, Annealing of arsenic and antimony implanted silicon single crystals using a CW xenon arc lamp, *Radiat. Eff. Lett.* **76**(1), 67–72 (1983).
91. S. J. Pennycook, J. Narayan, and O. W. Holland, Formation of partially coherent antimony precipitates in ion implanted thermally annealed silicon, *J. Appl. Phys.* **54**(12), 6875–6878 (1983).
92. S. J. Pennycook, J. Narayan, and O. W. Holland, Point defect trapping in solid-phase epitaxially grown silicon–antimony alloys, *J. Appl. Phys.* **55**(4), 837–840 (1984).
93. A. Nylandsted Larsen, F. T. Pedersen, G. Weyer, R. Galloni, R. Rizzoli, and A. Armigliato, The nature of electrically inactive antimony in silicon, *J. Appl. Phys.* **59**(6), 1908–1917 (1986).
94. D. G. Beanland, The behaviour of boron molecular ion implants into Si, *Solid-State Electron.* **21**(3), 537–547 (1978).
95. M. Y. Tsai, D. S. Day, B. G. Streetman, P. Williams, and C. Evans, Recrystallization of implanted amorphous Si layers. II. Migration of fluorine in BF_2^+-implanted silicon, *J. Appl. Phys.* **50**(1), 188–192 (1979).
96. A. Nylandsted Larsen and R. A. Jarjis, Investigation of fluorine-concentration profiles in pulsed-laser-irradiated BF_2-implanted silicon single crystals, *Appl. Phys. Lett.* **41**(4), 366–368 (1982).
97. J. S. Williams, Amorphisation, crystallization and related phenomena in silicon, in *Beam–Solid Interactions and Phase Transformation*, edited by H. Kurz, G. L. Olson, and J. M. Poate (North-Holland, Amsterdam, 1986), pp. 19–32.

98. W. O. Adekoya, J. C. Muller, P. Siffert, and L. Pedulli, Recrystallization kinetics during fast thermal annealing of PF_n^+ implanted silicon, in *Beam–Solid Interactions and Phase Transformation*, edited by G. L. Olson and J. M. Poate (North-Holland, Amsterdam, 1986), pp. 42–50.
99. F. F. Komarov, A. P. Novikov, V. S. Solovyev, and S. Y. Shiryaev, *Defekti Strukturi v Ionnolmplantirovannom Kremnii* (Izd. Universitetskoe, Minsk, 1989).
100. D. K. Sadana, Defect structures and electrical behavior of rapid thermally annealed ion implanted silicon, in *Rapid Thermal Processing of Electronic Materials*, edited by S. R. Wilson, R. Powell, and D. E. Davies (MRS, Pittsburgh, 1987), pp. 319–328.
101. K. S. Jones, S. Prussin, and E. R. Weber, A systematic analysis of defects in ion-implanted silicon, *Appl. Phys. A* **45**(1), 1–34 (1988).
102. D. J. Eaglesham, P. A. Stolk, H.-J. Gossmann, and J. M. Poate, Implantation and transient B diffusion in Si: The source of the interstitials, *Appl. Phys. Lett.* **65**(18), 2305–2307 (1994).
103. J. Huang and R. J. Jaccodine, Study of "reverse annealing" of boron under low temperature lamp anneals, in *Rapid Thermal Processing*, edited by T. O. Sedgwick, T. E. Seidel, and B.-Y. Tsaur (MRS, Pittsburgh, 1986), pp. 57–64.
104. C. R. Peter, J. P. De Souza, and C. M. Hasenack, Prolonged and rapid thermal annealing of boron implanted silicon, *J. Appl. Phys.* **64**(5), 2696–2699 (1988).
105. M. Tamura and K. Ohyu, Residual defects in high-energy B-, P- and As-implanted Si by rapid thermal annealing, *Appl. Phys. A* **49**(2), 143–155 (1989).
106. T. E. Seidel, D. J. Lischner, C. S. Pai, R. V. Knoell, D. M. Maher, and D. C. Jacobson, A review of rapid thermal annealing (RTA) of B, BF_2 and As ions implanted into silicon, *Nucl. Instrum. Methods Phys. Res. B* **7/8**(1), 251–260 (1985).
107. T. Sands, J. Washburn, E. Myers, and D. K. Sadana, On the origins of structural defects in BF_2^+-implanted and rapid-thermally-annealed silicon: Conditions for defect-free regrowth, *Nucl. Instrum. Methods Phys. Res. B* **7/8**(2), 337–341 (1985).
108. V. E. Borisenko, P. I. Gaiduk, V. V. Gribkovskii, F. F. Komarov, and V. S. Solov'yev, Structure modifications in As^+-ion implanted silicon layers during pulse thermal treatment, *Radiat. Eff. Lett.* **87**(3), 109–115 (1985).
109. D. M. Maher, R. V. Knoell, M. B. Ellington, and D. C. Jacobson, Extendet defects in amorphized and rapid-thermally annealed silicon, in *Rapid Thermal Processing*, edited by T. O. Sedgwick, T. E. Seidel, and B.-Y. Tsaur (MRS, Pittsburgh, 1986), pp. 93–105.
110. D. Baither, R. Koegler, D. Pankin, and E. Wieser, Residual defects in implanted silicon after annealing with incoherent light, *Phys. Status Solidi A* **94**(2), 767–772 (1986).
111. K. S. Jones, S. Prussin, and E. R. Weber, Enhanced elimination of implantation damage upon exceeding the solid solubility, *J. Appl. Phys.* **62**(10), 4114–4117 (1987).
112. Y. Kim, H. Z. Massoud, and R. B. Fair, Boron profile changes during low-temperature annealing of BF_2^+ -implanted silicon, *Appl. Phys. Lett.* **53**(22), 2197–2199 (1988).
113. S. Coffa, L. Calcagno, C. Spinella, S. U. Campisano, G. Foti, and E. Rimini, Rapid thermal annealing of hot implants in silicon, *Nucl. Instrum. Methods Phys. Res. B* **39**(1–4), 357–361 (1989).
114. T. Seidel and A. U. MacRae, The isothermal annealing of boron implanted silicon, *Radiat. Eff.* **7**(1/2), 1–6 (1971).
115. M. Y. Tsai and B. G. Streetman, Recrystallization of implanted amorphous silicon layers. 1. Electrical properties of silicon implanted with BF_2^+ or Si^++B^+, *J. Appl. Phys.* **50**(1), 183–187 (1979).
116. S. Mizuo, T. Kusaka, A. Shintani, M. Nanba, and H. Higuchi, Effect of Si and SiO_2 thermal nitridation on impurity diffusion and oxidation induced stacking fault in Si, *J. Appl. Phys.* **54**(7), 3860–3866 (1983).
117. P. Fahey, R. W. Dutton, and M. Moslehi, Effect of thermal nitridation process on boron and phosphorus diffusion in (100) silicon, *Appl. Phys. Lett.* **43**(7), 683–685 (1983).
118. S. Matsumoto and Y. Ishikawa, Oxidation enhanced and concentration dependent diffusions of dopants in silicon, *J. Appl. Phys.* **54**(9), 5049–5054 (1983).
119. R. M. Harris and D. A. Antoniadis, Silicon self-interstitial supersaturation during phosphorus diffusion, *Appl. Phys. Lett.* **43**(10), 937–939 (1983).
120. P. Fahey, R. W. Dutton, and S. M. Hu, Supersaturation of self-interstitials and undersaturation of vacancies during phosphorus diffusion in silicon, *Appl. Phys. Lett.* **44**(8), 777–779 (1984).

121. V. A. Panteleev, V. I. Okulich, A. S. Vasin, and V. A. Gusarov, Mekhanism diffuzii bora v kremnii, *Izv. Akad. Nauk SSSR Neorg. Mater.* **21**(8), 1253–1255 (1985).
122. P. Fahey, G. Barbuscia, M. Moslehi, and R. W. Dutton, Kinetics of thermal nitridation processes in the study of dopant diffusion mechanism in silicon, *Appl. Phys. Lett.* **46**(8), 784–786 (1985).
123. S. M. Hu, Modelling diffusion in silicon: Accomplishments and challenges, in *VLSI Science and Technology /1985/*, edited by W. M. Bullis and S. Broydo (Electrochemical Society, Princeton, NJ, 1985), pp. 465–506.
124. O. W. Holland, New mechanism for diffusion of ion-implanted boron in Si at high concentration, *Appl. Phys. Lett.* **54**(9), 798–800 (1989).
125. P. Fahey, Diffusion of shallow impurities in silicon, in *Shallow Impurities in Semiconductors*, Conf. Ser. No. 95 (Institute of Physics, London, 1989), pp. 483–492.
126. C. S. Nichols, C. G. Van de Walle, and S. T. Pantelides, Mechanisms of equilibrium and nonequilibrium diffusion of dopants in silicon, *Phys. Rev. Lett.* **62**(9), 1049–1052 (1989).
127. S. Hu, Diffusion in silicon and germanium, in *Atomic Diffusion in Semiconductors*, edited by D. Shaw (Plenum Press, New York, 1973), pp. 217–350.
128. R. B. Fair, Concentration-profiles of diffused dopants in silicon, in *Impurity Doping Processes in Silicon*, edited by F. F. G. Wang (North-Holland, Amsterdam, 1981), pp. 315–442.
129. P. M. Fahey, P. B. Griffin, and J. D. Plummer, Point defects and dopant diffusion in silicon, *Rev. Mod. Phys.* **61**(2), 289–384 (1989).
130. K. Nishiyama, M. Arai, and N. Watanabe, Radiation annealing of boron-implanted silicon with halogen lamp, *Jpn. J. Appl. Phys.* **19**(10), L563–L566 (1980).
131. A. Gat, Heat-pulse annealing of arsenic-implanted silicon with a CW arc lamp, *IEEE Electron Device Lett.* **2**(4), 85–87 (1981).
132. C. Drowley and C. Hu, Arsenic-implanted Si layers annealed using a CW Xe arc lamp, *Appl. Phys. Lett.* **38**(11), 876–878 (1981).
133. R. A. Powell, T. O. Yep, and R. T. Fulks, Activation of arsenic implanted silicon using an incoherent light source, *Appl. Phys. Lett.* **39**(2), 150–152 (1981).
134. R. T. Fulks, C. J. Russo, P. R. Hanley, and T. I. Kamins, Rapid isothermal annealing of ion implantation damage using a thermal radiation source, *Appl. Phys. Lett.* **39**(8), 604–606 (1981).
135. V. E. Borisenko and V. A. Labunov, Pulse electron beam annealing of phosphorus implanted silicon, in *Laser and Electron Beam Interactions with Solid*, edited by B. R. Appleton and G. K. Celler (North-Holland, Amsterdam, 1982), pp. 305–310.
136. V. E. Borisenko and V. A. Labunov, Annealing of antimony implanted silicon with halogen lamp irradiation, *Phys. Status Solidi A* **72**(2), K173–K176 (1982).
137. V. A. Labunov, V. E. Borisenko, V. V. Gribkovskii, and S. N. Kornilov, Otjig implantirovannih fosforom cloev kremniya sekundnimi impulsami nekogerentnogo sveta, *Elektronnaya Tekhnika, Ser. 2, Poluprovodnikovie Pribori* **3**, 73–77 (1984).
138. V. E. Borisenko, D. I. Zarovskii, V. I. Pachinin, and V. A. Ukhov, Otjig implantirovannih fosforom sloev kremniya impulsami elektronov sekundnoii dlitelnosti, *Elektronnaya Tekhnika, Ser. 6, Materiali* **6**, 28–31 (1984).
139. Y. Sasaki, K. Itoh, and S.-I. Tamura, Enhanced diffusion of implanted arsenic in silicon at an early stage of annealing, *Jpn. J. Appl. Phys.* **28**(8), 1421–1425 (1989).
140. D. A. Antoniadis, A. M. Lin, and R. W. Dutton, Oxidation-enhanced diffusion of arsenic and phosphorus in near-intrinsic (100) silicon, *Appl. Phys. Lett.* **33**(12), 1030–1033 (1978).
141. D. A. Antoniadis and I. Moskowitz, Diffusion of indium in silicon inert and oxidizing ambients, *J. Appl. Phys.* **53**(12), 9214–9216 (1982).
142. S. Mizuo and H. Higuchi, Effects of back-side oxidation of Si substrates on Sb diffusion at front side, *J. Electrochem. Soc.* **130**(9), 1942–1947 (1983).
143. M. Miyake, Oxidation-enhanced diffusion of ion-implanted boron in heavily phosphorus-doped silicon, *J. Appl. Phys.* **58**(2), 711–715 (1985).
144. S. M. Hu, Kinetics of interstitial supersaturation during oxidation of silicon, *Appl. Phys. Lett.* **43**(5), 449–451 (1983).
145. S. M. Hu, Kinetics of interstitial supersaturation and enhanced diffusion in short-time/low-temperature oxidation of Si, *J. Appl. Phys.* **57**(10), 4527–4532 (1985).

146. N. K. Chen and C. Lee, Oxynitridation-enhanced diffusion of phosphorus in ⟨100⟩ silicon, *J. Electrochem. Soc.* **142**(6), 2051–2054 (1995).
147. E. V. Nidaev and A. L. Vasiliev, Photon enhanced annealing of ion implantation amorphized silicon subjected to laser irradiation influence, in *International Conference on Ion Implantation in Semiconductors and Other Materials*, edited by P. Grigaitis and S. Tamulevichus (Kaunas Polytechnic Institute, Kaunas, 1985), pp. 217–218.
148. S. R. Wilson, R. B. Gregory, W. M. Paulson, and H. T. Dichl, Characterization of ion implanted silicon annealed with a graphite radiation source, *IEEE Trans. Nucl. Sci.* **30**(2), 1–4 (1983).
149. S. J. Pennycook, J. Narayan, and O. W. Holland, Transient-enhanced diffusion during furnace and thermal annealing of ion-implanted silicon, *J. Electrochem. Soc.* **132**(8), 1962–1968 (1985).
150. A. J. Walker, Rapid thermal diffusion of boron implanted as boron difluoride in preamorphized silicon, *J. Appl. Phys.* **71**(4), 2033–2035 (1992).
151. H. Kinoshita, T. H. Huang, D. L. Kwong, and P. E. Bakeman, Boron diffusion in fluorine preamorphized silicon during rapid thermal annealing, in *Rapid Thermal and Integrated Processing II*, edited by J. C. Gelpey, J. K. Elliott, J. J. Wortman, and A. Ajmera (MRS, Pittsburgh, 1993), pp. 253–258.
152. R. B. Fair, J. J. Wortman, and J. Lin, Modelling of rapid thermal diffusion of arsenic and boron in silicon, *J. Electrochem. Soc.* **131**(10), 2387–2394 (1984).
153. T. O. Sedgwick, R. Kalish, S. R. Mader, and S. C. Shatas, Short time annealing of As and B implanted Si using tungsten-halogen lamps, in *Energy Beam–Solid Interactions and Transient Thermal Processing*, edited by J. C. C. Fan and N. M. Johnson (North-Holland, Amsterdam, 1984), pp. 293–298.
154. K. Cho, M. Numan, T. G. Finstad, W. K. Chu, J. Lin, and J. J. Wortman, Transient enhanced diffusion during rapid thermal annealing of boron implanted Si, *Appl. Phys. Lett.* **47**(12), 1321–1323 (1985).
155. D. J. Godfrey, R. A. McMahon, D. G. Hasko, H. Ahmed, and M. G. Dowsett, Annealing and diffusion of boron in self-implanted silicon by furnace and electron beam heating, in *Impurity Diffusion and Gettering in Silicon*, edited by R. P. Fair (MRS, Pittsburgh, 1985), pp. 143–149.
156. R. Angelucci, P. Negrini, and S. Solmi, Transient enhanced diffusion of dopant in silicon induced by implantation damage, *Appl. Phys. Lett.* **49**(15), 1468–1470 (1986).
157. A. Armigliato, S. Guimaraes, S. Solmi, R. Kogler, and E. Wieser, Comparison of boron diffusivity during rapid thermal annealing in predamaged, preamorphized and crystalline silicon, *Nucl. Instrum. Methods Phys. Res. B* **19/20**(1), 512–515 (1987).
158. S. Guimaraes, E. Landi, and S. Solmi, Enhanced diffusion phenomena during rapid thermal annealing of preamorphized boron-implanted silicon, *Phys. Status Solidi A* **95**(2), 589–598 (1986).
159. A. E. Michel, Rapid annealing and the anomalous diffusion of ion implanted boron into silicon, *Appl. Phys. Lett.* **50**(7), 416–418 (1987).
160. V. F. Stelmakh, Y. R. Suprun-Belevich, and A. R. Chelyadinskii, Vliyanie radiatsionnih defektov i uprugih napryajenii nesootvetstviya na diffuziyu ionno-vnedrennogo bora v kremnii pri impulsnom otjige, *Izv. Vyssh. Uchebu. Zaved. Fiz.* **9**, 114–116 (1987).
161. T. O. Sedgwick, A. E. Michel, V. R. Deline, and S. A. Cohen, Transient boron diffusion in ion-implanted crystalline and amorphous silicon, *J. Appl. Phys.* **63**(5), 1452–1463 (1988).
162. M. Miyake and S. Aoyama, Transient enhanced diffusion of ion-implanted boron in Si during rapid thermal annealing, *J. Appl. Phys.* **63**(5), 1754–1757 (1988).
163. M. Servidori, R. Angelucci, F. Cembali, P. Negrini, S. Solmi, P. Zaumseil, and U. Winter, Retarded and enhanced dopant diffusion in silicon related to implantation-induced excess vacancies and interstitials, *J. Appl. Phys.* **61**(5), 1834–1840 (1987).
164. P. A. Packan and J. D. Plummer, Transient diffusion of low-concentration B in Si due to ^{29}Si implantation damage, *Appl. Phys. Lett.* **56**(18), 1787–1789 (1990).
165. E. Landi, S. Guimaraes, and S. Solmi, Influence of nucleation on the kinetics of boron precipitation in silicon, *Appl. Phys. A* **44**(2), 135–141 (1987).
166. K. Kugimiya and G. Fuse, Blink furnace annealing of ion-implanted silicon, *Jpn. J. Appl. Phys.* **21**(1), L16–L18 (1982).
167. R. A. Powell, Activation of shallow, high-dose BF_2^+-implants into silicon by rapid thermal processing, *J. Appl. Phys.* **56**(10), 2837–2843 (1984).
168. T. E. Seidel, R. Knoell, G. Poli, B. Schwartz, F. A. Stevie, and P. Chu, Rapid thermal annealing of dopants implanted into preamorphized silicon, *J. Appl. Phys.* **58**(2), 683–687 (1985).

169. R. Klabes, M. Voelskow, H. Woittennek, E. V. Nidaev, and L. S. Smirnov, Dopant redistribution after flash lamp annealing, *Phys. Status Solidi A* **71**(1), K127–K130 (1982).
170. H. B. Harrison, S. S. Iyer, G. A. Sai-Halasz, and S. A. Cohen, Highly activated shallow Ga profiles in silicon obtained by implantation and rapid thermal annealing, *Appl. Phys. Lett.* **51**(13), 992–994 (1987).
171. W. Katz, G. A. Smith, R. F. Reihl, and E. F. Koch, The effects of thermal and transient annealing on the redistribution of indium implanted silicon, in *Energy Beam–Solid Interactions and Transient Thermal Processing*, edited by J. C. C. Fan and N. M. Johnson (North-Holland, Amsterdam, 1984), pp. 299–302.
172. S. Y. Shiryaev, A. Nylandsted Larsen, and N. Safronov, Rapid thermal annealing of indium implanted silicon crystals, *J. Appl. Phys.* **65**(11), 4220–4224 (1989).
173. F. Marou, A. Claverie, P. Salles, and A. Martinez, The enhanced diffusion of boron in silicon after high-dose implantation and during rapid thermal annealing, *Nucl. Instrum. Methods Phys. Res. B* **55**, 655–660 (1991).
174. H. Kinoshita and D. L. Kwong, Physical model for defect mediated boron diffusion during rapid thermal annealing of ion implanted BF_2, *Appl. Phys. Lett.* **61**(1), 25–27 (1992).
175. S. Nishikawa, A. Tanaka, and T. Yamaji, Reduction of transient boron diffusion in preamorphized Si by carbon implantation, *Appl. Phys. Lett.* **60**(18), 2270–2272 (1992).
176. S. Nishikawa and T. Yamaji, Elimination of secondary defects in preamorphized Si by carbon implantation, *Appl. Phys. Lett.* **62**(3), 303–305 (1993).
177. H. Ryssel and I. Ruge, *Ionenimplantation* (Teubner, Stuttgart, 1978).
178. J. P. de Souza and H. Boudinov, Electrical activation of boron coimplanted with carbon in a silicon substrate, *J. Appl. Phys.* **74**(11), 6599–6602 (1993).
179. H. Boudinov and J. P. de Souza, Electrical activation of boron in B+C implanted Si during RTA with different heating rates, in *Rapid Thermal and Integrated Processing III*, edited by J. J. Wortman, J. C. Gelpey, M. L. Green, S. R. J. Brueck, and F. Roozeboom (MRS, Pittsburgh, 1994), pp. 357–362.
180. L. Correra and L. Pedulli, Incoherent-light-flash annealing of phosphorus-implanted silicon, *Appl. Phys. Lett.* **37**(1), 55–57 (1980).
181. L. D. Buiko, V. A. Gorushko, V. A. Pilipenko, V. V. Rojkov, N. I. Sterjanov, and V. N. Topchiii, Impulsniii nemonokhromaticheskiii opticheskiii otjig ionnolegirovannih sloev kremniya, *Elektronnaya Tekhnika, Ser. 3, Mikroelektronika* **3**, 47–51 (1982).
182. *Impurity Doping Processes in Silicon*, edited by F. F. G. Wang (North-Holland, Amsterdam, 1981).
183. N. E. B. Cowern and D. J. Godfrey, Transient enhanced diffusion of phosphorus in silicon, *Appl. Phys. Lett.* **49**(25), 1711–1713 (1986).
184. V. Borisenko, V. Gribkovskii, F. Korshunov, and N. Sobolev, Stability of phosphorus solid solution in silicon produced by ion implantation and transient thermal processing, in *Energy Pulse Modification of Semiconductors and Related Materials*, edited by K. Hennig (Akademie der Wissenschaften der DDR, Dresden, 1985), pp. 326–330.
185. S. R. Wilson, R. B. Gregory, and W. M. Paulson, Rapid isothermal anneal of ^{75}As implanted silicon, *Appl. Phys. Lett.* **41**(10), 978–980 (1982).
186. P. D. Scovell and E. J. Spurgin, Pulse thermal annealing of ion-implanted silicon, *J. Appl. Phys.* **54**(5), 2413–2418 (1983).
187. R. T. Hodgson, J. E. E. Baglin, and A. E. Michel, Rapid thermal annealing of silicon using an ultrahigh power arc lamp, in *Laser–Solid Interactions and Transient Thermal Processing of Materials*, edited by J. Narayan, W. L. Brown, and R. A. Lemons (North-Holland, Amsterdam, 1983), pp. 355–360.
188. R. Kalish, T. O. Sedgwick, and S. Mader, Transient enhanced diffusion in arsenic-implanted short time annealed silicon, *Appl. Phys. Lett.* **44**(1), 107–109 (1984).
189. R. Galloni, R. Rizzoli, and A. Nylandsted Larsen, Incoherent light annealing of high dose arsenic implantation in silicon, in *Energy Beam–Solid Interactions and Transient Thermal Processing*, edited by V. T. Nguyen and A. G. Cullis (MRS Europe, Strasbourg, 1985), pp. 331–336.
190. T. O. Sedgwick, A. E. Michel, S. A. Cohen, V. R. Deline, and G. S. Oehrlein, Investigation of transient diffusion effects in rapid thermally processed ion implanted arsenic in silicon, *Appl. Phys. Lett.* **47**(8), 848–850 (1985).
191. S. Y. Shiryaev, A. Nylandsted Larsen, E. S. Sorensen, and P. Tidemand-Petersson, Redistribution and activation of ion implanted As in Si during RTA for concentrations around solid solubility, *Nucl. Instrum. Methods Phys. Res. B* **19/20**(2), 507–511 (1987).

192. A. Nylandsted Larsen and L. Correra, Annealing of arsenic- and antimony-implanted silicon single crystals using a CW xenon arc lamp, *Radiat. Eff.* **76**(1), 67–72 (1983).
193. D. Pankin, E. Wieser, R. Klabes, and H. Syhre, Dose dependence of the flash lamp annealing of arsenic-implanted silicon, *Phys. Status Solidi A* **77**(1), 553–559 (1983).
194. A. Kamgar and F. A. Baiocchi, Metastable activation in rapid thermal annealed arsenic implanted silicon, in *Rapid Thermal Processing*, edited by T. O. Sedgwick, T. E. Seidel, and B.-Y. Tsaur (North-Holland, Amsterdam, 1986), pp. 23–29.
195. A. Kamgar, F. A. Baiocchi, and T. T. Sheng, Kinetics of arsenic activation and clustering in high dose implanted silicon, *Appl. Phys. Lett.* **48**(16), 1090–1092 (1986).
196. Y. Zysin, A. Nylandsted Larsen, G. Weyer, R. Galloni, and R. Rizzoli, Metastable phase of ion implanted antimony in silicon by incoherent light annealing, in *Energy Pulse Modification of Semiconductors and Related Materials*, edited by K. Hennig (Akademie der Wissenschaften der DDR, Dresden, 1985), pp. 291–295.
197. A. Nylandsted Larsen, F. T. Pedersen, G. Weyer, R. Galloni, and R. Rizzoli, Rapid thermal annealing of ion-implanted Sb in Si: A comparison of substitutional fractions from channeling-, electrical-, and Mossbauer measurements, in *Energy Beam–Solid Interactions and Transient Thermal Processing*, edited by V. T. Nguyen and A. G. Cullis (MRS Europe, Strasbourg, 1985), pp. 319–324.
198. O. W. Holland and D. Fathy, Annealing of Sb^+-ion-implanted Si, *J. Appl. Phys.* **63**(11), 5326–5330 (1988).
199. A. Nylandsted Larsen, P. Tidemand-Petersson, P. E. Andersen, and G. Weyer, The effect of heavy doping on complex formation and diffusivity of Sb in Si, in *Shallow Impurities in Semiconductors*, Inst. Phys. Conf. Ser. No. 95 (IOP Publishing, London, 1989), pp. 499–504.
200. Y. A. Kontsevoii, Y. M. Litvinov, and E. A. Fattakhov, *Plastichnost I Prochnost Poluprovodnikovih Materialov i Struktur* (Radio i svyaz, Moscow, 1982).
201. V. A. Labunov, N. T. Kvasov, and A. K. Polonin, Dinamicheskie effekti v kristallicheskoii reshetke pri ionnoii implantatsii, *Dokl. Akad. Nauk BSSR* **27**(7), 606–608 (1983).
202. A. P. Morozov, P. V. Pavlov, Y. A. Semin, V. D. Skupov, and D. I. Tetelbaum, Anomalno glubokoe proniknovenie defektov v kremnii pri ionnoii bombardirovke, in *Vzaimodeiistvie Atomnih Chastits s Tverdim Telom* (Minsk Radioengineering Institute, Minsk, 1984), pp. 14–15.
203. I. V. Vorob'eva, Y. E. Geguzin, and V. E. Monostyrenko, "Shock wave" mechanism of formation of surface tracks of heavy ions in solids, *Sov. Phys. Semicond.* **28**(1), 88–90 (1986).
204. A. I. Krestelev and A. N. Bekrenev, Mass transfer in metals under the effect of shock waves, *Phys. Chem. Mater. Treat.* **19**(2), 114–115 (1985).
205. V. S. Eremeev, *Diffusiya i Napryajeniya* (Energoatomizdat, Moscow, 1984).
206. V. A. Panteleev, M. I. Vasilevskii, and Y. L. Kalinkin, Microscopic study of the effect of elastic stresses on substitutional impurity atom migration in silicon, *Sov. Phys. Solid State* **25**(10), 1689–1692 (1983).
207. V. A. Labunov, N. T. Kvasov, and N. L. Ostroverkhov, O migratsii primesi pri impulsnoii svetovoli obrabotke legirovannogo kremniya, *Izv. Akad. Nauk BSSR Ser. Fiz. Mater. Nauk* **3**, 38–41 (1984).
208. M. I. Vasilevskii, G. M. Golemshtok, and V. A. Panteleev, Influence of intrinsic elastic stresses and distortions of the vibrational spectrum on formation of defects and diffusion in semiconductors, *Sov. Phys. Solid State* **27**(1), 73–77 (1985).
209. Y. A. Tkhorik and L. S. Hazan, *Plasticheskaya Deformatsia i Dislokatsii Nesootvetstviya v Geteroepitaksialnih Sistemah* (Naukova dumka, Kiev, 1983).
210. H. Siethoff and P. Haasen, *Lattice Defects in Semiconductors*, edited by R. R. Hasiguti (University of Tokyo Press, Tokyo; Pennsylvania State University Press, University Park, Pennsylvania; 1968).
211. H. Siethoff, K. Ahlborn, and W. Schroter, Two independent mechanisms of dynamical recovery in the high-temperature deformation of silicon and germanium, *Philos. Mag. A* **50**(1), L1–L6 (1984).
212. G. Charitat and A. Martinez, Stress evolution and point defect generation during oxidation of silicon, *J. Appl. Phys.* **55**(4), 909–913 (1984).
213. V. E. Borisenko, V. V. Gribkovskii, N. T. Kvasov, and A. K. Polonin, Mekhanicheskie napryajeniya i povedenie implantirovannih primeseii zamescheniya v kremnii pri sekundnoii termoobrabotke, *Izv. Akad. Nauk BSSR Ser. Fiz. Mater. Nauk* **1**, 106–108 (1988).
214. V. A. Labunov, N. T. Kvasov, and A. Polonin, Control of mechanical stresses in semiconductor wafers, USSR Invention Certificate No. 1087779 (1984).

215. N. L. Prokhorenko and Y. R. Suprun-Belevich, Internal mechanical stress and electrical activation of impurity in ion-implanted Si during pulse annealing, in *Energy Pulse and Particle Beam Modification of Materials*, edited by K. Hennig (Akademie-Verlag, Berlin, 1988), pp. 294–296.
216. V. F. Stelmakh, Y. R. Suprun-Belevich, and A. R. Chelyadinskii, Effect of radiation defects and elastic incompatibility stresses on the electrical activation and diffusion of boron in ion-implanted silicon, *Phys. Status Solidi A* **112**(1), 381–384 (1989).
217. V. F. Stelmakh, Y. R. Suprun-Belevich, and A. R. Chelyadinskii, Vliyanie uprugih napryajeniii nesootvetstviya na diffuziyu i elektricheskuyu aktivatsiyu ionno-vnedrennogo B v Si, *Poverkhnost* **10**, 123–127 (1989).
218. V. E. Borisenko and S. G. Yudin, Implanted impurity behavior in silicon subjected to transient heating, *Phys. Status Solidi A* **109**(1), 395–402 (1988).
219. V. E. Borisenko and S. G. Yudin, Diffuziya primeseii zamescheniya v poluprovodnikah v usloviyah raspada peresischennogo rastvora, *Dokl. Akad. Nauk BSSR* **32**(7), 613–616 (1988).
220. V. I. Fistul, *Raspad Peresischennih PoluDrovodnikovih Tverdih Rastvorov* (Metallurgiya, Moscow, 1977).
221. A. G. Tweet, Precipitation of Cu in Ge, *Phys. Rev.* **106**(2), 221–224 (1957).
222. F. S. Ham, Theory of diffusion limited precipitation, *J. Phys. Chem. Solids* **6**(4), 335–351 (1958).
223. N. T. Bendik, V. S. Garnyk, and L. S. Milevskii, Precipitation kinetics of solid solutions of chromium in silicon, *Sov. Phys. Solid State* **12**(1), 150–154 (1970).
224. B. I. Boltaks, M. K. Bakhadyrkhanov, and G. S. Kulikov, Electrical properties and decomposition of the solid solution of iron in silicon, *Sov. Phys. Solid State* **13**(9), 2240–2243 (1971).
225. A. Z. Badalov, Kinetics of precipitation of a solid solution of gold in n-type silicon, *Sov. Phys. Semicond.* **6**(5), 685–687 (1972).
226. S. A. Azimov, M. S. Yunusov, and B. V. Umarov, Kinetics of precipitation of solid solutions of iridium in silicon [photocapacitance and electrical resistivity], *Sov. Phys. Semicond.* **11**(9), 979–982 (1977).
227. P. Revesz, G. Farkas, and J. Gyulai, Behavior of antimony above solid solubility in silicon produced by implantation and laser annealing, *Radiat. Eff.* **47**(1/4), 149–152 (1980).
228. M. Miyao, K. Itoh, M. Tamura, H. Tamura, and T. Takuyama, Furnace annealing behavior of phosphorus implanted laser annealed silicon, *J. Appl. Phys.* **51**(8), 4139–4144 (1980).
229. A. Lietoila, J. F. Gibbons, and T. W. Sigmon, The solid solubility and thermal behavior of metastable concentrations of As in Si, *Appl. Phys. Lett.* **36**(9), 765–768 (1980).
230. K. Itoh, Y. Sasaki, T. Mitsushi, M. Miyao, and T. Tamura, Thermal behavior of B, P and As atoms in supersaturated Si produced by ion implantation and pulsed-laser annealing, *Jpn. J. Appl. Phys.* **21**(5), L245–L247 (1982).
231. J. Gotzlich, P. H. Tsien, G. Henghuber, and H. Ryssel, CO_2 laser annealing of ion-implanted silicon: Relaxation characteristics of metastable concentrations, in *Ion Implantation: Equipment and Techniques*, edited by H. Ryssel and H. Glawischnig (Springer-Verlag, Berlin, 1983), pp. 513–519.
232. V. P. Popov, A. V. Dvurechenskii, B. P. Kashnikov, and A. I. Popov, Defects in supersaturated solid solution of arsenic in silicon at rapid thermal annealing, *Phys. Status Solidi A* **94**(2), 569–572 (1986).
233. V. E. Borisenko, A. G. Dutov, K. E. Lobanova, and S. G. Yudin, Redistribution of phosphorus in implanted laser-annealed silicon during subsequent heat treatment lasting up to tens of seconds, *Sov. Phys. Semicond.* **21**(8), 891–892 (1987).
234. R. M. Bayazitov, V. E. Borisenko, D. A. Konovalov, I. B. Khaibullin, and S. G. Yudin, Precipitation of a supersaturated substitutional solution of phosphorus in silicon due to heat treatment for periods of seconds, *Sov. Phys. Semicond.* **21**(8), 917–918 (1987).
235. S. N. Ershov, V. A. Panteleev, S. N. Nagornykh, and V. V. Chernyakhovskii, Migration energy of intrinsic point defects in different charge states in silicon and germanium, *Sov. Phys. Solid State* **19**(1), 187 (1977).
236. G. D. Watkins, EPR studies of the lattice vacancy and low temperature damage processes in silicon, in *Lattice Defects in Semiconductors*, Inst. Phys. Conf. Ser. No. 23 (Institute of Physics, London, 1975), pp. 221–230.
237. G. B. Bronner and J. D. Plummer, Gettering of gold in silicon: A tool for understanding the properties of silicon interstials, *J. Appl. Phys.* **61**(12), 5286–5298 (1987).
238. V. E. Borisenko and S. G. Yudin, Steady-state solubility of substitutional impurities in silicon, *Phys. Status Solidi A* **101**(1), 123–127 (1987).

239. F. A. Trumbore, Solid solubilities of impurity elements in germanium and silicon, *Bell Syst. Tech. J.* **39**(1), 205–233 (1960).
240. N. H. Abrikosov, V. M. Glazov, and G.-Y. Lu, Issledovanie razdelnoii i sovmestnoii rastvorimosti allyuminiya i fosfora v germanii i kremnii, *G. Neorg. Khim.* **7**(4), 831–835 (1962).
241. S. Maekawa and T. Oshida, Diffusion of boron into silicon, *J. Phys. Soc. Jpn.* **19**(3), 253–267 (1964).
242. G. L. Vick and K. M. Whittle, Solid solubility and diffusion coefficients of boron in silicon, *J. Electrochem. Soc.* **116**(8), 1142–1144 (1969).
243. F. N. Schwettmann and D. L. Kendall, Carrier profile change for phosphorus diffused layers on low-temperature heat treatment, *Appl. Phys. Lett.* **19**(7), 218–220 (1971).
244. F. N. Schwettmann, Characterisation of incomplete activation of high-dose boron implants in silicon, *J. Appl. Phys.* **45**(4), 1918–1920 (1974).
245. P. Ostoja, D. Nobili, A. Armigliato, and R. Angelucci, Isochronal annealing of silicon–phosphorus solutions, *J. Electrochem. Soc.* **123**(1), 124–129 (1976).
246. A. Armigliato, D. Nobili, P. Ostoja, M. Servidori, and S. Solmi, Impurity solubility in silicon, in *Semiconductor Silicon*, edited by H. R. Huff and E. Sirtl (Electrochemical Society, Princeton, NJ, 1977), pp. 638–647.
247. G. Masetti, D. Nobili, and S. Solmi, Stability of impurity solutions in silicon, in *Semiconductor Silicon*, edited by H. R. Huff and E. Sirtl (Electrochemical Society, Princeton, NJ, 1977), pp. 648–654.
248. R. B. Fair and J. C. C. Tsai, A quantitative model for the diffusion of phosphorus in silicon and emitter dip effect, *J. Electrochem. Soc.* **124**(6), 1107–1115 (1977).
249. G. Wada and S. Nishimatsu, Grain growth mechanism of heavily phosphorus-implanted polycrystalline silicon, *J. Electrochem. Soc.* **125**(9), 1499–1504 (1978).
250. K. Nakamura and M. Kamoshida, Implanted As redistribution during annealing in oxidizing ambient, *J. Electrochem. Soc.* **125**(9), 1518–1521 (1978).
251. J. Murota, E. Arai, K. Kobayashi, and K. Kudo, Arsenic diffusion in silicon from doped polycrystalline silicon, *Jpn. J. Appl. Phys.* **17**(2), 457–458 (1978).
252. J. Murota, E. Arai, K. Kobayashi, and K. Kudo, Relationship between total arsenic and electrically active arsenic concentrations in silicon produced by the diffusion process, *J. Appl. Phys.* **50**(2), 804–808 (1979).
253. A. Lietoila, J. F. Gibbons, and T. W. Sigmon, The solid solubility and thermal behavior of metastable concentrations of As in Si, *Appl. Phys. Lett.* **36**(9), 765–768 (1980).
254. M. G. Tsai, F. F. Morehead, J. E. E. Baglin, and A. E. Michel, Shallow junctions by high-dose As implants in Si: Experiments and modelling, *J. Appl. Phys.* **51**(6), 3230–3234 (1980).
255. H. Ryssel, K. Muller, K. Haberger, K. Henkelman, and F. Jahnel, High concentration effects of ion implanted boron in silicon, *Appl. Phys. A* **22**(1), 35–38 (1980).
256. P. Cappelletti, G. F. Cerofolini, and G. U. Pignatel, A correlation between solid solubility and tetrahedral radius of III, IV and V group impurities in silicon, *Philos. Mag.* **46**(5), 863–868 (1982).
257. D. Nobili, Solid solubility and preçipitation of phosphorus and arsenic in silicon solar cells front layer, in *Fourth EC Photovoltaic Solar Energy Conference*, edited by W. H. Bloss and G. Gressi (Reidel, Dordrecht, 1982), pp. 410–420.
258. D. Nobili, A. Armigliato, M. Finetti, and S. Solmi, Precipitation as the phenomenon responsible for the electrically inactive phosphorus in silicon, *J. Appl. Phys.* **53**(3), 1484–1491 (1982).
259. D. Nobili, A. Carabelas, G. Gelotti, and S. Solmi, Precipitation as the phenomenon responsible for the electrically inactive arsenic in silicon, *J. Electrochem. Soc.* **130**(4), 322–328 (1983).
260. J. Gotzlich, P. H. Tsien, and H. Ryssel, Relaxation behavior of metastable As and P concentrations in Si after pulsed and CW laser annealing, *Mater. Res. Soc. Symp. Proc.* **23**, 235–240 (1984).
261. V. M. Glazov and L. M. Pavlova, *Khimicheskaya Termodinamika i Fazovie Ravnovesiya: Dvuhkomponentnie Poluprovodnikovie Sistemi* (Metallurgiya, Moscow, 1981).
262. V. A. Ryabinin, *Termodinamicheskie Svoistva Veschestv. Spravochnik* (Khimiya, Leningrad, 1977).
263. *Termodinamicheskie Svoistva Individualnikh Veschestv. Spravochnik* Izd. v 4-h t., sostaviteli L. V. Gurevich i dr. (Nauka, Moscow, 1978).
264. A. A. Brown, P. J. Rosser, P. B. Moynegh, D. J. Godfrey, D. de Cogan, and D. Nobili, Diffusion, solid solubility and implantation group III and group V impurities, in *Properties of Silicon* (INSPEC, London, 1988), pp. 325–408.

265. C. A. Londos, K. Eftaxias, and V. Hadjicontis, Correlation of solubilities of various elements in silicon, *Phys. Status Solidi A* **118**(1), K13–K16 (1990).
266. V. E. Borisenko and A. Nylandsted Larsen, Incoherent light induced diffusion of arsenic into silicon from spin-on source, *Appl. Phys. Lett.* **43**(6), 582–584 (1984).
267. A. Nylandsted Larsen, V. E. Borisenko, and L. D. Nielsen, Doping of silicon with arsenic and phosphorus from spin-on sources exposed to incoherent light, *J. Phys. (Paris)* **44**(10), c5/427–c5/431 (1983).
268. V. E. Borisenko and S. G. Yudin, Legirovanie kremniya fosforom iz poverkhnostnogo istochnika v processe secundnoii termoobrabotki svetom, *Elektronnaya Tekhnika, Ser. 6, Materiali* **5**, 32–35 (1987).
269. V. E. Borisenko, G. V. Litvinovich, and N. V. Gaponenko, Spekroskopicheskoe issledovanie plenok emulsionnih istochnikov primesi na kremnii, podvergnutih sekundnoii termoobrabotke nekogerentnim svetom, *Zh. Prikl. Spektrosk.* **47**(2), 313–315 (1981).
270. Y. Kato and Y. Ono, Phosphorus diffusion using spin-on phosphosilicate-glass source and halogen lamps, *J. Electrochem. Soc.* **132**(7), 1730–1732 (1985).
271. D. E. Davies and C. E. Ludington, Transient diffusion doping in Si, *J. Appl. Phys.* **59**(6), 2035–2037 (1986).
272. J. P. de Souza, C. M. Hasenack, and J. E. Swart, The doping of silicon with boron by rapid thermal processing, *Semicond. Sci. Technol.* **3**(4), 277–280 (1988).
273. M. Miyake, Diffusion of boron into silicon from borosilicate glass using rapid thermal processing, *J. Electrochem. Soc.* **138**(10), 3031–3039 (1991).
274. A. Usami, M. Ando, M. Tsunekane, and T. Wada, Shallow-junction formation of silicon by rapid thermal diffusion of impurities from a spin-on source, *IEEE Trans. Electron Devices* **39**(1), 105–110 (1992).
275. B. Hartiti, A. Slaoui, J. C. Muller, R. Stuck, and P. Siffert, Phosphorus diffusion into silicon from a spin-on source using rapid thermal processing, *J. Appl. Phys.* **71**(11), 5474–5478 (1992).
276. L. Ventura, A. Slaoui, B. Hartiti, J. C. Muller, R. Stuck, and P. Siffert, Shallow-junction formation by rapid thermal diffusion into silicon from doped spin-on glass films, in *Rapid Thermal and Integrated Processing III*, edited by J. J. Wortman, J. C. Gelpey, M. L. Green, S. R. J. Brueck, and F. Roozeboom (MRS, Pittsburgh, 1994), pp. 345–350.
277. M. Rastogi, W. Zagozdzon-Wosik, F. Romero-Borja, J. M. Heddleson, R. Beavers, P. Grabiec, and L. T. Wood, Boron doping using proximity rapid thermal diffusion from spin-on dopants, in *Rapid Thermal and Integrated Processing III*, edited by J. J. Wortman, J. C. Gelpey, M. L. Green, S. R. J. Brueck, and F. Roozeboom (MRS, Pittsburgh, 1994), pp. 369–374.
278. B. H. Justice, D. F. Harnish, and H. F. Jones, Diffusion processing of arsenic spin-on diffusion sources, *Solid State Technol.* **21**(7), 39–42 (1978).
279. V. V. Zadde, K. V. Zinoviev, D. S. Strebkov, and T. I. Suriyaninova, Ispolzovanie rastvornih kompozitsii pri nizkotemperaturnoii diffuzii fosfora i bora v kremnii, *Elektronnaya Promishlennost* **1**, 53–55 (1980).
280. *Borofilm and Phosphorofilm Diffusion Sources* (Emulsitone Company data, 1982).
281. *Arsenosilicafilm and Antimony Silicafilm* (Emulsitone Company data, 1982).
282. See for example K. Ghaderi and G. Hobler, Simulation of phosphorus diffusion in silicon using a pair diffusion model with a reduced number of parameters, *J. Electrochem. Soc.* **142**(5), 1654–1658 (1995), and references therein.
283. I. V. Antonova, K. B. Kadyrakinov, E. V. Nidaev, and L. S. Smirnov, Diffusion of iron and gold in silicon annealed with millisecond pulses, *Phys. Status Solidi A* **76**(2), K213–K215 (1983).
284. V. E. Borisenko, A. G. Dutov, K. E. Lobanova, and S. G. Yudin, Diffuziya zolota v kremnii pri impulsnom nagreve, *Izv. Akad. Nauk SSSR Neorg. Mater.* **26**(5), 1092–1094 (1990).
285. C. Boit, F. Lau, and P. Sitting, Gold diffusion in silicon by rapid optical annealing, *Appl. Phys. A* **50**(2), 197–205 (1990).
286. Y. Ishikawa, K. Yamanchi, and I. Nakamichi, The enhanced diffusion of low concentration phosphorus, arsenic and boron in silicon during IR-heating, *Jpn. J. Appl. Phys.* **28**(8), L1319–L1321 (1989).
287. V. A. Labunov, V. P. Bondarenko, and V. E. Borisenko, Poristii kremnii v poluprovodnikovoii elektronike, *Zarubejnaya Elektronnaya Tekhnika* **15**, 3–48 (1978).
288. L. T. Canham, Silicon quantum wire array fabrication by electrochemical and chemical dissolution of wafers, *Appl. Phys. Lett.* **57**(10), 1046–1048 (1990).

289. V. E. Borisenko, Nanostructure based informatics, in *Physics, Chemistry and Application of Nanostructures*, edited by V. E. Borisenko, A. B. Filonov, S. V. Gaponenko, and V. S. Gurin (BSUIR, Minsk, 1995), pp. 246–256.
290. V. E. Borisenko and A. M. Dorofeev, Gettering of impurities by incoherent light annealed porous silicon, in *Laser–Solid Interactions and Transient Thermal Processing of Materials*, edited by J. Narayan, W. L. Brown, and R. A. Lemons (North-Holland, Amsterdam, 1983), pp. 375–379.
291. V. P. Bondarenko, V. E. Borisenko, L. F. Gorskaya, A. M. Dorofeev, and A. G. Dutov, Redistribution of gold in single-crystal silicon during brief annealing by incoherent light, *Sov. Phys. Tech. Phys.* **29**(10), 1184–1186 (1984).
292. V. P. Bondarenko, V. E. Borisenko, and V. A. Labunov, Diffusion of arsenic through porous silicon as a result of heat treatment by exposure to incoherent light for tens of seconds, *Sov. Phys. Semicond.* **20**(5), 586–588 (1986).
293. T. Kimura, A. Yokoi, H. Horiguchi, R. Saito, T. Ikoma, and A. Sato, Electrochemical Er doping of porous silicon and its room-temperature luminescence at ~1.54 μm, *Appl. Phys. Lett.* **65**(8), 983–985 (1994).
294. A. M. Dorofeev, N. V. Gaponenko, V. P. Bondarenko, E. E. Bachilo, N. M. Kazuchits, A. A. Leshok, G. N. Troyanova, N. N. Vorozov, V. E. Borisenko, H. Gnaser, W. Bock, P. Becker, and H. Oechsner, Erbium luminescence in porous silicon doped from spin-on films, *J. Appl. Phys.* **77**(6), 2679–2683 (1995).
295. G. B. Larrabee and J. A. Keenan, Neutron activation analysis of epitaxial silicon, *J. Electrochem. Soc.* **118**(8), 1351–1355 (1971).
296. D. R. Sparks and R. G. Chapman, The use of rapid thermal annealing for studying transition metals in silicon, *J. Electrochem. Soc.* **133**(6), 1201–1205 (1986).
297. D. R. Sparks, R. G. Chapman, and N. S. Alvi, Anomalous diffusion and gettering of transition metals in silicon, *Appl. Phys. Lett.* **49**(9), 525–527 (1986).
298. B. I. Boltaks, *Diffusion in Semiconductors* (Academic Press, New York, 1963).
299. V. Labunov, V. Bondarenko, L. Glinenko, A. Dorofeev, and L. Tabulina, Heat treatment effect on porous silicon, *Thin Solid Films* **137**(1), 123–134 (1986).
300. V. A. Labunov, V. P. Bondarenko, V. E. Borisenko, and A. M. Dorofeev, High-temperature treatment of porous silicon, *Phys. Status Solidi A* **102**(1), 193–198 (1987).
301. H. Yamanaka, M. Kamoshida, and G. Haneta, Impurity diffusion in porous silicon formed by anodic reaction, *Jpn. J. Appl. Phys.* **13**(10), 1661–1662 (1974).
302. A. Polman, Erbium ion implantation for optical doping, *Mater. Res. Soc. Symp. Proc.* **316**, 385–395 (1994).
303. F. J. Morin and J. P. Maita, Electrical properties of silicon containing arsenic and boron, *Phys. Rev.* **96**(1), 28–35 (1954).
304. A. Zeeger, H. Foll, and W. Frank, Self-interstitials, vacancies and clusters in silicon and germanium, in *Radiation Effects in Semiconductors* (The Institute of Physics, Bristol, 1977), pp. 12–29.
305. G. B. Bronner and J. D. Plummer, Silicon interstitial generation by argon implantation, *Appl. Phys. Lett.* **46**(5), 510–512 (1985).
306. K. Taniguchi, D. A. Antoniadis, and Y. Matsushita, Kinetics of self-interstitial generated at the Si/SiO_2 interface, *Appl. Phys. Lett.* **42**(11), 961–963 (1983).
307. P. B. Griffin, P. M. Fahey, J. D. Plummer, and R. W. Dutton, Measurement of silicon interstitial diffusivity, *Appl. Phys. Lett.* **47**(3), 319–321 (1985).
308. D. Mathiot, Thermal donor formation in silicon: A new kinetic model based on self-interstitial aggregation, *Appl. Phys. Lett.* **51**(12), 904–906 (1987).
309. D. Maroudas and R. A. Brown, Atomistic calculation of the self-interstitial diffusivity in silicon, *Appl. Phys. Lett.* **62**(2), 172–174 (1993).
310. V. M. Gusev and V. V. Titov, Investigation of the kinetics of thermal annealing of radiation defects in silicon doped by the ion implantation method, *Sov. Phys. Semicond.* **3**(1), 1–6 (1969).
311. G. D. Watkins, Defect in irradiated silicon: EPR and electronnuclear double resonance of interstitial boron, *Phys. Rev. B* **12**(12), 5824–5839 (1975).
312. A. K. Tipping and R. C. Newman, An infrared study of the production, diffusion and complexing of interstitial boron in electron-irradiated silicon, *Semicond. Sci. Technol.* **2**(7), 389–398 (1987).

313. J. C. C. Tsai, D. G. Schimmel, R. B. Fair, and W. Maszara, Point defect generation during phosphorus diffusion in silicon. 1. Concentrations above solid solubility, *J. Electrochem. Soc.* **134**(6), 1508–1518 (1987).
314. A. M. Kosevich, *Fizicheskaya Mekhanika Realniikh Kristallov* (Nauka, Moscow, 1981).
315. P. V. Petrashen, Bragg diffraction of X-ray by crystals with impurities, *Sov. Phys. Solid State* **16**(8), 1417–1421 (1974).
316. S. P. Nikanorov and B. K. Kardashev, *Uprugost i Dislokatsionnaya Neuprugost Kristallov* (Nauka, Moscow, 1985).
317. A. N. Orlov, Tochechnie defektii v kristallakh i ih svoistva, in *Defekti v Kristallakh i Ih Modelirovanie na EVM* (Nauka, Leningrad, 1980), pp. 5–22.
318. G. Masetti, M. Severi, and S. Solmi, Modeling of mobility against carrier concentration in arsenic-, phosphorus-, and boron-doped silicon, *IEEE Trans. Electron Devices* **30**(5), 764–769 (1983).
319. J. C. Plunkett, J. L. Stone, and A. Leu, A computer algorithm for accurate and repeatable profile analysis using anodization and stripping of silicon, *Solid-State Electron.* **20**(5), 447–453 (1986).

CHAPTER 3

1. *Comparison of Thin Film Transistor and SOI Technologies,* edited by H. W. Lam and M. J. Thompson (North-Holland, Amsterdam, 1984).
2. *Polycrystalline Semiconductors. Physical Properties and Applications,* edited by G. Harbeke (Springer-Verlag, Berlin, 1985).
3. M. J. M. J. Josquin, P. R. Boudewijn, and Y. Tammiaga, Effectiveness of polycrystalline silicon diffusion sources, *Appl. Phys. Lett.* **43**(10), 960–962 (1983).
4. V. E. Borisenko and V. A. Samujlov, Tverdofaznie processi v polikristallicheskom kremnii pri impulsnoii termoobrabotke nekogerentnim svetom, *Zarubejnaya Elektronnaya Tekhnika* **1**, 45–68 (1987).
5. S. Solmi, M. Severy, and R. Angelucci, Electrical properties of thermally and laser annealed polycrystalline silicon films heavily doped with arsenic and phosphorus, *J. Electrochem. Soc.* **129**(8), 1811–1818 (1982).
6. J. Y. W. Seto, Annealing characteristics of boron- and phosphorus-implanted polycrystalline silicon, *J. Appl. Phys.* **47**(12), 5167–5170 (1976).
7. M. Kiselewicz, M. Zielinska-Szot, and W. Zuk, Ion implantation of impurities into polycrystalline silicon, *Acta Phys. Pol. A* **56**(5), 609–618 (1979).
8. J. R. Monkowski, J. Bloem, L. J. Giling, and M. W. M. Graef, Comparison of dopant incorporation into polycrystalline and monocrystalline silicon, *Appl. Phys. Lett.* **35**(5), 410–412 (1979).
9. S. Hasegawa, T. Kasajima, and T. Shimizu, Electrical activation process of phosphorus atoms with annealing for doped CVD poly-Si, *J. Appl. Phys.* **50**(11), 7256–7257 (1979).
10. T. Makino and H. Nakamura, Resistivity changes of heavily-boron-doped CVD-prepared polycrystalline silicon caused by thermal annealing, *Solid-State Electron.* **24**(1), 49–55 (1981).
11. N. Lifshitz, Solubility of implanted dopants in polysilicon: Phosphorus and arsenic, *J. Electrochem. Soc.* **130**(12), 2464–2467 (1983).
12. J. Murota and T. Sawai, Electrical characteristics of heavily arsenic and phosphorus doped polycrystalline silicon, *J. Appl. Phys.* **53**(5), 3702–3708 (1982).
13. M. Mandurah, K. C. Saraswat, C. R. Helms, and T. I. Kamins, Dopant segregation in polysilicon, *J. Appl. Phys.* **51**(11), 5755–5763 (1980).
14. M. E. Cowher and T. O. Sedgwick, Chemical vapor deposited polycrystalline silicon, *J. Electrochem. Soc.* **119**(11), 1565–1570 (1972).
15. T. I. Kamins, Structure and properties of LPCVD silicon films, *J. Electrochem. Soc.* **127**(3), 686–687 (1980).
16. M. M. Mandurah, K. C. Saraswat, and T. I. Kamins, Arsenic segregation in polycrystalline silicon, *Appl. Phys. Lett.* **36**(8), 683–685 (1980).
17. L. L. Kazmerski, P. J. Ireland, and T. F. Ciszek, Evidence for the segregation of impurities to grain boundaries in multigrained silicon using Auger electron spectroscopy and secondary ion mass spectroscopy, *Appl. Phys. Lett.* **36**(4), 323–325 (1980).

18. B. Swaminathan, E. Demoulin, T. W. Sigmon, R. W. Dutton, and R. Reif, Segregation of arsenic to the grain boundaries in polycrystalline silicon, *J. Electrochem. Soc.* **127**(10), 2227–2229 (1980).
19. A. Carabelas, D. Nobili, and S. Solmi, Grain boundary segregation in silicon heavily doped with phosphorus and arsenic, *J. Phys. (Paris)* **43**(10), c1/187–c1/192 (1982).
20. T. I. Kamins, J. Manolin, and R. N. Tucker, Diffusion of impurities in polycrystalline silicon, *J. Appl. Phys.* **43**(1), 83–91 (1972).
21. S. Horiuchi, Electrical characteristics of boron diffused polycrystalline silicon layers, *Solid-State Electron.* **18**(7/8), 659–665 (1975).
22. A. D. Buonaquisti, W. Carter, and P. H. Holloway, Diffusion characteristics of boron and phosphorus in polycrystalline silicon, *Thin Solid Films* **100**(3), 235–248 (1983).
23. P. H. Holloway, Grain boundary diffusion of phosphorus in polycrystalline silicon, *J. Vac. Sci. Technol.* **21**(1), 19–22 (1982).
24. H. F. Matare, Comments on: "Grain boundary diffusion of phosphorus in polycrystalline silicon," *J. Vac. Sci. Technol. B* **1**(1), 107 (1983).
25. J. L. Liotard, R. Biberian, and J. Cabane, La diffusion intergranulaire dans le silicium, *J. Phys. (Paris)* **43**(10), c1/213–c1/218 (1982).
26. H. Boumgart, H. J. Leamy, G. K. Celler, and L. E. Trimble, Grain boundary diffusion in polycrystalline silicon films on SiO_2, *J. Phys. (Paris)* **43**(10), c1/363–c1/369 (1982).
27. B. Swaminathan, K. C. Saraswat, R. W. Dutton, and T. I. Kamins, Diffusion of arsenic in polycrystalline silicon, *Appl. Phys. Lett.* **40**(9), 795–798 (1982).
28. Y. Sato, K. Murase, and H. Harada, A novel method to measure lateral diffusion length in polycrystalline silicon, *J. Electrochem. Soc.* **129**(7), 1635–1638 (1982).
29. M. Arienzo, Y. Komem, and A. E. Michel, Diffusion of arsenic in bilayer polycrystalline silicon films, *J. Appl. Phys.* **55**(2), 365–369 (1984).
30. H. Ryssel, H. Iberl, M. Bleier, G. Prinke, K. Haberger, and H. Kranz, Arsenic-implanted polysilicon layers, *Appl. Phys.* **24**(3), 197–200 (1981).
31. D. L. Losee, J. P. Lavine, E. A. Trabka, S.-T. Lee, and C. M. Jarman, Phosphorus diffusion in polycrystalline silicon, *J. Appl. Phys.* **55**(4), 1218–1220 (1984).
32. F. H. M. Spit, H. Albers, A. Lubbes, Q. J. A. Rijke, L. J. Van Ruijven, J. P. A. Westerveld, H. Bakker, and S. Radelaar, Diffusion of antimony (^{125}Sb) in polycrystalline silicon, *Phys. Status Solidi A* **89**(1), 105–115 (1985).
33. M. Takai, M. Izumi, K. Matunaga, K. Gamo, S. Namba, T. Minamisono, M. Miyauchi, and T. Hirao, Backscattering study of implanted arsenic distribution in poly-silicon on insulator, *Nucl. Instrum. Methods Phys. Res. B* **19/20**(1), 603–606 (1987).
34. K. Tsukamoto, Y. Akasaka, and K. Horie, Arsenic implantation into polycrystalline silicon and diffusion to silicon substrate, *J. Appl. Phys.* **48**(5), 1815–1821 (1977).
35. J. Murota and E. Arai, Relationship between total arsenic and electrically active arsenic concentrations in silicon produced by the diffusion process, *J. Appl. Phys.* **50**(2), 804–808 (1979).
36. H. C. De Graaff and J. G. De Groot, The SIS tunnel emitter: A theory for emitters with thin interface layers, *IEEE Trans. Electron Devices* **26**(11), 1771–1776 (1979).
37. P. Ashburn and B. Soerowirdjo, Arsenic profiles in bipolar transistors with polysilicon emitters, *Solid-State Electron.* **24**(5), 475–476 (1981).
38. W. J. M. J. Josquin, P. R. Boudewijn, and Y. Taminga, Effectiveness of polycrystalline silicon diffusion sources, *Appl. Phys. Lett.* **43**(10), 960–962 (1983).
39. S. P. Murarka, Phosphorus out-diffusion during high temperature anneal of phosphorus-doped polycrystalline silicon and SiO_2, *J. Appl. Phys.* **56**(8), 2225–2230 (1984).
40. Y. Wada and S. Nishimatsu, Grain growth mechanism of heavily phosphorus-implanted polycrystalline silicon, *J. Electrochem. Soc.* **125**(9), 1499–1504 (1978).
41. J. P. Colinge, E. Demoulin, F. Delannay, M. Lobet, and J. M. Temerson, Grain size and resistivity of LPCVD polycrystalline silicon films, *J. Electrochem. Soc.* **128**(9), 2009–2014 (1981).
42. C. H. Lee, Heat-treatment effect on boron implantation in polycrystalline silicon, *J. Electrochem. Soc.* **129**(7), 1604–1607 (1982).
43. L. Mei, M. Rivier, Y. Kwark, and R. W. Dutton, Grain-growth mechanisms in polysilicon, *J. Electrochem. Soc.* **129**(8), 1791–1795 (1982).

44. C. V. Thompson and H. I. Smith, Surface-energy-driven secondary grain growth in ultrathin (< 100 nm) films on silicon, *Appl. Phys. Lett.* **44**(6), 603–605 (1984).
45. L. R. Zheng, L. S. Hang, and J. W. Mayer, Grain growth in arsenic-implanted polycrystalline Si, *Appl. Phys. Lett.* **51**(25), 2139–2141 (1987).
46. C. Hill and S. Jones, Recrystallization of poly-Si, in *Properties of Silicon* (INSPEC, London, 1988), pp. 964–986.
47. R. Klabes, J. Matthai, M. Voelskow, and S. Mutze, Pulsed incoherent light annealing of arsenic and phosphorus implanted polycrystalline silicon, *Phys. Status Solidi A* **47**(1), K5–K7 (1982).
48. K. B. Kadyrakunov, E. V. Nidaev, A. E. Plotnicov, L. S. Smirnov, I. G. Melnik, and M. V. Makeev, Flash lamp annealing of ion-implanted polycrystalline silicon, *Phys. Status Solidi A* **75**(2), 483–488 (1983).
49. V. E. Borisenko, V. V. Gribkovskii, V. A. Labunov, V. A. Samuilov, and K. D. Yashin, Incoherent light annealing of phosphorus-doped polycrystalline silicon, *Phys. Status Solidi A* **75**(1), 117–120 (1983).
50. V. E. Borisenko, V. A. Samuilov, V. F. Stelmakh, and K. D. Yashin, Electrical properties of phosphorus doped polycrystalline silicon subjected to transient heating, in *Energy Pulse Modification of Semiconductors and Related Materials,* edited by K. Hennig (Akademie der Wissenschaften der DDR, Dresden, 1985), pp. 331–336.
51. J. Matthai, M. Voelskow, and R. Klabes, Electrical and structural properties of light pulse and thermally annealed polycrystalline silicon films, in *Energy Pulse Modification of Semiconductors and Related Materials,* edited by K. Hennig (Akademie der Wissenschaften der DDR, Dresden, 1985), pp. 337–342.
52. M. Voelskow, J. Matthai, and R. Klabes, Electrical properties of ion implanted and short time annealed polycrystalline silicon, *Phys. Status Solidi A* **86**(2), 781–788 (1984).
53. K. Takebayashi, T. Yokoyama, M. Yoshida, and M. Inoue, Infrared radiation annealing of ion-implanted polycrystalline silicon using a graphite heater, *J. Electrochem. Soc.* **130**(11), 2271–2274 (1983).
54. S. R. Wilson, W. M. Paulson, R. B. Gregory, J. D. Gressett, A. H. Hamdi, and F. D. McDaniel, Fast diffusion of As in polycrystalline silicon during rapid thermal annealing, *Appl. Phys. Lett.* **45**(4), 464–466 (1984).
55. S. J. Krause, S. R. Wilson, W. M. Paulson, and R. B. Gregory, Grain growth during transient annealing of As-implanted polycrystalline silicon, *Appl. Phys. Lett.* **45**(7), 778–780 (1984).
56. S. J. Krause, S. R. Wilson, R. B. Gregory, W. M. Paulson, J. A. Leavitt, L. C. McIntyre, J. L. Seerveld, and P. Stoss, Structural changes during transient post-annealing of preannealed and arsenic implanted polycrystalline silicon films, in *Rapid Thermal Processing*, edited by T. O. Sedgwick, T. E. Seidel, and B.-Y. Tsaur (MRS, Pittsburgh, 1986), pp. 145–152.
57. R. A. Powell and R. Chow, Dopant activation and redistribution in As-implanted polycrystalline silicon by rapid thermal processing, *J. Electrochem. Soc.* **132**(1), 194–198 (1985).
58. R. Chow and R. A. Powell, Activation and redistribution of implanted P and B in polycrystalline silicon by rapid thermal processing, *J. Vac. Sci. Technol. A* **3**(3), 892–895 (1985).
59. V. E. Borisenko, V. V. Gribkovskii, V. A. Samuilov, V. F. Stelmakh, and K. D. Yashin, Elektrofizicheskie parametri implantirovannih fosforom sloev polikremniya na kremnii, podvergnutih impulsnoii termoobrabotke, *Elektronnaya Tekhnika, Ser. 6, Materiali* **6**, 21–25 (1985).
60. H. B. Harrison, A. P. Pogany, and Y. Komem, Properties of gallium implanted furnace and rapidly annealed polycrystalline silicon, in *Rapid Thermal Processing of Electronic Materials*, edited by S. R. Wilson, R. Powell, and D. E. Davies (MRS, Pittsburgh, 1987), pp. 329–333.
61. E. F. Krimmel, A. G. Lutsch, and E. Doering, Contribution to electron beam annealing of high-dose ion-implanted polysilicon, *Phys. Status Solidi A* **71**(1), 451–456 (1982).
62. R. C. Cammarata, C. V. Thompson, and S. M. Garrison, Secondary grain growth during rapid thermal annealing of doped polysilicon films, in *Rapid Thermal Processing of Electronic Materials*, edited by S. R. Wilson, R. Powell, and D. E. Davies (MRS, Pittsburgh, 1987), pp. 335–339.
63. V. E. Borisenko, L. F. Gorskaya, A. G. Dutov, and V. A. Samuilov, Povedenie implantirovannogo fosfora v polikristallicheskom kremnii pri impulsnoii termoobrabotke, *Elektronnaya Tekhnika, Ser. 2, Poluprovodnikovie Pribori* **2**, 53–57 (1987).
64. J. L. Hoyt, E. Crabbe, J. F. Gibbons, and R. F. W. Pease, Epitaxial alignment of arsenic implanted polycrystalline silicon films on (100) silicon obtained by rapid thermal annealing, *Appl. Phys. Lett.* **50**(12), 751–753 (1987).
65. H. J. Bohm, H. Wendt, and H. Oppolzer, Diffusion of B and As from polycrystalline silicon during rapid optical annealing, *J. Appl. Phys.* **62**(7), 2784–2788 (1987).

66. J. L. Hoyt, E. F. Crabbe, J. F. Gibbons, and R. F. W. Pease, Epitaxial alignment of As implanted polysilicon emitters, in *Rapid Thermal Processing of Electronic Materials*, edited by S. R. Wilson, R. Powell, and D. E. Davies (MRS, Pittsburgh, 1987), pp. 47–52.
67. T. L. Alford, D. K. Yang, W. Maszara, V. H. Ozguz, J. J. Wortman, and G. A. Rozgonyi, Microstructure of implanted and rapid thermal annealed semi-insulating polycrystalline oxygen-doped silicon, *J. Electrochem. Soc.* **134**(4), 998–1003 (1987).
68. W. Andra, G. Gotz, H. Hobert, V. Misyuchenko, V. A. Samuilov, and V. Stelmakh, IR and RBS spectroscopy investigation of semi-insulating phosphorus-doped polycrystalline silicon layers, *Phys. Status Solidi A* **110**(1), 181–187 (1988).
69. A. Almaggoussi, J. Sicart, J. L. Robert, J. F. Joly, and A. Laugier, Enhanced mobility in SOI films annealed by rapid thermal annealing, *Appl. Surf. Sci.* **36**(1), 572–578 (1989).
70. M. Takai, M. Izumi, T. Yamamoto, S. Namba, and T. Minamisono, Rapid thermal annealing of arsenic-implanted poly-Si layers on insulator, *Nucl. Instrum. Methods Phys. Res. B* **39**(1–4), 352–356 (1989).
71. N. Natsuaki, M. Tamura, and T. Tokuyama, Transformation of CVD poly-Si films on Si substrates into single crystal during rapid thermal annealing, in *Layered Structures and Interface Kinetics*, edited by S. Furukawa (KTK Scientific Publishers, Tokyo, 1985), pp. 137–146.
72. H. B. Harrison, S. T. Johnson, Y. Komem, C. Wong, and S. Cohen, Using rapid thermal processing to induce epitaxial alignment of polycrystalline silicon films on (100) silicon, in *Materials Issues in Silicon Integrated Circuit Processing*, edited by M. Wittmer, I. Stimmell, and M. Strathan (MRS, Pittsburgh, 1986), pp. 455–458.
73. R. B. Fair, Concentration profiles of diffused dopants in silicon, in *Impurity Doping Processes in Silicon*, edited by F. F. Wang (North-Holland, Amsterdam, 1981), pp. 315–442.
74. V. E. Borisenko and S. G. Yudin, Steady-state solubility of substitutional impurities in silicon, *Phys. Status Solidi A* **101**(1), 123–127 (1987).
75. V. A. Samuilov and V. F. Stelmakh, Nonequilibrium impurity segregation in phosphorus implanted polycrystalline silicon subjected to transient heating, in *Energy Pulse and Particle Beam Modification of Materials*, edited by K. Hennig (Akademie-Verlag, Berlin, 1988), pp. 285–287.
76. L. Gerzberg and J. Meindl, A quantitative model of the effect of grain size on the resistivity of polycrystalline silicon, *IEEE Electron Device Lett.* **1**(3), 38–41 (1980).
77. V. E. Borisenko, S. N. Kornilov, V. A. Labunov, and Y. V. Kucherenko, Oborudovanie dlya impul'snoii termoobrabotki materialov poluprovodnikovoii elektroniki intensivnim nekogerentnim svetom, *Zarubejnaya Elektronnaya Tekhnika* **6**, 45–65 (1985).
78. G. A. Ruggles, S. N. Hong, J. J. Wortman, F. Y. Sorrel, and M. C. Ozturk, Sample geometry effects in rapid thermal annealing, *J. Vac. Sci. Technol. B* **8**(2), 122–127 (1990).
79. H. J. Kim and C. V. Thompson, The effects of dopants on surface-energy-driven secondary grain growth in silicon films, *J. Appl. Phys.* **67**(2), 757–767 (1990).
80. R. Angelucci, M. Severi, S. Solmi, and L. Baldi, Electrical properties of phosphorus-doped polycrystalline silicon films contaminated with oxygen, *Thin Solid Films* **103**(3), 275–281 (1983).
81. J. Bloem and W. A. P. Classen, Carbon in polycrystalline silicon, influence on resistivity and grain size, *Appl. Phys. Lett.* **40**(8), 725–726 (1982).

CHAPTER 4

1. *GaAs Microelectronics*, edited by N. G. Einspruch and W. R. Wisseman (Academic Press, New York, 1985).
2. B. L. Sharma, Implantation in InP technology, Solid State Technol. **32**(11), 113–117 (1989).
3. V. E. Borisenko, V. V. Gribkovskii, V. A. Labunov, and S. G. Yudin, Pulsed heating of semiconductors, *Phys. Status Solidi A* **86**(2), 573–583 (1984).
4. S. S. Gill and B. J. Sealy, Review of rapid thermal annealing of ion implanted GaAs, *J. Electrochem. Soc.* **133**(12), 2590–2596 (1986).
5. R. Singh, Rapid isothermal processing, *J. Appl. Phys.* **63**(8), R59–R114 (1988).
6. V. E. Borisenko, N. V. Gaponenko, and A. V. Nosenko, Impulsnaya termoobrabotka poluprovodnikovih soedinenii A^3B^5, *Zarubejnaya Elektronnaya Tekhnika* **7**, 3–61 (1990).

7. C. D. Thurmond, Phase equilibria in the GaAs and the GaP systems, *J. Phys. Chem. Solids* **26**(5), 785–802 (1965).
8. J. R. Arthur, Vapor pressures and phase equilibria in the Ga-As system, *J. Phys. Chem. Solids* **28**(11), 2257–2267 (1967).
9. C. T. Foxon, J. A. Harvey, and B. A. Joyce, The evaporation of GaAs under equilibrium and non-equilibrium conditions using a modulated beam technique, *J. Phys. Chem. Solids* **34**(10), 1693–1701 (1973).
10. C. Pupp, J. J. Murray, and R. F. Pottie, Vapour pressures of arsenic over InAs(c) and GaAs(c). The enthalpies of formation of InAs(c) and GaAs(c), *J. Chem. Thermodyn.* **6**(2), 123–134 (1974).
11. R. F. C. Farrow, The evaporation of InP under Knudsen (equilibrium) and Langmuir (free) evaporation conditions, *J. Phys. D* **7**(17), 2436–2448 (1974).
12. A. V. Nosenko, personal communication.
13. *Handbook of Thin Film Technology*, edited by L. I. Maissel and R. Glang (McGraw–Hill, New York, 1970).
14. J. T. A. Pollock and A. Rose, Surface temperature and dissociation loss during the pulsed laser annealing of GaAs, in *Energy Beam–Solid Interactions and Transient Thermal Processing*, edited by J. C. C. Fan and N. M. Johnson (North-Holland, Amsterdam, 1984), pp. 513–519.
15. T. E. Haynes, W. K. Chu, T. L. Aselage, and S. T. Picraux, Initial decomposition of GaAs during rapid thermal annealing, *Appl. Phys. Lett.* **49**(11), 666–668 (1986).
16. T. E. Haynes, W. K. Chu, T. L. Aselage, and S. T. Picraux, Initial evaporation rates from GaAs during rapid thermal processing, *J. Appl. Phys.* **63**(4), 1168–1176 (1988).
17. A. R. Von Neida, S. J. Pearton, M. Stavola, and R. Caruso, Effect of crystal stoichiometry on activation efficiency in Si implanted, rapid thermal annealed GaAs, *Appl. Phys. Lett.* **49**(25), 1708–1710 (1986).
18. M. Arai, K. Nishiyama, and N. Watanabe, Radiation annealing of GaAs implanted with Si, *Jpn. J. Appl. Phys.* **20**(2), L124–L126 (1981).
19. M. Kuzuhara, H. Kohzu, and Y. Takayama, Infrared rapid thermal annealing of Si-implanted GaAs, *Appl. Phys. Lett.* **41**(8), 755–758 (1982).
20. J. Grno, S. Bederka, and M. Vesely, Pulse radiation annealing of capless GaAs, *Phys. Status Solidi A* **79**(1), K41–K44 (1983).
21. P. Chambon, M. Berth, and B. Prevot, Shallow beryllium implantation in GaAs annealed by rapid thermal annealing, *Appl. Phys. Lett.* **46**(2), 162–164 (1985).
22. H. Kanber, R. J. Cipolli, W. B. Hellderson, and J. M. Whelan, A comparison of rapid thermal annealing and controlled atmosphere annealing of Si-implanted GaAs, *J. Appl. Phys.* **57**(10), 4732–4737 (1985).
23. J. A. Del Alamo and T. Mizutani, Rapid thermal annealing of InP using GaAs and InP proximity caps, *J. Appl. Phys.* **62**(8), 3456–3458 (1987).
24. T. Hara and C. Gelpey, Rapid thermal processing of silicon ion implanted channel layers in GaAs, in *Rapid Thermal Processing of Electronic Materials*, edited by S. R. Wilson, R. Powell, and D. E. Davies (MRS, Pittsburgh, 1987), pp. 417–424.
25. H. Kohzu, M. Kuzuhara, and Y. Takayama, Infrared rapid thermal annealing for GaAs device fabrication, *J. Appl. Phys.* **54**(9), 4998–5003 (1983).
26. J. S. Willims and S. J. Pearton, Rapid annealing of GaAs and related compounds, in *Energy Beam–Solid Interactions and Transient Thermal Processing*, edited by D. K. Biegelsen, G. A. Rozgonyi, and C. V. Shank (MRS, Pittsburgh, 1985), pp. 427–438.
27. N. Duhamel, E. V. K. Rao, M. Gauneau, H. Thibierge, and A. Mircea, Silicon implantation in semi-insulating bulk InP: Electrical and photoluminescence measurements, *J. Cryst. Growth* **64**(1), 186–193 (1983).
28. P. K. Bhattacharya, W. H. Goodman, and M. V. Rao, Photoluminescence in Si-implanted InP, *J. Appl. Phys.* **55**(2), 509–514 (1984).
29. M. Sacilotti, R. A. Masut, and A. P. Roth, Stabilization of InP substrate under annealing in the presence of GaAs, *Appl. Phys. Lett.* **48**(7), 481–483 (1986).
30. S. J. Pearton, K. D. Cummings, and G. P. Vella-Coleiro, Electrical activation of implanted Be, Mg, Zn, and Cd in GaAs by rapid thermal annealing, *J. Appl. Phys.* **58**(8), 3252–3254 (1985).

31. R. Blanchet, P. Viktorovich, J. Chave, and C. Santinelli, Reduction of fast interface states and suppression of drift phenomena in arsenic stabilized metal-insulator-InP structures, *Appl. Phys. Lett.* **46**(8), 761–763 (1985).
32. R. T. Blunt, M. S. M. Lamb, and R. Szweda, Crystallographic slip in GaAs wafers annealed using incoherent radiation, *Appl. Phys. Lett.* **47**(3), 304–306 (1985).
33. A. Tamura, T. Uenoyama, K. Nishii, K. Inoue, and T. Onuma, New rapid thermal annealing for GaAs digital integrated circuits, *J. Appl. Phys.* **62**(3), 1102–1107 (1987).
34. M. J. Goff, S. C. Wang, and T.-H. Yu, Elimination of slip lines in capless rapid thermal annealing of GaAs, *J. Mater. Res.* **3**(5), 911–913 (1988).
35. H. A. Lord, Thermal and stress analysis of semiconductor wafers in a rapid thermal processing oven, *IEEE Trans. Semicond. Manuf.* **1**(3), 105–114 (1988).
36. C. A. Armiento and F. C. Prince, Capless rapid thermal annealing of GaAs using an enhanced overpressure proximity technique, *Appl. Phys. Lett.* **48**(23), 1623–1625 (1986).
37. C. A. Armiento, L. L. Lehman, F. C. Prince, and S. Zemon, Capless rapid thermal annealing of GaAs implanted with Si^+ using an enhanced overpressure proximity method, *J. Electrochem. Soc.* **134**(8), 2010–2017 (1987).
38. J. D. Woodhouse, M. C. Gaidis, and J. P. Donnelly, Capless rapid thermal annealing of Si-implanted InP, *Appl. Phys. Lett.* **51**(3), 186–188 (1987).
39. S. J. Pearton and R. Caruso, Rapid thermal annealing of GaAs in graphite susceptor—Comparison with proximity annealing, *J. Appl. Phys.* **66**(2), 663–665 (1989).
40. S. J. Pearton, F. Ren, A. Katz, T. R. Fullowan, C. R. Abernathy, W. S. Hobson, and R. F. Kopf, Rapid isothermal processing for fabrication of GaAs-based electronic devices, *IEEE Trans. Electron Devices* **39**(1), 154–159 (1992).
41. H. Tews, R. Neumann, A. Hoepfner, and S. Gisdakis, Mg implant activation and diffusion in GaAs during rapid thermal annealing arsine ambient, *J. Appl. Phys.* **67**(6), 2857–2861 (1990).
42. T. Hiramoto, T. Saito, and T. Icoma, Rapid thermal annealing of Si^+ implanted GaAs in the presence of arsenic pressure by GaAs powder, *Jpn. J. Appl. Phys.* **24**(3), L193–L195 (1985).
43. T. E. Haynes, W. K. Chu, and S. T. Picraux, Direct measurement of evaporation during rapid thermal processing of capped GaAs, *Appl. Phys. Lett.* **50**(16), 1071–1073 (1987).
44. J. P. Donnelly, The electrical characteristics of ion implanted compound semiconductors, *Nucl. Instrum. Methods* **182/183**(Pt. 2), 553–571 (1981).
45. K. G. Stephens, Doping of III–V compound semiconductors by ion implantation, *Nucl. Instrum. Methods* **209/210**(Pt. 2), 589–614 (1983).
46. S. S. Gill and B. J. Sealy, Annealing of selenium implanted indium phosphide using a graphite strip heater, *J. Appl. Phys.* **56**(4), 1189–1194 (1984).
47. L. S. Vanasupa, M. D. Deal, and J. D. Plummer, Effects of stress on the electrical activation of implanted Si in GaAs, *Appl. Phys. Lett.* **55**(3), 274–276 (1989).
48. J. P. de Souza and D. K. Sadana, Ion implantation in gallium arsenide MESFET technology, *IEEE Trans. Electron Devices* **39**(1), 166–175 (1992).
49. N. J. Barrett, J. D. Grange, B. J. Sealy, and K. G. Stephens, Annealing of selenium-implanted GaAs, *J. Appl. Phys.* **56**(12), 3503–3507 (1984).
50. R. Bensalem, N. J. Barrett, and B. J. Sealy, AlN capped annealing of Se and Sn implanted semi-insulating GaAs, *Electron. Lett.* **19**(3), 112–113 (1983).
51. N. J. Barrett, J. D. Grange, B. J. Sealy, and K. G. Stephens, Annealing of zinc-implanted GaAs, *J. Appl. Phys.* **57**(12), 5470–5476 (1985).
52. H. Nishi, Ion implantation for high-speed III–V IC s, *Nucl. Instrum. Methods Phys. Res. B* **7/8**(Pt. 1), 395–401 (1985).
53. R. L. Chapman, J. C. C. Fan, J. P. Donnelly, and B.-Y. Tsaur, Transient annealing of selenium implanted gallium arsenide using a graphite strip heater, *Appl. Phys. Lett.* **40**(9), 805–807 (1982).
54. M. Kuzuhara, Rapid thermal processing for high-speed III–V compound devices, in *Rapid Thermal Processing of Electronic Materials*, edited by S. R. Wilson, R. Powell, and D. E. Davies (MRS, Pittsburgh, 1987), pp. 401–410.
55. J. P. Donnelly and C. E. Hurwitz, Ion-implanted n- and p-type layers in InP, *Appl. Phys. Lett.* **31**(7), 418–420 (1977).

56. N. Arnold, R. Schmitt, and K. Heime, Diffusion in III–V semiconductors from spin-on-film sources, *J. Phys. D* **17**(3), 443–474 (1984).
57. T. P. Ma and K. Miyauchi, MIS structures based on spin-on SiO_2 on GaAs, *Appl. Phys. Lett.* **34**(1), 88–90 (1979).
58. E. Arai and Y. Terunuma, Structural changes of arsenic silicate glasses with heat treatment, *Jpn. J. Appl. Phys.* **9**(6), 691–704 (1970).
59. M. Nishitsuji and A. Tamura, Rapid thermal annealing of Si-implanted GaAs using the Ga-doped spin-on glass films, *Appl. Phys. Lett.* **63**(10), 1384–1386 (1993).
60. J. M. Molarius, E. Kolawa, K. Morishita, M.-A. Nicolet, J. L. Tandon, J. A. Leavitt, and L. C. McIntyre, Jr., Tantalum-based encapsulants for thermal annealing of GaAs, *J. Electrochem. Soc.* **138**(3), 834–837 (1991).
61. M. Ghezzo and D. M. Brown, Diffusivity summary of B, Ga, P, As and Sb in SiO_2, *J. Electrochem. Soc.* **120**(1), 146–148 (1973).
62. A. H. Van Ommen, Diffusion of ion-implanted Ga in SiO_2, *J. Appl. Phys.* **57**(6), 1872–1879 (1985).
63. I. Ohdomari, S. Mizutani, H. Kume, M. Mori, I. Kimura, and K. Yoneda, High-temperature annealing of the SiO_2/GaAs system, *Appl. Phys. Lett.* **32**(4), 218–220 (1978).
64. M. Kuzuhara, H. Kohzu, and Y. Takayama, Rapid thermal annealing of III–V compound materials, in *Energy Beam–Solid Interactions and Transient Thermal Processing*, edited by J. C. C. Fan and N. M. Johnson (North-Holland, Amsterdam, 1984), pp. 651–662.
65. J. D. Oberstar, B. G. Streetman, J. E. Baker, N. L. Finnegan, E. A. Sammann, and P. Williams, Annealing encapsulants for InP. I. Auger electron and secondary ion mass spectrometric studies, *Thin Solid Films* **94**(2), 149–159 (1982).
66. J. D. Oberstar and B. G. Streetman, Annealing encapsulants for InP. II. Photoluminescence studies, *Thin Solid Films* **94**(2), 161–170 (1982).
67. J. S. Blakemore, Semiconducting and other major properties of gallium arsenide, *J. Appl. Phys.* **53**(10), R123–R181 (1982).
68. S. Adachi, GaAs, AlAs, $Al_xGa_{1-x}As$: Material parameters for use in research and device applications, *J. Appl. Phys.* **58**(3), R1–R29 (1985).
69. G. A. Slack and S. F. Bartram, Thermal expansion of some diamond-like crystals, *J. Appl. Phys.* **46**(1), 89–98 (1975).
70. R. Bisaro, P. Merenda, and T. P. Pearsall, The thermal-expansion parameters of some $Ga_xIn_{1-x}As_yP_{1-y}$ alloys, *Appl. Phys. Lett.* **34**(1), 100–102 (1979).
71. P. J. Burkhardt and R. F. Marvel, Thermal expansion of sputtered silicon nitride films, *J. Electrochem. Soc.* **116**(6), 864–866 (1969).
72. J. Wong, Thermal stress in CVD films: The case of binary arsenosilicate glasses [diffusion source], *J. Electrochem. Soc.* **119**(8), 1080–1084 (1972).
73. S. J. Pearton, J. M. Gibson, D. C. Jacobson, J. M. Poate, J. S. Williams, and D. O. Boerma, Transient thermal processing of GaAs, in *Rapid Thermal Processing*, edited by T. O. Sedgwick, T. E. Seidel, and B.-Y. Tsaur (MRS, Pittsburgh, 1986), pp. 351–360.
74. D. V. Lang, A. Y. Cho, A. C. Gossard, and M. Ilegems, Study of electron traps in n-GaAs grown by molecular beam epitaxy, *J. Appl. Phys.* **47**(6), 2558–2564 (1976).
75. G. M. Martin, A. Mitonneau, and A. Mircea, Electron traps in bulk and epitaxial GaAs crystals, *Electron. Lett.* **13**(7), 191–193 (1977).
76. J. H. Neave, P. Blood, and B. A. Joyce, A correlation between electron traps and growth process in n-GaAs prepared by molecular beam epitaxy, *Appl. Phys. Lett.* **36**(4), 311–312 (1980).
77. P. Blood and J. J. Harris, Deep states in GaAs grown by molecular beam epitaxy, *J. Appl. Phys.* **56**(4), 993–1007 (1984).
78. R. Y. DeJule, M. A. Hasse, G. E. Stillman, S. C. Palmateer, and J. C. M. Hwang, Measurements of deep levels in high-purity molecular beam epitaxial GaAs, *J. Appl. Phys.* **57**(12), 5287–5289 (1985).
79. M. O. Manasreh, D. W. Fischer, and W. C. Mitchel, The EL2 defect in GaAs: Some recent developments, *Phys. Status Solidi B* **154**(1), 11–41 (1989).
80. D. Vignaud and J. L. Farvacque, A quantitative study of the creation of EL2 defects in GaAs by plastic deformation, *J. Appl. Phys.* **65**(4), 1516–1520 (1989).

81. T. Wosinski, Evidence for the electron traps at dislocations in GaAs crystals, *J. Appl. Phys.* **65**(4), 1566–1570 (1989).
82. M. Kuzuhara, T. Nozaki, and T. Kamejima, Characterization of Ga out-diffusion from GaAs into SiO_xN_y films during thermal annealing, *J. Appl. Phys.* **66**(12), 5833–5836 (1989).
83. S. T. Lai, B. D. Never, D. Alexiev, F. Faraone, T. C. Ku, and N. Dytlewski, Comparison of neutron and electron irradiation on the EL2 defects in GaAs, *J. Appl. Phys.* **77**(7), 3088–3094 (1995).
84. F. A. Wang, M.-F. Rau, J. Kurz, D.-D. Liao, and R. Carter, Distinguishing between EL2 and dislocation formation mechanisms in GaAs by mapping topographies, *J. Cryst. Growth* **103**(1–4), 311–322 (1990).
85. J. E. Bisbee and N. C. Halder, Activation energy and distribution function of the EL2 defect level in Si-implanted GaAs, *Phys. Status Solidi A* **119**(2), 545–553 (1990).
86. M. Kuzuhara and T. Nozaki, Study of electron traps in n-GaAs resulting from infrared rapid thermal annealing, *J. Appl. Phys.* **59**(9), 3131–3136 (1986).
87. A. Kitagawa, A. Usami, T. Wada, Y. Tokuda, and H. Kano, Production of the midgap electron trap (EL2) in molecular-beam-epitaxial GaAs by rapid thermal processing, *J. Appl. Phys.* **61**(3), 1215–1217 (1987).
88. M. Katayama, A. Usami, T. Wada, and Y. Tokuda, Variations of electron traps in bulk n-GaAs by rapid thermal processing, *J. Appl. Phys.* **62**(2), 528–533 (1987).
89. W. R. Buchwald, N. M. Johnson, and L. P. Trombetta, New metastable defects in GaAs, *Appl. Phys. Lett.* **50**(15), 1007–1009 (1987).
90. F. P. Korshunov, N. A. Sobolev, N. G. Kolin, E. A. Kudryavtseva, and T. A. Prokhorenko, Pulsed annealing of neutron-transmutation-doped gallium arsenide, *Sov. Phys. Semicond.* **22**(10), 1169–1171 (1988).
91. A. Kitagawa, A. Usami, and W. Takao, Effects of rapid thermal processing on electron traps in molecular-beam-epitaxial GaAs, *J. Appl. Phys.* **65**(2), 606–611 (1989).
92. H. Y. Cho, E. K. Kim, S.-K. Min, J. H. Yoon, and S. H. Choh, Deep levels in GaAs grown on Si during rapid thermal annealing, *Appl. Phys. Lett.* **56**(8), 761–763 (1990).
93. S. Dhar, K. S. Seo, and P. K. Bhattacharya, Nature and distribution of electrically active defects in Si-implanted and lamp-annealed GaAs, *J. Appl. Phys.* **58**(11), 4216–4220 (1985).
94. A. Kitagawa, A. Usami, and T. Wada, Characteristics of electron traps in Si-implanted and rapidly thermal-annealed GaAs, *J. Appl. Phys.* **63**(2), 414–420 (1988).
95. H. Y. Cho, E. K. Kim, and S.-K. Min, Deep levels in Si- and Be-implanted GaAs, *J. Appl. Phys.* **70**(2), 661–664 (1991).
96. K. Wada and N. Inoue, Effects of heat treatments of GaAs on the near surface distribution of EL2 defects, *Appl. Phys. Lett.* **47**(9), 945–947 (1985).
97. H. J. Von Bardeleben, D. Stievenard, J. C. Bourgoin, and A. Huber, Identification of EL2 in GaAs, *Appl. Phys. Lett.* **47**(9), 970–972 (1985).
98. J. Lagowski, H. C. Gatos, J. M. Parsey, K. Wada, M. Kaminska, and W. Walukiewich, Origin of the 0.82-eV electron trap in GaAs and its annihilation by shallow donors, *Appl. Phys. Lett.* **40**(4), 342–344 (1982).
99. G. F. Wager and J. A. Van Vechten, Atomic model for the EL2 defect in GaAs, *Phys. Rev. B* **35**(5), 2330–2339 (1987).
100. F. S. Ham, Stress-assisted precipitation on dislocations, *J. Appl. Phys.* **30**(6), 915–926 (1959).
101. D. K. Sadana, Mechanisms of amorphization and recrystallization in ion implanted III–V compound semiconductors, *Nucl. Instrum. Methods Phys. Res. B* **7/8**(Pt. 1), 375–386 (1985).
102. P. Auvray, A. Guivarc'h, H. L'Haridon, G. Pelous, M. Salvi, and P. Henoc, Epitaxial regrowth of (100) InP layers amorphized by ion implantation at room temperature, *J. Appl. Phys.* **53**(9), 6202–6207 (1982).
103. L. A. Christel and J. F. Gibbons, Stoichiometric disturbances in ion implanted compound semiconductors, *J. Appl. Phys.* **52**(8), 5050–5055 (1981).
104. J. S. Williams, Transient annealing of ion implanted gallium arsenide, in *Energy Beam–Solid Interactions and Transient Thermal Processing*, edited by J. Narayan, W. L. Brown, and R. A. Lemons (North-Holland, Amsterdam, 1983), pp. 621–632.
105. W. Wesch and G. Gotz, Rapid annealing of ion-implanted GaAs, *Phys. Status Solidi A* **94**(2), 745–766 (1986).
106. J. S. Williams and M. W. Austin, Low-temperature epitaxial regrowth of ion implanted amorphous GaAs, *Appl. Phys. Lett.* **36**(12), 994–996 (1980).

107. R. S. Bhattacharya, P. P. Pronko, and S. C. Ling, Low temperature annealing behavior of Se-implanted GaAs studied by high resolution Rutherford backscattering channeling, *J. Appl. Phys.* **53**(3), 1804–1806 (1982).
108. S. I. Kwun, M. H. Lee, L. L. Liou, W. G. Spitzer, H. L. Dunlap, and K. V. Vaidyanathan, Solid phase regrowth of low temperature Be-implanted GaAs, *J. Appl. Phys.* **57**(4), 1022–1028 (1985).
109. C. Licoppe, Y. I. Nissim, and C. Meriadec, Direct measurement of solid-phase epitaxial growth kinetics in GaAs by time-resolved reflectivity, *J. Appl. Phys.* **58**(8), 3094–3096 (1985).
110. C. Licoppe, Y. I. Nissim, P. Krauz, and P. Henoc, Solid phase epitaxial regrowth of ion-implanted amorphized InP, *Appl. Phys. Lett.* **49**(6), 316–318 (1986).
111. C. Licoppe, Y. I. Nissim, C. Meriadec, and P. Henoc, Recrystallization kinetics pattern in III–V implanted semiconductors, *Appl. Phys. Lett.* **50**(23), 1648–1650 (1987).
112. T. Suzuki, H. Sakurai, and M. Arai, Infrared thermal annealing of Zn-implanted GaAs, *Appl. Phys. Lett.* **43**(10), 951–953 (1983).
113. S. J. Pearton, R. Hull, D. C. Jacobson, J. M. Poate, and J. S. Williams, Relationship between secondary defects and electrical activation in ion implanted, rapidly annealed GaAs, *Appl. Phys. Lett.* **48**(1), 38–40 (1986).
114. Y. I. Nissim, B. Joukoff, J. Sapriel, and P. Henoc, Annealing of high dose implanted GaAs with halogen lamps, in *Energy Beam–Solid Interactions and Transient Thermal Processing*, edited by J. C. C. Fan and N. M. Johnson (North-Holland, Amsterdam, 1984), pp. 675–680.
115. G. Bahir, J. L. Merz, J. R. Abelson, and T. W. Sigmon, Correlation of Rutherford backscattering and electrical measurements on Si implanted InP following rapid thermal and furnace annealing, in *Ion Beam Processes in Advanced Electronic Materials and Device Technology*, edited by B. R. Appleton, F. H. Eisen, and T. W. Sigmon (MRS, Pittsburgh, 1985), pp. 297–302.
116. D. Kirillov, J. L. Merz, R. Kalish, and A. Ron, Band-to-band luminescence of ion implanted InP after rapid lamp annealing, *Appl. Phys. Lett.* **44**(6), 609–610 (1984).
117. D. Kirillov, J. L. Merz, R. Kalish, and S. Shatas, Luminescence study of rapid lamp annealing of Si implanted InP, *J. Appl. Phys.* **57**(2), 531–536 (1985).
118. B. Tell, K. F. Brown-Goebeler, and C. L. Cheng, Rapid thermal annealing of elevated-temperature silicon implants in InP, *Appl. Phys. Lett.* **52**(4), 299–301 (1988).
119. D. E. Davies, J. K. Kennedy, and C. E. Ludington, Comparison of group IV and VI doping by implantation in GaAs, *J. Electrochem. Soc.* **122**(10), 1374–1377 (1975).
120. S. S. Kular, B. J. Sealy, Y. Onos, and K. G. Stephens, The electrical properties of zinc implanted GaAs, *Solid-State Electron.* **27**(1), 83–88 (1984).
121. P. Kringhoj, J. L. Hansen, and S. Y. Shiryaev, Structural and electric characteristics of Ge and Se implanted InP after rapid thermal annealing, *J. Appl. Phys.* **72**(6), 2249 –2255 (1992).
122. H. G. Robinson, T. E. Haynes, E. L. Allen, C. C. Lee, M. D. Deal, and K. S. Jones, Effect of implant temperature on dopant diffusion and defect morphology for Si implanted GaAs, *J. Appl. Phys.* **76**(8), 4571–4575 (1994).
123. B. I. Boltaks, *Diffusiva i Tochechnie Defekti v Poluprovodnikah* (Nauka, Leningrad, 1972).
124. H. C. Casey, Jr., Diffusion in the III–V compound semiconductors, in *Atomic Diffusion in Semiconductors*, edited by D. Shaw (Plenum Press, New York, 1973), pp. 351–430.
125. K. V. Vaidyanathan and H. L. Dunlap, Rapid thermal annealing of ion implanted GaAs and InP, in *Energy Beam–Solid Interactions and Transient Thermal Processing*, edited by J. C. C. Fan and N. M. Johnson (North-Holland, Amsterdam, 1984), pp. 687–691.
126. K. D. Cumming, S. J. Pearton, and G. P. Vella-Coleiro, Rapid thermal annealing of GaAs uniformity and temperature dependence of activation, *J. Appl. Phys.* **60**(1), 163–168 (1986).
127. N. J. Barrett, D. C. Bartle, R. Nicholls, and J. D. Grange, Optical furnace annealing of Be implanted GaAs, in *GaAs and Related Compounds*, Inst. Phys. Conf. Ser. No. 74, pp. 77–82 (1984).
128. K. Tabatabaie-Alavi, A. N. M. Masum Choudhury, C. G. Fonstad, and J. C. Gelpey, Rapid thermal annealing of Be, Si, and Zn implanted GaAs using an ultrahigh power argon arc lamp, *Appl. Phys. Lett.* **43**(5), 505–507 (1983).
129. M. D. Deal and H. G. Robinson, Diffusion of implanted beryllium in gallium arsenide as a function of anneal temperature and dose, *Appl. Phys. Lett.* **55**(10), 996–998 (1989).

130. A. C. T. Tang, B. J. Sealy, and A. A. Rezazadeh, Thermal stability of Be-, Mg-, and Zn-implanted layers in GaAs for high temperature device-processing technology, *J. Appl. Phys.* **66**(6), 2759–2761 (1989).
131. H. Baratte, D. K. Sadana, J. P. de Souza, P. E. Hallali, R. G. Schad, M. Norcott, and F. Cardone, Outdiffusion of Be during rapid thermal annealing of high-dose Be-implanted GaAs, *J. Appl. Phys.* **67**(10), 6589–6591 (1990).
132. A. N. M. Masum Choudhury, K. Tabatabaie-Alavi, C. G. Fonstad, and J. C. Gelpey, Rapid thermal annealing of Se and Be implanted InP using an ultrahigh power argon arc lamp, *Appl. Phys. Lett.* **43**(4), 381–383 (1983).
133. B. Molnar and H. B. Dietrich, Comparison of isothermal anneal techniques for Be or Si implanted S.I. InP, in *Rapid Thermal Processing*, edited by T. O. Sedgwick, T. E. Seidel, and B.-Y. Tsaur (MRS, Pittsburgh, 1986), pp. 417–422.
134. S. J. Perton, S. Nakahara, A. R. Von Neida, K. I. Short, and L. J. Oster, Implantation characteristics of InSb, *J. Appl. Phys.* **66**(5), 1942–1946 (1989).
135. M. Maier and J. Selders, Rapid thermal and furnace anneal of beryllium-implanted $Ga_{0.47}In_{0.53}As$, *J. Appl. Phys.* **60**(8), 2783–2787 (1986).
136. G. S. Lam and C. G. Fonstad, Rapid thermal annealing of Be implanted $In_{0.53}Ga_{0.47}As$, in *Rapid Thermal Processing*, edited by T. O. Sedgwick, T. E. Seidel, and B.-Y. Tsaur (MRS, Pittsburgh, 1986), pp. 397–402.
137. M. V. Rao, Rapid isothermal annealing of high- and low-energy ion-implanted InP and $In_{0.53}Ga_{0.47}As$, *IEEE Trans. Electron Devices* **39**(1), 160–165 (1992).
138. W. Lee and C. G. Fonstad, Rapid thermal annealing of Be^+- implanted $In_{0.62}Al_{0.48}As$, *J. Appl. Phys.* **61**(12), 5272–5278 (1987).
139. S. J. Pearton, W. S. Hobson, A. E. Von Neida, N. M. Haegel, K. S. Jones, N. Morris, and B. J. Sealy, Implant activation and redistribution in $Al_xGa_{1-x}As$, *J. Appl. Phys.* **67**(5), 2396–2409 (1990).
140. K. K. Patel and B. J. Sealy, Incoherent light annealing of Mg implanted GaAs, *Radiat. Eff.* **91**(1), 53–60 (1985).
141. K. K. Patel and B. J. Sealy, Rapid thermal annealing of Mg^+ + As^+ dual implants in GaAs, *Appl. Phys. Lett.* **48**(21), 1467–1469 (1986).
142. B. Descouts, N. Duhamel, S. Godefroy, and P. Krauz, Rapid thermal anneal in InP, GaAs and GaAs/GaAlAs, *Nucl. Instrum. Methods Phys. Res. B* **19/20**(1), 496–500 (1987).
143. A. N. M. Masum Choudhury and C. A. Armiento, Formation of device quality p-type layers in GaAs using co-implantation of Mg^+ and As^+ and capless rapid thermal annealing, in *Rapid Thermal Processing of Electronic Materials*, edited by S. R. Wilson, R. Powell, and D. E. Davies (MRS, Pittsburgh, 1987), pp. 425–430.
144. W. H. Van Berlo and G. Landgren, High dose magnesium implantation in InP activated by rapid thermal annealing, *J. Appl. Phys.* **66**(7), 3117–3120 (1989).
145. S. S. Kular, B. J. Sealy, K. G. Stephens, D. Sadana, and G. R. Booker, Electrical, Rutherford backscattering and transmission electron microscopy studies of furnace annealed zinc implanted GaAs, *Solid-State Electron.* **23**(8), 831–838 (1980).
146. D. E. Davies, Transient thermal annealing in gallium arsenide, *Nucl. Instrum. Methods Phys. Res. B* **7/8**(Pt. 1), 387–394 (1985).
147. D. E. Davies and P. J. McNally, Enhanced activation of Zn-implanted GaAs, *Appl. Phys. Lett.* **44**(3), 304–306 (1984).
148. K. Yokota, M. Kimura, H. Nakanishi, and S. Tamura, Halogen and mercury lamp annealing of Cd-implanted GaAs, *J. Electrochem. Soc.* **136**(11), 3450–3454 (1989).
149. C. S. Lam and C. G. Fonstad, Ion implantation and rapid thermal annealing of Mg, Cd, and Si in $Al_xGa_{1-x}As$ grown by molecular-beam epitaxy, *J. Appl. Phys.* **64**(4), 2103–2106 (1988).
150. J. H. Wilkie and B. J. Sealy, Rapid thermal annealing of 200 °C mercury implants into InP, *Electron. Lett.* **22**(24), 1308–1309 (1986).
151. J. H. Wilkie and B. J. Sealy, Implantation damage control of silicon indiffusion during rapid thermal annealing of InP using AlN/Si_3N_4 as encapsulant, *Thin Solid Films* **162**(1), 49–57 (1988).
152. J. H. Wilkie and B. J. Sealy, Redistribution of ion-implanted mercury during rapid thermal annealing of $Ga_{0.47}In_{0.53}As$ and InP, in *Solid State Devices*, edited by G. Soncini and P. U. Calzolari (North-Holland, Amsterdam, 1988), pp. 919–922.

153. J. C. Soares, A. A. Melo, E. Alves, and M. F. da Silva, Lattice site location and outdiffusion of mercury implanted in GaAs, *Nucl. Instrum. Methods Phys. Res. B* **59/60**(1), 1090–1093 (1991).
154. R. Kwor, Y. K. Yeo, and Y. S. Park, Electrical properties and distribution of sulfur implants in GaAs, *J. Appl. Phys.* **53**(7), 4786–4792 (1982).
155. K. Ito, M. Yoshida, M. Otsubo, and T. Murotani, Radiation annealing of Si- and S-implanted GaAs, *Jpn. J. Appl. Phys.* **22**(5), L299–L300 (1983).
156. M. Kuzuhara, H. Kohzu, and Y. Takayama, Electrical properties of S implants in GaAs activated by rapid thermal annealing, *J. Appl. Phys.* **54**(6), 3121–3124 (1983).
157. S. Banerjee and J. Baker, Proximity annealing of sulfur-implanted gallium arsenide using a strip heater, *Jpn. J. Appl. Phys.* **24**(5), L377–L379 (1985).
158. S. J. Pearton and K. D. Cummings, Diffusion phenomena and defect generation in rapidly annealed GaAs, *J. Appl. Phys.* **58**(4), 1500–1504 (1985).
159. B. J. Sealy, R. Bensalem, and K. K. Patel, Transient annealing for the production of n^+ contact layers in GaAs, *Nucl. Instrum. Methods Phys. Res. B* **6**(1/2), 325–329 (1985).
160. K. K. Patel, R. Bensalem, M. A. Shahid, and B. J. Sealy, Se^+ and Sn^+ implants for n^+ layers in GaAs, *Nucl. Instrum. Methods Phys. Res. B* **7/8**(Pt. 1), 418–422 (1985).
161. T. Penna, B. Tell, A. S. H. Liao, T. J. Bridges, and G. Burkhard, Ion implantation of Si and Se donors in $In_{0.53}Ga_{0.47}As$, *J. Appl. Phys.* **57**(2), 351–354 (1985).
162. S. T. Johnson, K. G. Orrman-Rositer, and J. S. Williams, Solid solubility of Sb and Te implanted GaAs following solid phase annealing, in *Energy Beam–Solid Interactions and Transient Thermal Processing*, edited by J. C. C. Fan and N. M. Johnson (North-Holland, Amsterdam, 1984), pp. 663–667.
163. Y. I. Nissim, B. Joukoff, J. Sapriel, and P. Henoc, Annealing of high dose implanted GaAs with halogen lamps, in *Energy Beam–Solid Interactions and Transient Thermal Processing*, edited by J. C. C. Fan and N. M. Johnson (North-Holland, Amsterdam, 1984), pp. 675–679.
164. V. M. Vorob'ev, V. A. Murav'ev, and V. A. Panteleev, Migration of amphoteric impurities in A^3B^5 compounds, *Sov. Phys. Solid State* **27**(9), 2568–2572 (1985).
165. A. T. Yuen, S. I. Long, and J. L. Merz, Rapid thermal anneal and furnace anneal of silicon and beryllium implanted gallium arsenide, in *Ion Beam Processes in Advanced Electronic Materials and Device Technology*, edited by B. R. Appleton, F. H. Eisen, and T. W. Sigmon (MRS, Pittsburgh, 1985), pp. 285–290.
166. J. Wagner, H. Seelewind, and W. Jantz, Dopant incorporation in Si-implanted and thermally annealed GaAs, *J. Appl. Phys.* **67**(4), 1779–1783 (1990).
167. J. Maguire, R. Murray, R. C. Newman, R. B. Beall, and J. J. Harris, Mechanism of compensation in heavily silicon-doped gallium arsenide grown by molecular beam epitaxy, *Appl. Phys. Lett.* **50**(9), 516–518 (1987).
168. R. Murray, R. C. Newman, M. J. L. Sangster, R. B. Beall, J. J. Harris, P. J. Wright, J. Wagner, and M. Ramsteiner, The calibration of the strength of the localized vibrational modes of silicon impurities in epitaxial GaAs revealed by infrared absorption and Raman scattering, *J. Appl. Phys.* **66**(6), 2589–2596 (1989).
169. H. Ono and R. C. Newman, The complexing of silicon impurities with point defects in plastically deformed and annealed GaAs, *J. Appl. Phys.* **66**(1), 141–145 (1989).
170. S. Sigitani, F. Hynga, and K. Yamasaki, Phosphorus coimplantation effects on optimum annealing temperature in Si-implanted GaAs, *J. Appl. Phys.* **67**(1), 552–554 (1990).
171. K. J. Soda, J. P. Lorenzo, D. E. Davies, and T. G. Ryan, Study of dielectric caps for incoherent lamp anneal of InP, in *Energy Beam–Solid Interactions and Transient Thermal Processing*, edited by J. C. C. Fan and N. M. Johnson (North-Holland, Amsterdam, 1984), pp. 693–698.
172. M. V. Rao, Electrical and optical nonuniformity of Si-implanted and rapid thermal annealed InP:Fe, *Appl. Phys. Lett.* **48**(22), 1522–1524 (1986).
173. M. V. Rao, Two-step rapid thermal annealing of Si-implanted InP:Fe, *Appl. Phys. Lett.* **50**(20), 1444–1446 (1987).
174. H. Kanber and J. M. Whelan, Substrate impurity migration during rapid thermal annealing of Si implanted GaAs, *J. Electrochem. Soc.* **134**(10), 2596–2599 (1987).
175. M. D. Deal and D. A. Stevenson, The solubility of chromium in gallium arsenide, *J. Electrochem. Soc.* **131**(10), 2343–2347 (1984).

176. A. Tamura, K. Inoue, and T. Onuma, Enhancement in activation efficiency for a SiF_3-implanted GaAs layer by a new annealing method, *Appl. Phys. Lett.* **54**(6), 503–504 (1988).
177. C. C. Lee, M. D. Deal, and J. C. Bravman, Eliminating dopant diffusion after ion implantation by surface etching, *Appl. Phys. Lett.* **64**(24), 3302–3304 (1994).
178. H. G. Robinson, M. D. Deal, G. Amaratunga, P. B. Griffin, D. A. Stevenson, and J. D. Plummer, Modeling uphill diffusion of Mg implants in GaAs using SUPREM-IV, *J. Appl. Phys.* **71**(6), 2615–2623 (1992).
179. P. Kringhoj, V. V. Gribkovskii, and A. Nylandsted Larsen, Rapid thermal annealing of Ge-implanted InP, *Appl. Phys. Lett.* **57**(15), 1514–1516 (1990).
180. M. C. Ridgway and P. Kringhoj, Rapid thermal annealing of Sn-implanted InP, *J. Appl. Phys.* **77**(6), 2375–2379 (1995).
181. S. J. Pearton and C. R. Abernathy, Carbon in GaAs: Implantation and isolation characteristics, *Appl. Phys. Lett.* **55**(7), 678–680 (1989).
182. S. J. Pearton, W. S. Hobson, A. P. Kinsella, J. Kovalchik, U. K. Chakvabarti, and C. R. Abernathy, Carbon implantation in InGaAs and AlInAs, *Appl. Phys. Lett.* **56**(13), 1263–1265 (1990).
183. A. J. Moll, K. M. Yu, W. Walukievicz, W. L. Hansen, and E. E. Haller, Coimplantation and electrical activity of C in GaAs: Stoichiometry and damage effects, *Appl. Phys. Lett.* **60**(19), 2383–2385 (1992).
184. W. H. van Berlo, Carbon implantation into gallium arsenide, *J. Appl. Phys.* **73**(6), 2765–2769 (1993).
185. A. J. Moll, E. E. Haller, J. W. Ager III, and W. Walukievicz, Direct evidence of carbon precipitates in GaAs and InP, *Appl. Phys. Lett.* **65**(9), 1145–1147 (1994).
186. D. M. Dobkin and J. F. Gibbons, Thermal pulse diffusion of Zn in GaAs from an elemental source, *J. Electrochem. Soc.* **131**(7), 1699–1702 (1984).
187. D. M. Dobkin and J. F. Gibbons, Monolayer surface doping of GaAs from a plated zinc source, *Appl. Phys. Lett.* **44**(9), 884–886 (1984).
188. S. K. Ghandhi, R. T. Huang, and J. M. Borrego, Fabrication of GaAs tunnel junctions by a rapid thermal diffusion process, *Appl. Phys. Lett.* **48**(6), 415–416 (1986).
189. T. S. Kalkur, Y. C. Lu, and C. A. Araujo, Non-alloyed ohmic contacts on rapid thermally Zn diffused GaAs, *Solid-State Electron.* **32**(4), 281–285 (1989).
190. D. L. Plumton, Tungsten silicide zinc as a high temperature zinc diffusion source, in *Rapid Thermal Processing of Electronic Materials*, edited by S. R. Wilson, R. Powell, and D. E. Davies (MRS, Pittsburgh, 1987), pp. 469–474.
191. G. Franz and M.-C. Amann, Reliable spin-on source for acceptor diffusion into III/V compound semiconductors, *J. Electrochem. Soc.* **136**(8), 2410–2413 (1989).
192. G. J. Gualtieri, G. P. Schwartz, G. J. Zydzik, and L. G. Van Uitert, Metal-p^+-n enhanced Schottky barriers on (100) InP formed by an open tube diffusion technique, *J. Electrochem. Soc.* **133**(7), 1425–1429 (1986).
193. R. Singh, F. Radpour, P. Chou, Q. Nguen, S. P. Joshi, H. S. Ullal, R. J. Matson, and S. Asher, Junction and ohmic contact formation in compound semiconductors by rapid isothermal processing, *J. Vac. Sci. Technol. A* **5**(4), 1819–1823 (1987).
194. K.-W. Wang, S. M. Parker, C.-L. Cheng, and J. Long, Diffusion in InP using evaporated Zn_3P_2 film with transient annealing, *J. Appl. Phys.* **63**(6), 2104–2109 (1988).
195. U. Konig and M. Kuisl, Simultaneous annealing for implantation activation and spin-on source diffusion into GaInAs: A novel approach for the formation of pn junctions, *J. Appl. Phys.* **60**(9), 3376–3378 (1986).
196. A. Usami, Y. Tokuda, H. Shiraki, H. Ueda, T. Wada, H. Kan, and T. Murakami, Diffusion of Zn into $GaAs_{0.6}P_{0.4}$: Te from Zn-doped oxide films by rapid thermal processing, in *Rapid Thermal Processing of Electronic Materials*, edited by S. R. Wilson, R. Powell, and D. E. Davies (MRS, Pittsburgh, 1987), pp. 393–398.
197. M. E. Greiner and J. F. Gibbons, Diffusion of silicon in gallium arsenide using rapid thermal processing: Experiment and model, *Appl. Phys. Lett.* **44**(8), 750–752 (1984).
198. S. D. Lester, C. W. Farley, T. S. Kim, B. G. Streetman, and J. M. Anthony, Pulse diffusion of Ge into GaAs, *Appl. Phys. Lett.* **48**(16), 1063–1065 (1986).
199. C. S. Hernandes, J. W. Swart, M. A. A. Pudenzi, G. T. Kraus, Y. Shacham-Diamand, and E. P. Giannelis, Rapid thermal diffusion of Sn from spin-on-glass into GaAs, *J. Electrochem. Soc.* **142**(8), 2829–2832 (1995).
200. B. Goldstein, Diffusion of cadmium and zinc in gallium arsenide, *Phys. Rev.* **118**(4), 1024–1027 (1960).

201. R. J. Field and S. K. Ghandhi, An open-tube method for diffusion of zinc into GaAs, *J. Electrochem. Soc.* **129**(7), 1567–1570 (1982).
202. H. Ando, N. Susa, and H. Kanbe, Carrier density profiles in Zn- and Cd-diffused in InP, *Jpn. J. Appl. Phys.* **20**(3), L197–L200 (1981).
203. F. Schmitt, L. M. Su, D. Franke, and R. Kaumanns, A new open diffusion technique using evaporated Zn_3P_2 and its application to a lateral p–n–p transistor, *IEEE Trans. Electron Devices* **31**(8), 1083–1085 (1984).
204. E. A. Montie and G. L. van Gurp, Photoluminescence of Zn-diffused and annealed InP, *J. Appl. Phys.* **66**(11), 5549–5553 (1989).
205. J. E. Bisberg, A. K. Chin, and F. P. Dabkowski, Zinc diffusion in III–V semiconductors using a cubic-zirconia protection layer, *J. Appl. Phys.* **67**(3), 1347–1351 (1990).
206. A. B. Y. Young and G. L. Pearson, Diffusion of sulfur in gallium phosphide and gallium arsenide, *J. Phys. Chem. Solids* **31**(3), 517–527 (1970).
207. V. A. Presnov, A. P. Mamontov, and L. L. Shirokov, Diffusion of sulfur into gallium arsenide through films of silicon dioxide, *Sov. Phys. Semicond.* **2**(2), 253–254 (1968).
208. J. Ohsawa, H. Kakinoki, H. Ikeda, and M. Migitaka, Diffusion of iron into GaAs from a spin-on source, *J. Electrochem. Soc.* **137**(8), 2608–2611 (1990).

CHAPTER 5

1. G. V. Samsonov and I. M. Vinitskii, *Handbook of Refractory Compounds* (IFI/Plenum Press, New York, 1980).
2. A. E. Gershinskii, A. V. Rzhanov, and E. I. Cherepov, Silicide thin films in microelectronics, *Sov. Microelectron.* **11**(2), 51–59 (1982).
3. K. N. Tu and J. W. Mayer, Silicide formation, in *Thin Films. Interdiffusion and Reactions*, edited by J. M. Poate, K. N. Tu, and J. W. Mayer (Wiley, New York, 1978), pp. 359–405.
4. S. P. Murarka, *Silicides for VLSI Applications* (Academic Press, New York, 1983).
5. G. Ottaviani, Metallurgical aspects of the formation of silicides, *Thin Solid Films* **140**(1), 3–21 (1986).
6. V. M. Koleshko, V. F. Belitsky, and A. A. Khodin, Thin films of rare-earth metal silicides in microelectronics, *Vacuum* **36**(10), 669–676 (1986).
7. T. Hirano and M. Kaise, Electrical resistivities of single-crystalline transition-metal disilicides, *J. Appl. Phys.* **68**(2), 627–633 (1990).
8. S. Mantl, *Ion Beam Synthesis of Epitaxial Silicides: Fabrication, Characterization and Application* (North-Holland, Amsterdam, 1992).
9. K. Maex, Silicides for integrated circuits: $TiSi_2$ and $CoSi_2$, *Mater. Sci. Eng.* **R11**(2–3), 53–153 (1993).
10. V. A. Labunov, V. E. Borisenko, D. I. Zarovskii, L. I. Ivanenko, V. V. Tokarev, and V. I. Hit'ko, Formirovanie silicidov impulsnoii termoobrabotkoii plenochnih struktur, *Zarubejnaya Elektronnaya Tekhnika* **8**, 27–53 (1985).
11. V. I. Hit'ko, V. A. Labunov, V. E. Borisenko, and T. T. Samojlyuk, Vliyanie termoobrabotki na povedenie vol'frama v pripoverkhnostnih sloyah monokristallicheskogo kremniya, *Poverkhnost* **7**, 134–137 (1984).
12. C. S. Wei, J. Van der Spiegel, and J. Santiago, Incoherent radiative processing of titanium silicides, *Thin Solid Films* **118**(2), 155–162 (1984).
13. I. C. Wu, J. J. Chu, and L. J. Chen, Local epitaxy of $TiSi_2$ on (111) Si: Effects due to rapid thermal annealing and to the annealing atmosphere, *J. Appl. Phys.* **60**(9), 3172–3175 (1986).
14. V. E. Borisenko and V. P. Parkhutik, RBS study of transient thermal and anodic oxidation of tantalum films, *Phys. Status Solidi A* **93**(1), 123–130 (1986).
15. A. E. Morgan, E. K. Broadbent, and A. H. Reader, Formation of titanium nitride/silicide bilayers by rapid thermal annealing in nitrogen, in *Rapid Thermal Processing*, edited by T. O. Sedgwick, T. E. Seidel, and B.-Y. Tsaur (MRS, Pittsburgh, 1986), pp. 279–287.
16. S. W. Sun, F. Pintchovski, P. J. Tobin, and R. L. Hance, Effects of $TiSi_x/TiN_x/Al$ contact metallization process on the shallow junction related properties, in *Rapid Thermal Processing of Electronic Materials*, edited by S. R. Wilson, R. Powell, and D. E. Davies (MRS, Pittsburgh, 1987), pp. 165–170.

17. B. Cohen and J. Nulman, Low temperature rapid thermal formation of TiN barrier layers, in *Rapid Thermal Processing of Electronic Materials*, edited by S. R. Wilson, R. Powell, and D. E. Davies (MRS, Pittsburgh, 1987), pp. 171–176.
18. T. Okamoto, M. Shimizu, A. Ohsaki, and Y. Mashiko, Simultaneous formation of TiN and $TiSi_2$ by lamp annealing in NH_3 ambient and its application to diffusing barriers, *J. Appl. Phys.* **62**(11), 4465–4470 (1987).
19. A. E. Morgan, E. K. Broadbent, K. N. Ritz, D. K. Sadana, and B. J. Burrow, Interactions of thin Ti films with Si, SiO_2, Si_3N_4 and SiO_xN_x under rapid thermal annealing, *J. Appl. Phys.* **64**(1), 344–353 (1988).
20. F. Richter, E. Bugiel, H. B. Erzgraber, and D. Panknin, Formation of titanium silicide during rapid thermal annealing: Influence of oxygen, *J. Appl. Phys.* **72**(2), 815–817 (1992).
21. A. T. S. Wee, A. C. H. Huan, W. H. Thian, K. L. Tan, and R. Hogan, Investigation of titanium silicide formation using secondary ion mass spectrometry, in *Rapid Thermal and Integrated Processing III*, edited by J. J. Wortman, J. C. Gelpey, M. L. Green, S. R. J. Brueck, and F. Roozeboom (MRS, Pittsburgh, 1994), pp. 117–122.
22. A. Kamgar, F. A. Baiocchi, A. B. Emerson, T. T. Sheng, M. J. Vasile, and R. W. Haynes, Self-aligned TiN barrier formation by rapid thermal nitration, *J. Appl. Phys.* **66**(6), 2395–2401 (1989).
23. M. P. Siegal, W. R. Graham, and J. J. Santiago, The formation of thin film tungsten silicide annealed in ultra-high vacuum, *J. Appl. Phys.* **66**(12), 6073–6076 (1989).
24. V. E. Borisenko, D. I. Zarovskii, and V. P. Lesnikova, Struktura i fazovie prevrascheniya v poverkhnostnih sloyah platini na kremnii pri sekundnoii termoobrabotke, *Poverkhnost* **4**, 96–100 (1989).
25. A. K. Pant, S. P. Murarka, C. Shepard, and W. Lanford, Kinetics of platinum silicide formation during rapid thermal processing, *J. Appl. Phys.* **72**(5), 1833–1836 (1992).
26. A. E. Morgan, E. K. Broadbent, and D. K. Sadana, Reaction of titanium with silicon nitride under rapid thermal annealing, Appl. Phys. Lett. **49**(19), 1236–1238 (1986).
27. L. J. Chen, I. W. Wu, J. J. Chu, and C. W. Nieh, Effects of backsputtering and amorphous silicon capping layer on the formation of $TiSi_2$ in sputtered Ti films on (001) Si by rapid thermal annealing, *J. Appl. Phys.* **63**(8), 2778–2782 (1988).
28. A. Kikuchi and T. Ishiba, Role of oxygen and nitrogen in the titanium–silicon reaction, *J. Appl. Phys.* **61**(5), 1891–1894 (1987).
29. H. Jiang, C. S. Petersson, and M.-A. Nicolet, Thermal oxidation of transition metal silicides, *Thin Solid Films* **140**(1), 115–129 (1986).
30. R. J. Nemanich, R. T. Fulks, B. L. Stafford, and H. A. Van der Plas, Reactions of thin-film titanium on silicon studied by Raman spectroscopy, *Appl. Phys. Lett.* **46**(7), 670–672 (1985).
31. H. Jiang, H. J. Whitlow, M. Ostling, E. Niemi, F. M. d'Heurle, and C. S. Petersson, A quantitative study of oxygen behavior during $CrSi_2$ and $TiSi_2$ formation, *J. Appl. Phys.* **65**(2), 567–574 (1989).
32. J. W. Rogers, K. L. Erickson, D. N. Belton, and S. J. Ward, Thermal diffusion of oxygen in titanium and titanium oxide films, *J. Vac. Sci. Technol. A* **4**(3), 1685–1687 (1986).
33. A. Anttila, J. Raisanen, and J. Keinonen, Diffusion of nitrogen in α-Ti, *Appl. Phys. Lett.* **42**(6), 498–500 (1983).
34. *Binary Alloy Phase Diagrams*, Vol. 2, edited by T. B. Massalski, J. L. Murray, C. H. Bennett, and H. Baker (American Society for Metals, Metals Park, 1986), pp. 1789–1790.
35. R. Beyers, R. Sinclair, and M. E. Thomas, Phase equilibria in the thin-film metallizations, *J. Vac. Sci. Technol. B* **2**(4), 781–784 (1984).
36. R. D. Thompson, H. Takai, R. A. Psaras, and K. N. Tu, Effect of substrate on the phase transformations of amorphous $TiSi_2$ thin films, *J. Appl. Phys.* **61**(2), 540–544 (1987).
37. M. Bartur, Thermal oxidation of transition metal silicides: The role of mass transport, *Thin Solid Films* **107**(1), 55–65 (1983).
38. F. M. d'Heurle, E. A. Irene, and C. Y. Ting, Oxidation of silicide thin films: $TiSi_2$, *Appl. Phys. Lett.* **42**(4), 361–363 (1983).
39. W. Y. Yang, H. Iwakuro, H. Yagi, T. Kuroda, and S. Nakamura, Study of oxidation of $TiSi_2$ thin film by XPS, *Jpn. J. Appl. Phys.* **23**(12), 1560–1567 (1984).
40. R. M. Tromp, G. W. Rubloff, and E. J. Van Loenen, Low temperature material reaction at the Ti/Si(111) interface, *J. Vac. Sci. Technol. A* **4**(3), 865–868 (1986).

41. S. A. Chambers, D. M. Hill, F. Xu, and T. H. Weaver, Silicide formation at Ti/Si interface: Diffusion parameters and behavior at elevated temperatures, *Phys. Rev. B* **35**(2), 634–640 (1987).
42. W. Lur and L. J. Chen, Growth kinetics of amorphous interlayer formed by interdiffusion of polycrystalline Ti thin-film and single-crystal silicon, *Appl. Phys. Lett.* **54**(13), 1217–1219 (1989).
43. S. P. Murarka and D. E. Fraser, Thin film interaction between titanium and polycrystalline silicon, *J. Appl. Phys.* **51**(1), 342–349 (1980).
44. P. Revesz, J. Gyimesi, L. Pogany, and G. Peto, Lateral growth of titanium silicide over a silicon dioxide layer, *J. Appl. Phys.* **54**(4), 2114–2115 (1983).
45. A. Guldman, V. Schuiller, A. Steffen, and P. Balk, Formation and properties of $TiSi_2$ films, *Thin Solid Films* **100**(1), 1–7 (1983).
46. L. S. Hung, J. Gyulai, J. W. Mayer, S. S. Lau, and M.-A. Nicolet, Kinetics of $TiSi_2$ formation by thin Ti films on Si, *J. Appl. Phys.* **54**(9), 5076–5080 (1983).
47. G. G. Bentini, R. Nipoti, A. Armigliato, M. Berti, A. V. Drigo, and C. Cohen, Growth and structure of titanium silicide phases formed by thin Ti films on Si crystals, *J. Appl. Phys.* **57**(2), 270–275 (1985).
48. A. E. Borzakovskii, V. G. Verbitskii, O. N. Visotskii, I. V. Gusev, A. A. Klyuchnikov, V. M. Koptenko, V. M. Pozdnyakova, and V. N. Scherbin, Kinetika fazovih izmenenii pri diffuzionnom vzaimodeistvii plenok titana s kremniem, *Poverkhnost* **7**, 95–100 (1988).
49. C. A. Pico and M. G. Lagally, Kinetics of titanium silicide formation on single-crystal Si: Experiment and modeling, *J. Appl. Phys.* **64**(10), 4957–4967 (1988).
50. E. A. Maydell-Ondrusz, P. L. F. Hemment, K. G. Stephens, and S. Moffat, Processing of titanium films on silicon using a multiscanned electron beam, *Electron. Lett.* **18**(17), 752–754 (1982).
51. D. Levy, J. P. Ponpon, J. J. Grob, and R. Stuck, Rapid thermal annealing and titanium silicide formation, *Appl. Phys. A* **38**(1), 23–29 (1985).
52. C. X. Dexin, H. B. Harrison, and G. K. Reeves, Titanium silicides formed by rapid thermal vacuum processing, *J. Appl. Phys.* **63**(6), 2171–2173 (1988).
53. M. Nathan, Solid phase reactions in free-standing layered M-Si (M = Ti,V,Cr,Co) films, *J. Appl. Phys.* **63**(11), 5534–5540 (1988).
54. R. Furlan and J. W. Swart, Titanium silicide formation and arsenic dopant behavior under rapid thermal treatment in vacuum, *J. Electrochem. Soc.* **136**(6), 1806–1810 (1989).
55. E. A. Maydell-Ondrusz, R. E. Harper, I. H. Wilson, and K. G. Stephens, Formation of $TiSi_2$ by electron beam annealing of arsenic implanted titanium films on silicon substrates, *Vacuum* **34**(10/11), 995–999 (1984).
56. J. J. Santiago, C. S. Wei, and J. Van der Spiegel, Fast radiative processing of titanium silicides under vacuum, *Mater. Lett.* **2**(6), 477–482 (1984).
57. E. A. Maydell-Ondrusz, R. E. Harper, A. Abid, P. L. F. Hemment, and K. G. Stephens, Formation of titanium disilicide by electron beam irradiation under non steady state conditions, in *Thin Films and Interfaces II*, edited by J. E. E. Baglin, D. R. Campbel, and W. K. Chu (North-Holland, Amsterdam, 1984), pp. 99–104.
58. V. E. Borisenko, D. I. Zarovskii, V. A. Labunov, and A. N. Larsen, $TiSi_2$ formation during brief heat treatment of titanium films on silicon, *Sov. Phys. Tech. Phys.* **31**(5), 599–600 (1986).
59. T. Brat, C. M. Osburn, T. Finstad, J. Liu, and B. Ellington, Self-aligned Ti silicide formed by rapid thermal annealing, *J. Electrochem. Soc.* **133**(7), 1451–1458 (1986).
60. D. Pramanik, A. N. Saxena, O. K. Wu, G. G. Peterson, and M. Tanielian, Influence of the interfacial oxide on titanium silicide formation by rapid thermal annealing, *J. Vac. Sci. Technol. B* **2**(4), 775–780 (1984).
61. R. K. Shukla, P. W. Davies, and B. M. Tracy, The formation of titanium silicide by arsenic ion beam mixing and rapid thermal annealing, *J. Vac. Sci. Technol. B* **4**(6), 1344–1351 (1986).
62. E. G. Colgan, L. A. Clevenger, and C. Cabral, Jr., Activation energy for C49 and C54 $TiSi_2$ formation measured during rapid thermal annealing, *Appl. Phys. Lett.* **65**(16), 2009–2011 (1994).
63. Y. H. Ku, S. K. Lee, D. K. Shih, D. L. Kwong, C.-O. Lee, and J. R. Yeargain, Suppression of lateral Ti silicide growth by ion beam mixing and rapid thermal annealing, *Appl. Phys. Lett.* **52**(11), 877–879 (1988).
64. M. Berti, A. V. Drigo, C. Cohen, J. Siejka, G. G. Bentini, R. Nipoti, and S. Guerri, Titanium silicide formation: Effect of oxygen distribution in the metal film, *J. Appl. Phys.* **55**(10), 3558–3565 (1984).
65. P. J. Rosser and G. Tomkins, Thermal pulse annealing of titanium and tantalum silicides, *J. Phys. (Paris)* **44**(10), c5/445–c5/448 (1983).

66. T. Tien, G. Ottaviani, and K. N. Tu, Temperature dependence of structural and electrical properties of Ta-Si thin alloy films, *J. Appl. Phys.* **54**(12), 7047–7057 (1983).
67. D. L. Kwong, Rapid thermal annealing of co-sputtered tantalum silicide films, *Thin Solid Films* **121**(1), 43–50 (1984).
68. B. Z. Li, R. E. Jones, K. Daneshvar, and J. Davis, Graphite strip rapid isothermal annealing of tantalum silicide, *J. Appl. Phys.* **56**(4), 1242–1244 (1984).
69. K. Daneshvar and R. E. Jones, The rapid isothermal annealing of tantalum silicide, *Nucl. Instrum. Methods Phys. Res. B* **10/11**(1), 529–531 (1985).
70. J. Schulte, R. Ferretti, W. Hasse, Y. Danto, and Y. Ousten, Quantitative Rutherford backscattering spectrometry in-depth composition analysis of sputtered $TaSi_2$ layers, *Thin Solid Films* **140**(1), 183–188 (1986).
71. I. C. Wu, J. J. Chu, and L. J. Chen, Localized epitaxial growth of $TaSi_2$ on (111) and (001) Si by rapid thermal annealing, *J. Appl. Phys.* **62**(3), 879–884 (1987).
72. M. Natan and S. C. Shatas, Structure and properties of rapidly thermal annealed Ta/Si multilayers, *J. Vac. Sci. Technol. B* **3**(6), 1707–1714 (1985).
73. G. D. Davis and M. Natan, Effect of impurities on the reaction of Ta and Si multilayers processed by rapid thermal annealing, *J. Vac. Sci. Technol. A* **4**(2), 159–167 (1986).
74. V. K. Raman, F. Mahmood, R. A. McMahon, H. Ahmed, C. Jeynes, and D. Sarcar, Rapid electron beam induced tantalum–silicon reactions, *Appl. Surf. Sci.* **36**(1), 654–663 (1989).
75. F. Mahmood, H. Ahmed, M. Suleman, and M. M. Ahmed, Measurement of the activation energy of tantalum silicide growth on silicon by rapid electron beam annealing, *Jpn. J. Appl. Phys.* **30**(8a), L1418–L1421 (1991).
76. L. D. Locker and C. D. Capio, Reaction kinetics of tungsten thin films on silicon (100) surfaces, *J. Appl. Phys.* **44**(10), 4366–4369 (1973).
77. J. Lajzerowicz, J. Torres, G. Goltz, and R. Pantel, Kinetics of WSi_2 formation at low and high temperatures, *Thin Solid Films* **140**(1), 23–28 (1986).
78. P. R. Gage and R. W. Bartlett, Diffusion kinetics affecting formation of silicide coatings on molybdenum and tungsten, *Trans. Metall. Soc.* **233**, 832–834 (1965).
79. R. W. Bartlett, P. R. Gage, and P. A. Larssen, Growth kinetics of intermediate silicides in the $MoSi_2$/Mo and WSi_2/W systems, *Trans. Metall. Soc.* **230**, 1528–1534 (1964).
80. T. S. Jaydev and A. Joshi, Conductivity changes in tungsten silicide films due to rapid thermal processing, *Electron. Lett.* **20**(14), 604–606 (1984).
81. T. O. Sedgwick, F. M. d'Heurle, and S. A. Cohen, Short time annealing of coevaporated tungsten silicide films, *J. Electrochem. Soc.* **131**(10), 2446–2451 (1984).
82. T. Hara, H. Takahashi, and S.-C. Chen, Ion implantation of arsenic in chemical vapor deposition tungsten silicide, *J. Vac. Sci. Technol. B* **3**(6), 1664–1667 (1985).
83. B.-Y. Tsaur, C. K. Chen, C. H. Anderson, and D. L. Kwong, Selective tungsten silicide formation by ion-beam mixing and rapid thermal annealing, *J. Appl. Phys.* **57**(6), 1890–1894 (1985).
84. D. L. Kwong, D. C. Meyers, N. S. Alvi, L. W. Li, and E. Norbeck, Refractory metal silicide formation by ion-beam mixing and rapid thermal annealing, *Appl. Phys. Lett.* **47**(7), 688–691 (1985).
85. J. Kato, M. Asahina, H. Shimura, and Y. Yamamoto, Rapid annealing of tungsten polycide films using halogen lamps, *J. Electrochem. Soc.* **133**(4), 794–798 (1986).
86. C. Nobili, M. Bosi, G. Ottaviani, G. Queirolo, and L. Bacci, Rapid thermal annealing of WSi_x, *Appl. Surf. Sci.* **53**(1), 219–223 (1991).
87. M. P. Siegal and J. J. Santiago, Effects of rapid thermal processing on the formation of uniform tetragonal tungsten disilicide films on Si (100) substrates, *J. Appl. Phys.* **63**(2), 525–529 (1988).
88. J. M. Liang and L. J. Chen, Autocorrelation function analysis of phase formation in the initial stage of interfacial reactions of molybdenum thin films on (111) Si, *Appl. Phys. Lett.* **64**(10), 1224–1226 (1994).
89. A. Guivar'h, P. Auvray, L. Berthou, M. Le Cun, J. P. Boulet, P. Henok, G. Pelous, and A. Martinez, Reaction kinetics of molybdenum thin films on silicon (111) surface, *J. Appl. Phys.* **49**(1), 233–237 (1978).
90. P. Urwank, E. Wieser, A. Hassner, C. Kaufmann, H. Lippmann, and I. Melzer, Formation of $MoSi_2$ by light pulse irradiation, *Phys. Status Solidi A* **90**(2), 463–468 (1985).
91. S. A. Agamy, V. Q. Ho, and H. M. Naguib, As^+ implantation and transient annealing of $MoSi_2$ thin films, *J. Vac. Sci. Technol. A* **3**(3), 718–722 (1985).

92. V. Q. Ho and H. M. Naguib, Characterization of rapidly annealed Mo-polycide, *Vac. Sci. Technol. A* **3**(3), 896–899 (1985).
93. R. Portius, D. Dietrich, A. Hassner, T. Raschke, E. Wieser, and W. Wolke, Formation of stable $MoSi_2$/poly-Si films by rapid thermal annealing, *Phys. Status Solidi A* **100**(1), 199–206 (1987).
94. D. L. Kwong, D. C. Meyers, and N. S. Alvi, Molybdenum silicide formation by ion beam mixing and rapid thermal annealing, in *VLSI Science and Technology/1985/*, edited by W. M. Bullis and S. Broydo (Electrochemical Society, Pennington, 1985), pp. 195–202.
95. K. N. Tu, W. K. Chu, and J. W. Mayer, Structure and growth kinetics of Ni_2Si on silicon, *Thin Solid Films* **25**(2), 403–413 (1975).
96. G. Ottaviani, Review of binary alloy formation by thin film interactions, *J. Vac. Sci. Technol.* **16**(5), 1112–1119 (1979).
97. F. d'Heurle, C. S. Peterson, J. E. E. Baglin, S. J. LaPlaca, and C. Y. Wong, Formation of thin films of NiSi. Metastable structure, diffusion mechanisms in intermetallic compound, *J. Appl. Phys.* **55**(12), 4208–4218 (1984).
98. C. D. Lien, M.-A. Nicolet, and S. S. Lau, Kinetiks of silicides on Si (100) and evaporated silicon substrates, *Thin Solid Films* **143**(1), 63–72 (1986).
99. A. J. Brunner, E. Ma, and M.-A. Nicolet, Silicide formation by furnace annealing of thin Si films on large-grained Ni substrates, *Appl. Phys. A* **48**(3), 229–232 (1989).
100. K. Nakamura, J. O. Olowolafe, S. S. Lau, M.-A. Nicolet, J. W. Mayer, and R. Shima, Interaction of metal layers with polycrystalline Si, *J. Appl. Phys.* **47**(4), 1278–1283 (1976).
101. L. R. Zheng, E. Zingu, and J. W. Mayer, Lateral silicide growth, in *Thin Films and Interfaces II*, edited by J. E. E. Baglin, D. R. Campbel, and W. K. Chu (North-Holland, Amsterdam, 1984), pp. 75–85.
102. L. A. Clevenger, C. V. Thompson, R. C. Cammarata, and K. N. Tu, Reaction kinetics of nickel/silicon multilayer films, *Appl. Phys. Lett.* **52**(10), 795–797 (1988).
103. J. C. Ciccariello, S. Poize, and P. Gas, Lattice and grain boundary self-diffusion in Ni_2Si: Comparison with thin-film formation, *J. Appl. Phys.* **67**(7), 3315–3322 (1990).
104. A. Singh and W. S. Khokle, Window size effect on lateral growth of nickel silicide, *J. Appl. Phys.* **66**(3), 1190–1194 (1989).
105. D. M. Scott and M.-A. Nicolet, Modification of nickel silicide formation by oxygen implantation, *Nucl. Instrum. Methods* **182/183**(Pt. 2), 655–660 (1981).
106. D. M. Scott and M.-A. Nicolet, Implanted oxygen in NiSi formation, *Phys. Status Solidi A* **66**(2), 773–778 (1981).
107. G. Majni, F. Della Velle, and G. Nobili, Growth kinetics of NiSi on (100) and (111) silicon, *J. Phys. D* **17**(5), L77–L81 (1984).
108. C.-D. Lien, M.-A. Nicolet, and S. S. Lau, Low temperature formation of $NiSi_2$ from evaporated silicon, *Phys. Status Solidi A* **81**(1), 123–128 (1984).
109. O. Nishikawa, M. Shibata, T. Yoshimura, and E. Nomura, Atom-probe study of silicide formation at Ni/Si interfaces, *J. Vac. Sci. Technol. B* **2**(1), 21–23 (1984).
110. V. Hinkel, L. Sorba, H. Haak, and K. Horn, Evidence for Si diffusion through epitaxial $NiSi_2$ grown on Si(111), *Appl. Phys. Lett.* **50**(18), 1257–1259 (1987).
111. M. Natan, Anomalous first-phase-formation in rapidly thermal annealed, thin-layered Si/Ni/Si films, *Appl. Phys. Lett.* **49**(5), 257–259 (1986).
112. E. Ma, M. Natan, B. S. Lim, and M.-A. Nicolet, Comparison of silicide formation by rapid thermal annealing and conventional furnace annealing, in *Rapid Thermal Processing of Electronic Materials*, edited by S. R. Wilson, R. Powell, and D. E. Davies (MRS, Pittsburgh, 1987), pp. 205–211.
113. E. Ma, B. S. Lim, M.-A. Nicolet, and M. Natan, Growth of Ni_2Si by rapid thermal annealing: Kinetics and moving species, *Appl. Phys. A* **44**(2), 157–160 (1987).
114. V. E. Borisenko, D. I. Zarovskii, A. I. Ivanov, V. V. Tokarev, and A. A. Tomchenko, Vliyanie implantacii bora na fazovie prevrascheniya v silicidnih sloyah nikelya, *Poverkhnost* **6**, 116–118 (1989).
115. V. V. Tokarev, A. I. Demchenko, A. I. Ivanov, and V. E. Borisenko, Influence of boron implantation on phase transformations in nickel silicides, *Appl. Surf. Sci.* **44**(1), 241–247 (1990).
116. A. Nylandsted Larsen, J. Chevallier, and G. Sorensen, Growth of nickel silicides on silicon by short duration incoherent light exposure, in *Energy Beam–Solid Interactions and Transient Thermal Processing*, edited by J. C. C. Fan and N. M. Johnson (North-Holland, Amsterdam, 1984), pp. 727–732.

117. J. Chevallier and A. Nylandsted Larsen, Epitaxial nickel and cobalt silicide formation by rapid thermal annealing, *Appl. Phys. A* **39**(2), 141–145 (1986).
118. R. E. Harper, E. A. Maydell-Ondrusz, I. H. Wilson, and K. F. Stephenes, Epitaxial silicide formation by electron beam annealing of Ni thin films, *Vacuum* **34**(10/11), 875–879 (1984).
119. V. V. Tokarev, V. E. Borisenko, T. M. Pyatkova, and D. I. Zarovskii, Argon ion implantation influence on phase-formation and structure of nickel silicides, *Phys. Status Solidi A* **111**(2), 499–505 (1989).
120. V. V. Tokarev, V. E. Borisenko, A. I. Demchenko, and T. M. Pyatkova, Growth of epitaxial nickel disilicide during rapid thermal processing of argon-implanted nickel films on silicon, *Phys. Status Solidi A* **116**(1), 331–336 (1989).
121. C.-D. Lien, M.-A. Nicolet, and P. Williams, Effects of ion irradiation on thermally formed silicides in the presence of interfacial oxide, *Thin Solid Films* **136**(1), 69–76 (1986).
122. G. J. Van Gurp and C. Langereis, Cobalt silicide layers on Si. Structure and growth. *J. Appl. Phys.* **46**(10), 4301–4307 (1975).
123. G. J. Van Gurp, D. Sigurd, and W. F. Van der Weg, Tungsten as a marker in thin-film diffusion studies, *Appl. Phys. Lett.* **29**(3), 159–161 (1976).
124. S. S. Lau and J. W. Mayer, Interaction in the Co/Si thin-film system. 1. Kinetics, *J. Appl. Phys.* **49**(7), 4005–4010 (1978).
125. G. J. Van Gurp, W. F. Van der Weg, and D. Sigurd, Interactions in the Co/Si thin-film system. 2. Diffusion marker experiments, *J. Appl. Phys.* **49**(7), 4011–4020 (1978).
126. C.-D. Lien, M.-A. Nicolet, C. S. Pai, and S. S. Lau, Growth of Co-silicides from single crystal and evaporated Si, *Appl. Phys. A* **36**(3), 13–17 (1984).
127. C.-H. Jan, C.-P. Chen, and Y. A. Chang, Growth of intermediate phases in Co/Si diffusion couples: Bulk versus thin-film studies, *J. Appl. Phys.* **73**(3), 1168–1179 (1993).
128. C.-D. Lien, M.-A. Nicolet, and S. S. Lau, Kinetics of $CoSi_2$ from evaporated silicon, *Appl. Phys. A* **34**(4), 249–251 (1984).
129. A. Appelbaum, R. V. Knoell, and S. P. Murarka, Study of cobalt-disilicide formation from cobalt monosilicide, *J. Appl. Phys.* **57**(6), 1880–1886 (1985).
130. F. M. d'Heurle and C. S. Petersson, Formation of thin films of $CoSi_2$: Nucleation and diffusion mechanisms, *Thin Solid Films* **128**(3/4), 283–297 (1985).
131. A. H. Reader, J. P. W. B. Duchateau, J. E. Crombeen, E. P. Naburgh, and M. A. J. Somers, The formation of epitaxial $CoSi_2$ thin films on (001) Si from amorphous Co-W alloys, *Appl. Surf. Sci.* **53**(1), 92–102 (1991).
132. H. Miura, E. Ma, and C. V. Thompson, Initial sequence and kinetics of silicide formation in cobalt/amorphous-silicon multilayer thin films, *J. Appl. Phys.* **70**(8), 4287–4294 (1991).
133. T. Barge, S. Poize, J. Bernardini, and P. Gas, Cobalt lattice diffusion in bulk cobalt disilicide, *Appl. Surf. Sci.* **53**(1), 180–185 (1991).
134. S. Vaidya, S. P. Murarka, and T. T. Sheng, Formation and thermal stability of $CoSi_2$ on polycrystalline Si, *J. Appl. Phys.* **58**(2), 971–978 (1985).
135. F. D. Schowengerdt, T. L. Lin, R. W. Fathauer, and P. J. Grunthaner, Diffusion of Si in thin $CoSi_2$ layers, *Appl. Phys. Lett.* **54**(14), 1314–1316 (1989).
136. B. S. Lim, E. Ma, M.-A. Nicolet, and M. Natan, Kinetics and moving species during Co_2Si formation by rapid thermal annealing, *J. Appl. Phys.* **61**(11), 5027–5030 (1987).
137. L. Van den hove, R. Wolters, K. Maex, R. De Keersmaecker, and G. Declerck, A self-aligned cobalt silicide technology using rapid thermal processing, *J. Vac. Sci. Technol. B* **4**(6), 1358–1363 (1986).
138. K. Maex, G. Brijs, J. Vanhellemont, and W. Vandervorst, Formation of thin films of monocrystalline $CoSi_2$ on (100) Si, *Nucl. Instrum. Methods Phys. Res. B* **59/60**(Pt. I), 660–665 (1991).
139. R. A. Collins, S. C. Edwards, and G. Dearnaley, Thermal spike mixing in cobalt silicide formation, *J. Phys. D* **24**(10), 1822–1831 (1991).
140. W. J. Freitas and J. W. Swart, The influence of impurities on cobalt silicide formation, *J. Electrochem. Soc.* **138**(10), 3067–3070 (1991).
141. E. G. Colgan, C. Cabral, Jr., and D. E. Kotecki, Activation energy for CoSi and $CoSi_2$ formation measured during rapid thermal annealing, *J. Appl. Phys.* **77**(2), 614–619 (1995).
142. S. A. Chambers, H. W. Anderson, S. B. Chen, and J. H. Weaver, High-temperature nucleation and silicide formation at the Co/Si(111)-7×7 interface: A structural investigation, *Phys. Rev. B* **34**(2), 913–919 (1986).

143. R. W. Bower and J. W. Mayer, Growth kinetics observed in the formation of metal silicides on silicon, *Appl. Phys. Lett.* **20**(9), 359–361 (1972).
144. J. O. Olowolafe, M.-A. Nicolet, and J. W. Mayer, Formation kinetics of $CrSi_2$ films on Si substrates with and without interposed Pd_2Si layer, *J. Appl. Phys.* **47**(12), 5182–5186 (1976).
145. A. P. Botha and R. Pretorius, Co_2Si, $CrSi_2$, $ZrSi_2$ and $TiSi_2$ formation studied by a radioactive ^{31}Si marker technique, *Thin Solid Films* **93**(1/2), 127–133 (1982).
146. N. I. Plyusnin, N. G. Galkin, A. N. Kamenev, A. G. Lifshits, and S. A. Lobachev, Atomnoe peremeshivanie na granitse razdela Si-Cr i nachal nie stadii epitaksii $CrSi_2$, *Poverkhnost* **9**, 55–61 (1989).
147. A. Martinez and D. Esteve, Metallurgical and electrical properties of chromium interfaces, *Solid-State Electron.* **23**(1), 55–64 (1980).
148. M. Natan and S. W. Duncan, Microstructure and growth kinetics of $CrSi_2$ on Si{100} studied using cross-sectional transmission electron microscopy, *Thin Solid Films* **123**(1), 69–85 (1985).
149. V. E. Borisenko and D. I. Zarovskii, Formation of chromium silicides by annealing for times < 1 minute, *Sov. Phys. Tech. Phys.* **30**(10), 1188 (1985).
150. V. E. Borisenko, D. I. Zarovskii, G. V. Litvinovich, and V. A. Samuilov, Opticheskaya spectroskopiya silitsidov khroma, sformirovannih secundnoii termoobrabotkoii cloev khroma na kremnii, *Zh. Prikl. Spektrosc.* **44**(2), 314–317 (1986).
151. V. E. Borisenko, L. I. Ivanenko, and S. Y. Nikitin, Semiconductor properties of chromium disilicide, *Microelectronics* **121**(2), 69–77 (1992), in Russian.
152. G. Ottaviani, Phase formation and kinetic processes in silicide growth, in *Thin Films and Interfaces II*, edited by J. E. E. Baglin, D. R. Campbel, and W. K. Chu (North-Holland, Amsterdam, 1984), pp. 21–31.
153. T. Yamauchi, S. Zaima, K. Mizuno, H. Kitamura, Y. Koide, and Y. Yasuda, Solid-phase reactions and crystallographic structures in Zr/Si systems, *J. Appl. Phys.* **69**(10), 7050–7056 (1991).
154. F. M. d'Heurle, Nucleation phenomena in transitions from two to three phases, diffusion couples, phase stability, the thin film formation of some silicides, in *Thin Films and Interfaces II*, edited by J. E. E. Baglin, D. R. Campbel, and W. K. Chu (North-Holland, Amsterdam, 1984), pp. 3–8.
155. V. E. Borisenko, D. I. Zarovskii, and V. V. Tokarev, Formirovanie silitsidov tsirkoniya sekundnoii termoobrabotkoii, *Dokl. Akad. Nauk BSSR* **32**(8), 703–705 (1988).
156. R. W. Bower, D. Sigurd, and R. E. Scott, Formation kinetics and structure of Pd_2Si films on Si, *Solid-State Electron.* **16**(12), 1461–1471 (1973).
157. F. Edelman, C. Cytermann, R. Brener, M. Eizenberg, and R. Weil, Interfacial reactions in the Pd/a-Si/c-Si system, *J. Appl. Phys.* **71**(1), 289–295 (1992).
158. D. J. Fertig and G. Y. Robinson, A study of Pd_2Si films on silicon using Auger electron spectroscopy, *Solid-State Electron.* **19**(5), 407–413 (1976).
159. N. W. Cheung, M.-A. Nicolet, M. Wittmer, C. A. Evans, and T. T. Sheng, Growth kinetics of Pd_2Si from evaporated and sputter-deposited films, *Thin Solid Films* **79**(1), 51–60 (1981).
160. M. Wittmer and K. N. Tu, Growth kinetics and diffusion mechanism in Pd_2Si, *Phys. Rev. B* **27**(2), 1173–1179 (1983).
161. T. W. Little and H. Chen, In situ x-ray diffraction measurement of Pd_2Si transformation kinetics using a linear position-sensitive detector, *J. Appl. Phys.* **63**(4), 1182–1190 (1988).
162. G. E. White and H. Chen, An in situ observation of the growth kinetics and stress relaxation Pd_2Si films on Si (111), *J. Appl. Phys.* **67**(8), 3689–3692 (1990).
163. D. Levy, A. Grob, J. J. Grob, and J. P. Ponpon, Formation of palladium silicide by rapid thermal annealing, *Appl. Phys. A* **35**(3), 141–144 (1984).
164. H. Ishiwara and H. Yamamoto, Epitaxial growth of Pd_2Si films on Si(111) substrates by scanning electron-beam annealing, *Appl. Phys. Lett.* **41**(8), 718–720 (1982).
165. H. Ishiwara and H. Yamamoto, Epitaxial silicide formation by scanning electron beam annealing, in *Laser and Electron–Beam Interactions with Solids*, edited by B. R. Appleton and G. K. Celler (North-Holland, Amsterdam, 1982), pp. 437–442.
166. A. Nylandsted Larsen and J. Chevallier, Epitaxial PtSi and Pd_2Si formed by rapid thermal annealing, *Mater. Lett.* **3**(5/6), 242–246 (1985).
167. N. S. Alvi, D. L. Kwong, C. G. Hopkins, and S. G. Bauman, Dopant redistribution during Pd_2Si formation using rapid thermal annealing, *Appl. Phys. Lett.* **48**(21), 1433–1435 (1986).

168. J. M. Poate and T. C. Tisone, Kinetics and mechanism of platinum silicide formation on silicon, *Appl. Phys. Lett.* **24**(8), 391–393 (1974).
169. C. Canali, C. Catellani, M. Prudenziati, W. H. Wadlin, and C. A. Evans, Pd_2Si and PtSi formation with high-purity Pt thin films, *Appl. Phys. Lett.* **31**(1), 43–45 (1977).
170. C. A. Crider, J. M. Poate, J. E. Rowe, and T. T. Sheng, Platinum silicide formation under ultrahigh vacuum and controlled impurity ambients, *J. Appl. Phys.* **52**(4), 2860–2868 (1981).
171. H. Muta and D. Shinoda, Solid-solid reactions in Pt-Si systems, *J. Appl. Phys.* **43**(6), 2913–2915 (1972).
172. M. Wittmer, Growth kinetics of platinum silicide, *J. Appl. Phys.* **54**(9), 5081–5086 (1983).
173. G. Majni, F. Panini, G. Sodo, and P. Cantoni, Lateral growth of platinum and palladium silicides on an SiO_2 layer, *Thin Solid Films* **125**(3/4), 313–320 (1985).
174. J.-S. Song and C.-A. Chang, Enhanced PtSi formation using a gold layer between Pt and Si, *Appl. Phys. Lett.* **50**(7), 422–424 (1987).
175. C.-A. Chang and J.-S. Song, Selectively enhanced silicide formation by a gold interlayer: Probing the dominant diffusing species and reaction mechanisms during thin-film reactions, *Appl. Phys. Lett.* **51**(8), 572–574 (1987).
176. D. I. Zarovskii and V. E. Borisenko, Platinum silicides formed by rapid thermal processing for Schottky barriers, in *Abstracts of MRS Fall Meeting 1989* (MRS, Boston, 1989), Abstract C9.36.
177. C. G. Hopkins, S. M. Baumann, and R. J. Blattner, The kinetics of platinum silicide formation using CW lamp annealing, in *Thin Films and Interfaces II*, edited by J. E. E. Baglin, D. R. Campbel, and W. K. Chu (North-Holland, Amsterdam, 1984), pp. 87–91.
178. J. Van der Spiegel, C. S. Wei, and J. J. Santiago, Fast radiative processing of platinum silicide, *J. Appl. Phys.* **57**(2), 607–609 (1985).
179. A. A. Naem, Platinum silicide formation using rapid thermal processing, *J. Appl. Phys.* **64**(8), 4161–4167 (1988).
180. V. E. Borisenko, D. I. Zarovskii, and V. V. Tokarev, Influence of argon implantation on the formation of platinum silicides, *Phys. Status Solidi A* **107**(1), K33–K35 (1988).
181. V. E. Borisenko, D. I. Zarovskii, V. P. Lesnikova, and V. V. Tokarev, Vliyanie implantatsii argona na obrazovanie silitsidov platini, *Izv. Akad. Nauk BSSR Ser. Fiz. Mater. Nauk* **3**, 85–87 (1989).
182. J. H. Lin and L. J. Chen, Growth kinetics of amorphous interlayers by solid–solid diffusion in ultrahigh-vacuum deposited polycrystalline silicon V thin films on (001) Si, *J. Appl. Phys.* **77**(9), 4425–4430 (1995).
183. R. M. Walser and R. W. Bene, First phase nucleation in silicon-transition-metal planar interfaces, *Appl. Phys. Lett.* **28**(10), 624–625 (1976).
184. R. Pretorius, Prediction of silicide first phase and phase sequence from heat of formation, in *Thin Films and Interfaces II*, edited by J. E. E. Baglin, D. R. Campbel, and W. K. Chu (North-Holland, Amsterdam, 1984), pp. 15–20.
185. R. W. Bene, A kinetic model for solid-state silicide nucleation, *J. Appl. Phys.* **61**(5), 1826–1833 (1987).
186. V. S. Eremeev, *Diffusiya i Napryajeniya* (Energoatomizdat, Moscow, 1984).
187. V. E. Borisenko and S. A. Kulyako, Diffusion Synthesis of Silicides. Personal computer code (Belorussian State University of Informatics and Radioelectronics, Minsk, 1994).
188. J. Philibert, Reactive interdiffusion, *Mater. Sci. Forum* **155–156**, 15–30 (1994).
189. S.-L. Zhang and F. M. d'Heurle, Modellization of the growth of three intermediate phases, *Mater. Sci. Forum* **155–156**, 59–70 (1994).
190. V. E. Borisenko, L. I. Ivanenko, and E. A. Krushevski, Combined reaction and diffusion controlled kinetics of silicidation, *Thin Solid Films* **250**(1), 53–55 (1994).
191. R. T. Tung, J. M. Poate, J. C. Bean, J. M. Gibson, and D. C. Jacobson, Epitaxial silicides, *Thin Solid Films* **93**(1/2), 77–90 (1982).
192. J. M. Gibson, J. C. Bean, J. M. Poate, and R. T. Tung, Direct determination of atomic structure at the epitaxial cobalt disilicide on (111) Si interface by ultrahigh resolution electron microscopy, *Appl. Phys. Lett.* **41**(9), 818–820 (1982).
193. L. J. Chen, J. W. Mayer, and K. N. Tu, Formation and structure of epitaxial $NiSi_2$ and $CoSi_2$, *Thin Solid Films* **93**(1/2), 135–141 (1982).
194. R. T. Tung, J. M. Gibson, and J. M. Poate, Growth of single crystal epitaxial silicides on silicon by the use of template layers, *Appl. Phys. Lett.* **42**(10), 888–890 (1983).

195. S. Furukawa and H. Ishiwara, Epitaxial silicide films for integrated circuits and future devices, *Jpn. J. Appl. Phys. Suppl.* **1**, 21–27 (1983).
196. G. J. Chang and J. L. Erskine, Diffusion layers and Schottky-barrier height in nickel silicide–silicon interfaces, *Phys. Rev. B* **28**(10), 5766–5773 (1983).
197. C. Pirri, J. C. Peruchetti, G. Gewinner, and J. Derrien, Cobalt disilicide epitaxial growth on the silicon (111) surface, *Phys. Rev. B* **29**(6), 3391–3397 (1984).
198. L. J. Chen, W. T. Lin, and M. B. Chang, Epitaxial growth of $NiSi_2$ on (011) Si, in *Thin Films and Interfaces II*, edited by J. E. E. Baglin, D. R. Campbel, and W. K. Chu (North-Holland, Amsterdam, 1984), pp. 447–452.
199. R. T. Tung, S. Nakahara, and T. Boone, Growth of single crystal $NiSi_2$ layers on Si (100), *Appl. Phys. Lett.* **46**(9), 895–897 (1985).
200. K. Ishibashi and S. Furukawa, Formation of uniform solid-phase epitaxial $CoSi_2$ films by patterning method, *Jpn. J. Appl. Phys.* **24**(8), 912–917 (1985).
201. C. Pirri, J. C. Peruchetti, D. Bolmont, and G. Gevinner, Surface structure of epitaxial $CoSi_2$ crystals grown on Si (111), *Phys. Rev. B* **33**(6), 4108–4113 (1986).
202. J. Zegenhagen, M. A. Kayed, K.-G. Huang, W. M. Gibson, J. C. Phillips, L. J. Schowalter, and B. D. Hunt, Determination of lattice mismatch in $NiSi_2$ overlayers on Si (111), *Appl. Phys. A* **44**(4), 365–369 (1987).
203. J. Zegenhagen, K.-G. Huang, B. D. Hunt, and L. J. Schowalter, Interface structure and lattice mismatch of epitaxial $CoSi_2$ on Si (111), *Appl. Phys. Lett.* **51**(15), 1176–1178 (1987).
204. J. Zegenhagen, K.-G. Huang, W. M. Gibson, B. D. Hunt, and L. J. Schowalter, Structure properties of epitaxial $NiSi_2$ on Si (111) investigated with x-ray standing waves, *Phys. Rev. B* **39**(14), 10254–10260 (1989).
205. M. C. Ridgway, R. G. Elliman, R. P. Thornton, and J. S. Williams, Thermally induced epitaxial recrystallization of $NiSi_2$ and $CoSi_2$, *Appl. Phys. Lett.* **56**(20), 1992–1994 (1990).
206. M. N. Bodyako, S. A. Astapchik, and G. B. Yaroshevich, *Termokinetika Rekristallizatsii* (Nauka i Tekhnika, Minsk, 1968).
207. T. J. Magel, G. R. Woolhouse, H. A. Kawayoshi, I. C. Niemeyer, B. Rodrigues, R. D. Ormond, and A. S. Bhandia, Microstructural investigations of refractory metal silicide films on silicon, *J. Vac. Sci. Technol. B* **2**(4), 756–761 (1984).
208. T. P. Nolan, R. Sinclair, and R. Beyers, Modeling of agglomeration in polycrystalline thin films: Application to $TiSi_2$ on a silicon substrate, *J. Appl. Phys.* **71**(2), 720–724 (1992).
209. H. J. W. Van Houtum and I. J. M. M. Raaijmakers, First phase nucleation and growth of titanium disilicide with an emphasis on the influence of oxygen, in *Thin Films—Interfaces and Phenomena*, edited by R. J. Nemanich, P. S. Ho, and S. S. Lau (MRS, Pittsburgh, 1986), pp. 37–42.
210. Z. Ma, L. H. Allen, and D. D. J. Allman, Microstructural aspects and mechanism of the C49-to-C54 polymorphic transformation in titanium silicide, *J. Appl. Phys.* **77**(9), 4384–4388 (1995).
211. R. A. Roy, L. A. Clevenger, C. Cabral, Jr., K. L. Saenger, S. Brauer, J. Jordan-Sweet, J. Bucchignano, G. B. Stephenson, G. Morales, and K. F. Ludwig, Jr., In situ x-ray diffraction analysis of the C49–C54 titanium silicide phase transformation in narrow lines, *Appl. Phys. Lett.* **66**(14), 1732–1734 (1995).
212. R. W. Mann, L. A. Clevenger, and Q. Z. Hong, The C49 to C54 $TiSi_2$ transformation in self-aligned silicide application, *J. Appl. Phys.* **73**(7), 3566–3568 (1993).
213. T. Yachi, Formation of $TiSi_2/n^+$poly-Si layer by rapid lamp heating and its application to MOS devices, *IEEE Electron Device Lett.* **5**(7), 217–220 (1984).
214. R. Ganapathiraman, S. Koh, Z. Ma, L. H. Allen, and S. Lee, Formation of $TiSi_2$ during rapid thermal annealing: In situ resistance measurements at heating rates from 1 °C/s to 100 °C/s, in *Rapid Thermal and Integrated Processing II*, edited by J. C. Gelpey, J. K. Elliott, J. J. Wortman, and A. Ajmera (MRS, Pittsburgh, 1993), pp. 63–68.
215. Y. Matsubara, T. Horiuchi, and K. Okumura, Activation energy for the C49-to-C54 transition of polycrystalline $TiSi_2$ films with arsenic impurities, *Appl. Phys. Lett.* **62**(21), 2634–2636 (1993).
216. L. A. Clevenger, J. M. E. Harper, C. Cabral, Jr., C. Nobili, G. Ottaviani, and R. Mann, Kinetic analysis of C49-$TiSi_2$ and C54-$TiSi_2$ formation at rapid thermal annealing rates, *J. Appl. Phys.* **72**(10), 4978–4980 (1992).
217. W. T. Lin and L. J. Chen, Localized epitaxial growth of hexagonal and tetragonal $MoSi_2$ on (111) Si, *Appl. Phys. Lett.* **46**(11), 1061–1063 (1985).

218. W. T. Lin and L. J. Chen, Localized epitaxial growth of tetragonal and hexagonal WSi_2 on (111) Si, *J. Appl. Phys.* **58**(4), 1515–1518 (1985).
219. W. T. Lin and L. J. Chen, Localized epitaxial growth of WSi_2 on silicon, *J. Appl. Phys.* **59**(10), 3481–3488 (1986).
220. W. T. Lin and L. J. Chen, Localized epitaxial growth of $MoSi_2$ on silicon, *J. Appl. Phys.* **59**(5), 1518–1524 (1986).
221. F. Hong, B. K. Patnaik, G. A. Rozgonyi, and C. M. Osburn, Self-aligned epitaxial $CoSi_2$ formation from multilayer Co/Ti-Si(100) by a two-step RTA process, in *Rapid Thermal and Integrated Processing II*, edited by J. C. Gelpey, J. K. Elliott, J. J. Wortman, and A. Ajmera (MRS, Pittsburgh, 1993), pp. 63–68.
222. A. Catana, P. E. Schmid, M. Heinze, F. Levy, P. Stadelmann, and R. Bonnet, Atomic scale study of local $TiSi_2$/Si epitaxies, *J. Appl. Phys.* **67**(4), 1820–1825 (1990).
223. M. S. Fung, H. C. Cheng, and L. J. Chen, Localized epitaxial growth of C54 and C49 $TiSi_2$ on (111) Si, *Appl. Phys. Lett.* **47**(12), 1312–1314 (1985).
224. R. Nipoti and A. Armigliato, On the epitaxial relationships of $TiSi_2$ on silicon, *Jpn. J. Appl. Phys.* **24**(11), 1421–1424 (1985).
225. K. H. Kim, J. J. Lee, D. J. Seo, C. K. Choi, S. R. Hong, J. D. Koh, S. C. Kim, J. Y. Lee, and M.-A. Nicolet, Growth of epitaxial C54 TiS_2 on Si (111) substrate by in situ annealing in ultrahigh vacuum, *J. Appl. Phys.* **71**(8), 3812–3815 (1992).
226. C. J. Chien, H. C. Cheng, C. W. Nieh, and L. J. Chen, Epitaxial growth of VSi_2 on (111) Si, *J. Appl. Phys.* **57**(6), 1887–1889 (1985).
227. F. Y. Shiau, H. C. Cheng, and L. J. Chen, Epitaxial growth of $CrSi_2$ on (111) Si, *Appl. Phys. Lett.* **45**(5), 524–526 (1984).
228. F. Y. Shiau, H. C. Cheng, and L. J. Chen, Localized epitaxial growth of $CrSi_2$ on silicon, *J. Appl. Phys.* **59**(8), 2784–2787 (1986).
229. H. C. Cheng, L. J. Chen, and T. R. Your, Epitaxial growth of $FeSi_2$ on (111) Si, in *Thin Films and Interfaces II*, edited by J. E. E. Baglin, D. R. Campbel, and W. K. Chu (North-Holland, Amsterdam, 1984), pp. 441–446.
230. J. E. Mahan, K. M. Geib, G. Y. Robinson, R. G. Long, Y. Xinghna, G. Bai, M.-A. Nicolet, and M. Nathan, Epitaxial films of semiconducting $FeSi_2$ on (001) silicon, *Appl. Phys. Lett.* **56**(21), 2126–2128 (1990).
231. K. Konuma, J. Vrijmoeth, P. M. Zagwijn, J. W. M. Frenken, E. Vlieg, and J. F. van der Veen, Formation of epitaxial β-$FeSi_2$ films on Si(100) as studied by medium energy ion scattering, *J. Appl. Phys.* **73**(3), 1104–1109 (1993).
232. D. R. Peale, R. Haight, and J. Ott, Heteroepitaxy of β-$FeSi_2$ on unstrained and strained Si(100) surfaces, *Appl. Phys. Lett.* **62**(12), 1402–1404 (1993).
233. J. E. Mahan, V. Le Thanh, J. Clevrier, I. Berbezier, J. Derrien, and R. G. Long, Surface electron-diffraction patterns of β-$FeSi_2$ films epitaxially grown on silicon, *J. Appl. Phys.* **74**(3), 1747–1761 (1993).
234. N. Jedrecy, Y. Zheng, A. Waldhaner, M. Sanvage-Simkin, and R. Pinchaux, Epitaxy of β-$FeSi_2$ on Si(111), *Phys. Rev. B* **48**(12), 8801–8808 (1993).
235. J. F. Morar and M. Wittmer, Growth of epitaxial $CaSi_2$ films on Si (111), *J. Vac. Sci. Technol. A* **6**(3), 1340–1342 (1988).
236. F. Arnaud d'Avitaya, P.-A. Badoz, Y. Campidelli, J. A. Chroboczek, J.-Y. Duboz, A. Perio, and J. Pierre, Growth, characterization and electrical properties of epitaxial erbium silicide, *Thin Solid Films* **184**(1), 283–293 (1990).
237. F. H. Kaatz, M. P. Siegal, W. R. Graham, J. Van der Spiegel, and J. J. Santiago, Epitaxial growth of $ErSi_2$ on (111) Si, *Thin Solid Films* **184**(1), 325–333 (1990).
238. D. Cherns, D. A. Smith, W. Krakow, and P. E. Batson, Electron microscope studies of the structure and propagation of the Pd_2Si/(111)Si interface, *Philos. Mag. A* **45**(1), 107–125 (1982).
239. Y. Yokota, R. Matz, and P. S. Ho, TEM study of Pt silicide formation on clean Si surfaces, in *Thin Films and Interfaces II*, edited by J. E. E. Baglin, D. R. Campbel, and W. K. Chu (North-Holland, Amsterdam, 1984), pp. 435–440.
240. C. W. T. Bulle Lieuwma, A. H. van Ommen, J. Hornstra, and N. A. M. Aussems, Observation and analysis of epitaxial growth of $CoSi_2$ on (100) Si, *J. Appl. Phys.* **71**(5), 2211–2224 (1992).
241. M. Wittmer, C.-Y. Ting, and K. N. Tu, Atomic motion of dopant during interfacial silicide formation, *Thin Solid Films* **104**(1/2), 191–195 (1983).

242. C. B. Cooper III, R. A. Powell, and R. Chow, Dopant redistribution in silicides: Materials and process issues, *J. Vac. Sci. Technol. B* **2**(4), 718–722 (1984).

243. I. Ohdomari, K. Konuma, M. Takano, T. Chikyow, H. Kawarada, J. Nakanishi, and T. Ueno, Dopant redistribution during silicide formation, in *Thin Films—Interfaces and Phenomena*, edited by R. J. Nemanich, P. S. Ho, and S. S. Lau (MRS, Pittsburgh, 1986), pp. 63–72.

244. C. J. Sofield, R. E. Harper, and P. J. Rosser, Boron redistribution during transient thermal metal silicide growth on Si, in *Energy Beam–Solid Interactions and Transient Thermal Processing*, edited by D. K. Biegelsen, G. A. Rozgonyi, and C. V. Shank (MRS, Pittsburgh, 1985), pp. 445–450.

245. A. A. Pasa, J. P. de Souza, I. J. R. Baumvol, and F. L. Freire, Dopant redistribution during titanium-disilicide formation by rapid thermal processing, *J. Appl. Phys.* **61**(3), 1228–1230 (1987).

246. D. S. Wen, P. L. Smith, C. M. Osburn, and G. A. Rozgonyi, Defect annihilation in shallow p^+ junctions using titanium silicide, *Appl. Phys. Lett.* **51**(15), 1182–1184 (1987).

247. W. Zhou, G. Yang, N. Yu, Z. Zhou, and S. Zou, RBS studies of As redistribution during silicide formation by RTA, *Vacuum* **39**(2–4), 153–157 (1989).

248. M. P. Siegal and J. J. Santiago, The effects of dopant and impurity redistribution on WSi_2 formation by rapid thermal processing, *J. Appl. Phys.* **65**(2), 760–766 (1988).

249. P. Revesz, J. Gyimesi, and E. Zsoldos, Growth of titanium silicide on ion-implanted silicon, *J. Appl. Phys.* **54**(4), 1860–1864 (1983).

250. J. Amano, P. Merchant, and T. Koch, Arsenie out-diffusion during $TiSi_2$ formation, *Appl. Phys. Lett.* **44**(8), 744–746 (1984).

251. M. Wittmer and K. N. Tu, Low-temperature diffusion of dopant atoms in silicon during interfacial silicide formation, *Phys. Rev. B* **29**(4), 2010–2020 (1984).

252. L. R. Zheng, L. S. Hung, and J. W. Mayer, Redistribution of dopant arsenic during silicide formation, *J. Appl. Phys.* **58**(4), 1505–1514 (1985).

253. J. Amano, P. Merchant, T. R. Cass, J. N. Miller, and T. Koch, Dopant redistribution during titanium silicide formation, *J. Appl. Phys.* **59**(8), 2689–2693 (1986).

254. S. M. Hu, Point defect generation and enhanced diffusion in silicon due to tantalum silicide overlayer, *Appl. Phys. Lett.* **51**(3), 308–310 (1987).

255. P. Pan, N. Hsieh, H. J. Geipel, and G. J. Slusser, Dopant diffusion in tungsten silicide, *J. Appl. Phys.* **53**(4), 3059–3062 (1982).

256. S. Solmi, R. Angelucci, G. Cicognani, and R. Canteri, Diffusion of boron and arsenic in molybdenum disilicide films, *Appl. Surf. Sci.* **53**(1), 186–189 (1991).

257. A. H. Van Ommen, H. J. W. Van Houtum, and A. M. L. Theunissen, Diffusion of ion implanted As in $TiSi_2$, *J. Appl. Phys.* **60**(2), 627–630 (1986).

258. P. Gas, V. Deline, F. M. d'Heurle, A. Michel, and G. Scilla, Boron, phosphorus, and arsenic diffusion in $TiSi_2$, *J. Appl. Phys.* **60**(5), 1634–1639 (1986).

259. P. Gas, G. Scilla, A. Michel, F. K. LeGoues, O. Thomas, and F. M. d'Heurle, Diffusion of Sb, Ga, Ge, and (As) in $TiSi_2$, *J. Appl. Phys.* **63**(11), 5335–5345 (1988).

260. J. Pelleg and S. P. Murarka, Phosphorus distribution in $TaSi_2$ films by diffusion from a polycrystalline silicon layer, *J. Appl. Phys.* **54**(3), 1337–1345 (1983).

261. O. Thomas, P. Gas, A. Charai, F. K. Le Goues, A. Michel, G. Scilla, and F. M. d'Heurle, The diffusion of elements implanted in films of cobalt disilicide, *J. Appl. Phys.* **64**(6), 2973–2980 (1988).

262. F. M. d'Heurle, R. T. Hodgson, and C. Y. Ting, Silicides and rapid thermal annealing, in *Rapid Thermal Annealing*, edited by T. O. Sedgwick, T. E. Seidel, and B.-Y. Tsaur (MRS, Pittsburgh, 1986), pp. 261–270.

263. R. A. Powell, R. Chow, C. Thridandom, R. T. Fulks, I. A. Blech, and J. T. Pan, Formation of titanium silicide films by rapid thermal processing, *IEEE Electron Device Lett.* **4**(10), 380–382 (1983).

264. V. Privitera, F. La Via, E. Rimini, and G. Ferla, Titanium silicide as a diffusion source for arsenic, *J. Appl. Phys.* **67**(11), 7174–7176 (1990).

265. C. B. Cooper III and R. A. Powell, The use of rapid thermal processing to control dopant redistribution during formation of tantalum and molybdenum silicide/n^+polysilicon bilayers, *IEEE Electron Device Lett.* **6**(5), 234–236 (1985).

266. J. Kato, A. Fujisawa, and M. Asahina, Diffusion of boron in Mo silicide films, *J. Appl. Phys.* **59**(12), 4186–4189 (1986).

267. V. Privitera, F. La Via, C. Spinella, V. Raineri, and E. Rimini, Titanium silicide as a diffusion source for phosphorus: Precipitation and activation, *Appl. Surf. Sci.* **53**(1), 190–195 (1991).
268. F. La Via, V. Privitera, S. Lombardo, C. Spinella, V. Raineri, E. Rimini, P. Baeri, and G. Ferla, Precipitation of arsenic diffused into silicon from a $TiSi_2$ source, *J. Appl. Phys.* **69**(2), 726–731 (1991).
269. V. Probst, H. Schaber, A. Mitwalsky, H. Kabza, B. Hoffmann, K. Maex, and L. Van den hove, Metal-dopant-compound formation in $TiSi_2$ and $TaSi_2$. Impact on dopant diffusion and contact resistance, *J. Appl. Phys.* **70**(2), 693–707 (1991).
270. W. Eichhammer, K. Maex, K. Elst, and W. Vandervorst, Boron outdiffusion from poly- and monocrystalline $CoSi_2$, *Appl. Surf. Sci.* **53**(1), 171–179 (1991).
271. V. Probst, H. Schaber, A. Mitwalsky, H. Kabza, L. Van den hove, and K. Maex, WSi_2 and $CoSi_2$ as diffusion sources for shallow-junction formation in silicon, *J. Appl. Phys.* **70**(2), 708–719 (1991).
272. J. Lin, W. Chen, A. Sultan, S. Banerjee, and J. Lee, Enhanced boron diffusion in silicon with BF_2-implanted $CoSi_2$ as diffusion source and under rapid thermal annealing, in *Rapid Thermal and Integrated Processing II*, edited by J. C. Gelpey, J. K. Elliott, J. J. Wortman, and A. Ajmera (MRS, Pittsburgh, 1993), pp. 265–270.
273. W. Chen, J. Lin, S. Banerjee, and J. Lee, Using $CoSi_2$/polysilicon polycide structure as a gate diffusion source in rapid thermal processing, in *Rapid Thermal and Integrated Processing II*, edited by J. C. Gelpey, J. K. Elliott, J. J. Wortman, and A. Ajmera (MRS, Pittsburgh, 1993), pp. 271–276.
274. E. C. Jones, S. Ogawa, P. Ameika, M. Lawrence, A. Dass, D. B. Fraser, P. Chu, and N. W. Cheung, Rapid thermal processing of shallow junctions using epitaxial $CoSi_2$ as a doping source, in *Rapid Thermal and Integrated Processing III*, edited by J. J. Wortman, J. C. Gelpey, M. L. Green, S. R. J. Brueck, and F. Roozeboom (MRS, Pittsburgh, 1994), pp. 339–343.
275. A. S. Pedersen, J. Chevallier, and A. Nylandsted Larsen, Electrical characteristics of PtSi formed by incoherent light annealing, *Thin Solid Films* **164**(1–4), 487–492 (1988).
276. N. Toyama, T. Takahashi, H. Murakami, and H. Koriyama, Variation of the effective Richardson constant of Pt-Si Schottky diode due to annealing treatment, *Appl. Phys. Lett.* **46**(6), 557–559 (1985).
277. C.-A. Chang, A. Segmuller, H. C. W. Huang, B. Cunningham, F. E. Turene, A. Sugerman, and P. A. Totta, PtSi contact metallurgy using sputtered Pt and different annealing processes, *J. Electrochem. Soc.* **133**(6), 1256–1260 (1986).
278. F. Nava, T. Tien, and K. N. Tu, Temperature dependence of semiconducting and structural properties of Cr-Si thin films, *J. Appl. Phys.* **57**(6), 2018–2025 (1985).
279. E. V. Buzaneva, V. V. Ilchenko, and V. I. Strikha, Electronnie svoistva granitsi razdela silitsida khroma i molibdena v kontaktah s kremniem, *Izv. Vyssh. Uchebu. Zaved. Fiz.* **8**, 105–106 (1984).
280. M. C. Bost and J. E. Mahan, Semiconducting silicides as potential materials for electro-optic very large scale integrated circuit interconnects, *J. Vac. Sci. Technol. B* **4**(6), 1336–1338 (1986).
281. A. B. Filonov, I. E. Tralle, N. N. Dorozhkin, D. B. Migas, V. L. Shaposhnikov, G. V. Petrov, A. M. Anishchik, and V. E. Borisenko, Semiconducting properties of hexagonal chromium, molybdenum, and tungsten disilicides, *Phys. Status Solidi B* **186**(1), 209–215 (1994).

CHAPTER 6

1. P. C. Frazan, A. Ditali, C. H. Dennison, H. E. Rhodes, H. C. Chan, and Y. C. Liu, Reliability and characterization of composite oxide/nitride dielectrics for multi-megabit dynamic random access memory stacked capacitors, *J. Electrochem. Soc.* **138**(7), 2052–2057 (1991).
2. T. Hori, S. Akamatsu, and Y. Odake, Deep-submicrometer CMOS technology with reoxidized or annealed nitrided-oxide gate dielectrics prepared by rapid thermal processing, *IEEE Trans. Electron Devices* **39**(1), 118–126 (1992).
3. H. Fukuda, M. Yasuda, T. Iwabuchi, and S. Ohno, Novel N_2O-oxynitride technology for forming highly reliable EEPROM tunnel oxide films, *IEEE Electron Device Lett.* **12**(11), 587–589 (1991).
4. C. d'Anterroches, High resolution T.E.M. study of the Si(001)/SiO_2 interface, *J. Microsc. Spectrosc. Electron.* **9**, 147–162 (1984).
5. J.-J. Ganem, S. Rigo, I. Trimaille, and G.-N. Lu, NRA characterization of pretreatment operations of silicon, *Nucl. Instrum. Methods Phys.* **B64**, 784–788 (1992).

6. M.-T. Tang, K. W. Evans-Lutterodt, G. S. Higashi, and T. Boone, Roughness of the silicon (001)/SiO_2 interface, *Appl. Phys. Lett.* **62**(24), 3144–3146 (1993).
7. M.-T. Tang, K. W. Evans-Lutterodt, M. L. Green, D. Brasen, K. Krisch, L. Manchanda, G. S. Higashi, and T. Boone, Growth temperature dependence of the Si(100)/SiO_2 interface width, *Appl. Phys. Lett.* **64**(6), 748–750 (1994).
8. J. Nulman, J. P. Krusius, and A. Gat, Rapid thermal processing of thin gate dielectrics, *IEEE Electron Device Lett.* **6**(5), 205–207 (1985).
9. R. Singh, S. Sinha, R. P. S. Thakur, and P. Chou, Some photoeffect roles in rapid isothermal processing, *Appl. Phys. Lett.* **58**(11), 1217–1219 (1991).
10. A. Kazor and I. W. Boyd, Growth and modeling of cw-UV induced oxidation of silicon, *J. Appl. Phys.* **75**(1), 227–230 (1994).
11. E. M. Young and W. A. Tiller, Photon enhanced oxidation of silicon, *Appl. Phys. Lett.* **42**(1), 63–65 (1983).
12. E. M. Young, Electron population factor in light enhanced oxidation of silicon, *Appl. Phys. Lett.* **50**(1), 46–48 (1987).
13. Y. Sato and K. Kiuchi, Oxidation of silicon using lamp light radiation, *J. Electrochem. Soc.* **133**, 652–654 (1986).
14. N. C. Tung and Y. Caratini, Rapid thermal oxidation of silicon for thin gate dielectric, *Electron. Lett.* **22**(13), 694–696 (1986).
15. M. M. Moslehi, S. C. Shatas, and K. C. Saraswat, Thin SiO_2 insulators grown by rapid thermal oxidation of silicon, *Appl. Phys. Lett.* **47**(12), 1353–1355 (1985).
16. J. P. Ponpon, J. J. Grob, A. Grob, and R. Stuck, Formation of thin silicon oxide films by rapid thermal heating, *J. Appl. Phys.* **59**(11), 3921–3923 (1986).
17. A. Slaoui, J. Ponpon, and P. Siffert, Characterization of thin silicon oxide obtained by lamp heating, *Appl. Phys. A* **43**, 301–304 (1987).
18. Y. L. Chiou, C. H. Sow, G. Li, and K. A. Ports, Growth characteristics of silicon dioxide produced by rapid thermal oxidation processes, *Appl. Phys. Lett.* **57**(9), 881–883 (1990).
19. B. E. Deal and A. S. Grove, General relationship for the thermal oxidation of silicon, *J. Appl. Phys.* **36**, 3770–3778 (1965).
20. H. Z. Massoud, J. D. Plummer, and E. A. Irene, Thermal oxidation of silicon in dry oxygen growth-rate enhancement in the thin regime I. Experimental results, *J. Electrochem. Soc.* **132**(11), 2685–2693 (1985).
21. N. C. Tung, Y. Caratini, C. d'Anterroches, and J. L. Buevoz, Characteristics of thin gate dielectric in a rapid thermal processing machine and temperature uniformity studies, *Appl. Phys. A* **47**, 237–247 (1988).
22. V. P. Parkhutik, V. A. Labunov, and G. G. Chigir, Kinetics and mechanisms of transient thermal oxidation of silicon, *Phys. Status Solidi A* **96**, 11–18 (1986).
23. J. J. Ganem, G. Battistig, S. Rigo, and I. Trimaille, A study of the initial stages of the oxidation of silicon using $^{18}O_2$ and RTP, *Appl. Surf. Sci.* **65/66**, 647–653 (1993).
24. F. C. Stedile, I. J. R. Baumvol, J. J. Ganem, S. Rigo, I. Trimaille, G. Battistig, W. H. Schulte, and H. W. Becker, IBA study of the growth mechanisms of very thin silicon oxide films: The effect of wafer cleaning, *Nucl. Instrum. Methods Phys. Res. B* **85**, 248–254 (1994).
25. A. Y. Messaoud, E. Scheid, G. Sarrabayrouse, A. Claverie, and A. Martinez, A comprehensive study of thin rapid thermal oxide films, *Jpn. J. Appl. Phys.* **32**(12A), 5805–5812 (1993).
26. K. Ohyu, Y. Wade, S. Iijima, and N. Natsuaki, Highly reliable thin silicon dioxide layers grown on heavily phosphorous doped poly-Si by rapid thermal oxidation, *J. Electrochem. Soc.* **137**(7), 2261–2265 (1990).
27. E. A. Irene, H. Z. Massoud, and E. Tierney, Silicon oxidation studies: Silicon orientation effects on thermal oxidation, *J. Electrochem. Soc.* **133**, 1253–1256 (1986).
28. S. C. Sun, L. S. Wang, F. L. Yeh, T. S. Lai, and Y. H. Lin, Rapid thermal oxidation of lightly doped silicon in N_2O, *Mater. Res. Soc. Symp. Proc.* **342**, 181–186 (1994).
29. M. M. Moslehi, S. C. Shatas, K. C. Saraswat, and J. D. Meindl, Interfacial and breakdown characteristics of MOS devices with rapidly grown ultrathin SiO_2 gate insulators, *IEEE Trans. Electron Devices* **34**(6), 1407–1410 (1987).
30. J. M. deLarios, D. B. Kao, B. E. Deal, and C. R. Helms, The effect of aqueous chemical cleaning on Si (100) dry oxidation kinetics, *J. Electrochem. Soc.* **138**(8), 2353–2361 (1991).
31. J. M. deLarios, D. B. Kao, C. R. Helms, and B. E. Deal, Effect of SiO_2 surface chemistry on oxidation of silicon, *Appl. Phys. Lett.* **54**(8), 715–717 (1989).

32. J. M. deLarios, C. R. Helms, D. B. Kao, and B. E. Deal, Effect of silicon surface cleaning procedures on oxidation kinetics and surface chemistry, *Appl. Surf. Sci.* **30**, 17–24 (1987).
33. E. A. Irene, The effects of trace amounts of water on the thermal oxidation of silicon in oxygen, *J. Electrochem. Soc.* **121**(12), 1613–1616 (1974).
34. E. A. Irene and R. Ghez, Silicon oxidation studies: The role of H_2O, *J. Electrochem. Soc.* **124**, 1757–1761 (1977).
35. M. A. George and D. A. Bohling, Hydrogen content of silicon and thermal oxidation induced moisture generation in an integrated rapid thermal processing reactor, *J. Vac. Sci. Technol. B* **11**(1), 86–91 (1993).
36. M. Glück, U. König, J. Hersener, Z. Nenyei, and A. Tillmann, Homogeneity of wet oxidation by RTP, *Mater. Res. Soc. Symp. Proc.* **342**, 215–225 (1994).
37. D. K. Nayak, K. Kamjoo, J. S. Park, J. C. S. Woo, and K. L. Wang, Rapid isothermal processing of strained GeSi layers, *IEEE Trans. Electron Devices* **39**(1), 56–62 (1992).
38. S. A. Ajuria, P. U. Kenkare, A. Nghiem, and T. C. Mele, Kinetic analysis of silicon oxidations in the thin regime by incremental growth, *J. Appl. Phys.* **76**(8), 4618–4624 (1994).
39. R. Deaton and H. Z. Massoud, Effect of thermally induced stresses on the rapid-thermal oxidation of silicon, *J. Appl. Phys.* **70**(7), 3588–3592 (1991).
40. Y. T. Kim and C. H. Jun, Effects of chlorine based gettering on the electrical properties of rapid thermal oxidation//nitridation dielectric films, *J. Vac. Sci. Technol. A* **11**(4), 1039–1043 (1993).
41. S. C. Chao, R. Pitchai, and Y. H. Lee, Enhancement of thermal oxidation of silicon by ozone, *J. Electrochem. Soc.* **136**(9), 2751–2752 (1989).
42. A. Kazor, R. Gwilliam, and I. W. Boyd, Growth rate enhancement using ozone during rapid thermal oxidation of silicon, *Appl. Phys. Lett.* **65**(4), 412–414 (1994).
43. H. Z. Massoud, J. D. Plummer, and E. A. Irene, Thermal oxidation of silicon in dry oxygen growth-rate enhancement in the thin regime II. Physical mechanisms, *J. Electrochem. Soc.* **132**(11), 2693–2700 (1985).
44. V. K. Samalam, Theoretical model for the oxidation of silicon, *Appl. Phys. Lett.* **47**(7), 736–737 (1985).
45. M. Orlowski and V. Pless, Analysis of rapid dry oxidation of silicon, *Appl. Phys. A* **46**, 67–71 (1988).
46. S. A. Schafer and S. A. Lyon, New model of the rapid initial oxidation of silicon, *Appl. Phys. Lett.* **47**(2), 154–156 (1985).
47. E. A. Irene, Silicon oxidation studies: A revised model for thermal oxidation, *J. Appl. Phys.* **54**(9), 5416–5420 (1983).
48. E. A. Irene, Silicon oxidation studies: Some aspects of the initial oxidation region, *J. Electrochem. Soc.* **125**(10), 1708–1714 (1978).
49. H. Z. Massoud and J. D. Plummer, Analytical relationship for the oxidation of silicon in dry oxygen in the thin-film regime, *J. Appl. Phys.* **62**(8), 3416–3423 (1987).
50. C.-J. Han and C. R. Helms, Parallel oxidation mechanism for Si oxidation in dry O_2, *J. Electrochem. Soc.* **134**(5), 1297–1301 (1987).
51. J. M. deLarios, C. R. Helms, D. B. Kao, and B. E. Deal, Parallel oxidation model for Si including both molecular and atomic oxygen mechanisms, *Appl. Surf. Sci.* **39**, 89–102 (1989).
52. P. J. Rosser, P. B. Moynagh, and C. N. Duckworth, Modelling of rapid thermal and furnace oxide growth, *Mater. Res. Soc. Symp. Proc.* **54**, 611–616 (1986).
53. V. Murali, Rapid thermal processing of electronic materials—Oxidation of silicon and annealing of silicon dioxide, Ph.D. dissertation, Department of Materials Engineering, Rensselaer Polytechnic Institute, Troy, NY (1987).
54. C. A. Paz de Araujo, R. W. Gallegos, and Y. P. Huang, Kinetics of rapid thermal oxidation, *J. Electrochem. Soc.* **136**(9), 2673–2676 (1989).
55. A. Kazor, Space-charge oxidant diffusion model for rapid thermal oxidation of silicon, *J. Appl. Phys.* **77**(4), 1477–1481 (1995).
56. B. E. Deal, Standard terminology for oxide charges associated with thermally oxidized silicon, *J. Electrochem. Soc.* **127**(4), 979–981 (1980).
57. S. Wolf and R. N. Tauber, *Silicon Processing for the VLSI Era* (Lattice Press, Sunset Beach, CA, 1986).
58. D. J. DiMaria, E. Cartier, and D. Arnold, Impact ionization, trap creation, degradation, and breakdown in silicon dioxide films on silicon, *J. Appl. Phys.* **73**(7), 3367–3384 (1993).

59. C. T. Sah, Models and experiments on degradation of oxidized silicon, *Solid-State Electron.* **33**(2), 147–167 (1990).
60. D. A. Buchanan and D. J. DiMaria, Interface and bulk trap generation in metal-oxide-semiconductor capacitors, *J. Appl. Phys.* **67**(12), 7439–7452 (1990).
61. D. Arnold, E. Cartier, and D. J. DiMaria, Theory of high-field electron transport and impact ionization in silicon dioxide, *Phys. Rev.* **49**(15), 10278–10297 (1994).
62. R. F. Pierret, *Field Effect Devices*, edited by G. W. Neudeck and R. F. Pierret (Addison–Wesley, Reading, MA, 1990).
63. S. M. Sze, *Physics of Semiconductor Devices* (Wiley, New York, 1981).
64. E. H. Nicollian and J. R. Brews, *MOS (Metal Oxide Semiconductor) Physics and Technology* (Wiley, New York, 1982).
65. W. R. Runyan and K. E. Bean, *Semiconductor Integrated Circuit Processing Technology* (Addison–Wesley, Reading, MA, 1990).
66. H. Fukuda, T. Iwabuchi, and S. Ohno, The dielectric reliability of very thin SiO_2 films grown by rapid thermal processing, *Jpn. J. Appl. Phys.* **27**(11), L2164–L2167 (1988).
67. J. H. Stathis, D. A. Buchanan, D. L. Quinian, and A. H. Parsons, and D. E. Kotecki, Interface defects of ultrathin rapid-thermal oxide on silicon, *Appl. Phys. Lett.* **62**(21), 2682–2684 (1993).
68. G. J. Gerardi, E. H. Poindexter, P. J. Caplan, and N. M. Johnson, Interface traps and P_b centers in oxidized (100) silicon wafers, *Appl. Phys. Lett.* **49**(6), 348–350 (1986).
69. G. Q. Lo, D. L. Kwong, K. J. Abbott, and D. Nazarian, Thickness dependence of charge-trapping properties in ultrathin thermal oxides prepared by rapid thermal oxidation, *J. Electrochem. Soc.* **140**(2), L16–L19 (1993).
70. G. Eftekhari, Characterization of thin SiO_2 grown by rapid thermal processing as influenced by processing parameters, *J. Electrochem. Soc.* **140**(3), 787–789 (1993).
71. L. Fonseca and F. Campabadal, Effect of RTP/furnace processing on the minority carrier lifetime in very thin oxide MOS capacitors, *Solid-State Electron.* **37**(1), 115–117 (1994).
72. L. Fonseca and F. Campabadal, Breakdown characteristics of RTO 10 nm SiO_2 films grown at different temperatures, *IEEE Electron Device Lett.* **15**(11), 449–451 (1994).
73. M. Depas, R. L. Van Meirhaeghe, W. H. Laflére, and F. Gardon, Electrical characteristics of Al/SiO_2/n-Si tunnel diodes with an oxide layer grown by rapid thermal oxidation, *Solid-State Electron.* **37**(3), 433–441 (1994).
74. R. Singh, A new growth enhancement in thin SiO_2 formed by rapid isothermal oxidation of silicon, *J. Vac. Sci. Technol. A* **6**(3), 1480–1483 (1988).
75. S. G. dos Santos Filho, C. M. Hansenack, M. C. V. Lopes, and V. Baranauskas, Rapid thermal oxidation of silicon with different thermal annealing cycles in nitrogen: Influence on surface microroughness and electrical characteristics, *Semicond. Sci. Technol.* **10**, 990–996 (1995).
76. T. Ohmi, M. Miyashita, M. Itano, T. Imaoka, and I. Kawanabe, Dependence of thin-oxide films quality of surface microroughness, *IEEE Trans. Electron Devices* **39**(3), 537–545 (1992).
77. G. Q. Lo, W. Ting, J. H. Ahn, D.-L. Kwong, and J. Kuehne, Thin fluorinated gate dielectrics grown by rapid thermal processing in O_2 with diluted NF_3, *IEEE Trans. Electron Devices* **39**(1), 148–153 (1992).
78. W.-S. Lu and J.-G. Hwu, Preparation of fluorinated gate oxides by liquid phase deposition following rapid thermal oxidation, *Appl. Phys. Lett.* **66**(24), 3322–3324 (1995).
79. G. A. Hames, J. J. Wortman, S. E. Beck, and D. A. Bohling, Degradation of rapid thermal oxides due to the presence of nitrogen in the oxidation ambient, *Appl. Phys. Lett.* **64**(8), 980–982 (1994).
80. H. Z. Massoud, Rapid thermal growth and processing of dielectrics, in *Rapid Thermal and Integrated Processing II, Rapid Thermal Processing*, edited by R. B. Fair (Academic Press, New York, 1993), pp. 45–77.
81. M. M. Moslehi and K. C. Saraswat, Thermal nitridation of Si and SiO_2 for VLSI, *IEEE Trans. Electron Devices* **32**(2), 106–123 (1985).
82. J. Nulman and J. P. Krusius, Rapid thermal nitridation of thin thermal silicon dioxide films, *Appl. Phys. Lett.* **47**(2), 148–150 (1985).
83. T. Hori, Y. Naito, H. Iwasaki, and H. Esaki, Interface states and fixed charges in nanometer range thin nitrided oxides prepared by rapid thermal annealing, *IEEE Electron Device Lett.* **7**(12), 669–671 (1986).

84. T. Hori, H. Iwasaki, Y. Naito, and H. Esaki, Electrical and physical characteristics of thin nitrided oxides prepared by rapid thermal nitridation, *IEEE Trans. Electron Devices* **34**(11), 2238–2245 (1987).
85. D. K. Shih, W. T. Chang, S. K. Lee, Y. H. Ku, D. L. Kwong, and S. Lee, Metal-oxide-semiconductor characteristics of rapid thermal nitrided thin oxides, *Appl. Phys. Lett.* **52**(20), 1698–1700 (1988).
86. Zhi Hong Liu, P. T. Lai, and Y. C. Cheng, Characterization of charge trapping and high-field endurance for 15-nm thermally nitrided oxides, *IEEE Trans. Electron Devices* **38**(2), 344–354 (1991).
87. P. J. Wright, A. Kermani, and K. C. Saraswat, Nitridation and post-nitridation anneals of SiO_2 for ultrathin dielectrics, *IEEE Trans. Electron Devices* **37**(8), 1836–1841 (1990).
88. Y.-L. Wu and J.-G. Hwu, Characterization of metal-oxide-semiconductor capacitors with improved gate oxides prepared by repeated rapid thermal annealings in N_2O, *J. Vac. Sci. Technol. B* **12**(4), 2400–2404 (1994).
89. T. Hori and H. Iwasaki, Excellent charge-trapping properties of ultrathin reoxidized nitrided oxides prepared by rapid thermal processing, *IEEE Electron Device Lett.* **9**(41), 168–170 (1988).
90. T. Hori and H. Iwasaki, Correlation between electron trap density and hydrogen concentration in ultrathin rapidly reoxidized nitrided oxides, *Appl. Phys. Lett.* **52**(9), 736–738 (1988).
91. D. K. Shih and D. L. Kwong, Study of the SiO_2/Si interface endurance property during rapid thermal nitridation and reoxidation processing, *Appl. Phys. Lett.* **54**(9), 822–824 (1989).
92. T. Hori, H. Iwasaki, and K. Tsuji, Electrical and physical properties of ultrathin reoxidized nitrided oxides prepared by rapid thermal processing, *IEEE Trans. Electron Devices* **36**(2), 340–349 (1989).
93. L. K. Han, J. Kim, G. W. Yoon, J. Yan, and D. L. Kwong, High quality oxynitride gate dielectrics prepared by reoxidation of NH_3-nitrided SiO_2 in N_2O ambient, *Electron. Lett.* **31**(14), 1196–1198 (1995).
94. J. T. Yount, P. M. Lenahan, and J. T. Krick, Comparison of defect structure in N_2O- and NH_3-nitrided oxide dielectrics, *J. Appl. Phys.* **76**(3), 1754–1758 (1994).
95. Y. Okada, P. J. Tobin, P. Rushbrook, and W. L. DeHart, The performance and reliability of 0.4 micron MOSFET's with gate oxynitrides grown by rapid thermal processing using mixtures of N_2O and O_2, *IEEE Trans. Electron Devices* **41**(2), 191–197 (1994).
96. J. S. Cable, R. A. Mann, and J. C. S. Woo, Impurity barrier properties of reoxidized nitrided oxide films for use with p^+-doped polysilicon gates, *IEEE Electron Device Lett.* **12**(3), 128–130 (1991).
97. T. Hori, T. Yasui, and S. Akamatsu, Hot-carrier effects in MOSFET's with nitrided-oxide gate-dielectrics prepared by rapid thermal processing, *IEEE Trans. Electron Devices* **39**(1), 134–147 (1992).
98. A. B. Joshi and D.-L. Kwong, Excellent immunity of GIDL to hot-electron stress in reoxidized nitrided gate oxide MOSFET's, *IEEE Electron Device Lett.* **13**(1), 47–49 (1992).
99. H. C. Cheng, H. W. Liu, H. P. Su, and G. Hong, Superior low-pressure-oxidized Si_3N_4 films on rapid-thermal-nitrided poly-Si for high-density DRAM's, *IEEE Electron Device Lett.* **16**(11), 509–511 (1995).
100. Y. Naito, T. Hori, H. Iwasaki, and H. Esaki, Effect of nitrogen distribution in nitrided oxide prepared by rapid thermal annealing on its electrical characteristics, *J. Vac. Sci. Technol. B* **5**(3), 633–637 (1987).
101. F. Iberl, P. Ramm, and W. Lang, ONO structures investigated by SIMS, RBS and NRA, *Nucl. Instrum. Methods Phys. Res. B* **64**, 650–653 (1992).
102. M. Dutoit, P. Letourneau, J. Mi, and M. Novkovski, Optimization of thin Si oxynitride films produced by rapid thermal processing for applications in EEPROMs, *J. Electrochem. Soc.* **140**(2), 549–555 (1993).
103. M. Dutoit, D. Bouvet, J. Mi, N. Novkovski, and P. Letourneau, Thin SiO_2 films nitrided by rapid thermal processing in NH_3 or N_2O for applications in EEPROMs, *Microelectron. J.* **25**, 539–551 (1994).
104. W.-S. Lu, C.-H. Chan, and J.-G. Hwu, A systematic study of the initial electrical and radiation hardness properties of reoxidized nitrided oxides by rapid thermal processing, *IEEE Trans. Nucl. Sci.* **42**(3), 167–173 (1995).
105. G. Q. Lo, A. W. Cheung, and D.-L. Kwong, Current conduction and charge trapping in thin interpoly dielectrics prepared by *in situ* multiple rapid thermal processing, *J. Electrochem. Soc.* **139**(5), 1424–1430 (1992).
106. G. Q. Lo, A. W. Cheung, D.-L. Kwong, and N. S. Alvi, Polarity asymmetry of electrical characteristics of thin nitrided polyoxides prepared by *in-situ* multiple rapid thermal processing, *Solid-State Electron.* **34**(2), 181–184 (1991).

107. S. Itoh, G. Q. Lo, D.-L. Kwong, V. K. Mathews, and P. C. Frazan, Rapid thermal oxidation of thin nitride dielectrics deposited on rapid thermal nitrided polycrystalline silicon, *Appl. Phys. Lett.* **61**(11), 1313–1315 (1992).
108. H.-W. Liu, H.-P. Su, and H.-C. Cheng, High-performance superthin oxide/nitride/oxide stacked dielectrics formed by low-pressure oxidation of ultrathin nitride, *Jpn. J. Appl. Phys.* **34**(4A), Pt. 1, 1713–1715 (1995).
109. Y. Ma, T. Yasuda, and G. Lucovsky, Formation of device-quality metal-insulator-semiconductor structures with oxide–nitride–oxide dielectrics by low-temperature plasma-assisted processing, combined with high-temperature rapid thermal annealing, *J. Vac. Sci. Technol. A* **11**(4), 952–958 (1993).
110. K. Yoneda, Y. Todokoto, and M. Inoue, Thin silicon dioxide and nitrides oxide using rapid thermal processing for trench capacitors, *J. Mater. Res.* **6**(11), 2362–2370 (1991).
111. B. Fröschle, H. P. Bruemmer, W. Lang, K. Neumeier, and P. Ramm, Fabrication of high qualitity MOS devices for application in hazardous environments based on RTP gate dielectrics with in situ RTCVD of polysilicon gates, *Mater. Res. Soc. Symp. Proc.* **303**, 331–338 (1993).
112. H. Fukuda, T. Arakawa, and S. Ohno, Highly reliable thin nitrided SiO_2 films formed by rapid thermal processing in an N_2O ambient, *Electron. Lett.* **26**(18), 1505–1506 (1990).
113. H. Fukuda, T. Arakawa, and S. Ohno, Highly reliable thin nitrided SiO_2 thermal processing in an N_2O ambient, *Jpn. J. Appl. Phys.* **29**(12), 233–236 (1990).
114. G. W. Yoon, A. B. Joshi, J. Kim, and D.-L. Kwong, High-field-induced leakage in ultrathin N_2O oxides, *IEEE Electron Device Lett.* **14**(5), 231–233 (1993).
115. G. W. Yoon, A. B. Joshi, J. Kim, and D.-L. Kwong, High quality ultra-thin tunneling N_2O oxides fabricated by RTP, *Mater. Res. Soc. Symp. Proc.* **303**, 291–296 (1993).
116. N. Bellafiore, F. Pio, and C. Riva, Thin SiO_2 films nitrided in N_2O, *Microelectron. J.* **25**, 495–500 (1994).
117. H. Fukuda and S. Nomura, Effect of Fowler–Nordheim stress on charge trapping properties of ultrathin N_2O-oxynitrided SiO_2 films, *Jpn. J. Appl. Phys.* **34**(1), 87–88 (1995).
118. H. Fukuda, T. Arakawa, and S. Ohno, Thin-gate SiO_2 films formed by in situ multiple rapid thermal processing, *IEEE Trans. Electron Devices* **39**(1), 127–132 (1992).
119. H. Fukuda, M. Yasuda, and S. Ohno, Electrical properties of thin oxynitrided SiO_2 films formed by rapid thermal processing in an N_2O ambient, *Electron. Lett.* **27**(5), 440–441 (1991).
120. H. Hwang, W. Ting, B. Maiti, D.-L. Kwong, and J. Lee, Electrical characteristics of ultrathin oxynitride gate dielectric repaired by rapid thermal oxidation of Si in N_2O, *Appl. Phys. Lett.* **57**(10), 1010–1011 (1990).
121. W. Ting, H. Hwang, J. Lee, and D.-L. Kwong, Composition and growth kinetics of ultrathin SiO_2 films formed by oxidizing Si substrates in N_2O, *Appl. Phys. Lett.* **57**(26), 2808–2810 (1990).
122. Y. Okada, P. J. Tobin, V. Lakhotia, W. A. Feil, S. A. Ajuria, and R. I. Hege, Relationship between growth conditions, nitrogen profile, and charge to breakdown of gate oxynitrides grown from pure N_2O, *Appl. Phys. Lett.* **63**(2), 194–196 (1993).
123. A. E. T. Kuiper, H. G. Pomp, P. M. Asveld, W. A. Bik, and F. H. O. M. Habraken, Nitrogen and oxygen incorporation during rapid thermal processing of Si in N_2O, *Appl. Phys. Lett.* **61**(9), 1031–1033 (1992).
124. H. T. Tang, W. N. Lennard, M. Zinke-Allmang, I. V. Mitchell, L. C. Feldman, M. L. Green, and D. Brasen, Nitrogen content of oxynitride films on Si(100), *Appl. Phys. Lett.* **64**(25), 3473–3475 (1994).
125. M. L. Green, D. Brasen, K. W. Evans-Lutterodt, L. C. Feldman, K. Krisch, W. Lennard, H.-T. Tang, L. Manchanda, and M.-T. Tang, Rapid thermal oxidation of silicon in N_2O between 800 and 1200°C: Incorporated nitrogen and interfacial roughness, *Appl. Phys. Lett.* **65**(7), 848–850 (1994).
126. G. Weidner and D. Krüger, Nitrogen incorporation in SiO_2 by rapid thermal processing of silicon and SiO_2 in N_2O, *Appl. Phys. Lett.* **62**(3), 294–296 (1993).
127. Z.-Q. Yao, The nature and distribution of nitrogen in silicon oxynitride grown on silicon in a nitric oxide ambient, *J. Appl. Phys.* **78**(5), 2906–2912 (1995).
128. Z.-Q. Yao, H. B. Harrison, S. Dimitrijev, and Y. T. Yeow, The electrical properties of sub-5nm oxynitride dielectrics prepared in a nitric oxide ambient using rapid thermal processing, *IEEE Electron Device Lett.* **15**(12), 516–518 (1995).
129. E. C. Carr and R. A. Buhrman, Role of interfacial nitrogen in improving thin silicon oxides grown in N_2O, *Appl. Phys. Lett.* **63**(1), 54–56 (1993).
130. N. S. Saks, D. I. Ma, and W. B. Fowler, Nitrogen depletion during oxidation in N_2O, *Appl. Phys. Lett.* **67**(3), 374–376 (1995).

131. T. S. Chao, W. H. Chen, S. C. Sun, and H. Y. Chang, Characterizations of oxide grown by N_2O, *J. Electrochem. Soc.* **140**(10), 2905–2908 (1993).
132. T. Y. Chu, W. Ting, J. H. Ahn, S. Lin, and D.-L. Kwong, Study of the composition of the dielectrics grown on Si in a pure N_2O ambient, *Appl. Phys. Lett.* **59**(12), 1412–1414 (1991).
133. R. P. Vasquez and A. Madukar, Strain-dependent defect formation kinetics and a correlation between voltage and nitrogen distribution in thermally nitrided SiO_xN_y structures, *Appl. Phys. Lett.* **47**(9), 998–1000 (1985).
134. E. C. Carr, K. A. Ellis, and R. A. Buhrman, N depth profiles in thin SiO_2 grown or processed in N_2O: The role of atomic oxygen, *Appl. Phys. Lett.* **66**(12), 1492–1494 (1995).
135. Z.-Q. Yao, H. B. Harrison, S. Dimitrijev, D. Sweatman, and Y. T. Yeow, High quality ultrathin dielectric films grown on silicon in a nitric oxide ambient, *Appl. Phys. Lett.* **64**(26), 3584–3586 (1994).
136. Y. Ishikawa, Y. Takagi, and I. Nakamichi, Low-temperature thermal oxidation of silicon in N_2O by UV-irradiation, *Jpn. J. Appl. Phys.* **28**(8), L1453–L1455 (1989).
137. W. Ting, H. Hwang, J. Lee, and D.-L. Kwong, Growth kinetics of ultrathin SiO_2 films fabricated by rapid thermal oxidation of Si substrates in N_2O, *J. Appl. Phys.* **70**(2), 1072–1074 (1991).
138. G. W. Yoon, A. B. Joshi, J. Ahn, and D.-L. Kwong, Thickness uniformity and electrical properties of ultrathin gate oxides grown in N_2O ambient by rapid thermal processing, *J. Appl. Phys.* **72**(12), 5706–5710 (1992).
139. Y. Okada, P. J. Tobin, V. Lakhotia, S. A. Ajuria, R. I. Hege, J. C. Liao, P. Rushbrook, and L. J. Arias, Jr., Evaluation of interfacial nitrogen concentration of RTP oxynitrides by reoxidation, *J. Electrochem. Soc.* **140**(6), L87–L89 (1993).
140. J. M. Grant and T.-Y. Hsieh, Ultrathin gate dielectrics grown in mixtures of N_2O and O_2 using rapid thermal oxidation, *Mater. Res. Soc. Symp. Proc.* **342**, 163–168 (1994).
141. R. Mcintosh, C. Galewski, and J. Grant, Ultrathin gate dielectric growth at reduced pressure, *Mater. Res. Soc. Symp. Proc.* **342**, 209–214 (1994).
142. H. Hwang, W. Ting, D.-L. Kwong, and J. Lee, A physical model for boron penetration through an oxynitride gate dielectric prepared by rapid thermal processing in N_2O, *Appl. Phys. Lett.* **59**(13), 1581–1582 (1991).
143. J. Ahn, W. Ting, T. Chu, S. N. Lin, and D.-L. Kwong, High quality ultrathin gate dielectrics formation by thermal oxidation of Si in N_2O, *J. Electrochem. Soc.* **138**(9), L39–L41 (1991).
144. W. Ting, G. Q. Lo, J. Ahn, T. Y. Chu, and D.-L. Kwong, MOS characteristics of ultrathin SiO_2 prepared by oxidizing Si in N_2O, *IEEE Electron Device Lett.* **12**(8), 416–418 (1991).
145. R. Bienek and Z. Nényei, Uniform durable thin dielectrics prepared by rapid thermal processing in an N_2O ambient, *IEEE Trans. Electron Devices* **40**(9), 1706–1708 (1993).
146. T. Y. Chu, W. T. Ting, J. Ahn, and D.-L. Kwong, Thickness and compositional nonuniformities of ultrathin oxides grown by rapid thermal oxidation of silicon in N_2O, *J. Electrochem. Soc.* **138**(6), L13–L16 (1991).
147. Y. Okada, P. J. Tobin, R. I. Hege, J. Liao, and P. Rushbrook, Oxynitride gate dielectrics prepared by rapid thermal processing using mixtures of nitrous oxide and oxygen, *Appl. Phys. Lett.* **61**(26), 3163–3165 (1992).
148. A. J. Bauer and E. P. Burte, Structural and electrical properties of thin SiO_2 layers grown by RTP in a mixture of N_2O and O_2, *J. Non-Cryst. Solids* **187**, 361–364 (1995).
149. G. Eftekhari, Effects of oxide thickness and oxidation parameters on the electrical characteristics of thin oxides grown by rapid thermal oxidation of Si in N_2O, *J. Electrochem. Soc.* **141**(11), 3222–3225 (1994).
150. P. Lange, H. Bernt, E. Hartmannsgruber, and F. Naumann, Growth rate and characterization of silicon oxide films grown in N_2O atmosphere in a rapid thermal processor, *J. Electrochem. Soc.* **141**(1), 259–263 (1994).
151. R. Wrixon, A. Twomey, P. O'Sullivan, and A. Mathewson, Enhanced thickness uniformity and electrical performance of ultrathin dielectrics grown by RTP using various N_2O-oxynitridation processes, *J. Electrochem. Soc.* **142**(8), 2738–2742 (1995).
152. Y.-L. Wu and J.-G. Hwu, Characterization of metal-oxide-semiconductor capacitors with improved gates prepared by repeated rapid thermal annealings in N_2O, *J. Vac. Sci. Technol. B* **12**(4), 2400–2404 (1994).
153. Y.-L. Wu and J.-G. Hwu, Improvement in radiation hardness of gate oxides in metal-oxide semiconductor devices by repeated rapid thermal oxidations in N_2O, *Appl. Phys. Lett.* **64**(23), 3136–3138 (1994).

154. G. Eftekhari, Comparison of the electrical characteristics of Si metal-insulator-semiconductor tunnel diodes with interfacial layer grown by rapid thermal oxidation of Si in O_2 and in N_2O, *J. Vac. Sci. Technol. B* **13**(2), 390–395 (1995).
155. T. Arakawa, Y. Yamashita, H. Hoga, S. Noda, and H. Fukuda, Effect of synchrotron radiation on electrical characteristics of SiO_xN_y thin films formed by rapid thermal processing in a N_2O ambient, *Appl. Phys. Lett.* **63**(24), 3364–3366 (1993).

CHAPTER 7

1. J. F. Gibbons, C. M. Gronet, and K. E. Williams, Limited reaction processing: Silicon epitaxy, *Appl. Phys. Lett.* **47**(7), 721–723 (1985).
2. J. L. Hoyt, Rapid thermal processing-based epitaxy, in *Rapid Thermal Processing*, edited by R. B. Fair (Academic Press, New York, 1993), pp. 13–44.
3. A. Sherman, *Chemical Vapor Deposition for Microelectronics, Principles, Technology, and Applications (Noyes Publication, Park Ridge, NJ, 1987).*
4. S. Sivaron, *Chemical Vapor Deposition* (Van Nostrand–Reinhold, Princeton, NJ, 1995).
5. J. Bloem, Nucleation and growth of silicon by CVD, *J. Cryst. Growth* **50**, 581–604 (1980).
6. C. M. Gronet, J. C. Sturm, K. E. Williams, J. F. Gibbons, and S. D. Wilson, Thin, highly doped layers of epitaxial silicon deposited by limited reaction processing, *Appl. Phys. Lett.* **48**(15), 1012–1014 (1986).
7. D. G. Schimmel, A comparison of chemical etches for revealing ⟨100⟩ silicon crystal defects, *J. Electrochem. Soc.* **123**(5), 734–741 (1976).
8. M. L. Green, D. Brasen, H. Luftman, and V. C. Kannan, High-quality homoepitaxial silicon films deposited by rapid thermal chemical vapor deposition, *J. Appl. Phys.* **65**(6), 2558–2560 (1989).
9. B. S. Meyerson, Low temperature silicon epitaxy by ultrahigh vacuum/chemical vapor deposition, *Appl. Phys. Lett.* **48**(12), 797–799 (1986).
10. J. E. Turner, J. Amano, C. M. Gronet, J. F. Gibbons, and S. K. Lee, Secondary ion mass spectrometry of hyper-abrupt doping transitions fabricated by limited reaction processing, *Appl. Phys. Lett.* **50**(22), 1601–1603 (1987).
11. T. Y. Hsieh, K. H. Jung, and D.-L. Kwong, Silicon homoepitaxy by rapid thermal chemical vapor deposition (RTCVD)—Review, *J. Electrochem. Soc.* **138**(4), 1188–1207 (1991).
12. M. K. Sanganeria, K. E. Violette, M. C. Ozturk, G. Harris, C. A. Lee, and D. M. Maher, Low thermal budget in situ cleaning and passivation for silicon epitaxy in a multichamber rapid thermal processing cluster tool, *Mater. Lett.* **21**, 137–141 (1994).
13. M. L. Green, D. Brasen, M. Geva, W. Reents, Jr., F. Stevie, and H. Temkin, Oxygen and carbon incorporation in low temperature epitaxial Si films grown by rapid thermal chemical vapor deposition (RTCVD), *J. Electron Mater.* **19**(10), 1015–1019 (1990).
14. S. K. Lee, Y. H. Ku, and D.-L. Kwong, Silicon epitaxial growth by rapid thermal processing chemical vapor deposition, *Appl. Phys. Lett.* **54**(18), 1775–1777 (1989).
15. M. Liehr, C. M. Greenlief, S. R. Kasi, and M. Offenberg, Kinetics of silicon epitaxy using SiH_4 in a rapid thermal chemical vapor deposition reactor, *Appl. Phys. Lett.* **56**(7), 629–631 (1990).
16. M. Hierlemann, A. Kersch, C. Werner, and H. Schäfer, A gas-phase and surface kinetics model for silicon epitaxial growth with SiH_2C_{12} in an RTCVD reactor, *J. Electrochem. Soc.* **143**(1), 259–266 (1995).
17. T. Y. Hsieh, In situ cleaning and in situ doping for silicon epitaxy by rapid thermal processing chemical vapor deposition, Ph.D. dissertation, Electrical and Computer Engineering, University of Texas at Austin (1991).
18. T. Y. Hsieh, K. H. Jung, D.-L. Kwong, C. J. Hitzman, and R. Brennan, Role of dopant incorporation in low-temperature Si epitaxial growth by rapid thermal processing chemical vapor deposition, *IEEE Trans. Electron Devices* **39**(1), 203–205 (1992).
19. T. Y. Hsieh, K. H. Jung, Y. M. Kim, and D.-L. Kwong, Dopant-enhanced low-temperature epitaxial growth of in situ doped silicon by rapid thermal processing chemical vapor deposition, *Appl. Phys. Lett.* **58**(1), 80–82 (1991).

20. T. Y. Hsieh, K. H. Jung, D.-L. Kwong, C. J. Hitzman, and R. Brennan, Role of dopant incorporation in low-temperature Si epitaxial growth by rapid thermal processing chemical vapor deposition, *IEEE Trans. Electron Devices* **39**(1), 203–205 (1992).
21. Y. J. Jeon, M. F. Becker, and R. M. Walser, Concentration dependence of arsenic on solid phase epitaxial regrowth of amorphous silicon, *Mater. Res. Soc. Symp. Proc.* **157**, 745–750 (1990).
22. T. Y. Hsieh, K. H. Jung, D.-L. Kwong, Y. M. Kim, and R. Brennan, Boron-enhanced low-temperature Si epitaxy by rapid thermal processing chemical vapor deposition, *Appl. Phys. Lett.* **61**(4), 474–476 (1992).
23. M. K. Sanganeria, K. E. Violette, M. C. Öztürk, G. Harris, and D. M. Maher, Boron incorporation in epitaxial silicon using Si_2H_6 and B_2H_6 in an ultrahigh vacuum rapid thermal chemical vapor deposition reactor, *J. Electrochem. Soc.* **142**(1), 285–289 (1995).
24. J. C. Sturm, C. M. Gronet, and J. F. Gibbons, Minority-carrier properties of thin epitaxial silicon films fabricated by limited reaction processing, *J. Appl. Phys.* **59**(12), 4180–4182 (1986).
25. C. A. King, C. M. Gronet, J. F. Gibbons, and S. D. Wilson, Electrical characterization of in-situ epitaxially grown Si p–n junctions fabricated using limited reaction processing, *IEEE Electron Device Lett.* **9**(5), 229–231 (1988).
26. D. Mathiot, J. L. Regolini, and D. Dutartre, Electrical characterization of silicon epitaxial layers grown by limited reaction processing, *J. Appl. Phys.* **69**(1), 358–361 (1991).
27. J. L. Regolini, D. Bensahel, J. Mercier, and E. Scheid, Silicon selective epitaxial growth at reduced pressure and temperature, *J. Cryst. Growth* **96**, 505–512 (1989).
28. M. K. Sanganeria, K. E. Violette, and M. C. Öztürk, Low temperature silicon epitaxy in an ultrahigh vacuum rapid thermal chemical vapor deposition reactor using disilane, *Appl. Phys. Lett.* **63**(9), 1225–1227 (1993).
29. X. Ren, M. C. Öztürk, J. J. Wortman, B. Zhang, D. M. Maher, and D. Batchelor, Deposition and characterization of polysilicon films deposited by rapid thermal processing, *J. Vac. Sci. Technol. B* **10**(3), 1081–1086 (1992).
30. X.-L. Xu, V. Misra, M. C. Öztürk, J. J. Wortman, G. S. Harris, D. Maher, L. Spanos, and E. A. Irene, Effects of oxygen doping on properties of microcrystalline silicon film grown using rapid thermal chemical vapor deposition, *J. Electron. Mater.* **22**(11), 1345–1351 (1993).
31. X. Xu, V. Misra, G. S. Harris, L. Spanos, M. C. Öztürk, J. J. Wortman, D. M. Maher, and E. A. Irene, Characterization of oxygen-doped and non-oxygen-doped polysilicon films prepared by rapid thermal chemical vapor deposition, *Mater. Res. Soc. Symp. Proc.* **303**, 49–54 (1993).
32. L. L. Tedder, G. W. Rugloff, I. Shareed, M. Anderle, D.-H. Kim, and G. N. Parsons, Real-time process and product diagnostics in rapid thermal chemical vapor deposition using *in situ* mass spectrometric sampling, *J. Vac. Sci. Technol. B* **13**(4), 1924–1927 (1995).
33. J. L. Regolini, D. Bensahel, E. Scheid, and J. Mercier, Selective epitaxial silicon growth in the 650–1100°C range in a reduced pressure chemical vapor deposition reactor using dichlorosilane, *Appl. Phys. Lett.* **54**(7), 658–659 (1989).
34. J. Mercier, J. L. Regolini, D. Bensahel, and E. Scheid, Kinetic aspects of selective epitaxial growth using a rapid thermal processing system, *J. Cryst. Growth* **94**, 885–894 (1989).
35. S. K. Lee, Y. H. Ku, T. Y. Hsieh, K. Jung, and D.-L. Kwong, Selective epitaxial growth by rapid thermal processing, *Appl. Phys. Lett.* **57**, 273–275 (1990).
36. A. Ishitani, H. Kitajima, N. Endo, and N. Kasai, Silicon selective epitaxial growth and electrical properties of epi/sidewall interfaces, *Jpn. J. Appl. Phys.* **28**(5), 841–848 (1989).
37. L. Ye, B. M. Armstrong, and H. S. Gamble, The study of selectivity in silicon selective epitaxial growth, *Microelectron. Eng.* **25**, 153–158 (1994).
38. K. E. Violette, M. K. Sanganeria, M. C. Öztürk, G. Harris, and D. M. Maher, Growth kinetics, silicon nucleation on silicon dioxide, and selective epitaxy using disilane and hydrogen in an ultrahigh vacuum rapid thermal chemical vapor deposition reactor, *J. Electrochem. Soc.* **141**(11), 3269–3273 (1994).
39. T. Y. Hsieh, H. G. Chun, and D.-L. Kwong, Selective deposition of *in situ* doped polycrystalline silicon by rapid thermal processing chemical vapor deposition, *Appl. Phys. Lett.* **55**(23), 2408–2410 (1989).
40. J. C. Sturm, C. M. Gronet, C. A. King, S. D. Wilson, and J. F. Gibbons, In situ epitaxial silicon-oxide-doped polysilicon structures for MOS field-effect transistors, *IEEE Electron Device Lett.* **7**(10), 577–579 (1986).

41. B. Fröschle, H. P. Bruemmer, W. Lang, K. Neumeier, and P. Ramm, Fabrication of high quality MOS devices for application in hazardous environments based on RTP gate dielectrics with in situ RTCVD of polysilicon gates, *Mater. Res. Soc. Symp. Proc.* **303**, 331–338 (1993).
42. S. C. Jain and W. Hayes, Structure, properties and applications of Ge_xSi_{1-x}, strained layers and superlattices, *Semicond. Sci. Technol.* **6**, 547–576 (1991).
43. C. M. Gronet, C. A. King, W. Opyd, J. F. Gibbons, S. D. Wilson, and R. Hull, Growth of GeSi/Si strained-layer superlattices using limited reaction processing, *J. Appl. Phys.* **61**(6), 2407–2409 (1987).
44. M. L. Green, B. E. Weir, D. Brasen, Y. F. Higashi, A. Feygenson, L. C. Feldman, and R. L. Headrick, Mechanically and thermally stable Si-Ge films and heterojunction bipolar transistors grown by rapid thermal chemical vapor deposition at 900°C, *J. Appl. Phys.* **69**(2), 745–751 (1991).
45. G. F. A. Van De Walle, L. J. Ijzendoorn, A. A. Van Gorkum, R. A. Van den Heuvel, A. M. L. Theunissen, and D. J. Gravesteijn, Germanium diffusion and strain relaxation in Si/Si_{1-x} Ge_x/Si structures, *Thin Solid Films* **183**, 183 (1989).
46. B. Höllander, S. Mant, B. Stritzker, H. Jorke, and E. Kasper, Strain measurements and thermal stability at Si_{1-x} Ge_x/Si strained layers, *J. Mater. Res.* **4**, 163 (1989).
47. J. C. Sturm, P. V. Schwartz, E. J. Prince, and H. Manoharan, Growth of $Si_{1-x}Ge_x$ by rapid thermal chemical vapor deposition and application to heterojunction bipolar transistors, *J. Vac. Sci. Technol. B* **9**(4), 2011–2016 (1991).
48. D. Dutartre, P. Warren, I. Berbezier, and P. Perret, Low temperature silicon and Si_{1-N} epitaxy by rapid thermal chemical vapour deposition using hydrides, *Thin Solid Films* **222**, 52–56 (1992).
49. D. J. Howard, W. E. Bailey, and D. C. Paine, Observation of open-ended stacking fault tetrahedra in $Si_{0.85}$ $Ge_{0.15}$ grown on V-grooved (001) Si and planar (111) Si substrates, *Appl. Phys. Lett.* **63**(21), 2893–2895 (1993).
50. K. H. Jung, Y. M. Kim, and D.-L. Kwong, Relaxed Ge_xSi_{1-x} films grown by rapid thermal processing chemical vapor deposition, *Appl. Phys. Lett.* **56**(18), 1775–1777 (1990).
51. K. H. Jung, Y. M. Kim. T. Y. Hsieh, and D.-L. Kwong, Defect study of GeSi alloy multilayer structures grown by RTCVD, *J. Electrochem. Soc.* **138**(8), 2387–2392 (1991).
52. G. P. Watson, E. A. Fitzgerald, Y.-H. Xie, and D. Monroe, Relaxed, low threading defect density $Si_{0.7}Ge_{0.3}$ epitaxial layers grown on Si by rapid thermal chemical vapor deposition, *Appl. Phys. Lett.* **75**(1), 263–268 (1994).
53. K. H. Jung, T. Y. Hsieh, D.-L. Kwong, H. Y. Liu, and R. Brennan, In situ doping of Ge_xSi_{1-x} with arsenic by rapid thermal processing chemical vapor deposition, *Appl. Phys. Lett.* **60**(6), 724–726 (1992).
54. D. B. Noble, J. L. Hoyt, J. F. Gibbons, M. P. Scott, S. S. Landerman, S. J. Rosner, and T. I. Kamins, Thermal stability of $Si/Si_{1-x}Ge_x/Si$ heterojunction bipolar transistor structures grown by limited reaction processing, *Appl. Phys. Lett.* **55**(19), 1978–1980 (1989).
55. F. Glowacki and Y. Campidelli, Single wafer epitaxy of Si and SiGe using UHV-CVD, *Elsevier Science B. V. SSDI* Vol. 0167-9317, No. 94, 161–170 (1994).
56. J. L. Regolini, D. Bensahel, and J. Mercier, Reduced pressure and temperature epitaxial silicon CVD kinetics and applications, *J. Electron. Mater.* **19**, 1025–1080 (1990).
57. D. Dutartre, P. Warren, I. Sagnes, P. A. Badoz, J. C. Dupuis, G. Prudon, and A. Perio, Low temperature Si and $Si_{1-x}Ge_x$ epitaxy by RT-CVD in the system SiH_4, GeH_4, B_2H_6 and H_2, Final Program, American Vacuum Society National Symposium in Chicago, Nov. 9–13, 1992 (American Vacuum Society, New York), p. 164.
58. S. Gu, R. Wang, R. Zhang, L. Qin, Y. Shi, S. Zhu, and Y. Zheng, Substrate temperature and Ge concentration dependence of the microstructure of strained Si-Ge alloys, *J. Phys. Condens. Matter* **6**, 6163–6168 (1994).
59. S. Gu, Y. Zheng, R. Zhang, R. Wang, and P. Zhong, Ge composition and temperature dependence of the deposition of SiGe layers, *J. Appl. Phys.* **75**(10), 5382–5384 (1994).
60. S. Gu, R. Zhang, P. Han, R. Wang, P. Zhong, and Y. Zheng, Raman study of strained SiGe layers, *Elsevier Science B.V. SSDI* Vol. 0169-4332 No. 94, 431–434 (1994).
61. Y. Zhong, M. C. Öztürk, D. T. Grinder, J. J. Wortman, and M. A. Littlejohn, Selective low-pressure chemical vapor deposition of $Si_{1-x}Ge_x$ alloys in a rapid thermal processor using dichlorsilane and germane, *Appl. Phys. Lett.* **57**(20), 2092–2094 (1990).
62. J. L. Regolini, F. Gisbert, G. Dolino, and P. Boucaud, Growth and characterization of strain compensated $Si_{1-x-y}Ge_xC_y$ epitaxial layers, *Mater. Lett.* **18**, 57–60 (1993).

63. G. G. Fischer and P. Zaumseil, *In situ* x-ray investigation of the high-temperature behaviour of strained $Si_{1-x}Ge_x$/Si and $Si_{1-y}C_y$/Si heterostructures, *J. Phys. D* **28**, A109–A113 (1995).
64. J. Mi, P. Letourneau, J.-D. Ganiére, M. Gailhanou, M. Dutoit, C. Dubois, J. C. Dupuy, and G. Brémond, Improvement of crystal quality of epitaxial silicon-germanium alloy layers by carbon additions, *Helv. Phys. Acta* **67**(2), 219–220 (1994).
65. J. Mi, P. Warren, P. Letourneau, M. Judelewicz, M. Gailhanou, M. Dutoit, C. Dubois, and J. C. Dupuy, High quality $Si_{1-x-y}Ge_xC_y$ epitaxial layers grown on (100) Si by rapid thermal chemical vapor deposition using methylsilane, *Appl. Phys. Lett.* **67**(2), 259–261 (1995).
66. Z. Atzmon, A. E. Bair, E. J. Jaquez, J. W. Mayer, D. Chandrasekhar, D. J. Smith, R. L. Hervig, and McD. Robinson, Chemical vapor deposition of heteroepitaxial $Si_{1-x-y}Ge_xC_y$ films on (100) Si substrates, *Appl. Phys. Lett.* **65**(20), 2559–2561 (1994).
67. S. Bodnar and J. L. Regolini, Growth of ternary alloy $Si_{1-x-y}Ge_xC_y$ by rapid thermal chemical vapor deposition, *J. Vac. Sci. Technol. A* **13**(5), 2336–2340 (1995).
68. M. Sanganeria, D. T. Grider, M. C. Öztürk, and J. J. Wortman, Rapid thermal chemical vapor deposition of in-situ boron doped polycrystalline Si_xGe_{1-x}, *J. Electron. Mater.* **21**(1), 61–64 (1992).
69. M. C. Öztürk, D. T. Grider, J. J. Wortman, M. A. Littlejohn, Y. Zhong, D. Batchlor, and P. Russell, Rapid thermal chemical vapor deposition of germanium on silicon and silicon dioxide and new applications of Ge in ULSI technologies, *J. Electron. Mater.* **19**(10), 1129–1134 (1990).
70. A. St. Amour and J. C. Sturm, Deposition of monolayer-scale germanium/silicon heterostructures by rapid thermal chemical vapor deposition, *Mater. Res. Soc. Symp. Proc.* **342**, 31–36 (1994).
71. C. W. Liu, J. C. Sturm, Y. R. J. Lacroix, M. L. W. Thewalt, and D. D. Perovic, Growth and photoluminescence of strained ⟨110⟩ Si/$Si_{1-x}Ge_x$/Si quantum wells, *Mater. Res. Soc. Symp. Proc.* **342**, 37–42 (1994).
72. P. Boucaud, C. Francis, F. H. Julien, J.-M. Lourtioz, D. Bouchier, S. Bodnar, B. Lambert, and J. L. Regolini, Band-edge and deep level photoluminescence of pseudomorphic $Si_{1-x-y}Ge_xC_y$ alloys, *Appl. Phys. Lett.* **64**(7), 875–877 (1994).
73. G. Bermond, A. Souifi, T. Benyattou, and D. Dutarte, Photoluminescence and electrical characterization of SiGe/Si heterostructures grown by rapid thermal chemical vapour deposition, *Thin Solid Films* **222**, 60–68 (1992).
74. S. C. Jain and W. Hayes, Structure, properties and applications of Ge_xSi_{1-x-} strained layers and superlattices, *Semicond. Sci. Technol.* **6**, 547–576 (1991).
75. P. Warren, I. Sagnes, D. Dutartre, P. A. Badoz, J. M. Berroir, Y. Guldner, J. P. Vieren, and M. Voos, France Telecom-Cnet, BP 98, F-38243 Meylan Cedex France, Charge transfer in p^+-Si/$Si_{1-x}Ge_x$ modulation doped heterostructures grown by RTCVD, *Elsevier Science B.V. SSDI* Vol. 0167-9317, No. 94, 171–176 (1994).
76. R. L. Jiang, J. L. Liu, Y. D. Zheng, H. F. Li, and H. Z. Zheng, Transport property of Si/$Si_{1-x}Ge_x$ /Si p-type modulation doped double heterostructure, *J. Appl. Phys.* **76**(4), 2544–2546 (1994).
77. R. L. Jiang, J. L. Liu, and Y. D. Zheng, Hole transport properties of Si/$Si_{1-x}Ge_x$ modulation-doped heterostructures, *Superlattices Microstruct.* **16**(4), 375–377 (1994).
78. W. Ting, J. H. Ahn, and D.-L. Kwong, Ultrathin stacked Si_3N_4/SiO_2 gate dielectrics prepared by rapid thermal processing, *Electron. Lett.* **27**(12), 1046–1047 (1991).
79. S. H. Lee and M. G. So, Characteristics of thin nitrided oxides prepared by an *in situ* process, *J. Mater. Sci.* **24**, 6645–6649 (1993).
80. X.-L. Xu, R. T. Kuehn, J. J. Wortman, and M. C. Öztürk, Rapid thermal chemical vapor deposition of thin silicon oxide films using silane and nitrous oxide, *Appl. Phys. Lett.* **60**(24), 3063–3065 (1992).
81. C. Cobianu, J. B. Rem, J. H. Klootwijk, M. H. H. Weusthof, J. Holleman, and P. H. Woerlee, LPCVD SiO_2 layers prepared from SiH_4 and O_2 at 450°C in a rapid thermal processing reactor, *J. Phys. IV Suppl. Colloq. C5* **5**, 1005–1011 (1995).
82. X.-L. Xu, P. K. McLarty, H. Brush, V. Misra, J. J. Wortman, and G. S. Harris, Characterization of thin silicon oxynitride films prepared by low pressure rapid thermal chemical vapor deposition, *J. Electrochem. Soc.* **140**(10), 2970–2974 (1993).
83. P. K. McLarty, W. L. Hill, X.-L. Xu, V. Misra, J. J. Wortman, and G. S. Harris, Thin oxynitride film metal-oxide-semiconductor transistors prepared by low-pressure rapid thermal chemical vapor deposition, *Appl. Phys. Lett.* **63**(26), 3619–3621 (1993).

84. P. K. McLarty, W. L. Hill, X.-L. Xu, J. J. Wortman, and G. S. Harris, Thin oxynitride films prepared by low pressure rapid thermal chemical vapor deposition, *Mater. Res. Soc. Symp. Proc.* **303**, 43–48 (1993).
85. V. Misra, S. Hattangady, X.-L. Xu, M. J. Watkins, B. Hornung, G. Lucovsky, J. J. Wortman, U. Emmerichs, C. Meyer, K. Leo, and H. Kurz, Integrated processing of stacked-gate heterostructures: Plasma assisted low temperature processing combined with rapid thermal high temperature processing, *Microelectron. Eng.* **25**, 209–214 (1994).
86. F. Lebland, Z. Z. Wang, J. Flicstein, C. Licoppe, and Y. I. Nissan, Bulk and surface properties of RTCVD Si_3N_4 films for optical device applications, *Appl. Surf. Sci.* **69**, 198–203 (1993).
87. W. L. Hill, E. Vogel, P. K. McLarty, V. Misra, J. J. Wortman, and V. Watt, N-channel and P-channel MOSFETs with oxynitride gate dielectrics formed using low pressure rapid thermal chemical vapor deposition, *Microelectron. Eng.* **28**, 269–272 (1995).
88. G. W. Yoon, G. Q. Lo, J. Kim, L. K. Han, and D.-L. Kwong, Formation of high quality storage capacitor dielectrics by *in-situ* rapid thermal reoxidation of Si_3N_4 films in N_2O ambient, *IEEE Electron Device Lett.* **15**(8), 266–268 (1994).
89. A. J. Steckl and J. P. Li, Epitaxial growth of β-SiC on Si by RTCVD with C_3H_8 in SiH_4, *IEEE Trans. Electron Devices* **39**(1), 64–74 (1992).
90. J. P. Li and A. J. Steckl, Nucleation and void formation mechanism in SiC thin film growth on Si by carbonization, *J. Electrochem. Soc.* **142**(2), 634–641 (1995).
91. J. P. Li, A. J. Steckl, I. Golecki, F. Reidinger, L. Wang, X. J. Ning, and P. Pirouz, Structural characterization of nanometer SiC films grown on Si, *Appl. Phys. Lett.* **62**(24), 3135–3137 (1993).
92. C. J. Mogab and H. J. Leamy, Conversion of Si to epitaxial SiC by reaction with C_2H_2, *J. Appl. Phys.* **45**(3), 1075–1084 (1974).
93. S. P. Chiew, G. McBride, B. M. Armstrong, J. Grimshaw, H. S. Gamble, and J. Trocha-Grimshaw, Growth of β-SiC layers by rapid thermal chemical vapor deposition, *Microelectron. Eng.* **15**, 177–182 (1994).
94. P. H. Yih, J. P. Li, and A. J. Steckl, SiC/Si heterojunction diodes fabricated by self-selective and by blanket rapid thermal chemical vapor deposition, *IEEE Trans. Electron Devices* **41**(3), 281–287 (1994).
95. J. D. Hwang, Y. K. Fang, K. H. Chen, and D. N. Yaung, A novel β-SiC/Si heterojunction backward diode, *IEEE Electron Device Lett.* **16**(5), 193–195 (1995).
96. S. K. Reynolds, Application of rapid thermal-processing to epitaxy and doping of gallium-arsenide (semiconductors), Ph.D. dissertation, Stanford University (1988).
97. A. Katz, A. Feingold, N. Moriya, S. J. Pearton, M. Geva, F. A. Baiocchi, L. C. Luther, and E. Lane, Rapid thermal low pressure metalorganic chemical vapor deposition of $In_{0.53}Ga_{0.47}As$ films using tertiarybutylarsine, *Appl. Phys. Lett.* **63**(19), 2679–2681 (1993).
98. A. Katz, A. El-Roy, A. Feingold, M. Geva, N. Moriya, S. J. Pearton, E. Lane, T. Keel, and C. R. Abernathy, W(Zn) selectively deposited and locally diffused ohmic contacts to *p*-InGaAs/InP formed by rapid thermal low pressure metalorganic chemical vapor deposition, *Appl. Phys. Lett.* **62**(21), 2652–2654 (1993).
99. A. Katz, A. Feingold, S. J. Pearton, N. Moriya, C. J. Baiocchi, and M. Geva, Low-temperature rapid thermal low pressure metalorganic chemical vapor deposition of Zn-doped InP layers using tertiarybutylophosphine, *Appl. Phys. Lett.* **63**(18), 2546–2548 (1993).
100. A. Katz, A. Feingold, N. Moriya, S. Nakahara, C. R. Abernathy, S. J. Pearton, A. El-Roy, M. Geva, F. A. Baiocchi, L. C. Luther, and E. Lane, Growth of InP epitaxial layers by rapid thermal low pressure metalorganic chemical vapor deposition, using tertiarybutylophosphine, *Appl. Phys. Lett.* **63**(21), 2959–2960 (1993).
101. J. L. Regolini, D. Bensahel, G. Bomchil, and J. Mercier, Selective layers of $TiSi_2$ deposited without substrate consumption in a cold wall LPCVD reactor, *Appl. Surf. Sci.* **38**, 408–415 (1989).
102. J. Mercier, J. L. Regolini, and D. Bensahel, Selective $TiSi_2$ deposition with no silicon substrate consumption by rapid thermal processing in a LPCVD reactor, *J. Electron. Mater.* **19**(3), 253–258 (1990).
103. J. Mercier, J. L. Regolini, D. Bensahel, and E. Scheid, Kinetic aspects of selective epitaxial growth using a rapid thermal processing system, *J. Cryst. Growth* **94**, 885–894 (1989).
104. H. H. Lee, Silicon growth at low temperatures: SiH_4-HCl-H_2 system, *J. Cryst. Growth* **69**, 82–90 (1984).
105. J. L. Regolini, E. Mastromatteo, M. Gauneau, J. Mercier, D. Dutartre, G. Bomchil, C. Bernard, R. Madar, and D. Bensahel, Aspects of the selective deposition of $TiSi_2$ by LRP-CVD for use in ULSI submicron technology, *Appl. Surf. Sci.* **53**, 18–23 (1991).

106. J. L. Regolini, F. Trincat, I. Berbezier, J. Palleau, J. Mercier, and D. Bensahel, Selective and epitaxial deposition of β-$FeSi_2$ onto silicon by RTP–CVD, *J. Phys. III France* **2**, 1445–1452 (1992).
107. J. L. Regolini, F. Trincat, I. Berbezier, and Y. Shapira, Selective and epitaxial deposition of β-$FeSi_2$ on silicon by rapid thermal processing-chemical vapor deposition using a solid iron source, *Appl. Phys. Lett.* **60**(8), 956–958 (1992).
108. I. Berbezier, J. L. Regolini, and C. d'Anterroches, Epitaxial orientation of β-$FeSi_2$/Si heterojunctions obtained by RTP chemical vapor deposition, *Microsc. Microanal. Microstruct.* **4**, 5–21 (1993).
109. A. Katz, A. Feingold, S. J. Pearton, S. Nakahara, M. Ellington, U. K. Chakrabarti, M. Geva, and E. Lane, Properties of titanium nitride thin films deposited by rapid-thermal-low-pressure-metalorganic-chemical-vapor-deposition technique using tetrakis (dimethylamido) titanium precursor, *J. Appl. Phys.* **70**(7), 3666–3676 (1991).
110. R. Singh, S. Sinha, N. J. Hsu, J. T. C. Ng, P. Chou, R. P. S. Thakur, and J. Narayan, Reduced thermal budget processing of Y-Ba-Cu-O high temperature superconducting thin films by metalorganic chemical vapor deposition, *J. Vac. Sci. Technol. A* **9**(3), 401–404 (1991).
111. R. Singh, J. T. C. Ng, S. Sinha, and V. Dhall, Design and fabrication of a computer-controlled rapid-isothermal-processing-assisted metalorganic chemical-vapor-deposited system for high-temperature superconducting thin films and related materials, *Rev. Sci. Instrum.* **64**(2), 514–523 (1993).

Index